Maximum-Entropy and Bayesian Methods in Inverse Problems

Fundamental Theories of Physics

A New International Book Series on The Fundamental Theories of Physics: Their Clarification, Development and Application

Maximum-Entropy and Bayesian Methods in Inverse Problems

edited by

C. Ray Smith

and

W. T. Grandy, Jr.

Department of Physics and Astronomy,
The University of Wyoming, Laramie, Wyoming, U.S.A.

D. Reidel Publishing Company

A MEMBER OF THE KLUWER ACADEMIC PUBLISHERS GROUP

Dordrecht / Boston / Lancaster

Library of Congress Cataloging in Publication Data
Main entry under title:

Maximum-entropy and Bayesian methods in inverse problems.

Papers presented at two workshops held at the University of Wyoming from June 8 to 10, 1981 and from August 9 to 11, 1982.
Includes index.
1. Entropy (Information theory)–Congresses. 2. Spectral analysis–Congresses. 3. Bayesian statistical decision theory–Congresses. 4. Inverse problems (Differential equations)–Congresses. I. Smit, C. Ray, 1933- . II. Grandy, Walter T., 1933- . III. University of Wyoming.
Q370.M37 1985 001.53′9 85-11765
ISBN 90-277-2074-6

Published by D. Reidel Publishing Company,
P.O. Box 17, 3300 AA Dordrecht, Holland

Sold and distributed in the U.S.A. and Canada
by Kluwer Academic Publishers,
190 Old Derby Street, Hingham, MA 02043, U.S.A.

In all other countries, sold and distributed
by Kluwer Academic Publishers Group,
P.O. Box 322, 3300 AH Dordrecht, Holland

Printed in The Netherlands

To Edwin T. Jaynes

CONTENTS

PREFACE

This volume contains the text of the twenty-five papers presented at two workshops entitled Maximum-Entropy and Bayesian Methods in Applied Statistics, which were held at the University of Wyoming from June 8 to 10, 1981, and from August 9 to 11, 1982. The workshops were organized to bring together researchers from different fields to critically examine maximum-entropy and Bayesian methods in science, engineering, medicine, oceanography, economics, and other disciplines. An effort was made to maintain an informal environment where ideas could be easily exchanged. That the workshops were at least partially successful is borne out by the fact that there have been two succeeding workshops, and the upcoming Fifth Workshop promises to be the largest of all.

These workshops and their proceedings could not have been brought to their final form without the substantial help of a number of people. The support of David Hofmann, the past chairman, and Glen Rebka, Jr., the present chairman of the Physics Department of the University of Wyoming, has been strong and essential. Glen has taken a special interest in seeing that the proceedings have received the support required for their completion. The financial support of the Office of University Research Funds, University of Wyoming, is gratefully acknowledged. The secretarial staff, in particular Evelyn Haskell, Janice Gasaway, and Marce Mitchum, of the University of Wyoming Physics Department has contributed a great number of hours in helping C. Ray Smith organize and direct the workshops.

Preliminary versions of the manuscripts were typed by J. Dob, and final camera-ready copy was prepared by Martha Stockton. But Martha's contributions went far beyond typing; her help in editing, proofreading, checking references, redrawing figures, and much more was essential to this work. Thanks, Martha.

The Second Workshop was held shortly after Professor Edwin T. Jaynes' sixtieth birthday. The sheer number of citations of his works indicates the considerable influence he has had on this subject. His influence through personal interaction is even greater. We take pleasure in dedicating this volume to Professor Jaynes.

Laramie, Wyoming
March 1985

C. Ray Smith
W. T. Grandy, Jr.

INCOMPLETE INFORMATION AND GENERALIZED INVERSE PROBLEMS

W. T. Grandy, Jr.

Department of Physics and Astronomy, University of Wyoming,
Laramie, Wyoming 82071

The status of approaches to the solution of generalized inverse problems is reviewed in a broad context, and an apparent divergence in viewpoint is discussed. Both those of the mathematical school of deductive inversion, and their colleagues who concentrate on the extraction of solid information from limited data, often stand in ignorance of the work of one another while addressing different aspects of the same general problem. It is argued that these differing views are not in opposition but are actually complementary, and that each approach may well benefit from the ideas of the other. In particular, analysis of incomplete data must logically take place before, and provide input to, the application of any powerful mathematical machinery employed for, say, reconstructing an object.

C. Ray Smith and W. T. Grandy, Jr. (eds.),
Maximum-Entropy and Bayesian Methods in Inverse Problems, 1–19.

1. Introduction

Several years ago John Wheeler observed that the Galapagos Islands are not the only place where a single species of bird has now separated into distinct species that no longer interbreed [1]. This remarkable example of evolutionary biology has a counterpart in scattering theory, and in the field of inverse problems generally, where a similar speciation is very close to completion. One flock concerns itself with a limited amount of corrupted data, while the other thrives on the mathematical exactitude of uniqueness theorems, both turning away at the approach of the other. As in biology, the phenomenon can be understood only through detailed examination of the divergent lines of evolution. Our long-range goal is somewhat different, however, in that perhaps survival of both ultimately depends on their interbreeding once again.

In a very real sense all life consists of a continual confrontation with generalized inverse problems, the first abstract formulation possibly being found in Plato's allegory of shadows on the cave walls [2]. Much of Kant's epistemological writing is concerned implicitly with inverse problems [3], albeit on a philosophical level. In this paper we shall focus on more concrete realizations.

A more specific, yet still abstract, expression of the problem can be written in the form

$$u = KU, \tag{1}$$

where U describes a 'true' state of being and u its manifestation or perceived image after being corrupted by the mechanism of observation, or by otherwise unavoidable circumstances. The direct problem is primarily one of deductive prediction: Given prior knowledge of U and the operator K, deduce the effect u. Of a deeper nature is the inverse question: Given u and a specific K, what is the 'true' state U? Connection with the general life scenario mentioned above can be made by reading 'human brain' for K, a much more difficult problem because the nature of K itself remains poorly understood.

A less abstract and more familiar form of the problem posed by Eq. (1) is the direct mathematical inversion carried out by construction of the inverse operator K^{-1}. An ubiquitous example of major importance is the linear Fredholm integral equation of the first kind,

$$u(x) = \int_a^b K(x,y)\, U(y)\, dy\ , \tag{2}$$

which is to be solved for U given a definite class of functions u(x) on a specific interval. One major difficulty with the model of K as the human brain is that the kernel is almost certainly a nonlinear functional of U itself. Here we shall consider only linear inverse problems, principally because the theory of nonlinear integral equations of this kind is not well developed.

Under that restriction Eq. (2) encompasses many mathematical inverse problems, from the inversion of integral transforms to the use of Cauchy's theorem for determining a function in a region from its values on the bounding contour.

More generally, as a pure integral equation the problem may have no solutions, and when it does they may not be uniquely determined. For example, if K is degenerate, K(x,y) = A(x) B(y), there is no solution unless u(x) is proportional to A(x). These scenarios constitute a field of mathematical analysis that is both beautiful and challenging.

On the other side of the coin, though, is the more general class of problems in which the left-hand side of Eq. (1) is poorly known or specified, and usually only in a discontinuous manner. The members of this class most often correspond to the bulk of everyday physical problems, typified by receipt of a coded message, weak or blurred speech patterns on a noisy background, discrete frequency data, . . . or the seductive smile of a beautiful woman. The associated inversion problems, when viewed in a strict mathematical sense, are often unstable, possess nonunique solutions, or, in the last example, sometimes lead to disaster. Yet such problems must be faced squarely in daily life, as well as in science. An important psychological prerequisite for studying such problems, and one that explains much about the differences between our two bird species, is the realization that the actual problem can rarely be solved exactly, in the sense of a complete mathematical inversion. One almost never has information complete enough to effect such a solution. Similar observations much earlier led to the invention of statistical mechanics, just because one never knows sufficient initial conditions for the microscopic equations of motion. Workers in that field are possibly saddened by, but appreciate, the fact that second-virial-coefficient data are insufficient to uniquely determine the two-body potential, and that data on the required higher-order coefficients are next to impossible to obtain.

In the following we first focus on the pure inverse problem, meaning that the data are specified precisely, to the extent they are given. After learning something of the evolutionary tracks of the two species, we examine the ways in which each acknowledges the presence of noise and, perhaps, shall be able to reintroduce them to one another.

2. Direct Mathematical Inversion

Examples of the pure mathematical inversion problem have been mentioned above, along with some of their intrinsic difficulties. Although these problems are interesting in their own right, we here examine only the application of the related mathematical techniques to the solution of more physical inverse problems. In this respect we point out two further features of the integral equation (2) which have physical implications. If K(x,y) has a number of continuous derivatives in x, for example, these properties also accrue to u(x), thereby limiting the class of given functions. Along with this "smoothing" property the kernel often attenuates the contributions from

$U(y)$ near the upper limit, thus introducing possible instabilities into the solution. A physical example of this will appear presently.

As an aside, let us note at least one mathematical inverse problem of interest to physicists that does not fall into this mold. In both classical and quantum physics one encounters the Sturm-Liouville equation for an operator L:

$$[L + \lambda^2 + q(\mathbf{x})]\, \psi(\mathbf{x}) = 0\,, \tag{3}$$

where ψ satisfies specified boundary conditions. It was probably Rayleigh [4] who first framed the following question: If the spectrum is given, can $q(\mathbf{x})$ be determined? That is, can you hear the shape of a drum [5]? Generally the answer is "no" although complete results seem to be available in only one dimension: q is uniquely determined from knowledge of two complete spectral sets generated by two different sets of homogeneous boundary conditions [6]. For finite intervals, recent work indicates that uniqueness can actually be achieved with a single spectrum [7].

Although Heisenberg's program to encompass all of physics within the S-matrix has been found flawed [8,9], when most physicists think of inverse problems they usually still have in mind inverse scattering. Given scattering data in the far field, the scattering amplitude, say,

$$f(E, \theta) = \sum_{\ell=0}^{\infty} (2\ell + 1)\, S_\ell(E)\, P_\ell(\cos\theta)\,, \tag{4a}$$

$$S_\ell(E) \sim E^{-1}\, [\exp(2i\eta_\ell(E)) - 1]\,, \tag{4b}$$

can one determine the target or potential? This question has generated a great deal of effort by a relatively large number of theoreticians since it was first asked [10] and has led to some of the most beautiful mathematical results in physics. For spherically-symmetric potentials the inverse problem is a mapping of the two-parameter phase shifts $\eta_\ell(E)$ to the single-variable potential $V(r)$, so that there are several forms in which the data can be collected. It is now known that if the s-wave phase shift $\eta_0(E)$ is given for all energies, along with the values of all bound-state energies and the range constants a_n of their associated wavefunctions, the potential is uniquely determined by a single differentiation [11,12]. Or, if one knows all the phase shifts η_ℓ at a fixed energy, the class of possible potentials is uniquely determined [13]. Most of these remarkable results are already to be found in Faddeev's classic review article [14], and a recent analysis of the entire mathematical problem has been provided by Newton [15].

How, though, is one to obtain such data? On the one hand, all phase shifts can never be found for a single attainable energy, for then all angular momenta cannot be explored. On the other hand, determination of a single phase shift at all energies approaches the impossible; aside from the kind of surprises that emerged from polarized pp scattering in the form of strong spin dependence, higher energies will eventually change the problem. Thus, these pretty mathematical results are rather impracticable for actually determining a potential from scattering data, and physicists fortunately have the sense to correlate a great deal of data from different types of experiments.

There are, of course, many other important problems of scattering and radiation, aside from those of particle physics, and we classify these broadly under the heading of object reconstruction. Figure 1 describes a general optical imaging process envisaging the scattering of a plane wave from an object, subsequent emersion of the scattered wave from the aperture of the detection system, and the record imprinted on the two-dimensional image plane. In reconstruction of the object from the image, three distinct problems must be solved: (i) reconstruction of the actual far-field wave from data on the image plane, (ii) inverse propagation of the wave from a far surface to a near surface, and (iii) reconstruction of the object from the object wave under the presumption that this is precisely what was radiated. In accordance with the general topic it is appropriate to discuss these different aspects in inverse order.

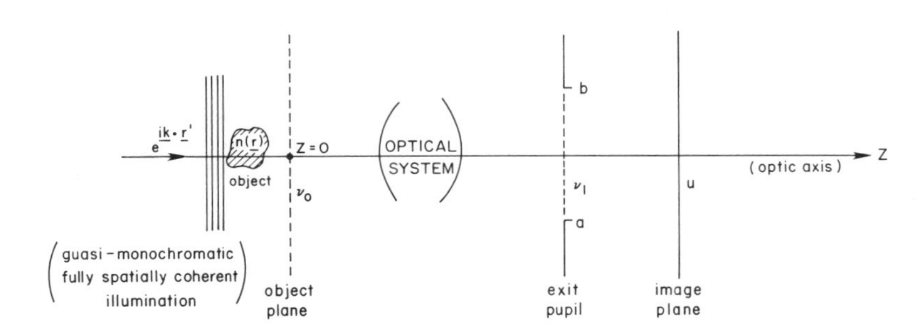

Figure 1. Idealization of a general optical imaging process.

Reconstruction of an object from the field immediately outside its support is not possible in general, for the result is often not unique. It has long been known that there exist charge-current distributions that do not radiate, and some objects can produce zero field when excited by a plane wave. Far fields from a scattered plane wave are insufficient to uniquely determine a target. In analogy with the quantum case, though, it is possible that an infinite number with different energies or directions can do the trick. But, again, we wonder whence cometh this knowledge. The need for prior information about the target is compelling.

Inverse diffraction concerns propagation of the object wave in the far field to a region just outside the scatterer. It is now known [16] that knowledge of a scalar or vector wave on any closed surface or infinite plane, which satisfies the asymptotic radiation condition, is sufficient to reconstruct the wave on any other surface or plane outside the scatterer.

How that knowledge is to be acquired is not included in the statements of the theorems.

Finally, problem (i) brings us to the realities of dealing with the data actually available. Even in the mathematical attempts at direct inversion, the more careful workers have recognized the limitations of observational band limits and the fact that one almost never has phase information. Usually only intensities are recorded. Although it is also noted that data in the form $u(x)$ are available over only a finite range, it is usually presumed that this is a precise continuous function. Nevertheless, when band-limited data of finite extent are specified in the form of a well-defined intensity distribution over part of the image plane, one can now obtain a mathematical reconstruction of the object wave outside this plane [17]. We shall examine briefly the results for one-dimensional objects, not only to simplify the discussion but also because rigorous theorems are available at present only for this case.

For either classical or quantum cases, the wavefunction in the image plane can be written as a finite Fourier transform of the object wave:

$$u(x) = \int_a^b U(r)\, e^{-ikrx}\, dr\,, \qquad -x_0 \le x \le x_0\,, \tag{5}$$

where (a,b) denotes the smaller of the object size or beam, and data are given over the interval $[-x_0,x_0]$. The fundamental approximation is the presumption that $u(x) = 0$ for $|x| > x_0$. This is, of course, an unwarranted extrapolation, and we return to a discussion of it below. When only intensities are measured, the data take the form

$$g(x) \equiv u(x)\, u^*(x) = \int_{-R}^{R} G(r)\, e^{-ikrx}\, dr\,, \tag{6a}$$

$$G(r) = 2\mathrm{Re} \int_a^{b-|r|} U(r')\, U^*(r' + |r|)\, dr'\,, \tag{6b}$$

and $R \equiv b - a$. In the absence of phase information, direct inversion can only yield $G(r)$, and it is generally not possible to obtain $U(r)$ from Eq. (6b). But if $U(r)$ is integrable over (a,b), it is possible in principle to extract the phase from the intensity alone, as follows.

Formal inversion of Eq. (5) yields

$$U(r) = \int_{-x_0}^{x_0} u(x)\, e^{ikrx}\, \frac{dx}{2\pi}\,, \qquad a \le r \le b\,, \tag{7}$$

and a theorem of Paley and Wiener [18] implies that both u(x) and U(r) can be continued into the complex plane as entire functions, with appropriate growth properties. The function u(z) can then be characterized completely by its complex zeros. For example, if a = -b and there is no zero at the origin, the Hadamard product is

$$u(z) = B \prod_{j=-\infty}^{\infty} \left[1 - \frac{z}{z_j}\right], \tag{8}$$

where B is a scale factor. If only the amplitude is known, we can write

$$u(z) = |u(z)|\, e^{i\phi(z)}, \tag{9a}$$

$$\ell n\, u(z) = \ell n|u(z)| + i[\phi(z) \pm 2n\pi], \tag{9b}$$

and use the theory of Hilbert transforms to relate $\phi(z)$ to $|u(z)|$. In this case the logarithmic Hilbert transform yields

$$\phi(x) = \frac{x}{\pi} P \int_{-\infty}^{\infty} \frac{\ell n|u(x')|}{x'(x'-x)}\, dx', \tag{10}$$

where the denominator emerges from a subtraction procedure needed to render the integrand square integrable. Evaluation of this last integral by contour integration, however, requires one to avoid the singularities at the zeros of u(z). Although there exist a number of methods for solving this problem, such as modifying the Hilbert transform [17,19] or shifting the zeros [20], eventually one is really forced to locate the zeros of u(z). In that eventuality, one may as well just use the Hadamard product, (8), to construct u(z).

In the most common problem one knows only g(x), which can also be continued into the complex plane as an entire function. Then g(z) = u(z) $u^*(z^*)$ contains zeros of both functions. Because

$$\left[1 - \frac{z^*}{z_j}\right]^* = \left[1 - \frac{z}{z_j^*}\right], \tag{11}$$

the zeros of g(z) are isolated and distributed symmetrically about the x-axis. Unfortunately, this leads to a basic ambiguity: Because g(x) is invariant with respect to "flipping" zeros from z_j to z_j^*, N zeros present in the interval of measurement will yield 2^N different functions u(z), all compatible with the observed g(x) and all having the same band limit [21]. Zeros on the real axis can be located by inspection, and the relative minima in the intensity will occur at the x-coordinate of a pair of complex-conjugate zeros. These assertions are summarized in a theorem of Titchmarsh [22], from which we can also conclude that the zeros tend to lie asymptotically along the real axis at the Nyquist frequency [23].

A restoration method begins with the approximate inversion of Eq. (6a):

$$G(r) = \int_{-x_0}^{x_0} g(x)\, e^{ikrx} \frac{dx}{2\pi}, \qquad a \le r \le b. \tag{12}$$

This result is then substituted into Eq. (6a) itself, and we study

$$g(x + iy) = \int_{-R}^{R} G(r)\, e^{kry}\, e^{-ikrx}\, dr, \qquad -x_0 \le x \le x_0, \tag{13}$$

as a function of the parameter y. Owing to symmetry, we can sweep just half the complex plane within the strip defined by the measurement interval and thus locate the zeros of g(z). Resolution of the 2^N ambiguity then consists of determining the sign of y_j for the correct U(r). If, in making the far-field approximation, we had retained the next term in the phase, we would merely replace the object wave U(r) by U(r) exp($ikr^2/2d$), where d is the distance from object to image plane. We then make intensity measurements at two different distances. In principle we can calculate the 2^N possible solutions for the first distance and, for each, form $|u_1(x)|$ from the Hadamard product [22]

$$|u_1(x)| = |u_1(0)| \prod_{n=1}^{\infty} \left|1 - \left(\frac{x}{r_n}\right) e^{-i\theta_n}\right| . \tag{14}$$

Next calculate the 2^N associated object waves U(r)exp($ikr^2/2d_1$) by Fourier inversion and for each determine the corresponding intensity $|u_2(x)|$ at the second distance from Eq. (5), again using the above replacement. Owing to the presence of the quadratic phase factor, and the fact that the distances are different, the set of 2^N functions $u_2(x)$ will be distinct, including their moduli. Now compare this set with the data, $g_2(x) = |u_2(x)|^2$, and determine the correct set of zeros. Having determined u(x), we now obtain the correct U(r) from the inversion formula, Eq. (7).

By clever use of the way in which the given information is encoded in the zeros of u(z), the enormous number of calculations can be reduced from 2^N to ~2N and the entire procedure made reasonable. Simulated computer experiments have produced some rather impressive results in this way [20]. Similar procedures have been developed for a time series when only samples of its phase or magnitude are given [24].

The difficulties with these techniques are almost obvious. Extension to two and three dimensions introduces <u>lines</u> of zeros, and any noise in the data will severely complicate the location of zeros. Most serious is the cutoff of

$u(x)$ for $|x| > x_0$, giving rise to problems only too familiar from time-series analyses. There may exist many zeros outside this interval and there seems to be no way to control the approximation. Zeros of higher order, of course, will further complicate the calculations.

In reality one generally has data on $g(x)$ at only a few points, and these are often noisy. What one would like to do is construct an optimal representation $\hat{g}(x)$ from the data and then extrapolate it as far as possible beyond x_0. One might then apply the preceding methods using $\hat{u}(x)$ to obtain $\hat{U}(r)$. Of course, it will still be necessary to unfold the desired object from $\hat{U}(r)$ eventually. A beginning has been made on such a program by Schick and Halloran [25].

As a final example in the direct-inversion context, consider the restoration of data that have been degraded by a linear band-limited system. In the absence of noise, and in a convenient single dimension, the noiseless image is given by

$$u(x) = \int_{-b}^{b} K(x-y)\, U(y)\, dy\,, \tag{15}$$

the convolution indicating a certain uniformity in the physical situation. The point spread function $K(x)$ has vanishing Fourier transform outside the inverval $[-\omega,\omega]$, and $U(y)$ is presumed to vanish outside $[-b,b]$. Hence, the Fourier transform of $U(y)$ is an entire function, and one might expect analytic continuation to provide a restoration with unlimited detail. But the object wave is rarely smooth enough to neglect its Fourier transform outside the band, and significant errors can be generated [26]. Nevertheless, when noise is neglected, the inversion is usually pursued, leading to predictably unsatisfactory results.

3. Confronting Reality

Although the practitioners of direct mathematical inversion have recognized that data are specified only over a finite range, and most often without phase information, a great deal of idealization necessarily remains in this approach. For example, the presence of noise in any realistic problem is too pervasive to ignore for long, and the resulting instabilities in the mathematical formulation have also had to be recognized.

A well-known example illustrating the mathematical problem is that of the integral equation (2). Let $U(y)$ be a solution for given $u(x)$, and add to this solution a function $U_n(y) = C \sin(ny)$, where C is an arbitrary constant. Then with $\overline{U}(y) = U(y) + U_n(y)$,

$$\int_{a}^{b} K(x,y)\, \overline{U}(y) dy = u(x) + C \int_{a}^{b} K(x,y) \sin(ny)\, dy\,. \tag{16}$$

But the Riemann-Lebesque lemma assures us that the second term on the right-hand side of Eq. (16) can be made as small as desired by taking n sufficiently large. Thus, small changes in the data can lead to large changes in the solution. Other examples are readily constructed by means of Tauberian theorems, and discretization of the Fredholm equation leads to similar difficulties.

Let us consider the inverse diffraction problem as a further example, in which the filtration properties of space itself are exhibited. Plane waves scattered by a sphere of radius a produce a scattering amplitude in the far field that can be represented by the partial-wave expansion (4). One knows that agreement with measurement can be obtained by keeping only ka terms, roughly. The higher-order partial waves apparently are attenuated before reaching the detector, so the measured scattering amplitude may not contain all the information originally encoded in the scattered wave. A radiative solution to the Helmholtz equation in a half-plane $z \geq z_0$ has a Fourier transform [27]

$$u(p,q;z) = \int_{-\infty}^{\infty}\int_{-\infty}^{\infty} e^{ik(px+qy)}\, U(x,y,z)\, dx\, dy\,, \tag{17a}$$

and so can be propagated from a plane at $z = z_0$ to any other plane by

$$u(p,q;z) = e^{isk(z-z_0)}\, u(p,q;z_0)\,,$$
$$s \equiv (1 - p^2 - q^2)^{1/2}\,,\ \mathrm{Im}(s) \geq 0. \tag{17b}$$

The inverse problem is the following: Given u on the plane $z = z_1 \geq z_0$, obtain u on any other plane between z_1 and z_0. But, because all modes for which $(p^2 + q^2) > 1$ are attenuated in the direct problem, any inaccuracies in $u(z_1)$ owing to noise will lead to instabilities in the inversion.

The initial study of unstable mathematical problems of the kind described above is generally attributed to Hadamard [28], who called them *ill-posed problems*, but they were actually studied and so named much earlier [29]. Only in the past 20 years have they been thought to be of physical importance, namely as an interpretation of the effects of noise on reconstruction problems. The major mathematical tool developed for treating these instabilities is that of *regularization*, which effectively restricts the class of admissible solutions to Eq. (2), say. That is, when u(x) is not specified with complete certainty, Eq. (2) admits a larger class of solutions than otherwise, and a measure of this increase is the magnitude of the uncertainty in u(x). The regularization procedure is designed to bring this class back down to size by imposing further constraints on the problem. Although the choice of constraints is somewhat arbitrary—in fact, ad hoc—one can be guided by using physical conditions on the problem as prior information. One may have a realistic upper bound on the energy of a signal, for example, which can be used to impose a global bound on the solutions; or, smoothness conditions can provide bounds on derivatives. Much of the work on regularization techniques has been pioneered by Tikhonov and his collaborators [30], and the application to inverse problems has been reviewed recently [31]. Indeed, an entire industry in regularization has sprung up.

If nothing else, these analyses of instabilities have contributed a great deal to our understanding of linear Fredholm equations of the first kind and linear algebraic systems. But one has the impression that the quintessential aspect of the data-inversion problem has been missed, for the data function is limited not only in range and phase, but also in continuity and precision. On the one hand, data received on an image plane, or on any other record, usually consist of isolated points confused by noise. All of these limitations are seen to be different aspects of the same problem: incomplete information. Without a reliable representation of the data on the image plane, say, no other phase of the restoration problem can even get started in a rational way. On the other hand, the regularization idea contains the seeds of the solution by recognizing the need to employ all the prior knowledge available about the physical problem.

Having followed the evolution of one species to this point, we now turn our attention to the other—to the group that nourishes itself on limited data from the beginning and seeks first to obtain the "true" data scene on the image plane.

4. Analysis of Incomplete Information

It appears that Laplace was the first to formulate the inverse problem in a careful scientific context [32]. If the event u can arise from a number of mutually exclusive possible causes U_i, $i = 1,2,\ldots,N$, and there exists a conceptual model giving the direct probabilities $P(u \mid U_i)$, what is the probability $P(U_i \mid u)$ that the cause of u was U_i? Realizing that often there is prior information I available that would yield prior probabilities $P(U_i \mid I)$, Laplace obtained the complete and correct solution to this problem for the posterior probabilities:

$$P(U_i \mid u, I) = \frac{P(U_i \mid I) P(u \mid U_i)}{\sum P(u \mid U_j) P(U_j \mid I)} , \tag{18}$$

which is usually referred to as Bayes' theorem. Unfortunately, Laplace had no way to calculate the priors, so he took them to be uniform, in which case $P(U_i \mid u) \propto P(u \mid U_i)$. Even so, he went on to use this form to make some of the most important discoveries of his time in astronomy. We note, also, that if there is no a priori conceptual model, the prior probabilities are all one has.

Although Laplace's solution to this inverse problem was ignored for almost 150 years, owing principally to disputes over the definition of probability, some professional statisticans did appreciate the need for statistical analyses of data in inverse problems. But they saw—and to large extent still see—only the random aspects of the data, in terms of the noise. To fit noisy data exactly, of course, is to introduce structure due to noise rather than information. In retrospect, the reason for this attitude is now clear: lacking a means by which to encode prior information into a probability distribution, they simply disregarded such prior information. Since the work of Shannon [33] and Jaynes [34], however, such an attitude is no longer tenable, and to ignore the essential incompleteness of data is sheer folly.

Jaynes has provided a detailed critique of the frequentist point of view toward inverse problems which clearly illustrates the "intuitive inadequacy" of orthodox statistical principles in this respect [35]. Given the scenario described by Eq. (1), in which the data set u is in one-to-one correspondence with the set U and the operator K is nonsingular, one expects the problem to be solvable exactly by direct mathematical inversion. In general, though, the set is underdetermined, and K is singular in the sense that the set u may result from more than one distinct set U. Or, if K is a point spread function, there will be some elements that are very small, and these may give rise to elements of K^{-1} that are very large. In these cases one can only obtain an estimate

$$\hat{U} = R u , \tag{19}$$

where the "resolvent" R is to be determined. At the very least R must be chosen such that $\hat{U}$ lies in the class C of <u>possible</u> states that could have produced u in Eq. (1). The mathematical statement of this requirement is

$$KRK = K , \tag{20}$$

so that R is a generalized inverse [36].

We see again that the basic problem is not one of randomness, but of incomplete information. The observed data, as well as orthodox statistics, can tell us only that the "true" scene must lie in class C, but there is no guidance as to what unique choice is to be made within that class. Indeed, without prior information the problem is not solvable, yet there exist common problems in economics (marketing decisions) and medicine (symptoms related nonuniquely to diseases), to name only two areas, in which such decisions must be faced daily. It is ironic that the one species, while almost blind to the essential incompleteness of information, nevertheless recognizes the need for prior information, whereas the other appreciates the imprecision in the data but sometimes fails to see that the crucial ingredient for effecting a solution lies with the prior information. Evolution works in strange ways.

The first step in solving any inverse problem, then, is to reconstruct the "true" data scene on the image plane, say. The total information consists of <u>both</u> the data <u>and</u> the physical knowledge contained in the statement of the problem. One must begin with the question: What is the actual information available? It was just for the purpose of answering this question, after all, that information theory was invented. Perhaps the first true resolution of an inverse problem in a modern context was presented by Boltzmann and Gibbs in precisely this way: Given several pieces of macroscopic information about a thermodynamic system in equilibrium, the minimum prior knowledge needed for a complete macroscopic description is a specification of its energy levels.

Until 25 years ago this evolutionary track for the inverse problem had been somewhat faint, but it was strengthened considerably in 1967 with Burg's now-classic work in geophysics [37]. This date is, in fact, a landmark in the evolutionary history of generalized inverse problems, for it begins the journey toward full recognition that the fundamental problem is one of inference, not deductive mathematics. That is, with reference to Laplace's result (18), and focusing for the moment only on the pure problem which presumes no noise, we see that a definite solution can be obtained only from prior probabilities. But we now have the principle of maximum entropy (PME) for encoding prior information into a probability distribution [34] and, in the event that information is given in the form of expectation values, the problem was solved once and for all by Gibbs [38].

Given M pieces of data in the form of expectation values,

$$F_k \equiv \sum_{i=1}^{m} p_i\, f_k(x_i)\,, \qquad 1 \le k \le M < m\,, \tag{21}$$

maximization of the entropy $S = -\Sigma_i\, p_i\, \ell n\, p_i$ subject to these constraints and the mutual exclusivity of the exhaustive set $\{x_i\}$ yields the distribution

$$p_i = Z^{-1} \exp[-\lambda_1 f_1(x_i) - \cdots - \lambda_M f_M(x_i)]\,, \tag{22a}$$

$$Z(\lambda_1 \ldots \lambda_M) = \sum_{i=1}^{m} \exp[-\lambda_1 f_1(x_i) - \cdots - \lambda_M f_M(x_i)]\,, \tag{22b}$$

where the Lagrange multipliers $\{\lambda_k\}$ are determined from the coupled set of differential equations

$$F_k = -\frac{\partial}{\partial \lambda_k} \log Z\,, \quad 1 \le k \le M\,. \tag{22c}$$

The expectation value of any other function $q(x)$ is then

$$\langle q(x) \rangle = \sum_{i=1}^{m} p_i\, q(x_i)\,, \tag{22d}$$

which is the best prediction we can make in the sense of minimum expected square of the error. Equations (22) provide the least-biased solution for the priors based on the data set (21), and only on that. Note that Eqs. (21) delineate the possible distributions, whereas maximization of S selects the optimal $\{p_i\}$ within that set.

As an elementary exercise in the use of this formalism, and one that contains important elements of the inverse problem, consider a die to be tossed a large number of times. With an honest die we would expect the average number of spots "up" to be 3.5, but suppose we find instead that the average is

$$4.5 = \sum_{i=1}^{6} i\, p_i . \tag{23}$$

If a wager is suggested as to the outcome of the next toss, we would do well to calculate the probabilities based on the evidence (23), and the above formalism yields the probability distribution [39]

$$\{p_1 \ldots p_6\} = 0.054, 0.079, 0.114, 0.166, 0.240, 0.348. \tag{24}$$

It must be emphasized that these are the probabilities of obtaining a particular number of spots on the next toss, a single trial.

But we now see that it is possible to reinterpret this scenario as an inverse problem, providing the answer to a different question. If we view the data set (23) as providing a mean-value constraint on the frequencies with which different numbers of spots turned up in the experiment, then, as we shall see below, Eq. (24) also gives the optimal estimate of what those frequencies were that led to Eq. (23). It depends on what questions we ask. These probability-frequency relations have been discussed at some length elsewhere [39], and here serve to illustrate that when frequency data are available the PME is quite capable of treating them as legitimate prior information. Quite often, the frequency distribution that can be realized in the greatest number of ways is numerically equal to the probability distribution that maximizes the entropy. In addition, Jaynes [40] has used his entropy concentration theorem to show that in a very large number of tosses the overwhelming majority of possible distributions compatible with Eq. (23) have entropy very close to the maximum.

The preceding discussion describes the simplest example of a class of generalized inverse problems for which we now have the complete solutions, in the noiseless case. More specifically, let us rewrite Eq. (1) in the discrete form

$$u_j = \sum_{i=1}^{m} K_{ji}\, f_i , \qquad 1 \leq j \leq M < m . \tag{25}$$

The class of problems we consider is described as follows: A random experiment has m possible results at each trial, so that there are m^n conceivable outcomes in n trials, and each outcome yields a set of sample numbers $\{n_i\}$ along with frequencies $f_i = n_i/n$, $1 \leq i \leq n$. Included in this class of problems is the above dice problem, as well as statistical mechanics, time series analysis, image reconstruction, etc. In the case of image reconstruction, for example, one considers n elements of luminance to be distributed over m pixels to form a scene, such that the ith pixel receives a fraction f_i of the total luminance. Each problem merely requires the proper interpretation of the relevant quantities in Eq. (25).

Although the data constraints (25) only delineate the class C of possible solutions, the very character of the problem provides a good deal of cogent prior information. If in n trials the ith result occurs $n_i = nf_i$ times, $1 \leq i \leq n$, then out of the m^n possible outcomes the number of those yielding a particular set of frequencies $\{f_i\}$ is given by the multiplicity factor

$$W = \frac{n!}{(nf_1)!\ldots(nf_m)!} \, . \tag{26}$$

If we specify the set $\{f_i\}$ to be uniform, this can be made into a probability distribution. But as $n \to \infty$, Stirling's approximation yields

$$n^{-1} \log W \equiv H = -\sum_i f_i \log f_i \, , \tag{27}$$

and in asking for the set $\{f_i\}$ that can be realized in the greatest number of ways, we have just reformulated the maximum entropy problem. Indeed, the entropy concentration theorem tells us that this is overwhelmingly the greatest number of ways. The frequencies are readily found to be

$$f_i = Z^{-1} \exp\left[-\sum_j \lambda_j K_{ji}\right] , \tag{28}$$

and the Lagrange multipliers are determined from the constraints in the usual way. Other multiplicity factors have been considered by Frieden [41].

We see, then, that whenever the pure inverse problem can be formulated in a frequency context, the solution is given immediately by Eq. (28). In the scenario of optical image reconstruction, for example, the set $\{f_i\}$ describes the optimal scene provided by the data set (25). But now we note that we have obtained even more, for the algorithm may permit us to interpolate missing data, as well as to possibly construct an optimal extrapolation to the remainder of the image plane based only on those data. This then provides the necessary input to the other phases of reconstruction. In the case of time series analysis one obtains an exact prescription for extrapolating the autocovariance beyond the data [40].

5. Including the Noise

The PME is designed to encode data and prior information into a prior probability distribution whenever this information can be put into an appropriate 'testable' form [42], and when we have known multiplicity factors the prescription outlined above is optimal for a certain class of problems. If noise is present in the data, however, an element of randomness is introduced that does not, at first glance, come under the purview of maximum entropy. Indeed, for a pure noise problem the methods of orthodox statistics often deal with the noise optimally, so that in the general case one anticipates the necessity of a full Bayesian solution which would incorporate both modes of reasoning.

It appears that this full solution has yet to be developed, although its possible form is suggested by Laplace's result, (18), in which $P(u|U)$ can be interpreted from a sampling model for the noise. A correct general procedure would be to obtain $P(U|u,I)$ and then calculate the expectation values $\langle U \rangle$ with this probability distribution. Although this prescription can be worked out [43], the data often are not in a form that allows us to do this. If the frequency interpretation is valid, however, some possibilities suggest themselves through modification of Eq. (25) to account for linear noise:

$$u_j = \sum_i K_{ji} f_i + e_j , \quad 1 \leq j \leq M \leq m, \tag{29}$$

where e_j are traditional noise terms.

As an example, consider the case of Gaussian independent noise and define the quadratic form

$$Q(f_1 \dots f_m) \equiv \frac{1}{2} \sum_{j=1}^{M} \sigma_j^{-2} \left[\sum_{i=1}^{m} K_{ji} f_i - u_j \right] \tag{30}$$

and thus a likelihood factor $\exp(-Q)$. The latter can be interpreted as a probability in the sense of Gauss' theory of errors. Following Bernoulli, we also derive from Eq. (26) the prior probability $P(U|I) = m^{-n}W$, and from Bayes' theorem we obtain the probability for a scene $\{f_i\}$:

$$P \propto e^{nH-Q} , \tag{31}$$

up to a normalization factor. We now follow Gauss' dictum and find the most probable set $\{\hat{f}_i\}$ as that which maximizes P. Although the point at which P is a maximum does not always coincide with the expectation value, in these cases the maximum is unique and very sharp as $n \to \infty$.

It is important to note that the likelihood factor behaves precisely as we would expect from an examination of Bayes' theorem. In the limit of zero noise it behaves almost like a δ-function, telling us only that the correct solution must be in the class C reproducing the data, and also imposing the constraints of the pure maximum-entropy problem. As the noise becomes overwhelming, the uniform gray scene emerges.

Jaynes has provided a nice geometric picture of this procedure [40], and observes that a reinterpretation is immediately suggested. As Lagrange noted long ago, imposition of a new constraint in a variational calculation cannot change the solution if the old solution already satisfied that constraint. Hence, because the noise is distributed in χ^2 in the above model, one can simply constrain this to some desirable value Q_0 and rewrite Eq. (31) as

$$P \propto e^{nH-\lambda Q} . \tag{32}$$

The unconditional Bayesian maximum is therefore reinterpreted as a constrained maximum, and there is only one Lagrange multiplier to be determined. This is precisely what Gull and Daniell did in a beautiful application of maximum entropy reconstruction in radio astronomy [44]. Undoubtedly, John Wheeler would classify this result as "magic without magic!"

But reflection upon the mathematical theory of regularization reveals that what we have just described is qualitatively just what is suggested there. In the present context the noiseless data delineate the class C of solutions as well as provide constraints on the multiplicity factor (prior information) enabling us to make a choice within that class. The presence of noise enlarges that class of possible solutions and, because it is noise, the class is no longer defined with precision; its boundaries are "fuzzy." Our prior information is now a bit more qualitative but must nevertheless be used to restrict the class of variations when maximizing W or H. This requirement is somewhat similar to that employed in deriving stationary-state ensembles in statistical mechanics [45], where the density matrix is constrained to commute with the Hamiltonian.

Of course, the semi-qualitative nature of the prior information about noise renders the choice of constraint somewhat arbitrary. A rather weak form might be to place a reasonable upper bound on the noise. One can do much better than this, however, for the experimentalist can usually supply a more detailed description of the noise. Owing to the availability of tables of χ^2, the choice described by Eq. (32) is quite reasonable, and possesses as well the desirable convexity property of the geometrical description. But one cannot "squeeze" χ^2 too closely in such a scheme, and moreover the choice of statistics is not necessarily unique. Bryan and Skilling [46] have utilized another with equally good results.

Although the practical applications of these choices have produced remarkable results [44,46], there obviously remains an element of ambiguity in the handling of noise. Yet in these constructions possibly lies the key to an eventual full Bayesian solution to the general inversion problem. Such a solution, of course, should encompass also those scenarios that can be interpreted neither in a frequency context nor in a sense of expectation-value data.

Having taken some notice of one another, our two species of practitioners may yet bring the best ideas of each to bear further on the problem. One can only hope. In the meantime, the frustrations undoubtedly felt by both may be slightly relieved by noting they are in good company [47]:

Oh, happy he who still hopes he can
Emerge from Error's boundless sea!
—Faust

Acknowledgments

A large portion of this article was written while the author was a Visiting Professor at the Instituto de Física e Química de São Carlos, Universidade de São Paulo, São Carlos, SP, Brasil. That support is much appreciated.

References

1. Wheeler, J. A. (1976) in E. H. Lieb, B. Simon, and A. S. Wightman, eds., Studies in Mathematical Physics (Princeton, N.J.: Princeton University Press).
2. Plato, The Republic, Book VII (Great Books of the Western World, Vol. 7, Encyclopaedia Britannica, Inc., Chicago, 1952).
3. Kant, I. (1964) Critique of Pure Reason (London: Macmillan).
4. Lord Rayleigh (1877) The Theory of Sound (New York: Dover, 1945).
5. Kac, M. (1966) Am. Math. Mon. **73**, 1, Part II.
6. Borg, G. (1946) Acta Math. **78**, 1.
7. Isaacson, E. L., and E. Trubowitz (1983) Commun. Pure Appl. Math. **36**, 767; Isaacson, E. L., H. P. McKean, and E. Trubowitz (1984) Commun. Pure Appl. Math. **37**, 1.
8. Ma, S. T. (1946) Phys. Rev. **69**, 668; (1947) **71**, 195, 210.
9. Bargmann, V. (1949) Phys. Rev. **75**, 301; Rev. Mod. Phys. **21**, 488.
10. Hoyt, F. C. (1939) Phys. Rev. **55**, 664.
11. Gel'fand, I. M., and B. M. Levitan (1951) Izv. Akad. Nauk SSSR, Mat. **15**, 309 [1955, Am. Math. Soc. Translations (2) **1**, 253].
12. Marchenko, V. A. (1950) Dokl. Akad. Nauk SSSR **72**, 457; (1955) **104**, 695.
13. Sabatier, P. C. (1972) J. Math. Phys. **13**, 675.
14. Faddeev, L. D.; translated by B. Seckler (1963) J. Math. Phys. **4**, 72.
15. Newton, R. G. (1980) J. Math. Phys. **21**, 1698; (1981) **22**, 631, 2191; (1982) **23**, 594.
16. Hoenders, B. J. (1978) in H. P. Baltes, ed., Inverse Source Problems in Optics (Berlin: Springer-Verlag).
17. Burge, R. E., M. A. Fiddy, A. H. Greenaway, and G. Ross (1976) Proc. R. Soc. London **A350**, 191.
18. Paley, R. E. A. C., and N. Wiener (1934) Fourier Transforms in the Complex Domain (Ann Arbor, Michigan: American Math. Soc.).

19. Toll, J. S. (1956) Phys. Rev. **104**, 1760.
20. Ross, G., M. A. Fiddy, and N. Nieto-Vesperinas (1980) in H. P. Baltes, ed., Inverse Scattering Problems in Optics (Berlin: Springer-Verlag).
21. Walther, A. (1963) Opt. Acta **10**, 41.
22. Titchmarsh, E. C. (1926) Proc. London Math. Soc. (2) **25**, 283.
23. Bond, F. E., and C. R. Cahn (1958) IRE Trans. Inf. Theory **IT-4**, 110.
24. Hayes, M. H., J. S. Lim, and A. V. Oppenheim (1980) IEEE Trans. Acoust. Speech Signal Process. **ASSP-28**, 672.
25. Schick L. H., and J. Halloran (personal communication).
26. Bertero, M., C. De Mol, and G. A. Viano (1979) J. Math. Phys. **20**, 509.
27. Shewell, J. R., and E. Wolf (1968) J. Opt. Soc. Am. **58**, 1596.
28. Hadamard, J. (1923) Lectures on the Cauchy Problem in Linear Partial Differential Equations (New Haven, Conn.: Yale University Press).
29. Bertrand, J. (1889) Calcul des probabilités (Paris: Gautier-Villars), pp. 4-5.
30. Tikhonov, A. N., and V. Y. Arsenin (1977) Solutions of Ill-Posed Problems (New York: Wiley).
31. Bertero, M., C. De Mol, and G. A. Viano (1980) in H. P. Baltes, ed., Inverse Scattering Problems in Optics (Berlin: Springer-Verlag).
32. Laplace, P. S. (1774) Mémoire sur la probabilité des causes par les événements [Oeuvr. compl. **8**, 27 (1891)].
33. Shannon, C. E. (1948) Bell Syst. Tech. J. **27**, 379, 623.
34. Jaynes, E. T. (1957) Phys. Rev. **106**, 620; **108**, 171.
35. Jaynes, E. T. (1981) International Convention on Fundamentals of Probability Theory and Statistics, Luino, Italy, September 1981.
36. Nashed, M. Z., ed. (1976) Generalized Inverses and Applications (New York: Academic Press).
37. Burg, J. P. (1967) Maximum entropy spectral analysis, 37th Annual International Meeting, Society of Exploration Geophysicists, Oklahoma City.
38. Gibbs, J. W. (1902) Elementary Principles in Statistical Mechanics (New Haven, Conn.: Yale University Press).
39. Jaynes, E. T. (1978) in R. D. Levine and M. Tribus, eds., The Maximum Entropy Formalism (Cambridge, Mass.: MIT Press).
40. Jaynes, E. T. (1981) What is the problem? First ASSP Workshop on Spectral Estimation, McMaster Univ., Canada, August 1981.
41. Frieden, B. Roy (1980) Comp. Graphics Image Process. **12**, 40.
42. Jaynes, E. T. (1968) IEEE Trans. Syst. Sci. Cybern. **SSC-4**, 227.
43. Grandy, W. T., Jr. (unpublished).
44. Gull, S. F., and G. J. Daniell (1978) Nature **272**, 686; (1980) IEE Proc. (E) **5**, 170.
45. Grandy, W. T., Jr. (1980) Phys. Rep. **62**, 175.
46. Bryan, R. K., and J. Skilling (1980) Mon. Not. R. Astron. Soc. **191**, 69.
47. von Goethe, J. W., Faust (Great Books of the Western World, Vol. 47, Encyclopaedia Britannica, Inc., Chicago, 1952).

WHERE DO WE GO FROM HERE?

E. T. Jaynes

Arthur Holly Compton Laboratory of Physics, Washington University, St. Louis, Missouri 63130

C. Ray Smith and W. T. Grandy, Jr. (eds.),
Maximum-Entropy and Bayesian Methods in Inverse Problems, 21–58.

1. Introduction

With this meeting, we enter a new era for the Principle of Maximum Entropy. Gathered in one room are many people who have been working largely in isolation from each other, while thinking very similar thoughts. Each has discovered, in his own way and in his own context, that this principle solves real, nontrivial problems in a way that cannot be approached by other statistical methods.

But future progress will be more rapid if we can pool the experience gained thus far in many different applications, and recognize common problems still in need of solution. That is the main purpose of these meetings.

My own work has been concerned mostly with development of the general theory of irreversible thermodynamics, although attempts have been made also to point out other possible applications. The history of the principle of maximum entropy, as it applies to thermodynamics, has been told recently (Jaynes, 1978) at such great length that hardly anyone could wish to have it all told again here. But as applied to problems far removed from thermodynamics, a few recollections about early experiences may help to set the record straight and explain to present students why the method is only now coming into its own.

Section 2 of this paper recalls a little history, but very briefly because our concern today should be with the future. It appears, in retrospect, that the development of computer programs capable of dealing with dozens to thousands of simultaneous constraints was the key factor in establishing the power of this method, in a way that transcended all philosophical arguments.

Sections 3 through 6 deal with technical problems, first in generality, then as applied specifically to spectrum analysis and image reconstruction. In these discussions we try to formulate some of the common problems that need to be dealt with in the immediate future in connection with presently established applications. Finally, we speculate a little about new applications that might develop in the more distant future.

2. Setting the Record Straight

Of course, the maximum-entropy algorithm originated with Boltzmann and Gibbs. But in their writings they did not make its meaning crystal clear—Boltzmann because he was not very clear in his own mind, repeatedly changing his position (Klein, 1973), and Gibbs because his work was left unfinished (Jaynes, 1967). In Gibbs' "Heterogeneous Equilibrium" (1875-1878), in fact, we find a much clearer and deeper explanation of the properties of entropy than in his final work, Statistical Mechanics (1902).

Therefore, as far as the rationale of the method was concerned, Boltzmann and Gibbs left a kind of vacuum that was filled, as such vacua always are, by followers whose thinking had little in common with that of Boltzmann, and nothing in common with that of Gibbs. For 60 years after Gibbs, the "official" statistical mechanics of our textbooks stood on the premise that the canonical (maximum entropy) distribution was an actual physical fact, the ultimate result of the mechanical equations of motion operating over long times, and that the fundamental unfinished goal of statistical mechanics was to prove this rigorously by "ergodic" theorems.

In the early 1950s—just when Enders Robinson was doing some of the first hand-run predictive deconvolutions of geophysical data at MIT—I was at Stanford, studying not only statistical mechanics but also the problem of detecting hidden regions of different dielectric or acoustical properties (nonmetallic land mines in Korea, crevasses in Greenland, brain tumors, flaws in structural materials, etc.).

In those days nobody even dreamed of being able to use digital signal processing hardware for such purposes although we did dream of analog calculation, recording the observed signal s(x,y) as a function of the antenna position, and correlating it with those to be expected from every possible target by endless-loop tape recorders (which would have been so bulky and slow as to be almost useless).

A conceivable fast solution was to try to design the antenna so that its radiation pattern (or more accurately, the dot product $\mathbf{E}_t \cdot \mathbf{E}_r$ of transmitter and receiver fields) maximized the ratio (peak signal)^2/(mean square response to soil anomalies), that is, so that the antenna itself approximated a two-dimensional matched filter. But of course the equations defining the optimal matched filter contradicted Maxwell's equations or the acoustical wave equations; and even if the optimal antenna could have been realized, it would have been optimal only for soil of one particular autocorrelation function. Today, of course, all this sounds trivial; an electronic package the size of a matchbook could do digital processing of these signals, according to any algorithm we please, in real time.

It was realized that if we had the capability to deconvolve any antenna response pattern, the antenna design problem would be radically changed. As those familiar with the theory of integral equations will see at once, the only important antenna design parameter is then the size of the "window" that it opens up in Fourier transform space. That is, over how large a region (k_1,k_2) is the transform

$$F(k_1,k_2) = \int dx \int dy(\mathbf{E}_r \cdot \mathbf{E}_t) \exp[i(k_1x+k_2y)]$$

above the thermal noise level, thus delivering relevant information to the computer?

These theoretical conclusions were summed up in a classified report in 1954, but at the time the technology to exploit them did not exist and no actual hardware resulted. Still, this association was an important influence on my general thinking, making me see a close relationship between the rationale of these engineering problems and that of Gibbsian statistical mechanics.

To see why a relationship exists, let us ask: What determines our probability distribution for soil anomalies? Conventional thinking—then and now—would suppose this is an actual physical fact, a "real physical property" of the soil, like density, chemical composition, etc., with different kinds of soil having different intrinsic autocorrelation functions. The more I held lumps of soil in my hand, the less I believed this.

The probability distribution for the soil dielectric constant that we use in our detection theory can be useful only to the extent that it tells us something about the one sample of soil that exists under our antenna. Whether variations in other samples do or do not follow this distribution in the frequency sense cannot be relevant to our problem. Indeed, if we had prior information telling us the actual condition of our sample of soil, the frequency distributions in other samples that we are not looking at would be completely irrelevant.

Realizing this seemed to me an important new insight into the nature of statistical inference. Orthodox thinking—then and now—wants us to define probabilities only as physical frequencies, and deplores any other criterion as not "objective." Yet when confronted with a (literally) dirty, objective real problem, common sense overrides orthodox teaching and tells us that to make the most reliable inferences about the special case before us, we ought to take into account all the information we have, whatever its nature; a rational person does not throw away cogent evidence merely because it does not fit into a preconceived pattern. But this means that our probabilities can be equal to frequencies only when (a) we actually have frequency data and (b) we have no other "prior information" beyond those frequencies. (We should add here that R. A. Fisher, the great proponent of frequency interpretations, eventually recognized this also, and in his final book (1956) he acknowledged that fiducial inference is valid only when we have no prior information; unfortunately, many writers of statistics textbooks still have not recognized this.)

The probability distributions that we use for inference must in general represent not merely frequencies, but our total state of knowledge, of which frequencies are only a part. Indeed, some of the most important real problems of inference are unrelated to frequencies in any "random experiment." The problem is that our information is incomplete; there is nothing "random" about it. But strict adherence to orthodox principles would deny us the use of probability theory in such problems.

So there is the connection: the only way known to set up a probability distribution that honestly represents a state of incomplete knowledge is to maximize the entropy, subject to all the information we have. Any other distribution would necessarily either assume information that we do not have, or contradict information that we do have. But that is just the procedure that Boltzmann and Gibbs advocated in statistical mechanics, long before Shannon showed us how to interpret it.

The principle itself was not new at all, but the realization that it applied as well to problems far removed from thermodynamics, and that these engineering problems, when finally solved correctly, would be seen as having the same rationale and the same algorithm as statistical mechanics, was so new and startling that it could not be conveyed to others. Attempts to explain this to other physicists, engineers, mathematicians, and statisticians at Stanford and Berkeley, and to the Army Engineers, met with strictly zero success—with only one exception.

David Blackwell saw the point at once, and put his finger on the basic "fairness" property of the MAXENT algorithm: It does not allow you to assign zero probability to any situation unless your information really rules out that situation. Nobody else was able to free his mind from frequentist preconceptions; and so from this engineering episode I returned to statistical mechanics with nothing to show for it except a new appreciation of the generality of the MAXENT principle.

But before one can see this generality it is necessary to realize that it is a principle of reasoning, not a principle of mechanics, and thus free it from the supposed dependence on Hamiltonian equations of motion and ergodic theorems.

What I did (Jaynes, 1957) was only to suggest a different interpretation of Gibbs' "canonical ensemble" method. If we regard it as representing, not mechanical prediction from the equations of motion, but only the process of inference (make the best predictions you can from the information you have), then in that sense his algorithm can be justified in great generality as a principle of probability theory—essentially just the process of rational thinking—without any appeal to ergodicity.

To this day, our ergodist critics seem to have a hang-up over the distinction between mechanical prediction and inference; they continue to complain about the latter because it is not the former. Nevertheless, for thermal equilibrium (the only case where ergodic theorems could ever have applied) we obtain the same actual predictions that the ergodists wanted; and indeed, any other predictions would be in conflict with experiment. Ergo: the most that ergodic theorems could ever have accomplished is to confirm what we already know, namely that in equilibrium problems taking the equations of motion into account does not alter the predictions that we obtain directly by maximizing the entropy.

In fact, it is rather elementary that equations of motion can tell us only how probabilities change with time, and not what probabilities should be assigned initially. Therefore, in any problem where the given information (set of constraints) tells us only something about constants of the motion, the equations of motion can provide zero additional information about the state of the system. In John Burg's happy phrase, then, all the philosophical arguments "do not change a single number" in the calculations.

The pragmatic advantage of the inference viewpoint (which hardly needs pointing out to this audience) is that we then see Gibbs' method as something of far greater generality than the ergodists ever dreamed of. It applies equally well not only to nonequilibrium situations but to any problem of inference, in or out of physics, in which the information at hand can be described by enumerating a set of conceivable hypotheses ("prior information") and specifying whatever else we know ("data") in the form of constraints that narrow down the set of possibilities. In such applications, it has some easily proved optimality properties.

However, in my papers of 1957 I was still under the influence of textbooks that attributed the ergodic view to Gibbs; therefore I thought I was suggesting something new. Three years later, finally getting around to a really careful reading of Gibbs, I realized that I had been brainwashed. Gibbs' actual thinking was utterly different from that commonly attributed to him, and I had only rediscovered his original viewpoint! (There was, however, a technical advance in that we now had Shannon's theorems that filled a gap in Gibbs' argument.) In the trauma of this discovery, a book review I wrote just then (Jaynes, 1961) came down hard on two other poor souls who had only been victims of this same brainwashing, but who had had the misfortune to write books before realizing it. Expressions of sympathy would have been more appropriate.

For some years (1958-1964) I traveled around a circuit of talks at universities and industrial research laboratories (including those of three well known oil companies, at Dallas, Tulsa, and Bartlesville), pointing out the existence and generality of the MAXENT algorithm to hundreds of people who might have profited by using it in all kinds of applications. But the reaction was mostly negative. Few saw the point, and most (under the influence of conventional physical and/or statistical teaching) denied vehemently—even angrily—that useful predictions could come from a method that expressed only "subjective" probabilities, not "objective" frequencies.

On one of these occasions a member of the audience was moved to express these doubts in poetry. I do not recall his name, or even the year and place, but the poem has been preserved because I was in turn moved to compose a counter-poem and read them both to my students. They were rediscovered this year in some yellowed old course notes.

Here is the original, whose author is presumably just as happy to remain anonymous (if not, he may come forward and claim it):

MAXIMUM ENTROPY

There's a great new branch of science which is coming to the fore,
And if you learn its principles you'll need to learn no more,
For though others may be doubtful (which can make them fuss and fret)
You'll be certain your uncertainty's as large as it can get.

This is no toy of theory, to invite the critic's jeering
But a powerful new method used in modern engineering.
For uncertainty, which often in the past just gave us fidgets
Has proved to be a vital tool in mass-producing Widgets.

The procedure's very simple (it is due to Dr. Jaynes)
You just maximize the entropy, which doesn't take much brains.
Then you form a certain function which we designate by Z,
Differentiate its log by every lambda that you see.

And LO!—we see before us (there is nothing more to do)
All the laws of thermal physics, and decision theory too.
Possibilities are endless, no frontier is yet in sight,
And regardless of your ignorance, you'll always know you're right!

So, are you faced with problems you can barely understand?
Do you have to make decisions, though the facts are not at hand?
Perhaps you'd like to win a game you don't know how to play.
Just apply your lack of knowledge in a systematic way.

Handed a copy of this upon leaving the meeting, I composed a reply during my return flight:

MAXIMUM ENTROPY REVISITED

There's a fine old branch of science coming back now to the fore
And if you learn its principles you'll need to grope no more.
For though others may be blinded (which can make them twist and turn)
You'll be certain that you see whatever is there to discern.

This is no cut-and-try device, empirical ad hockery
But a principle of reasoning, of proven optimality.
For prior information, which empiricism spurns
Has proved to be the pivot point on which decision turns.

The procedure's very simple (it is due to Willard Gibbs)
First define your prior knowledge without telling any fibs
On this space of possibilities, a constraint is then applied
For every piece of data, until all are satisfied.

If the states you thought were possible, in setting up this game
And the real possibilities in Nature, are the same
Then LO! you see before you (there is nothing more to grind)
Reproducible connections that experimenters find.

But the principles of logic are the same in every field
And regardless of your ignorance, you'll always know they yield
What your information indicates; and (whether good or bad)
The best predictions one could make, from data that you had.

This reply tried to correct the historical record and the mistaken emphasis by pointing out the positive nature of the principle, which seemed obviously true to me—and obviously false to nearly everybody else.

Not quite everybody, however. During the early 1960s several rumors circulated, too vague to leave a trail to fact, but hinting guardedly at uses of maximum entropy in oil exploration. Finally, with the landmark paper of Burg (1967), it became public knowledge that a few people had indeed taken the trouble to work out the detailed algorithms and computer programs needed to try the method on some real problems far removed from thermodynamics. They found, of course, that it worked just as I had said it would (or else we would not be meeting here today).

From this point on, I shall not presume to recount a history of these applications, so much better known to those here who made it. But, for that goal of setting the record straight, perhaps other speakers will be willing to share with us some of their own recollections of how it all started.

3. General Remarks

Maximum entropy represents an entirely different kind of thinking from what has been taught in statistics courses for 50 years. Many of those trained in orthodox statistics find the difference so mind-wrenching that the rationale of MAXENT remains incomprehensible to them after repeated attempts at explanation. Yet beginning students see the point at once with no difficulty because MAXENT is just the natural, common-sense way in which anybody does think about his problems of inference—unless his mind has been warped by orthodox teaching. This difference will be less troublesome if we explain it first in very general terms.

ORTHODOX APPROACH. You are given a set of observed data $D_{obs} = \{d_1 \ldots d_m\}$ from which you are to decide whether some hypothesis H about the real world is true (or, put more cautiously, whether to act as if it were true). For example, H might be the statement that some unobserved quantity θ has a value in a specified interval ($a \leq \theta \leq b$); in fact, by imaginative use of language, almost any hypothesis could be stated in this form.

Given this problem, the first thing orthodox statistics does is to imbed the observed data set in a "sample space," which is an imaginary collection containing other data sets $\{D_1 \ldots D_N\}$ that one thinks might have been observed but were not. Then one introduces the probabilities

$$p(D_i | H)\,, \qquad 1 \leq i \leq N, \tag{1}$$

that the data set D_i would be observed if H were true. This is called the "sampling distribution," and $p(D_i | H)$ is interpreted as the frequency with which the data set D_i would be observed in the long run if the measurement were made repeatedly with H constantly true.

When one asserts the long-run results of an arbitrarily long sequence of measurements that have not been performed, it would appear that he is drawing either on a vivid imagination or on a rather large hidden fund of prior knowledge about the phenomenon. If we are not told what that knowledge is and how it was obtained, we might be excused for doubting its existence. But, for the sake of argument, let us suppress these doubts and accept the orthodox interpretation of $p(D_i | H)$ at its face value.

The sampling distribution is the only probability distribution we are allowed to use. For that reason, orthodox statistics is often called "sampling theory." In principle, the merits of any proposed method of data analysis are to be judged by its sampling properties. That is, in the long run, how often would it lead us to a correct conclusion, or how large would the average error of estimation be? But practice sometimes ignores precept (as when one insists on using an unbiased estimator $(n-1)^{-1} \Sigma(x_i-\bar{x})^2$ of variance, even though the biased estimator $n^{-1} \Sigma(x_i-\bar{x})^2$ would yield a smaller mean square error).

PURE MAXENT APPROACH. By contrast, in the pure maximum-entropy (noiseless Bayes) method, our reasoning format is almost the opposite. Instead of considering the class of all data sets $\{D_1...D_N\}$ consistent with a hypothesis H, we consider the class of all hypotheses $\{H_1...H_n\}$ consistent with the one data set D_{obs} that was actually observed. In addition we use prior information I that represents our knowledge (from physical law, usually expressed as combinatorial multiplicity factors W_i) of the possible ways in which Nature could have generated the various H_i. Out of the class C of hypotheses consistent with our data, we pick the one favored by the prior information I—which means, usually, having the greatest multiplicity W (that is, greatest entropy log W).

Now, which kind of reasoning do you and I use in our everyday problems of inference? An automobile driver must make a decision (stop, slow down, go through, speed up, turn) at every intersection, based on what he can see. He does not think about the class of all things he might have seen but doesn't see. He thinks about the class of all contingencies that are consistent with what he does see, and acts according to which ones, in that class, seem a priori most likely from his previous experience.

If you go to a doctor and tell him your symptoms, he does not start thinking about the class of all symptoms you might have had but don't have. He thinks about the class of all disorders that might cause the symptoms you do have. The first one he will test for is the one which, in that class, appears to be a priori most likely from your medical history.

The same common-sense reasoning format is used by a TV repairman, a detective, an analytical chemist—and a player of the game "Twenty Questions." Each successive piece of data that one obtains is a new constraint that, if cogent, restricts the possibilities permitted by our previous information. By Blackwell's principle, at any stage an honest description of what we know must take into account (that is, assign nonzero probability to) every possibility that is not ruled out by our prior information and data.

Clearly, this is the way any rational person does—and should—think about such problems of inference. Only a warped mentality could see anything peculiar or illogical about it. Yet for 25 years MAXENT methods have been rejected—in some cases attacked—by persons who did not comprehend this simple rationale.

Even when we come to the proof of the pudding and present our superior numerical results (such as those of Burg, Currie, Montgomery, Frieden, Skilling, Gull and Daniell, and Shore as presented or recalled at this meeting), the issue is not always resolved. MAXENT results are sometimes regarded with suspicion because those familiar only with orthodox methods and results simply cannot believe it is possible to extract so much detailed information from the data.

In fact, their instincts are quite correct. It is not possible to extract all that detail from the data alone, or else orthodox methods might have done so. MAXENT gives us more information only because we have put more information into it. In image reconstruction, MAXENT is taking into account not only the data, but also our prior information about the multiplicity of different scenes. This information is, as we shall see, of the highest relevance to the problem, but orthodox ideology does not recognize it because it does not consist of frequencies in any random experiment.

In the problems where pure MAXENT is appropriate (which include all of statistical mechanics, equilibrium or nonequilibrium) we are concerned, not with frequencies in any random experiment, but with rational thinking in a situation where our information is incomplete. We are trying to do the best reasoning we can about the real situation that exists here and now—and not about the long run in some other situations that exist only in a statistician's imagination. In such applications, a probability distribution is not an assertion about frequencies but only a means of describing our state of knowledge.

In orthodox thinking, a frequency is considered "objective" and therefore respectable, while a mere state of knowledge is "subjective" and unscientific. But in the real-world problems of inference faced by every engineer, scientist, economist, business man, or administrator, it is evidently his state of knowledge that determines the quality of the decisions he is able to make in situations that will never be repeated; and it is the frequencies that are figments of the imagination.

Interestingly, the sampling distribution that orthodox theory does allow us to use is nothing more than a way of describing our prior knowledge about the "noise" (measurement errors). Thus, orthodox thinking is in the curious position of holding it decent to use prior information about noise, but indecent to use prior information about the "signal" of interest. Yet one man's signal is another man's noise. This arbitrary rejection of part of the information is the reason orthodox methods are incapable of dealing with "generalized inverse" problems of the aforementioned kind, which arise constantly in the real world.

FULL BAYES METHOD. We have expounded two more or less opposite extremes of reasoning, each of which is appropriate in a certain class of problems. The orthodox sampling distribution $p(D_i|H)$ is described only loosely as "noise." Stated more carefully, it might represent two different things:

(A) The variability of the data to be expected if the experiment were repeated under seemingly identical conditions.
(B) The likely deviations of the actually observed data D_{obs} from the truth.

Orthodox statistics, failing to distinguish between meanings (A) and (B), makes a single sampling distribution $p(D_i|H)$ serve two different functions. But, as I realized while holding those lumps of soil, (A) and (B) are logically quite different things. Indeed, it is (B) that one needs to know in order to make sound inferences that make allowance for the unreliability of the data D_{obs} actually obtained. Given the probability distribution appropriate for (B), it may or may not give also the variability (A) of data on other measurements that are not made, but the question is obviously irrelevant to our inference from the data we do have.

However, if the sole information we have pertaining to question (B) is derived from knowledge of variability in sense (A), then one may confuse the meanings (A) and (B) without harm. If in addition there is no prior information about the quantities being estimated, orthodox methods will be at least useful, and perhaps (if we have sufficient statistics and no nuisance parameters) even optimal. Likewise, if we have relevant prior information but no appreciable noise, the pure MAXENT method will be appropriate and near-optimal.

But just as prior information can make an orthodox analysis invalid (a maximum likelihood estimate $\hat{\theta} = 8$ is useless and misleading if we know in advance that $\theta < 6$), so appreciable noise can make the pure MAXENT method inappropriate and misleading.

Fortunately, in most of our everyday problems of inference, noise is not an important factor, so the MAXENT reasoning format applies. But if that automobile driver at the intersection had to look out through an optically turbulent medium, and so saw something different every time he looked even though the factual situation was unchanging, then he would have a more complicated problem and the MAXENT reasoning would have to be modified to make allowances for the unreliability of his data. We all know how much more cautiously we drive when a heavy rainstorm converts the windshield into just that turbulent medium. We cannot be sure what lies before us and must first guess at that before trying to decide what action to take.

If we have both noise and prior information, neither of the above methods is adequate. But both are only limiting special cases of a more general method that applies in all cases known to the writer. Adding prior information capabilities to orthodox methods, or noise capabilities to MAXENT, we arrive in either case at the Bayes method, which is actually simpler conceptually and older historically than either of these special cases.

To define the Bayesian method, we introduce some rudimentary Boolean algebra. We denote various propositions by A, B, C, etc., "not A" by ~A, "A and B" by AB. Then the probability symbols $p(A|{\sim}B)$ and $p(AB|C)$ are read as "the probability that A is true, given that B is false," and "the probability that both A and B are true, given that C is true." Mathematically, then, probability theory consists of nothing but the sum and product rules

$$p(A|B) + p(\sim A|B) = 1 \tag{2a}$$

$$p(AB|C) = p(A|C)\, p(B|AC) \tag{2b}$$

and the unending stream of consequences that can be deduced from them.

All schools of thought accept these rules as mathematically correct. It is over their interpretation—their relationship to the real world—that the 150-year-old philosophical controversies swirl. Orthodox doctrine, as noted, interprets p as a frequency, and it is a trivial observation that frequencies do indeed combine according to these rules. But orthodoxy also takes a militant stand, declaring all other interpretations to be metaphysical nonsense. In some 35 years of perusing the literature, I have found no orthodox writer who has advanced a logical reason for this position; it is merely asserted. But there is a vast literature indicating that this position was ill-advised, in effect putting orthodox statistics in a kind of straitjacket that makes it incapable of dealing with the current real problems of science, engineering, and economics.

A broader—and, we believe, far more useful—view, interprets $p(A|B)$ as a measure of the degree of plausibility of A, on a (0,1) scale. Then the equations of probability theory are not merely rules for calculating frequencies; they are also rules for conducting inference. In fact, R. T. Cox showed many years ago (Jaynes, 1976, 1978) that any set of rules for conducting inference, in which we represent degrees of plausibility by real numbers, is necessarily either equivalent to Eqs. (2) or inconsistent (in the sense that with any other rules one can find two different methods of calculation, both obeying the rules, that yield different results).

Of course, since Gödel one does not expect probability theory to provide a proof of its own consistency. The rules (2) as derived by Cox refer to finite discrete sets, and when they are applied in problems of inference on such sets, no inconsistency has ever been found. But there are many "paradoxes" in the literature—the marginalization paradox, the ambiguity of posterior odds ratios, the Borel-Kolmogorov paradox, the nonconglomerability paradox, etc.—all of which are caused by trying to apply these rules directly and indiscriminately on infinite sets. The common technique by which all these paradoxes, and any number of others, can be manufactured, is: (1) start from a well behaved finite result; (2) pass to a limit without specifying how the limit is to be approached; (3) ask a question whose answer depends on how the limit was carried out.

Clearly, the paradoxes of infinite sets have nothing to do with real problems of inference. After all, the number of atoms in our galaxy is only $G < 10^{70}$, a safely finite number, and it is hard to believe that a real problem will ever require a larger set than G! We propose a conjecture: All correct results in probability theory are either combinatorial theorems on finite sets generated by Eqs. (2), or well-behaved limits of such theorems. If this is correct, probability theory may be able to feed for a century on the combinatorial work of Rota (1975).

Let us see some consequences of Eq. (2), interpreted as rules for conducting inference. Take, as above, I = prior information, H = some hypothesis, D = data. Then, since the product rule (2b) is symmetric in A and B, we have

$$p(DH|I) = p(D|I)\,p(H|DI) = p(H|I)\,p(D|HI)\,. \tag{3}$$

If $p(D|I) \neq 0$ (that is, the data set is a possible one), this yields Bayes' theorem:

$$p(H|DI) = p(H|I)\,\frac{p(D|HI)}{p(D|I)}\,, \tag{4}$$

which represents in a very explicit form the process of learning. It shows how the "prior probability" $p(H|I)$ changes to the "posterior probability" $p(H|DI)$ as a result of acquiring new information D. This is exactly the kind of rule we need for inference. Let us note some of its properties.

Suppose we obtain data D_1 today, then additional data D_2 tomorrow. Today's data D_1 will of course be part of tomorrow's prior information, so our inferences about H on the two successive days will take the form

$$p(H|D_1I) = p(H|I)\,\frac{p(D_1|HI)}{p(D_1|I)} \tag{5}$$

$$p(H|D_2D_1I) = p(H|D_1I)\,\frac{p(D_2|HD_1I)}{p(D_2|D_1I)}\,. \tag{6}$$

But substituting Eq. (5) into Eq. (6), and using the product rule (3) in numerator and denominator, reduces Eq. (6) to

$$p(H|D_2D_1I) = p(H|I)\,\frac{p(D_2D_1|HI)}{p(D_2D_1|I)}\,, \tag{6'}$$

which is just of the form (4) of the original Bayes' theorem with $D = D_2D_1$, the total data now available. The result extends by induction to any number of additional data sets D_3, D_4, etc. Thus, when we take into account many different pieces of information, Bayes' theorem updates our state of knowledge at each step, but in such a way that our conclusions always depend only on the total information at hand, and would be the same if we got the information all at once. This is a most welcome consistency property, without which we would be in real trouble. Bayes' theorem has dozens of other nice features that express just the properties that a rational person would demand of a method of inference. Another important one is noted below [Eq. (56)].

But how does orthodoxy view this? Orthodoxy never gets to Eq. (6') because at the start it rejects Bayes' theorem (4) out of hand, as a method of inference, without ever bothering to examine the kind of results it gives. In orthodox ideology, probabilities can be assigned only to "random variables" because a probability, to be respectable, must be also a frequency.

But a hypothesis H is not a "random variable," so $p(H|I)$ and $p(H|DI)$ are held to be meaningless! At this point, orthodox statistics denies itself the use of the single most powerful and useful principle in probability theory, and it is left with no way to take the prior information I into account. This failure of orthodox statistics is just the reason why the new methods to be discussed at this workshop had to be developed.

Further details about the workings of Bayes' theorem are in Section 6, where we shall see that, in the Gull-Daniell image reconstruction problem, both MAXENT and orthodox results are contained in the Bayes solution, in the limits of zero noise and zero prior information respectively.

4. The Maximum-Entropy Formalism

Like any fairly general method, MAXENT can be approached in various ways. It can be based on information theory, on combinatorial theorems, or as a limiting form of Bayes' theorem. That is, all these approaches lead to the same algorithm although conceptually they express three slightly different problems. Arguing over these slight differences may solve the unemployment problems of philosophers for the next 50 years, but we need not dwell on them here.

We introduce "The Maximum-Entropy Formalism" by the information-theory approach. A real variable x can take on the values $\{x_1 \ldots x_n\}$ and we are to assign corresponding probabilities $\{p_1 \ldots p_n\}$ so as to represent our partial information about x. The information theory basis noted that the Shannon entropy

$$H = - \sum_{i=1}^{n} p_i \log p_i \tag{7}$$

is a measure of the "amount of uncertainty" in the distribution $\{p_1 \ldots p_n\}$, uniquely determined by certain very elementary consistency and additivity requirements (Shannon, 1948). Intuitively, then, the distribution that most honestly describes what we know, without assuming anything else (that is, is as noncommittal about x as it can be without violating what we know) is the one that maximizes H subject to the constraints imposed by our information.

But this is not a mathematically well-posed problem until we write down the explicit form of the constraints. Some kinds of information are too vague to use in a mathematical theory by any presently known methods, although our intuitive common sense may be able to make some use of them. Various aspects of this are discussed more fully elsewhere (Jaynes, 1968, 1978). With experience one becomes more adept at converting verbal information into mathematical constraints. For present purposes it is enough to give only one special case, which is by far the most useful one so far.

What does it mean to say that a probability distribution $\{p_1 \ldots p_n\}$ "contains" certain information? Presumably, this ought to mean that we can extract that information back out of it. Suppose we had a probability distribution and were asked to make the best estimate $\hat{A}$ of some function $A(x)$.

The exact value of A is not determined by the distribution, so we must introduce some criterion of what we mean by "best."

Since the time of Gauss, the mean square error criterion has been the most popular and easily implemented. Other criteria could be used if there were any advantage in using them; but for the applications we have in mind they would only make the computations longer, while leading to final results practically indistinguishable from the ones given below. If we make the estimate $\hat{A}$, the expected square of the error will be

$$\langle (A - \hat{A})^2 \rangle = \sum_i p_i [A(x_i) - \hat{A}]^2$$

$$= [\langle A^2 \rangle - \langle A \rangle^2] + (\hat{A} - \langle A \rangle)^2 , \qquad (8)$$

where the brackets $\langle\ \rangle$ denote averages over the distribution p_i, often called "expectations." The first term, the variance of the distribution of A,

$$\mathrm{var}(A) = \langle A^2 \rangle - \langle A \rangle^2 , \qquad (9)$$

is fixed by the distribution p_i, so the mean square error is minimized by choosing as our estimate the average

$$\hat{A} = E(A) = \langle A \rangle = \sum_i p_i\, A(x_i) . \qquad (10)$$

Statisticians denote expectations by E(A), while physicists, having already preempted E for energy and electric field, use the bracket notation $\langle A \rangle$. Writing for both, we use both notations.

Conversely, if we are asked to adjust the distribution $\{p_1 \ldots p_n\}$ to incorporate given information about A, we shall understand this as meaning that, by applying the usual prediction rule, (10), we can get that information back out of $\{p_1 \ldots p_n\}$. Thus our mathematical constraints, in the problems considered here, will take the form of fixing expectations of various quantities about which we have some information.

If we wish to incorporate information, not only about the value of A, but also about the accuracy with which A is known, we merely add another constraint, fixing $\langle A^2 \rangle$ as well as $\langle A \rangle$. The general formalism below automatically includes this possibility. But in practice this usually makes no appreciable difference in our conclusions. An exception can arise in the limit $n \to \infty$, where a constraint on $\langle A^2 \rangle$ may be needed to get a convergent solution, $\Sigma p_i = 1$.

More generally, whenever we find we are not getting a normalizable solution, that is how the theory tells us we have not yet specified enough information to justify any definite inferences. As in any situation of insufficient information, we cannot hope to get something for nothing, and the only remedy is to get more information. Almost always, it turns out that the person using the theory actually did have some more information that he had failed to put into the equations, not realizing that it was essential.

To proceed to "The Maximum-Entropy Formalism," there are m functions $\{A_1(x)\ldots A_m(x)\}$ for which we are given, in the statement of the problem, our "data," that is, a set of numbers $\{A'_1\ldots A'_m\}$ which are the values we want our probability distribution $\{p_1\ldots p_n\}$ to predict. To fit the distribution to our data, we impose the m simultaneous constraints:

$$\sum_{i=1}^{n} p_i\, A_k(x_i) = A'_k, \qquad 1 \le k \le m. \tag{11}$$

Much ink has been spilled over the relationship between the numbers A'_k appearing in Eq. (11) and real data. A persistent misconception is that MAXENT requires our data to consist of "mathematical expectations." This does not make sense, and is an egregious case of putting the cart before the horse. If we adjust a mathematical expectation to fit our new state of knowledge after getting some data, it does not follow that the data consisted of expectations. If I adjust my belt to fit my new state of girth after a heavy meal, it does not follow that my meal consisted of belt leather.

In various problems the A'_k could be generated in various ways. For present purposes we note that A'_k is simply a number given to us in the statement of the problem, and our present task is the mathematical one of incorporating the conditions (11) into our probability distribution. For further discussion, see Jaynes (1978, pp. 72-77 and 95-96).

The solution was given by Gibbs (in, of course, different notation):

Define the partition function

$$Z(\lambda_1\ldots\lambda_m) = \sum_{i=1}^{n} \exp[-\lambda_1 A_1(x_i) - \cdots - \lambda_m A_m(x_i)]\,. \tag{12}$$

Then the MAXENT distribution is

$$p_i = \frac{1}{Z(\lambda_1\ldots\lambda_m)} \exp[-\lambda_1 A_1(x_i) - \cdots - \lambda_m A_m(x_i)] \tag{13}$$

with the λ's determined, as our poet noted, by

$$A'_k = -\frac{\partial}{\partial\lambda_k} \log Z\,, \qquad 1 \le k \le m\,, \tag{14}$$

a set of m simultaneous equations for the m unknowns $\{\lambda_1\ldots\lambda_m\}$.

The result has also a combinatorial basis, a special case of which was given by Boltzmann (1877). Some process happens N times, and each time a result is that one of the x_i is chosen. This includes many different scenarios:

(I) BERNOULLI TRIALS. A random experiment is repeated N times, each trial yielding one of the values $\{x_i\}$.

(II) COMMUNICATION. We receive a message of N letters, chosen from the alphabet $\{x_1 \ldots x_n\}$.

(III) KINETIC THEORY. A gas contains N molecules, each in one of the quantum states $\{x_1 \ldots x_n\}$.

(IV) IMAGE RECONSTRUCTION. N elements of luminance are distributed over n pixels, labeled $\{x_1 \ldots x_n\}$, to form an image.

In any of these, the particular result x_i is chosen N_i times; that is, it is generated with frequency $f_i = N_i/N$. Needing a short name for the set $F = \{f_1 \ldots f_n\}$ whatever the scenario, we call it simply "the scene," which seems particularly appropriate in image reconstruction. Any given scene F has a certain multiplicity W(F):

$$W(F) = \frac{N!}{(Nf_1)! \ldots (Nf_n)!}, \tag{15}$$

which is equal to the number of ways in which Nature could have generated it. When N becomes large we have, by Stirling's rule,

$$\frac{1}{N} \log W(F) \to -\sum_{i=1}^{n} f_i \log f_i = H(f), \tag{16}$$

just the Shannon entropy again, only now depending on "objective" frequencies f_i instead of "subjective" probabilities p_i. For this reason, the combinatorial approach is the one most easily understood by those with orthodox training although the first approach is more general.

Now we do not know the specific sequence of $\{x_i\}$ that was generated; all we know is the resulting averages of m quantities $\{A_1(x) \ldots A_m(x)\}$:

$$\sum_{i=1}^{n} f_i A_k(x_i) = A'_k, \qquad 1 \leq k \leq m. \tag{17}$$

If we are asked which scene we consider most likely, it seems reasonable to favor the one that Nature could have generated in the greatest number of ways consistent with what we know, that is, the one that has maximum multiplicity W(F) subject to the constraints (17). In view of Eq. (16), then, we have formulated the same mathematical problem as in Eqs. (7) to (11), and the same MAXENT algorithm Eqs. (12) to (14) will solve it. We need only write f_i in place of p_i.

For those who have not seen elementary examples of numerical solutions by this method, the extensive analysis of Rudolph Wolf's dice-tossing experiments in Jaynes (1978) is a recommended tutorial introduction that explains many further points of rationale. A still more efficient procedure for testing hypotheses in the light of data by MAXENT appears in Jaynes (1983, Chapter 10). But let us turn now to some of the current nontrivial applications.

5. The No-Noise Spectral Analysis Problem

In the field of spectral analysis, this case is rather special but nevertheless real. Many time series that arise in econometrics and geophysics may be considered essentially uncontaminated by noise, because the variability of the phenomenon being observed greatly exceeds the error of the measurements. Sales on the New York Stock Exchange vary greatly from day to day, yet the value for any one day can be determined exactly. The air temperature in Tulsa may vary by 50 degrees in a few hours, yet its value at any one time can be measured to a fraction of a degree.

We have, then, some discrete time series $\{y_0, y_1, \ldots, y_N\}$ that has a Fourier transform

$$Y_\omega \equiv (N+1)^{-1} \sum_{k=0}^{N} y_k \, e^{i\omega k} \tag{18}$$

and a power spectrum

$$S_\omega \equiv |Y_\omega|^2 = \sum_{k=-N}^{N} R_k \, e^{i\omega k} \tag{19}$$

where $R_{-k} = R_k^*$ and

$$R_k \equiv \sum_{j=0}^{N-k} y_j^* \, y_{k+j} \, , \qquad 0 \leq k \leq N \, , \tag{20}$$

is usually called the 'autocovariance' or 'autocorrelation' although both terms seem unfortunate and inconsistent with other established usage. The Fourier inversion of Eq. (19) is

$$\frac{1}{2\pi} \int_{-\pi}^{\pi} S_\omega \, e^{-i\omega k} \, d\omega = R_k \, , \quad -N \leq k \leq N \, , \tag{21}$$

where $\omega_N = \pi$ is the Nyquist frequency, above which S_ω repeats itself periodically. Note that we have defined these quantities so that the last four equations are exact for finite N, without 'end-effects.' There is no mathematical advantage, in either precision or simplicity, to be gained by passing to the physically nonexistent limit $N \to \infty$.

The difficulty is that not all the R_k are known. Our data D, although exact, comprise only a subset $D = \{R'_{-m} \ldots R'_m\}$ where $m < N$. The problem we shall consider is how to estimate S_ω from this incomplete information. (Of course, practically everything we say can be carried over mutatis mutandis to the similar problem of estimating Y_ω from a subset of the y_k.)

This no-noise spectrum analysis problem is, then, an example of the standard generalized inverse difficulty; the data cannot distinguish between two spectra S_ω and S'_ω that differ by a solution of the homogeneous equation

$$\int_{-\pi}^{\pi} (S_\omega - S'_\omega)\, e^{-i\omega k}\, d\omega = 0\,, \qquad |k| \le m\,. \tag{22}$$

Therefore, by Eq. (19) the class C of possible spectra compatible with our data consists of all functions of the form

$$\hat{S}_\omega = \sum_{k=-m}^{m} R'_k\, e^{i\omega k} + \sum_{k=m+1}^{m} (\hat{R}_k\, e^{i\omega k} + \hat{R}_{-k}\, e^{-i\omega k})\,, \tag{23}$$

where the $\hat{R}_k$ $(m < k \le N)$ may be chosen arbitrarily but for the condition $\hat{S}_\omega \ge 0$ that any power spectrum, by definition, must satisfy. Clearly, the $\hat{R}_k$ represent estimates of the autocovariance, extrapolated beyond the data. Thus the problem of choosing, out of the class C of possible spectra, one "best" estimate of S_ω is equivalent to the problem of extrapolating the autocovariance to all lags $(-N \le k \le N)$.

Now, because there is no noise (the R'_k are considered known exactly for lags $|k| \le m$), there is no sampling distribution to describe it. It appears, then, that sampling theory can provide no basis for choosing one spectral estimate $\hat{S}_\omega$ over another; a pure generalized inverse problem lies entirely outside the domain of sampling theory statistical methods.

Indeed, in the Blackman-Tukey (1958) method of spectrum analysis the problem is never seen as one of extrapolating R_k at all. Blackman and Tukey state unequivocally that, "Surely, no estimate can be made for lags longer than the record." Nevertheless, because of the mathematical connections (19) and (21), any method of estimating S_ω is necessarily also a rule for estimating R_k beyond the record.

A spectral estimate $(\hat{S}_\omega)_{BT}$ found by the Blackman-Tukey method is, in fact, the spectrum of a time series whose autocovariance is zero at all lags beyond the data. As Burg (1975) pointed out, this is tantamount to making an extrapolation of R_k that is almost certainly wrong and that may even stand in violation of the nonnegativity condition $S_\omega \ge 0$, thus making $(\hat{S}_\omega)_{BT}$ outside the class C of logically possible spectra.

Even if this implied extrapolation fails to get us outside the class C, the following difficulty remains. The $\hat{S}_\omega$ that we obtain from Eq. (21) by setting

$\hat{R}_k = 0$, $k > m$, is the spectrum of a time series whose autocovariance is truncated abruptly at $k > m$. It is therefore the convolution of the true spectrum with the "Dirichlet kernel" $D(\omega) = (2\pi)^{-1}\sin[(2m+1)\omega/2]/\sin(\omega/2)$ and will exhibit spurious "side lobe" maxima that are not in the spectrum of the true time series $\{y_k\}$ but are separated from the true spectral lines by odd harmonics of $\omega_{SL} = \pi/(2m+1)$. As Burg realized, these side lobes are only an artifact of the method, caused by the failure to extrapolate R_k in a reasonable way.

However, the BT method proceeded to treat the symptom rather than the disease. To get rid of the abrupt truncation, instead of raising $\hat{R}_k$ for $k > m$, it lowers $\hat{R}_k$ for $k \leq m$ by introducing a "lag window" function W_k $(-m \leq k \leq m)$. It thus replaces Eq. (23) by

$$(\hat{S}_\omega)_{BT} = \sum_{k=-m}^{m} W_k R'_k \, e^{i\omega k} , \tag{24}$$

and indeed, by various choices of window functions, one can reduce the side lobes greatly, at the cost of losing about half the resolution.

But this does violence to Eq. (23), which is the most general estimate that lies in the class C of logically possible spectra. $(\hat{S}_\omega)_{BT}$ is the spectrum of a time series that is known to disagree with our data at every data point k where $W_k \neq 1$. We know, not as a plausible inference but as a logical deduction, that a time series with the spectrum (24) could not have produced our data!

In the noiseless case this criticism by Burg was crystal clear and unanswerable. Then does the use of windows become more defensible if the data are contaminated by noise? Our examination of the theoretically simpler image reconstruction problem below will provide a clue suggesting that allowing for noise must take us not toward window solutions, but away from them, that is, to solutions of still higher entropy.

But the claims of sampling theory are not yet disposed of. One might think that in a pure (noiseless) generalized inverse problem there is no place for sampling theory at all. Yet sampling theory has managed to work itself into the problem anyway, by a remarkable conceptual feat. If the real world has no sampling distribution, then we shall invent one, not by imbedding our data in a set of possible data, but by imbedding the whole real world in a set of possible worlds (just as Everett (1957) did in quantum theory). The one real, finite time series $Y = \{y_0 \ldots y_N\}$ that actually exists is regarded as only a "sample" drawn from some hypothetical ensemble of other infinitely long series—and we are back in business!

Of course, from this standpoint R'_k—which was by definition the exact autocovariance of Y—then appears to be "biased." So it is replaced by the "unbiased estimator"

$$R''_k = \frac{N+1}{N+1-k} R'_k , \qquad 0 \leq k \leq m . \tag{25}$$

However, we stress again that Eqs. (18) to (21) are exact as they stand for finite N, and so we have made a second error of reasoning. But pragmatically, this helps by canceling out some of the effect of the first error. Before falsifying our data R'_k downward by a window function W_k, if we first falsify them upward by using Eq. (25), we end up a little closer to being in the class C of possible spectra.

Here are some of the things that bother me about orthodox reasoning. In the real world no infinitely long time series exist—much less any ensemble of them. What does it mean to say that we are "estimating" something that is only a figment of our imagination? What determines this sampling distribution according to which the real world is now to be selected from this figment? If we are only imagining this ensemble, then it would seem that we are free to imagine it as having any properties we please. Then whose imagination is to take precedence—yours or mine?

If to achieve the "objectivity" of a frequency interpretation of our probabilities we must sacrifice the "objectivity" of dealing with the real world, then it seems to me that we have paid too high a price. True "objectivity" ought to address itself not to estimating figments of our imagination, but rather to representing the state of knowledge about the one real world that actually exists, based on the one data set that actually exists.

Now consider the generalized inverse problem (23) from the standpoint of maximum entropy. The given data

$$\{R'_{-m} \ldots R'_0 \ldots R'_m\}$$

represent, in complex notation where we treat R_k and $R_{-k} = R_k^*$ independently, (2m+1) constraints. (If we revert to the real notation $R_{-k} = R_k$ and treat only $(R_0 \ldots R_m)$ independently, then we have only (m+1) independent conditions.)

The maximum entropy distribution subject to these constraints is a Gibbsian generalized canonical ensemble

$$P(y_0 \ldots y_N) \propto \exp\left[-\sum_{k=-m}^{m} \lambda_k R_k\right]. \tag{26}$$

Writing out the exponent in full, we have

$$\begin{aligned}\sum \lambda_k R_k &= \lambda_0(y_0^2 + y_1^2 + \ldots + y_N^2)\\ &+ \lambda_1(y_0 y_1 + y_1 y_2 + \ldots + y_{N-1} y_N)\\ &+ \lambda_2(y_0 y_2 + y_1 y_3 + \ldots + y_{N-2} y_N)\end{aligned}$$

$$+ \dots$$

$$+ \lambda_m(y_0 y_m + \dots + y_{N-m} y_N)$$

$$+ \lambda_{-1}(y_1 y_0 + y_2 y_1 + \dots + y_N y_{N-1})$$

$$+ \dots$$

$$+ \lambda_{-m}(y_m y_0 + \dots + y_N y_{N-m}) . \tag{27}$$

But this can be rearranged into a matrix product:

$$\sum \lambda_k R_k = \sum_{ij=0}^{N} \Lambda_{ij} y_i y_j \tag{28}$$

where Λ is a Toeplitz matrix:

$$\Lambda_{ij} = \begin{cases} \lambda_{j-i}, & |j-i| \leq m \\ 0, & |j-i| > m \end{cases} \tag{29}$$

with the Lagrange multipliers repeated down bands parallel to the main diagonal.

Thus, without our having assumed any "Gaussian random process," the maximum entropy principle constructs the Gaussian form for us, as the distribution that most honestly represents our data:

$$P(y_0 \dots y_N) \propto \exp\left[- \sum_{ij} \Lambda_{ij} \, y_i y_j \right] . \tag{30}$$

The partition function (12) is then proportional to $[\det(\Lambda)]^{-1/2}$, and so

$$\log Z = -\frac{1}{2} \sum_{j=0}^{N} \log \ell_j + \text{constant} ,$$

where $\{\ell_j\}$ are the eigenvalues of Λ. Determination of the λ's from Eq. (14) then involves solving a number of simultaneous equations. In the limit $N \to \infty$, the eigenvalues $\{\ell_j\}$ are given by known theory of Toeplitz matrices.

The Levinson-Burg numerical algorithm proceeds by a Wiener-Hopf factorization of the polynomial

$$\sum_{k=-m}^{m} \lambda_k z_k = \left| \sum_{k=0}^{m} a_k z^k \right|^2 \tag{31}$$

and evaluating the new coefficients $\{a_k\}$ recursively. The $\{a_k\}$ may be interpreted as the coefficients of a Wiener prediction filter, or as the coefficients of an autoregressive model, if they are chosen to correspond to a minimum delay wavelet. However, from the standpoint of maximum entropy this choice is not mandatory, only the λ's appearing in our final results.

The details of this solution are so well known, and so well covered by others at this workshop, that it would be a needless duplication to repeat them here (they are, however, given from the writer's point of view in Jaynes, 1982). Suffice it to say that maximum entropy uses the form (23), thus guaranteeing that our estimated spectrum $\hat{S}_\omega$ lies in the class C of possible spectra—but we use the optimal extrapolation $\{\hat{R}_{m+1} \ldots \hat{R}_N\}$ of the covariance beyond our data, determined as expectations over the maximum-entropy distribution (30). By a bit of algebraic magic, the final maximum-entropy spectrum estimate turns out unexpectedly simple:

$$(\hat{S}_\omega)_{MAXENT} = \frac{1}{\sum_{k=-m}^{m} \lambda_k e^{-i\omega k}} , \tag{32}$$

which is the now classic result of Burg (1967).

The virtues of this solution are many. It not only gets us back into the class C of logically possible spectra, but also, instead of forcing us to compromise between side lobes and resolution, it gives us the best of both worlds—eliminating side lobes while giving much greater resolution. The practice of spectrum analysis was revolutionized by this discovery, which made all previous thinking and methods obsolete.

NEW PROBLEMS. Of course, this does not mean that the theory is now finished, and Eq. (32) with the Levinson-Burg evaluation of the $\{\lambda_k\}$ is the optimal solution for every conceivable problem. Some assumptions were made in the derivation of Eq. (32), and although our result is as robust with respect to small departures from those assumptions as were any previous solutions, we are now looking for higher standards of performance than were expected from previous methods. So, instead of being finished, now that our eyes have been opened we can see a mass of new problems in need of solution, some of which could not even be formulated in terms of previous notions.

Two problems now pressing for immediate attention are the variability of Eq. (32) in the presence of noise, pointed out to the writer by David Brillinger in December 1980, and the 'line-splitting' phenomenon pointed out to the writer by John Tukey just before this meeting. These can be traced to assumptions made in the derivation of Eq. (32) that are not always satisfied in real problems.

We assumed that our data were noiseless; otherwise the class C of possible spectra would not be sharply defined. With noise, the boundary of C becomes smeared out into a smooth transition region whose thickness represents the noise level. From the analogous image reconstruction solution to be noted in Section 6, we expect this will lead us to modify the pure MAXENT spectrum (32) in the direction of slightly higher entropy, by an amount of the order of the thickness of the transition region. However, the analytical details for the spectrum analysis case are not yet available.

Also it was assumed implicitly in the algorithm for evaluation of the λ_k that the autocovariance data $\{R_0 \ldots R_m\}$ were from a very long sample of length $N \gg m$, since one used the asymptotic distribution of eigenvalues for the limit of an infinitely large Toeplitz matrix. If our data are from a time series $\{y_0 \ldots y_{43}\}$ of only 44 observations, then the exact MAXENT spectrum should use the λ_k's computed from Eqs. (14) and (31) with the exact eigenvalues of the finite Toeplitz matrix with 44 rows and columns.

Again, the analytical details are not yet available, but I am confident, for reasons to be discussed more fully elsewhere, that this small correction of the algorithm will remove the line-splitting difficulty. The point is that when we have only a finite time series $\{y_0 \ldots y_N\}$, the present algorithm gives us only an approximate solution of the real problem; but that happens to be the exact solution for a circular time series that repeats itself forever: $y_0 = y_{N+1}$, $y_1 = y_{N+2}$, etc. A circular time series with the same autocovariances $\{R'_0 \ldots R'_m\}$ would indeed have a spectrum with split lines close together.

These two problems appear to be of high priority but also straightforward in principle, and surely solvable. But the details are not trivial, and some honest labor will be required to find the new algorithms. Fortunately, some of the analytical machinery needed has been provided by Gohberg and Fel'dman (1974) and Bleher (1981).

6. Image Reconstruction

The spectacular recent successes of Frieden, Gull and Daniell, Skilling, and others in the area of image reconstruction have done more than anything else to convert doubters into believers, because here the demonstration of what MAXENT can do stands, literally, before your eyes. Also, at its present stage of development, the problem is simpler theoretically than that of spectrum analysis because correlations in luminance of adjacent pixels are not yet incorporated into the MAXENT equations (whereas in spectrum analysis the whole problem lies in correlations).

This relative simplicity allows us to use the combinatorial approach to MAXENT, Eqs. (15) to (17) above, and to discuss the full Bayes solution. If N equal elements of luminance are distributed over n pixels to form a scene $F = \{f_1 \ldots f_n\}$, $f_i = N_i/N$, the number K of conceivable different scenes that could result is equal to the number of terms in the multinomial expansion of $(f_1 + \ldots + f_n)^N$:

$$K = \frac{(N + n-1)!}{N!(n-1)!}, \tag{33}$$

and when N is large, a given scene of entropy $H(f_1 \ldots f_n)$ has a multiplicity (number of ways it can be realized) given asymptotically by Eq. (16):

$$W(F) \sim e^{NH(F)} . \tag{34}$$

Our data $D \equiv \{d_1 \dots d_m\}$ consist of the luminances of m pixels of our blurred image, which we suppose determined by

$$d_k = \sum_{i=1}^{n} A_{ki} N_i + e_k , \qquad 1 \leq k \leq m < n , \tag{35}$$

where $N_i = Nf_i$ is the number of elements of luminance in the ith pixel, A_{ki} is the digitized point spread function of our telescope, and e_k are the traditional "Gaussian noise" terms, which are to have independently the probability distributions

$$p(e_k | \sigma) \propto \sigma^{-1} \exp(-e^2{}_k/2\sigma^2) , \qquad 1 \leq k \leq m , \tag{36}$$

and σ is the RMS noise level.

Let us proceed directly to the application of Bayes' theorem, (4). Given prior information I and data D, the probability that any specific scene F is the true one is

$$p(F | D,\sigma,I) = p(F | I) \frac{p(D | F\sigma I)}{p(D | \sigma I)} . \tag{37}$$

The denominator, being independent of F, is just a normalizing constant and is not needed in most applications. We have, then, to get the prior probability $p(F | I)$ and the sampling distribution $p(D | F\sigma I)$. The latter is determined from Eqs. (35) and (36): Given that the true scene is $F = \{f_1 \dots f_n\}$, the probability (density) that we shall obtain the data $D = (d_1 \dots d_m)$ is just the probability that the noise terms e_k will make up the difference:

$$p(D | F,\sigma,I) = (2\pi\sigma^2)^{-m/2} \exp(-Q/\sigma^2) , \tag{38}$$

where

$$Q(D,F) \equiv \frac{1}{2} \sum_{k=1}^{m} \left[d_k - \sum_{i=1}^{n} A_{ki} f_i \right]^2 \tag{39}$$

is the basic quadratic form of the kind that always gets into problems with Gaussian noise.

Assigning the prior probabilities $p(F | I)$ is not so straightforward because there is no end to the variety of different kinds of prior information that one might have in various problems. Clearly, any cogent prior information about which scenes are a priori likely ought to be taken into account and will increase the reliability of our reconstructed scene. For example, say

$$I = I_G \equiv \text{'The scene is a distant galaxy'} ;$$

we know in advance, in a general way, what galaxies look like; and a reconstruction depicting a horse or a New Jersey license plate is so improbable that we would not hesitate to assign

$$p(F_H | I_G) = p(F_{NJLP} | I_G) = 0 .$$

Equally clearly, however, it would be a task of unlimited complexity to try to characterize every conceivable nongalactic scene and eliminate all of them from our prior probability assignment, although that would undeniably improve our reconstructions.

Therefore, for the present we are going to do what is feasible, and pretend that we have no prior knowledge at all about what is being looked at, and so we cannot exclude any scene. If N elements of luminance are distributed, one by one, to the n pixels, this can be done in n^N different ways, and we shall look at the consequences of a state of primitive ignorance I_0 that assigns equal probability n^{-N} to each of them.

Indeed, the impressive results achieved thus far correspond to this I_0 although they are only the 'minimal performance' of the theory, and we know that they could be improved with a more elaborate treatment of prior information. We do not know how much further improvement is possible with a feasible amount of further analysis, but the present results prove to be quite good enough for many purposes.

Out of the n^N conceivable ways Nature could have made all possible scenes, any particular one $\{f_1 \ldots f_n\}$ can be realized in W(F) different ways, so our prior probability assignment is

$$p(F | I_0) = n^{-N} W(F) . \tag{40}$$

But since N is large and constant factors are irrelevant, it will suffice to take, from Eq. (34),

$$p(F | I_0) \propto e^{NH(F)} . \tag{41}$$

Then Bayes' theorem, (37), reduces to

$$p(F | D, \sigma, I_0) \propto e^{N[H(F) - wQ(F)]} , \tag{42}$$

where

$$w \equiv \frac{1}{N\sigma^2} . \tag{43}$$

The factor exp(NH) represents our prior information about the multiplicities of different scenes. The factor exp(-NwQ) in its dependence on $\{f_1 \ldots f_n\}$ is the 'likelihood function' that tells what we have learned from the data, making allowance for the noise.

The most probable scene $\{\hat{f}_1 \ldots \hat{f}_n\}$ is then at the peak of this distribution. To locate it we maximize (H - wQ) subject to the constraint $\Sigma f_i = 1$. By Lagrange multipliers, this tells us immediately that, at the peak, the f_i are proportional to

$$\hat{f}_i \propto \exp\left[-w \frac{\partial Q}{\partial f_i}\right], \qquad 1 \leq i \leq n \, . \tag{44}$$

This relationship, although interesting, is implicit only because all the f_i are still hidden in Q. Still, from Eq. (44) we can understand the transition from full Bayes to pure MAXENT. When w is small (large noise), the data provide only a rather "soft" constraint on the possible f_i, and Eq. (44) allows the solution point to wander up the entropy hill (that is, in the direction of uniform f_i), away from the hyperplane HP whose equation is Q = 0. But as the noise decreases and $w \to \infty$, Eq. (44) goes into a rigid "stone wall" constraint, forcing the solution point back to HP. In the zero-noise limit, the term wQ in the Bayes solution thus puts in just the constraint of the pure MAXENT solution.

Conversely, if all scenes had the same multiplicity, H(F) = constant, the most probable scene would revert to the maximum likelihood estimate corresponding to Q = 0. But if $m < n$, Q = 0 is the equation of the hyperplane HP, not a point, and there is no unique solution, the likelihood is flat at all points of HP and we have a generalized inverse problem of just the kind that orthodox statistics is unable to deal with. Without prior information, there is no criterion for preferring any point of HP satisfying $f_i \geq 0$, $\Sigma f_i = 1$, to any other. The Bayes solution thus contains the MAXENT and orthodox solutions as special cases.

But we have glossed over some conceptual difficulties in getting this result. There are quite a few unspecified quantities (n, m, N, σ) still flapping about loose. How do we decide on their values?

WHAT ABOUT n,m,N? In some cases the circumstances of the problem may determine one or more of these in a way not under our control, but in many problems it is up to us to choose them. That is how we define our "hypothesis space"—the conceptual field on which our game is to be played.

One might question why we bring in all these discrete integers at all. In the real world, Nature seldom divides up her scenes into neat little rectangular cells. The true scene has a continuous structure, and it would seem more realistic to replace Eq. (35) by an integral equation

$$d(x) = \int A(x,y) f(y) d^2y + e(x) \tag{45}$$

to be inverted. The discrete version, (35), that we have adopted is only a crude, inelegant approximation to Eq. (45).

Now if we were going to find an analytical solution, this would be the right way of looking at it, and in principle the general Bayes solution of

Eq. (45) might be given once and for all. Indeed, any specific galaxy, horse, or New Jersey license plate can be portrayed by some analytic function f(y). But an analytic solution that contains enough parameters to include all those possibilities does not seem feasible, and so we are being pragmatists, after useful numbers rather than formal elegance.

It is a practical necessity that our reconstructions be found, not by substituting into some grand and glorious general solution, but in each specific case by numerical processing of the data. Any data set that we can actually record is a collection of a finite number of integers. Any numerical calculation that we can actually perform—whether by hand or by the most powerful computer—is nothing but a finite number of manipulations of a finite number of integers. So if our problem is not solvable in finite, discrete terms, we shall not be able to solve it at all. Introducing discrete integers gives us a solvable problem.

Our choice of (n,m) is of course to be guided by our prior information about the nature of the problem, the quantity and quality of data, and the computing facilities available. For example, regardless of what the theory says, it would be fatal to chose (n,m) so large that our computer memory could not hold all the (f_i, d_k). Likewise, it seems irrational to choose n or m so small that we are obviously throwing away resolution that the data are capable of giving.

It seems, then, that the choice of (n,m) will involve some compromise between performance and computation cost. Larger values increase the potential resolution and squeeze more information out of our data; but beyond a certain point we have extracted essentially all that the data have to tell us, and further increases would only increase the amount of computation without useful return. To make an intelligent choice, we must understand how the resolution depends on our choice. Fortunately, this is not critical, and wide variations—within reason—make little difference in the quality of the reconstruction.

The choice of N is more subtle. As a first orientation, let us get some idea of the rather interesting numbers involved. Gull and Daniell (1978) considered an early example with $n = 128 \times 128 = 16384$ pixels (determined, obviously, by computer considerations). How many primitive elements of luminance N should we think of as strewn over them to make the unknown true scene F? This is something one can argue about for a long time, and I am sure John Skilling will agree with me that more thinking about it is also needed. But for purposes of illustration and to start the discussion, the following seems a reasonable first guess.

The question is important because at issue here is the relative weighting of the prior information (entropy) factor exp(NH) and the likelihood factor $\exp(-Q/\sigma^2)$. Just as the choice of the number n of pixels to use in our reconstructed scene is a pragmatic one (make it large enough that we achieve all the spatial resolution the data are capable of giving, but not much larger), so the choice of the number N of elements of luminance we suppose to be in the scene is a way of stating how fine-grained an intensity resolution the data are capable of giving. (Perhaps we should add, if it were not already obvious, that the term "luminance" is being used in a loose colloquial sense, not as a technical term of optometrics.)

The little elements of luminance represent basically the smallest increment that we could detect. Conceivably, at very high frequencies and very low intensities they might have something to do with individual photons, but in radio astronomy we are surely very far from that case, and "quantum considerations" are irrelevant. In optical cases they might be identified with individual developed grains in a photograph. If our RMS error σ decreases, then we can detect a smaller increment, and so N increases.

However, looking at it that way may be putting the cart before the horse. Presumably, if our apparatus is reasonably well designed, the "noise" σ is not coming from the apparatus itself but rather is generated by the phenomenon under observation. So perhaps we should state it the other way around: If N increases, then we can detect a smaller relative increment, and so σ decreases.

In any event, the result is that N and σ are linked in some way. Various relationships between them may be appropriate in different problems; we indicate one such connection that appears "crudely reasonable" in some cases.

If we think of our noise σ as generated by variability of the scene to be expected if it were created anew, then if the ith pixel got N_i elements of luminance, we should expect this might vary by about $\pm\sqrt{N_i}$, and so f_i might be uncertain to about

$$\delta f_i \simeq \sqrt{N_i}/N = \sqrt{f_i/N} \,. \tag{46}$$

The uncertainty σ in our data resulting from this uncertainty in f_i is then

$$\sigma = \Sigma/\sqrt{N} \,, \tag{47}$$

where Σ depends on details of the smearing matrix A_{ki} and the values of f_i, but is independent of N. But, comparing with Eq. (43), we see that $w = \Sigma^{-2}$. This is the "weight" of the log likelihood factor Q relative to the entropy factor H.

Obviously, many refinements in detail are possible and desirable—in particular, pursuing this line of reasoning will teach us, very quickly, that the errors e_k in Eq. (35) should not be taken as independent and identically distributed, which means that Q in Eq. (39) should be written as a more general quadratic form. But our important conclusion that w is independent of N will remain as long as the errors σ are generated more in the phenomenon than in the measuring instrument.

In the posterior distribution (42), increasing N and decreasing σ^2 so as to keep $w^{-1} = N\sigma^2$ constant has the effect of increasing the sharpness of the peak without affecting its actual location. We do not change our reconstructed scene, but only become more confident about its accuracy.

For most purposes it is w, not N, that is of primary interest. If we know that our reconstructed scene is the best one that can be made on the information used, then whether the peak at $\{\hat{f}_1 \ldots \hat{f}_n\}$ is sharp or broad we shall not be induced to seek another. Some nontrivial analysis may be needed to find the best value of w, but fortunately it appears from the results of Gull and Daniell that this too is not very critical, variations within reason not making a great deal of difference in the reconstructed scene.

Of course, if w is taken unreasonably low, that is tantamount to saying that the data are nearly worthless, and the reconstruction goes to the uninformative uniform gray scene of absolute maximum entropy. If w is taken unreasonably high, we are claiming that every tiny bit of detail in the data is real and not noise, and so the reconstruction starts showing spurious details. Even then, however, the pure MAXENT reconstruction, like the MAXENT spectrum estimate, has at least the merit that it keeps us in the class C of possible scenes and cannot yield a negative f_i for any pixel, as did some previous reconstruction algorithms.

Now we are ready for those interesting numbers. With the value n = 16384 of the Gull-Daniell 1978 work, let us choose arbitrarily N = 10n = 163840. This is probably unfair to Gull and Daniell, since it supposes the data were not very accurate, but even so it gives us big enough numbers to make our point. The number K of conceivable scenes from Eq. (33) is then

$$K = \frac{180332!}{163840!\ 16383!} = 3 \times 10^{23840} . \qquad (48)$$

Their multiplicities (15) range from

$$W_{min} = 1 \qquad (49)$$

up to

$$W_{max} = \frac{163840!}{(10!)^{16384}} = 4 \times 10^{675703} . \qquad (50)$$

Therefore, we could partition the scenes into 67571 categories, those in category c having multiplicity in

$$10^{10c} \leq W \leq 10^{10(c+1)}, \qquad 0 \leq c \leq 67570 . \qquad (51)$$

Higher W (higher entropy) means a smoother reconstructed scene, but the eye does not distinguish 30000 different degrees of smoothness. Two scenes, as alike as possible except that one is in category (c), the other in (c+2), will be virtually undistinguishable to the eye. But every scene in (c+2) has a greater multiplicity than any scene in (c) by a factor of more than 10^{10}.

So not only do we have variations

$$W_1/W_2 \simeq 10^{10} \tag{52}$$

between scenes, we have chains of thousands of such comparisons, with a factor of 10^{10} at each step.

These numbers should give us an appreciation of the importance of multiplicity factors in inference, and explain why methods that ignore them are at such a disadvantage. The success of MAXENT reconstructions appears almost magical until we realize that, as an elementary combinatorial result, the overwhelming majority of all possible scenes compatible with our data have entropy very close to the maximum. This statement can be refined into an 'entropy concentration theorem' (Jaynes, 1982, 1983).

Before closing this discussion, let us note one other aspect of Bayes' theorem. Suppose our problem was different from what was assumed in Eqs. (46) and (47) and we knew N in advance, but this told us nothing about σ. (Perhaps σ is instrumental noise in a new instrument never before used.)

WHAT ABOUT σ? In all the above, we have supposed the noise level σ known. What if it isn't? The answer will give us a good example of the power of Bayesian methods. If σ is unknown, then it too becomes a parameter to which we assign some prior probability $p(\sigma \mid I)$, and Bayes' theorem will give us, instead of Eq. (37), a joint posterior distribution for F and σ:

$$p(F,\sigma \mid D,I_0) = p(F,\sigma \mid I_0) \frac{p(D \mid F,\sigma,I_0)}{p(D \mid I_0)} . \tag{53}$$

As we have argued very extensively elsewhere (Jaynes 1968, 1978, 1980), 'complete ignorance' of any scale parameter such as σ corresponds to the Jeffreys prior $p(\sigma \mid I_0) \propto d\sigma/\sigma$ (strictly, the limit of a sequence of such distributions over finite intervals ($0 < a < \sigma < b < \infty$), but our present problem has such good convergence that this nit-picking isn't needed). On prior information I_0, the prior probabilities of F and σ are independent: $p(F,\sigma \mid I_0) = p(F \mid I_0)p(\sigma \mid I_0)$; so in place of Eq. (38) we have

$$p(F,\sigma \mid D,I) \propto \sigma^{-(m+2)/2} \exp(NH - Q/\sigma^2) . \tag{54}$$

But if we care only about estimating F, then σ is what is called a 'nuisance parameter' in statistics. To get rid of it, we integrate it out to get the marginal posterior distribution of various scenes:

$$p(F \mid DI) = \int_0^\infty p(F,\sigma \mid DI)\, d\sigma . \tag{55}$$

The integration yields

$$p(F \mid DI) \propto e^{NH}/Q^{m/2} . \tag{56}$$

So now, in order to find the most probable scene, instead of maximizing $(NH - \sigma^{-2}Q)$ we should maximize $[NH - (m/2) \log Q]$. This gives, in place of Eq. (44), the implicit relationship

$$f_i \propto \exp\left[-\frac{m}{2N}\frac{\partial}{\partial f_i}\log Q\right]. \tag{57}$$

But this is formally the same as Eq. (44), if we put into Eq. (44)

$$\sigma^2 = \frac{2Q}{m}, \tag{58}$$

which is just the estimate of σ^2 that any rational person would make if he knows the (d_k, f_i) and nothing else.

To make a long story short, then, Bayes' theorem tells us that, if σ is unknown, then: (A) we should choose the scene $\{\hat{f}_1 \ldots \hat{f}_n\}$ as the point of "self-consistency" where the previous connection (44) still holds, with σ^2 replaced by our best estimate of σ^2; (B) the penalty we pay for this ignorance is that the peak of the distribution (56) is not as sharp (no longer exponential in Q) as that of Eq. (42). We still get a definite reconstruction but are not quite so confident of it. This is a good example of what I meant back in Section 3 [Eq. (6)], in saying that Bayes' theorem has other nice features that a rational person would demand of a method of inference.

7. Speculations—Realizable Fantasies

Already in the third issue of Dr. Dobb's Journal of Computer Calisthenics and Orthodontia (March 1976), it seemed time for an editorial with our same title, "Where do we go from here?" Dr. Dobb noted the obvious extensions and more powerful implementations of things already under way, and added: "We will continue the active pursuit of 'realizable fantasies.'" By this was meant projects not yet under way but that appear to be within the bounds of present technology and knowledge, and achievable within the next few years.

Doubtless, each of us has his own favorite list of realizable fantasies: those we would like to do ourselves, and those we would like to persuade others to do. Each of us would like to achieve some new breakthrough that is conceptually difficult but technically easy, and leave to others problems that are technically difficult. Problems that are both conceptually and technically easy are going to be solved automatically.

From a historical perspective, the great achievements of the past always appear to be new advances in conceptual understanding, rather than "mere" technical accomplishments. But we are seeing this through a filter; although technical accomplishment is often a prerequisite for conceptual advance, it tends to be forgotten. Today, everybody knows about the concept of ellipses, but very few living people know even the formulation, much less the technical details, of Kepler's analysis that led to this concept.

But there is also truth in the opposite emphasis. As we have learned in the theory of irreversible statistical mechanics, problems that seemed horrendously difficult technically may become, if not exactly "easy," at least many orders of magnitude less difficult, as a result of deeper conceptual insight.

There are many problems that may become technically easy as soon as the conceptual cobwebs are swept away. Decades of discussion from the standpoint of sampling theory may have obscured completely the fact that there is no sampling distribution, and the real problem is logically a generalized inverse type where MAXENT, rather than orthodox statistics, is the appropriate method. But the actual implementation of MAXENT may still be very difficult conceptually, because it requires us to define an explicit "hypothesis space" on which our counting of multiplicities is done.

In statistical mechanics we feel we know how to do this: The set of linearly independent "global" quantum states of a system is, according to all present knowledge, the proper field on which our game is to be played. But in forecasting economic time series, it may be perfectly clear that the essence lies not in the "randomness" of our data but in its incompleteness, yet still very unclear just what is the proper hypothesis space of elementary possibilities on which we should define multiplicities W and entropies log W.

There are many problems of this type, in which it seems clear that substantial improvements should be realizable, since these multiplicities must be highly cogent evidence, but present methods do not even recognize their existence. Even in areas such as spectrum analysis and image reconstruction, which are in a sense already "established," there is still much room for creative thinking about what is the appropriate hypothesis space for different kinds of problems. In this respect, I don't think any of our present solutions are optimal except in some very special kinds of problems. Burg's solution is now the classic example, but in time it may be seen as only the first of several MAXENT solutions for various spectral analysis problems.

Let us illustrate the last three paragraphs by the example of turbulence, which has indeed been discussed for decades from the standpoint of sampling theory. Is it a realizable fantasy to find a different conceptual standpoint which makes it clear that the problem is logically a generalized inverse one rather than a sampling theory exercise?

TURBULENCE. In thermodynamics, the first law for closed systems, or a "generalized first law" for open systems (conservation of energy plus the other conservation laws for number of atoms of each type, charge, etc.) determines a class of possible macroscopic states that is permitted by those constraints. Out of all these possibilities allowed by the first law, the one chosen by Nature (second law) is the one that can be realized in the greatest number of ways, that is, that has maximum entropy. This is the principle given by Gibbs (1875-1878), governing heterogeneous equilibrium. For 100 years most of physical chemistry has been based on it.

A similar principle must govern the spectrum of turbulence; a stirred fluid has energy added to it in motions of small wavenumber k, such that $kL \simeq 1$, where L is a typical dimension of the system. This is transferred to successively smaller eddies, of greater wavenumber, until it finally degrades into the uniform thermal energy of the highest possible wavenumbers, such that $kd \simeq 1$, where d is the molecular separation.

Imagining this degradation process in a momentary steady state, suppose the energy in turbulent eddies in the wavenumber range dk varies as $(1/k^n)$. As far as conservation of mass, energy, and momentum are concerned, many different values of n are allowed. Which one is chosen by Nature? This must be the value that provides the greatest number of channels (sequences of microstates) by which the process can take place. The logarithm of the number of channels W(P) by which a macroscopic process P can take place is a kind of entropy, which we have ventured (Jaynes, 1980) to call the "caliber" of P.

As an analogy, even if individual drivers have no particular preference for one traffic lane over another, still the six-lane road will end up carrying more traffic than the two-lane road to the same place. However, we are concerned here with ratios, not of 3:1, but perhaps of 10^{20}:1.

Among the possible macroscopic paths by which low-k kinetic energy may be degraded into high-k thermal energy, there will be one that opens up overwhelmingly more "traffic lanes" than any other—or indeed, than all others. This is surely the one that Nature will choose. Since the caliber of a path is not determined by the Navier-Stokes equation, but by the multiplicity factors that the N-S equation ignores, we can understand why attempts to base turbulence theory on the N-S equation met with little success, and degenerated into attempts to "guess the right statistical assumptions." Again, the problem is not one of mechanics, but of inference, of the generalized inverse type.

Many years ago, on the basis of some "statistical assumptions" and reasoning that I have been unable to follow, Kolmogorov proposed the value $n = 5/3$, which does seem to have some experimental support. But its theoretical justification and range of validity do not seem at all well understood as yet. This is a problem for the future, on which I think we have a good chance of succeeding.

Now turbulence theory is in so many respects like quantum field theory that a realizable fantasy in one may point to a realizable fantasy in the other. The high-caliber paths in turbulence are undoubtedly "lumpy" in time, small bifurcations suddenly transforming a lump of energy, which would appear as a suddenly generated wavelet rather than a continuous mode-transfer process. This immediately reminds us of quantum theory.

QUANTUM WAVELETS. Having started out as a conventionally trained, if somewhat apostate, quantum theorist, I was of course well indoctrinated in the quantum theorists' view of the world, in which they see themselves at the top of the pecking order in science, and almost disdain to recognize a geophysicist.

Just in the past year I have been shocked by the discovery that geophysicists, uninhibited by the kind of logic taught in quantum theory courses, have answered questions that quantum theorists were afraid to ask. Geophysicists have learned how to deal with wave phenomena in a more complete and sophisticated manner, and are in possession of some fundamental knowledge that quantum physicists need to have, if they could only bring themselves to admit it.

A recent review (Jaynes, 1973) concludes with a very incomplete preliminary sketch, suggesting that today the interesting, relevant physics of blackbody radiation may be contained in the intermode correlations that conventional theory neglects at the outset. To the best of my knowledge, only von Laue (1915) had ever recognized that these correlations might be important. His ideas were not pursued by others because Einstein (1915) argued against them, and, for that reason, the issue remains unresolved to this day.

The ultimate verdict may be that this is the only error Einstein ever made. Apparently, an earlier analysis (Einstein and Hopf, 1910) had convinced him that equilibrium of radiation necessarily leads to independent Gaussian probability distributions for the mode amplitudes. But here we meet again that constant plague of probability theory: What is the Problem? Einstein and Hopf had of course given a correct solution—but to a different problem.

Clearly, in the present problem, every act of emission or absorption must affect not only the field energy but also the intermode correlations. If our probability distribution is independent Gaussian immediately before an emission, it cannot be independent Gaussian immediately afterward.

It is curious that Einstein's theory (1917) of blackbody radiation insisted on, from the standpoint of the atoms, instantaneous transfer of energy and momentum—but blandly ignored, from the standpoint of the field, the intermode correlations that this would necessarily entail. Just at this point, causality was lost, and we have never learned how to restore it.

The probability theory of the early 1900s was not attuned to these realities, because Maxwell was dead and Jeffreys had not yet been heard from. If we define the probability of some condition as the fraction of a very long time in which the condition holds, as Einstein did, then we have sacrificed the very concept of a time-dependent probability. A theory of probability so curtailed cannot be used for inference about situations where we have information referring to specific times. Yet the change in our knowledge about a radiation field when we are told that an emission has occurred at time t must clearly affect our subsequent predictions about it, in a way that depends on t.

Now a correlation in the amplitudes of many modes close together in k-space is what we should describe in ordinary coordinate space (x-space) as a "wave packet" or "wavelet." It is not essentially different from the wavelets envisaged by Huygens. An emitting atom generates a spherical wavelet that propagates outward at the speed of light. Until it has been scattered, reflected, or otherwise perturbed, it retains its integrity and represents an additional contribution to the amplitudes of many modes, perfectly correlated.

These wavelets are the entities that figure so prominently in the work of H. Wold and have been exploited so well in the work of Enders Robinson and other geophysicists. It is surprising that physicists outside geophysics have made almost no use of the concept of wavelets, and had virtually no awareness of their existence, until the recent interest in solitons, which are wavelets that have become somewhat more indestructible thanks to nonlinearities in the wave equation.

For this reason, conventional physical theory does not use causal mechanisms to account for radiative equilibrium. Indeed, it cannot until it is reformulated in wavelet terms. Emission from an atom does not go into just one mode at a time as one might think from the conventional "Fermi Golden Rule" approximations of quantum theory; it goes into a coherent superposition of many modes simultaneously.

Likewise, absorption of energy by an atom is accomplished by emission of a wavelet of such phase that it cancels out, partially and momentarily, the incident wave in the forward direction (incipient shadow formation). Thus we conjecture that, in both turbulence and blackbody radiation, it is these wavelets that provide the physical mechanism by which the equilibrium energy distribution is brought about and maintained.

This is thrown out as a half-baked idea to ponder: When equilibrium is not maintained by a continuous process, is it possible that the temperature might be related more closely to the average energy of a wavelet than to the average energy of a mode?

After all, when the wavelet gets "hard" and indestructible, we call it a "particle," and then there is no doubt that the temperature relates to its kinetic energy. Then at what degree of hardness does the transition from mode to wavelet take place? Is localization in space a necessary part of that indestructibility? Or does this require only phase persistence for a time long compared to some characteristic interaction time? Indeed, if our concern is with the spectral distribution of the energy, why should its spatial distribution matter? It seems to me that it is the coherent organization of the energy—whether it occupies a cubic micron or a cubic mile—that makes a wavelet a unique "object."

So, to take the final deep plunge into fantasy: Is it possible that wavelets are the missing link, which physics has needed all these years, to resolve the contradictions of the "wave-particle duality"? In quantum theory, two generations of physicists have been taught that waves and particles are so antithetical that when we think about them we must forego using even the notion of an "objectively real" world. That is, present quantum theory cannot, as a matter of principle, answer any question of the form: "What is really happening when . . . ?" It can answer only: "What are the possible results of this experiment and their probabilities?" Yet after being taught that waves and particles form a fundamental, unbridgeable dichotomy, do not wavelets provide a simple, continuous interpolation between them?

References

Blackman, R. B., and J. W. Tukey (1958) The Measurement of Power Spectra (New York: Dover).

Bleher, P. M. (1981) Inversion of Toeplitz matrices, Trans. Moscow Math. Soc., Issue 2.

Boltzmann, L. (1877) Wien. Ber. **76**, 373.

Burg, J. P. (1967) Maximum entropy spectral analysis, Proc. 37th Meeting, Society of Exploration Geophysicists: Reprinted in D. G. Childers, ed. (1978) Modern Spectrum Analysis (New York: Wiley).

Burg, J. P. (1975) Ph.D. thesis, Stanford University.

Einstein, A. (1915) Ann. Phys. **47**, 879-885.

Einstein, A. (1917) Phys. Zeit. **18**, 121-128.

Einstein, A., and L. Hopf (1910) Ann. Phys. **33**, 1096-1104.

Everett, H. (1957) Relative state formulation of quantum mechanics, Rev. Mod. Phys. **29**, 454-462.

Fisher, R. A. (1956) Statistical Methods and Scientific Inference (New York: Hafner).

Gibbs, J. W. (1875-1878) Heterogeneous equilibrium, Conn. Academy of Science (reprinted 1928, New York: Longmans, Green and Co.; 1961, New York: Dover).

Gibbs, J. W. (1902) Statistical Mechanics (reprinted 1928, New York: Longmans, Green and Co.; 1961, New York: Dover).

Gohberg, I. C., and I. A. Fel'dman (1974) Convolution Equations and Projection Methods for Their Solution, Trans. Math. Monographs **41** (Providence, R.I.: Am. Math. Soc.), p. 86.

Gull, S. F., and G. J. Daniell (1978) Nature **272**, 686-690.

Jaynes, E. T. (1957) Information theory and statistical mechanics, Phys. Rev. **106**, 620; **108**, 171.

Jaynes, E. T. (1961) Review of Y. W. Lee, Statistical Theory of Communication, Am. J. Phys. **29**, 276.

Jaynes, E. T. (1967) Foundations of probability theory and statistical mechanics, in M. Bunge, ed., Delaware Seminar in the Foundations of Physics (Berlin: Springer-Verlag).

Jaynes, E. T. (1968) Prior probabilities, IEEE Trans. Syst. Sci. Cybern. **SSC-4**, 227-241. Reprinted in V. M. Rao Tummala and R. C. Henshaw, eds. (1976) Concepts and Applications of Modern Decision Models (Michigan State University Business Studies Series).

Jaynes, E. T. (1973) Survey of the present status of neoclassical radiation theory, in L. Mandel and E. Wolf, eds., Coherence and Quantum Optics (New York: Plenum), pp. 35-81.

Jaynes, E. T. (1976) Confidence intervals vs. Bayesian intervals, in W. L. Harper and C. A. Hooker, eds., Foundations of Probability Theory, Statistical Inference, and Statistical Theories of Science (Dordrecht, Holland: D. Reidel).

Harper and C. A. Hooker, eds., Foundations of Probability Theory, Statistical Inference, and Statistical Theories of Science (Dordrecht, Holland: D. Reidel).

Jaynes, E. T. (1978) Where do we stand on maximum entropy? in R. D. Levine and M. Tribus, eds., The Maximum Entropy Formalism (Cambridge, Mass.: M.I.T. Press).

Jaynes, E. T. (1980) The minimum entropy production principle, in B. S. Rabinovitch et al., eds., Ann. Rev. Phys. Chem. (Palo Alto, Calif.: Annual Reviews, Inc.), pp. 579-602.

Jaynes, E. T. (1982) On the rationale of maximum entropy methods, Proc. IEEE **70**, 939-952.

Jaynes, E. T. (1983) Papers on Probability, Statistics and Statistical Physics (Dordrecht, Holland: D. Reidel).

Klein, M. J. (1973) The development of Boltzmann's statistical ideas, in E. G. D. Cohen and W. Thirring, eds., The Boltzmann Equation (Berlin: Springer-Verlag), pp. 53-106.

Rota, G.-C. (1975) Finite Operator Calculus (New York: Academic Press).

Shannon, C. E. (1948) Bell Syst. Tech. J. **27**, 379-423, 623-656.

von Laue, M. (1915) Ann. Phys. **47**, 853; **48**, 668.

INTRODUCTION TO MINIMUM-CROSS-ENTROPY SPECTRAL ANALYSIS OF MULTIPLE SIGNALS

Rodney W. Johnson and John E. Shore

Computer Science and Systems Branch, Naval Research Laboratory, Washington, D.C. 20375

This paper outlines a new information-theoretic method for simultaneously estimating a number of power spectra when a prior estimate of each is available and new information is obtained in the form of values of the autocorrelation function of their sum. One application of this method is the separate estimation of the spectra of a signal and additive noise, based on autocorrelations of the signal plus noise. The basic estimation equations are given, and an application to speech noise suppression is described.

C. Ray Smith and W. T. Grandy, Jr. (eds.),
Maximum-Entropy and Bayesian Methods in Inverse Problems, 59–66.

We outline here an information-theoretic method for simultaneously estimating a number of power spectra when a prior estimate of each is available and new information is obtained in the form of values of the autocorrelation function of their sum. The method applies, for instance, when one obtains autocorrelation measurements for a signal with independent additive interference, and one has some prior knowledge concerning the signal and noise spectra; the result is signal- and noise-spectrum estimates that take both the prior estimates and the autocorrelation information into account. One thus obtains a procedure for noise suppression that offers some advantages over more traditional procedures, such as those based on spectral subtraction.

The method described here, multi-signal minimum-cross-entropy spectral analysis (multi-signal MCESA), is a generalization of minimum-cross-entropy spectral analysis (MCESA) [1], which is in turn a generalization of maximum-entropy (or linear-predictive or autoregressive) spectral analysis (MESA) [2,3]. All these methods proceed from autocorrelation values. MCESA differs from MESA in that it explicitly uses a prior estimate of the power spectrum; it reduces to MESA as a special case when the prior estimate is uniform and one of the given autocorrelation values is for zero lag. Multi-signal MCESA differs from MCESA in that it treats an arbitrary number of independent spectra simultaneously; in the special case of a single spectrum, it becomes identical to MCESA.

MESA may be regarded as an application of the principle of maximum entropy [4,5]; single- and multi-signal MCESA are applications of a generalization of that principle, the principle of minimum cross entropy (also called minimum discrimination information, directed divergence, I-divergence, relative entropy, or Kullback-Leibler number) [6-8].

MESA addresses the following problem: Estimate the power spectrum $S(f)$ of a real, band-limited, stationary process, given values of the autocorrelation function

$$R(t) = 2 \int_0^W df\, S(f) \cos 2\pi ft$$

for finitely many lags $t = t_r$, $r = 0,\ldots,M$. (Here W is the bandwidth.) The solution proposed by Burg [2,3] is to choose the estimate Q of S that maximizes

$$\int_0^W df \log Q(f) \tag{1}$$

subject to the constraint that the autocorrelation function assume the given values:

$$R(t_r) = 2 \int_0^W df\, Q(f) \cos 2\pi f t_r \,. \tag{2}$$

The resulting estimator has the form

$$Q(f) = \frac{1}{\sum_r 2\beta_r \cos 2\pi f t_r} \,, \tag{3}$$

where the coefficients β_r $(r = 0,\ldots,M)$ are chosen so that Q satisfies Eq. (2).

MCESA is applicable to the problem of estimating S(f) when, in addition to the autocorrelation values, a prior estimate P of S is given; P may be thought of as the best guess at S we could make in the absence of autocorrelation data. The MCESA estimator has the form [1]

$$Q(f) = \frac{1}{\frac{1}{P(f)} + \sum_r 2\beta_r \cos 2\pi f t_r} \,, \tag{4}$$

where again the β_r are chosen so that Q satisfies the constraints in Eq. (2). We call Q the posterior estimate of S based on the prior estimate P and constraints in Eq. (2). This estimator can be obtained directly from the minimum-cross-entropy principle [1]; it can also be obtained by minimizing the Itakura-Saito distortion measure [9]

$$\int df \left[\frac{Q(f)}{P(f)} - \log \frac{Q(f)}{P(f)} - 1 \right]$$

subject [1] to Eq. (2). When P(f) is uniform, and one of the autocorrelation values is at lag zero (say $t_0 = 0$), we can write Eq. (4) in the form of Eq. (3), since the constant 1/P can be absorbed into the coefficient β_0. Thus in this case MCESA reduces to MESA.

For multi-signal MCESA, the problem is to estimate the power spectra $S_i(f)$ of a number of independent processes, given values of the total autocorrelation

$$R(t) = 2 \sum_i \int_0^W df\, S_i(f) \cos 2\pi f t$$

and a prior estimate P_i for each S_i. The estimator has the form

$$Q_i(f) = \frac{1}{\dfrac{1}{P_i(f)} + \sum_r 2\beta_r \cos 2\pi f t_r}, \tag{5}$$

where the β_r are chosen so that the constraint equations

$$\dot{R}(t_r) = 2 \sum_i \int_0^W df\ Q_i(f) \cos 2\pi f t_r \tag{6}$$

are satisfied. Note that the summation term in the denominator in Eq. (5) is independent of i. Elsewhere [10] we derive the estimates (5) directly from the principle of minimum cross entropy, and we also show that they can be obtained by minimizing the sum

$$\sum_i \int df \left[\frac{Q_i(f)}{P_i(f)} - \log \frac{Q_i(f)}{P_i(f)} - 1 \right]$$

of Itakura-Saito distortions subject to the constraints in Eq. (6). Equations (5) and (6) reduce to Eqs. (4) and (2) when there is only one spectrum S_i; thus multi-signal MCESA reduces to ordinary MCESA in case there is only one signal. Properties of the estimator (5) are also discussed in Ref. 10.

We now consider a numerical example based on time-domain samples of voiced speech and noise. The speech comprises a portion of an English sentence spoken by a male speaker and includes the first word, "Sue," of the sentence together with silent segments before and after it. The noise consists of a segment of helicopter noise equal in duration to the speech. These were separately filtered, sampled, and digitized at 8000 samples per second. The speech and noise data were then added sample by sample, resulting in samples of noisy speech. These samples were segmented into analysis frames of 180 samples, and 11 autocorrelations R_r ($r = 0,1,\ldots,10$) were estimated for each frame by the formula

$$R_r = \frac{1}{180} \sum_{j=1}^{180-r} s_j s_{j+r},$$

where s_j is the jth sample in the frame. This is a biased estimate but guarantees positive-definiteness. No additional windowing or filtering was used.

The last frame before the actual beginning of the word was selected; this frame of "noisy speech" thus consisted entirely of noise. From the autocorrelation estimates for this frame, a conventional MESA (that is, uniform-prior MCESA) spectral estimate was computed for use as a prior estimate of the noise spectrum in subsequent frames. A uniform spectrum was used as a prior estimate for the speech spectrum in the subsequent frames. These two priors are shown in Fig. 1. Much of the noise power is concentrated in a peak near 2780 Hz.

From the two priors and the autocorrelation estimates, multi-signal MCESA estimates of the speech and noise spectra were computed for later frames. From the autocorrelation estimates, MESA (LPC) spectral estimates were computed for the noisy speech. We present the results for a selected frame of voiced speech—the second of seven frames that span the vowel "u." For comparison with these results, we present in Fig. 2 a MESA estimate of the uncorrupted speech. This was computed exactly like the MESA

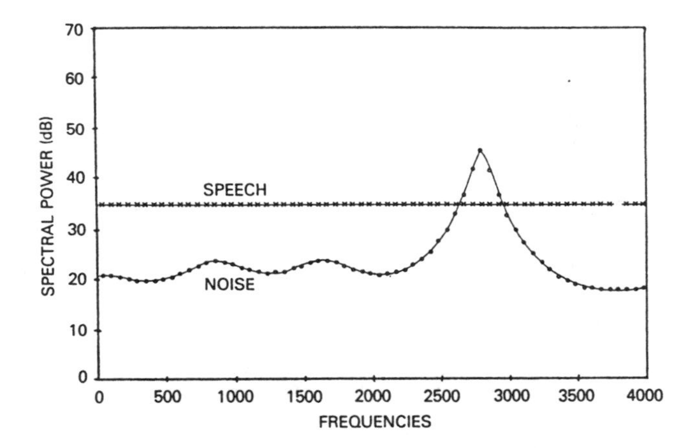

Figure 1. Prior estimates of speech and noise spectra.

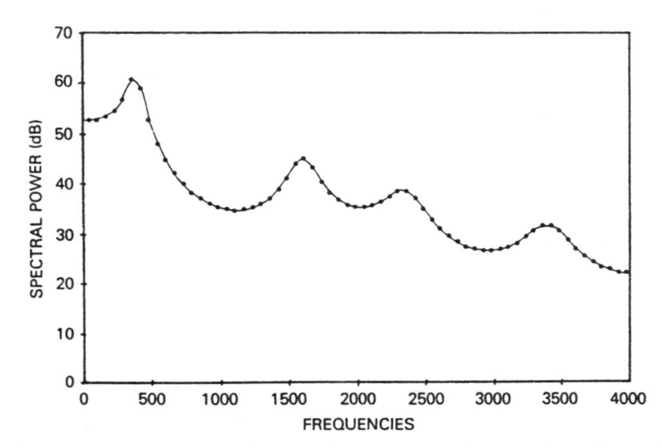

Figure 2. MESA estimate of speech spectrum from noise-free data.

estimate for the noisy speech except that the R_r were estimated from the speech samples only, not from the sums of speech and noise samples.

The MESA estimate for the noisy speech is shown in Fig. 3. This spectrum agrees rather well with the noise-free estimate in the band from 0 to about 2000 Hz, which includes the first two formants. Above 2000 Hz, however, there is only a single maximum; the third and fourth formants have merged with the peak in the noise spectrum to form a single peak at about 2690 Hz.

We subtracted the noise prior (Fig. 1) from this result (Fig. 3). The difference, shown in Fig. 4, represents an attempt to estimate the speech

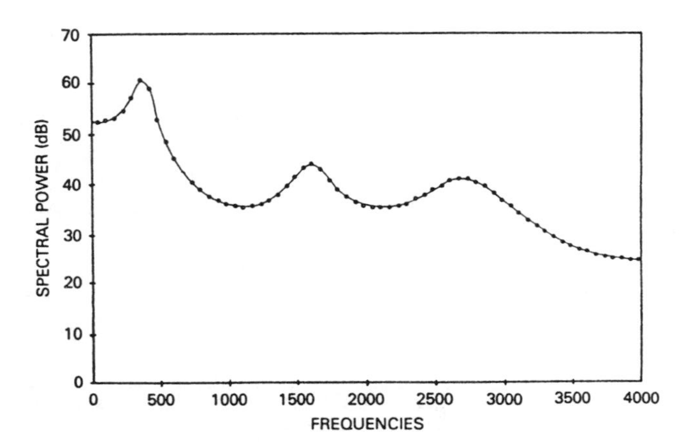

Figure 3. MESA estimate of total spectrum.

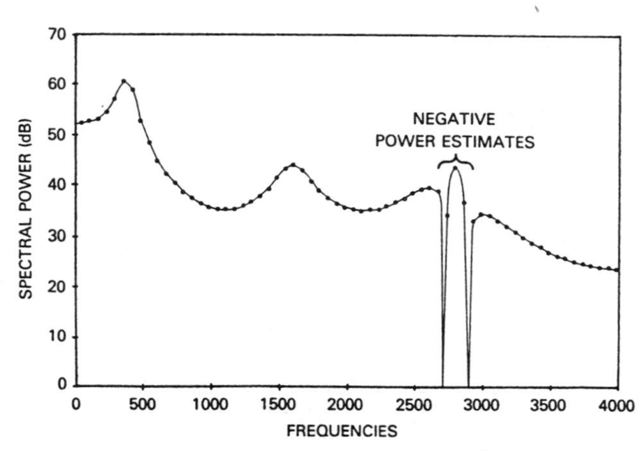

Figure 4. Result of subtracting noise prior from spectrum.

spectrum by a MESA analysis and spectral subtraction. The subtracted MESA spectrum is fairly close to the unsubtracted MESA spectrum except in the neighborhood of the noise peak at 2780 Hz. Near that frequency, the subtraction so far overcompensates that the difference actually assumes rather large negative values. (Absolute values are plotted in the figure.)

The multi-signal MCESA posteriors are shown in Figs. 5 and 6; Fig. 5 is the speech, and Fig. 6 is the noise. Figure 6 shows a maximum near 2440 Hz, about 130 Hz higher than the third formant, and a suggestion of the fourth formant is discernible. Except for frequencies near the noise peak, the multi-signal speech spectrum (Fig. 5) and the subtracted MESA result

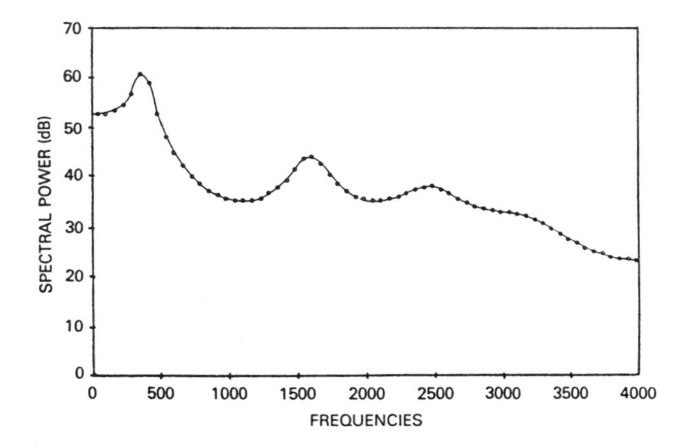

Figure 5. Multi-signal MCESA posterior estimate of speech spectrum.

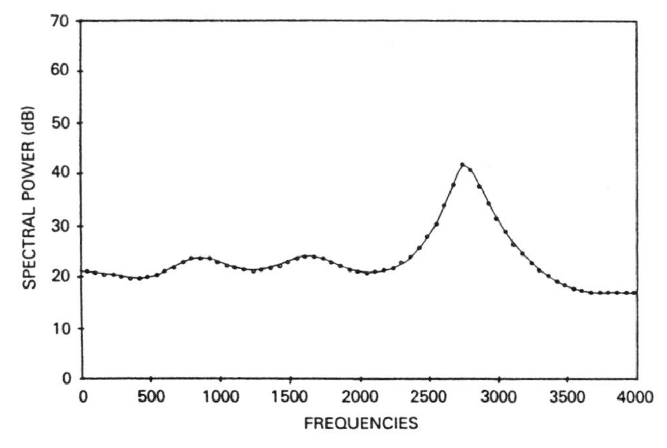

Figure 6. Multi-signal MCESA posterior estimate of noise spectrum.

(Fig. 4) are quite close, the multi-signal result being usually the closer of the two to the estimate based on noise-free data (Fig. 2). Near 2780 Hz, the multi-signal result is substantially closer, and where the subtracted MESA becomes negative, the multi-signal estimate takes only physically meaningful positive values. Both methods underestimate the total power near 2780 Hz (cf. Fig. 6); however, the multi-signal method apportions the total between speech and noise in a somewhat reasonable way, whereas the other does not.

Recently, we have synthesized speech from the outputs of multi-signal MCESA computed in this way. There was a clear reduction in the intensity of the background noise in comparison with speech synthesized from MESA (LPC) analysis of noisy speech.

References

1. J. E. Shore (1981) Minimum cross-entropy spectral analysis, IEEE Trans. Acoust. Speech Signal Process. **ASSP-29**, 230.
2. J. P. Burg (1967) Maximum entropy spectral analysis, Presented at the 37th Annual Meeting, Society of Exploration Geophysicists, Oklahoma City, Okla.
3. J. Burg (1975) Maximum Entropy Spectral Analysis, Ph.D. thesis, Stanford University (University Microfilms No. 75-25, 499).
4. W. M. Elsasser (1937) On quantum measurements and the role of the uncertainty relations in statistical mechanics, Phys. Rev. **52**, 987.
5. E. T. Jaynes (1968) Prior probabilities, IEEE Trans. Syst. Sci. Cybern. **SSC-4**, 227.
6. S. Kullback (1959) Information Theory and Statistics (New York: Wiley) (reprinted 1969, New York: Dover).
7. J. E. Shore and R. W. Johnson (1980) Axiomatic derivation of the principle of maximum entropy and the principle of minimum cross-entropy, IEEE Trans. Inf. Theory **IT-26**, 26.
8. J. E. Shore and R. W. Johnson (1981) Properties of cross-entropy minimization, IEEE Trans. Inf. Theory **IT-27**, 472.
9. F. Itakura and S. Saito (1968) Analysis synthesis telephone based upon the maximum likelihood method, in Y. Yonsai, ed., Reports of the 6th International Congress on Acoustics, Tokyo.
10. R. W. Johnson and J. E. Shore (1983) Minimum cross-entropy spectral analysis of multiple signals, IEEE Trans. Acoust. Speech Signal Process. **ASSP-31**, 574-582.

ON AN ALLEGED BREAKDOWN OF THE MAXIMUM-ENTROPY PRINCIPLE

E. Rietsch

Texaco USA, Bellaire Research Laboratories, Bellaire, Texas 77401

During the first workshop on Maximum Entropy and Bayesian Methods in Applied Statistics it was claimed that the maximum entropy principle cannot be applied if the entropy is a datum. It is shown here by means of an example that this situation can be handled by introducing a probability of a probability. The solution illustrates how different ways of incorporating constraints lead to different results, in particular if the constraints are non-linear functions of the parameters to be estimated.

C. Ray Smith and W. T. Grandy, Jr. (eds.),
Maximum-Entropy and Bayesian Methods in Inverse Problems, 67–82.

1. Introduction

In this paper I address a problem that came up during the First Workshop on Maximum Entropy and Bayesian Methods in Applied Statistics, held at the University of Wyoming. In spite of the fact that its solution probably has little practical significance, I nevertheless believe it is worth being discussed since it also illustrates other questions that must be faced by anyone seriously trying to solve inverse problems by means of the maximum-entropy method.

2. Formulation of the Problem

Instead of presenting the problem in general terms, I will explain it by means of an example.

Assume we are in Las Vegas in front of a slot machine. It is a simple model. Feeding one dollar to this machine leads to one of four different events E_n,

E_1: machine returns nothing—$1 lost
E_2: machine returns $1—nothing lost, nothing won
E_3: machine returns $2—$1 won
E_4: machine returns $10—$9 won

Each of these events E_n occurs with a probability p_n satisfying the normalization condition

$$\sum_n p_n = 1 , \tag{1}$$

which simply means that one of the four possible events will occur. (Unless indicated otherwise, summations and products over n, m, μ, ν extend from 1 to 4).

Standing in front of this machine, we are facing a situation of uncertainty. We know that one of the four events will occur whenever we insert one dollar, but we do not know which of them it is going to be.

A measure of the uncertainty is afforded by the entropy

$$h(\mathbf{p}) \equiv h(p_1, p_2, p_3, p_4) = - \sum_n p_n \log p_n , \tag{2}$$

which depends on the probabilities p_n. It is customary to use 2 as the basis of the logarithm in Eq. (2), and then the entropy is measured in units of bits.

The maximum entropy principle tells us that the "most honest guess" of the probabilities of the four events is the one that maximizes the entropy.

Subject to only the normalization condition (1), the entropy (and thus the uncertainty) has a maximum if all events are equally likely—in this case $p_n = 1/4$. The entropy is then 2 bits.

Equal probability of all outcomes maximizes the entropy but not the profits of the machine's owner. In fact, his expected profit $\bar{r}$ per game, which is given by

$$\bar{r} = \sum r_n p_n , \tag{3}$$

where

$$r_1 = 1, \quad r_2 = 0, \quad r_3 = -1, \quad r_4 = -9 , \tag{4}$$

is negative ($-2.25).

We know, of course, that the owner of the machine wants to make money, and therefore the expected return $\bar{r}$ should be positive. This is information. In a simple case like this it can be used to derive expectation values of inequalities in the sense that we expect that

$$\begin{aligned} p_1 &> 0.4272 \\ p_2 &< 0.3167 \\ p_3 &< 0.2347 \\ p_4 &< 0.0214 \end{aligned} \tag{5}$$

These inequalities have been derived with the assumption that $r = 0$, noting that the probabilities are monotonic functions of $\bar{r}$ (Appendix A). The same arguments lead to

$$h(\mathbf{p}) < 1.6590 . \tag{6}$$

In contrast to Eqs. (5), which are the "most honest guesses," inequality (6) is exact: No set of probabilities has a nonnegative expected return $\bar{r}$ and an entropy greater than 1.6590.

Let us assume the owner of the machine is cooperative. He tells us that in the long run the machine returns only $0.90 for each dollar inserted. Thus

$$\bar{r} = 0.1 . \tag{7}$$

Maximizing the entropy subject to the normalization condition (1) and the constraint (3) leads to (Appendix A)

$$\begin{aligned} p_1 &= 0.4473 \\ p_2 &= 0.3158 \\ p_3 &= 0.2231 \\ p_4 &= 0.0138 \end{aligned} \tag{8}$$

The entropy turns out to be

$$h(\mathbf{p}) = 1.6124 . \tag{9}$$

Obviously, the inequalities (5) and (6) are satisfied.

Without additional information, nothing more can be said about the probabilities. But then the owner volunteers one more piece of information: the entropy is 1.5. And now, I seem to have a problem. The entropy that I maximized to find the probabilities is now a known datum.

3. The Entropy as Datum

It has been claimed that the maximum entropy principle is not applicable if the entropy is a datum rather than a quantity to be maximized. It is one of the purposes of this paper to show that this is not the case. According to Jaynes (1968), all it requires is a "split personality." In following Jaynes' lead, I introduce a probability

$$P(\mathbf{p})d\mathbf{p} = P(p_1,p_2,p_3,p_4)dp_1dp_2dp_3dp_4$$

of the probabilities $p_1,\ldots,p_4$. Of course, $P(\mathbf{p})$ must satisfy the normalization condition

$$\int P(\mathbf{p})d\mathbf{p} = 1\,, \tag{10}$$

and its entropy is defined as

$$H = -\int P(\mathbf{p}) \log\left[\frac{P(\mathbf{p})}{W(\mathbf{p})}\right] d\mathbf{p}\,. \tag{11}$$

This presents the next stumbling block—the prior distribution, $W(\mathbf{p})$, describing "complete ignorance" (Jaynes, 1968). A quick look through the literature conveys the impression that this prior distribution is generally set equal to 1. To simplify further analysis I will also make this assumption here and proceed to the next problem: In which way should I incorporate the available information about the probabilities p_n?

$$\begin{aligned} \sum_n p_n &= 1 \\ \sum_n r_n p_n &= 0.1 \\ h(\mathbf{p}) = -\sum_n p_n \log p_n &= 1.5 \end{aligned} \tag{12}$$

One approach is to request that the expectation of conditions (12) be satisfied. Thus Eqs. (12) are used in the form

$$\int (\sum p_n)\, P(\mathbf{p})\, d\mathbf{p} = 1$$

$$\int (\sum r_n p_n)\, P(\mathbf{p})\, d\mathbf{p} = 0.1 \qquad (13)$$

$$\int h(\mathbf{p})\, P(\mathbf{p})\, d\mathbf{p} = 1.5 \,.$$

Now we can compute the probability density $P(\mathbf{p})$ that maximizes the entropy H defined in Eq. (11) subject to Eqs. (10) and (13) (Appendix B). From this probability density we can compute the expectation values

$$\overline{p_n} = \int p_n P(\mathbf{p})\, d\mathbf{p} \,. \qquad (14)$$

The result is

$$\begin{aligned} \overline{p_1} &= 0.4828 \\ \overline{p_2} &= 0.3027 \\ \overline{p_3} &= 0.1935 \\ \overline{p_4} &= 0.0210 \end{aligned} \qquad (15)$$

These expectation values differ from the probabilities computed previously without the condition on the entropy.

Computing the entropy of the p_n by substituting Eqs. (15) into Eq. (2) leads to

$$h(\mathbf{p}) = 1.6047 \,, \qquad (16)$$

which is lower than Eq. (9) but by no means close to the desired 1.5. The reason for this discrepancy is, of course, the fact that

$$\int h(\mathbf{p})\, P(\mathbf{p})\, d\mathbf{p} = \overline{h(\mathbf{p})} \neq h(\overline{\mathbf{p}}) \,; \qquad (17)$$

the expectation value of the entropy is not equal to the entropy of the expectation values. And I just computed the latter.

It is important to pay attention to this problem. In seismic inversion, for example, the condition that the expectation value of a synthetic seismogram must agree with the recorded seismic data will lead to somewhat different estimates of the elastic parameters than will the condition that a synthetic seismogram computed from the expectation values of the elastic parameters must agree with the recorded seismic data.

To show the difference in this simple case, let me replace the last equation in Eqs. (13) by

$$h(\overline{\mathbf{p}}) = 1.5 . \qquad (18)$$

Numerical evaluation of the resultant equations (Appendix C) leads to

$$\begin{aligned} \overline{p}_1 &= 0.5996 \\ \overline{p}_2 &= 0.2317 \\ \overline{p}_3 &= 0.1273 \\ \overline{p}_4 &= 0.0414 \end{aligned} \qquad (19)$$

These probabilities differ markedly from those in Eqs. (15) and do indeed satisfy condition (18).

Let me make one more modification. Owing to the special kind of our constraints, the probability density $P(\mathbf{p})$ has the general form

$$P(\mathbf{p}) = (\text{constant}) \exp(-\sum u_n p_n) , \qquad (20)$$

a truncated exponential distribution. Obviously, it does not exclude the possibility

$$p_1 = p_2 = p_3 = p_4 = 1 .$$

Defining

$$p = p_1 + p_2 + p_3 + p_4 , \qquad (21)$$

we see that $p = 4$ is possible (though very unlikely). The probability density of p is proportional to

$$\int \delta(p - \sum p_n) P(\mathbf{p}) d\mathbf{p} , \qquad (22)$$

where $\delta(\cdot)$ denotes the δ-function (Roos, 1969, p. 291). As shown in Appendix D, this integral can be evaluated in closed form, leading to the probability density shown in Fig. 1. It has a maximum for $p \simeq 1.1$, and then drops off rapidly. The case $p = 4$ is therefore extremely unlikely. Of potentially more concern is the fairly wide peak of the distribution.

The reason for this deficiency is that the first constraint in Eqs. (13) requires only that the expectation value of the sum of the probabilities be 1. To those who find this disturbing it might seem to be advisable to drop this constraint and, instead, perform the integration in Eqs. (10), (11), and (13) only over the hyperplane $\Sigma p_n = 1$. This is equivalent to introducing a prior distribution of the form

$$W(\mathbf{p}) = \delta(1 - \sum p_n) . \qquad (23)$$

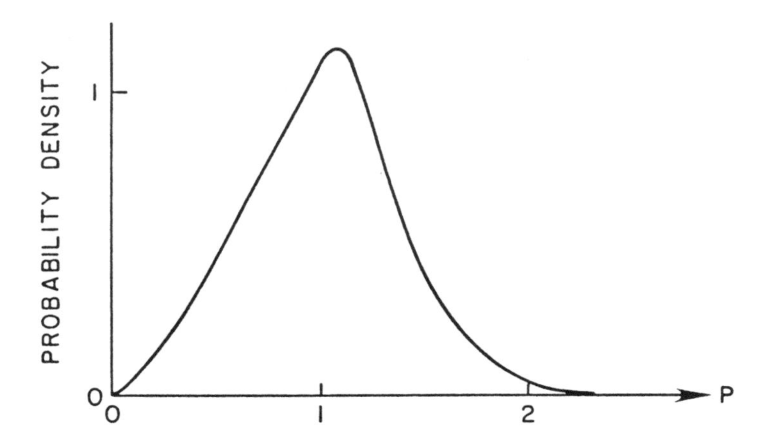

Figure 1. The probability density of p, defined by Eq. (21).

Numerical evaluation of the equations derived in Appendix D leads to the following expectation values of the probabilities:

$$\begin{aligned} \overline{p}_1 &= 0.6044 \\ \overline{p}_2 &= 0.2161 \\ \overline{p}_3 &= 0.1389 \\ \overline{p}_4 &= 0.0406 \end{aligned} \tag{24}$$

These turn out to be fairly close to those obtained with the 'softer' condition

$$\sum \overline{p}_n = 1 .$$

It is, of course, possible to also replace the second condition in Eqs. (13) by a restriction on the domain of integration thus giving $W(\mathbf{p})$ the form

$$W(\mathbf{p}) = \delta(1 - \sum p_n)\, \delta(0.1 - \sum r_n p_n) . \tag{25}$$

However, this change does not appear to be significant. Its influence would furthermore decrease with increasing number of variables. If hundreds of (rather than four) variables were being estimated, the discrepancy would be completely negligible for all practical purposes.

4. Conclusion

Why did I show this variety of ways to include information into the maximum-entropy formalism? It may have been confusing, though this has not been my intention. What I wanted to show is that the maximum-entropy approach is very flexible, even more flexible than many people seem to think. One other point at least as important is that the way the conditions or constraints are incorporated may have an important bearing on the result.

Appendix A. Computation of the Probabilities

The probabilities are solutions of the following extremal problem:

Maximize

$$h = -\sum_n p_n \log p_n \tag{A1}$$

subject to

$$\sum_n p_n = 1 \tag{A2}$$

$$\sum_n r_n p_n = \bar{r} . \tag{A3}$$

The solution is well known (Rietsch, 1977):

$$p_n = \frac{\exp(\lambda r_n)}{\sum_m \exp(\lambda r_m)} . \tag{A4}$$

It obviously satisfies Eq. (A2) for any value of the Lagrange multipliers λ that can be determined by substituting Eq. (A4) for p_n in Eq. (A3). This leads to the equation

$$\sum_n r_n \exp(\lambda r_n) = \bar{r} \sum_n \exp(\lambda r_n) , \tag{A5}$$

which can be solved by a variety of techniques.

Equation (A5) defines λ as a function of $\overline{r}$. Its differentiation shows that λ increases with increasing $\overline{r}$,

$$\left[\sum_n r_n(r_n - \overline{r}) \exp(\lambda r_n)\right] d\lambda = \left[\sum_n \exp(\lambda r_n)\right] d\overline{r}, \tag{A6}$$

since both terms in brackets are positive for $0 \leq \overline{r} \leq 1$ and the r_n defined in Eq. (4). Hence,

$$dp_n/d\overline{r} = (r_n - \overline{r}) p_n \, d\lambda/d\overline{r} \tag{A7}$$

increases for increasing $\overline{r}$ if $n = 1$ and decreases for increasing $\overline{r}$ if $n = 2, 3,$ or 4. Choosing $\overline{r} = 0$ therefore provides us with the lower and upper bounds, respectively, for p_1 and p_2, p_3, p_4 shown in Eqs. (5).

Appendix B. Computation of the Probability Density $P(\mathbf{p})$ with Constraints of the Form $\overline{f_j(\mathbf{p})} = \overline{f_j}$

Maximizing the entropy (using the natural logarithm for computational convenience)

$$H = - \int P(\mathbf{p}) \ln \left[\frac{P(\mathbf{p})}{W(\mathbf{p})}\right] d\mathbf{p} \tag{B1}$$

subject to the normalization condition

$$\int P(\mathbf{p}) d\mathbf{p} = 1 \tag{B2}$$

and J additional conditions of the form

$$\overline{f_j(\mathbf{p})} \equiv \int f_j(\mathbf{p}) P(\mathbf{p}) \, d\mathbf{p} = \overline{f_j} \tag{B3}$$

leads to the equation

$$\ln \left[\frac{P(\mathbf{p})}{W(\mathbf{p})}\right] + 1 + \mu + \sum_j \lambda_j f_j(\mathbf{p}) = 0 . \tag{B4}$$

Here μ and λ_j are Lagrange multipliers. The volume element $d\mathbf{p}$ indicates that the integration is performed over the support of $P(\mathbf{p})$. Sums over j, k, or ℓ extend from 1 to J. Choosing μ in such a way that Eq. (B2) is satisfied leads to

$$P(\mathbf{p}) = \frac{W(\mathbf{p})}{Z(\boldsymbol{\lambda})} \exp\left[-\sum_j \lambda_j f_j(\mathbf{p})\right] \tag{B5}$$

with

$$Z(\boldsymbol{\lambda}) = \int W(\mathbf{p}) \exp\left[-\sum_j \lambda_j f_j(\mathbf{p})\right] d\mathbf{p} \,. \tag{B6}$$

Substitution of Eq. (B5) into Eq. (B3) yields a system of equations for λ_j that can be written in the form

$$-\frac{\partial}{\partial \lambda_j} \ln Z(\boldsymbol{\lambda}) = \overline{f_j} \,. \tag{B7}$$

With Eq. (B5) substituted into Eq. (B1) we get the entropy

$$H = \ln Z(\boldsymbol{\lambda}) + \sum_j \lambda_j \overline{f_j} \,. \tag{B8}$$

Equation (B7) can therefore be written as

$$\frac{\partial H}{\partial \lambda_j} = 0 \,, \tag{B9}$$

defining a maximum of the entropy, since

$$\frac{\partial^2 H}{\partial \lambda_j \partial \lambda_k} = \frac{\left[\frac{\partial^2 Z(\boldsymbol{\lambda})}{\partial \lambda_j \partial \lambda_k} Z(\boldsymbol{\lambda}) - \frac{\partial Z(\boldsymbol{\lambda})}{\partial \lambda_j} \frac{\partial Z(\boldsymbol{\lambda})}{\partial \lambda_k}\right]}{Z^2(\boldsymbol{\lambda})}$$

$$= -\frac{1}{2Z^2(\boldsymbol{\lambda})} \int W(\mathbf{p}) W(\mathbf{p}') [f_j(\mathbf{p}) - f_j(\mathbf{p}')] [f_k(\mathbf{p}) - f_k(\mathbf{p}')]$$

$$\times \exp\left[-\sum_\ell \lambda_\ell [f_\ell(\mathbf{p}) - f_\ell(\mathbf{p}')]\right] d\mathbf{p}\, d\mathbf{p}' \tag{B10}$$

is obviously negative definite. Therefore the Lagrange multipliers can be computed by minimizing -H. Once the λ_j are known, the expectation values

$$\overline{p_n} = \int p_n P(\mathbf{p})\, d\mathbf{p} \tag{B11}$$

can be computed.

For the case discussed in the main body of this paper, J = 3 and

$$\begin{aligned} f_1(\mathbf{p}) &= \sum_n p_n \\ f_2(\mathbf{p}) &= p_1 - p_3 - 9\, p_4 \\ f_3(\mathbf{p}) &= -\sum_n p_n \log p_n \end{aligned} \tag{B12}$$

$$\overline{f_1} = 1, \quad \overline{f_2} = \overline{r} = 0.1, \quad \overline{f_3} = 1.5 \tag{B13}$$

$$W(\mathbf{p}) = 1\,. \tag{B14}$$

With Eqs. (B12) substituted for $f_j(\mathbf{p})$ in Eq. (B6), $Z(\boldsymbol{\lambda})$ can be computed by numerical integration for any $\lambda_1, \lambda_2, \lambda_3$ (within a reasonable range, of course). Minimizing -H with respect to the λ_j leads to the values of the Lagrange multipliers that are required to compute $P(\mathbf{p})$. Numerical integration of Eq. (B11) yields the expectation values of the probabilities p_n shown in Eqs. (15).

Appendix C. Computation of the Probability Density $P(\mathbf{p})$ with Constraints of the Form $f_j(\overline{\mathbf{p}}) = \overline{f_j}$

Maximizing, as in Appendix B, the entropy

$$H = -\int P(\mathbf{p}) \ln \left[\frac{P(\mathbf{p})}{W(\mathbf{p})}\right] d\mathbf{p} \tag{C1}$$

subject to the normalization condition

$$\int P(\mathbf{p})\, d\mathbf{p} = 1 \tag{C2}$$

and J additional conditions that now have the form

$$f_j(\overline{\mathbf{p}}) = \overline{f_j}\,, \tag{C3}$$

with the expectation value of $\mathbf{p}$ denoted by

$$\overline{\mathbf{p}} = \int \mathbf{p}\,P(\mathbf{p})\,d\mathbf{p} ,$$

leads to the following condition for $P(\mathbf{p})$:

$$\ln P(\mathbf{p}) + 1 + \mu + \sum_j \lambda_j \sum_n \frac{\partial f_j}{\partial p_n} p_n = 0 , \tag{C4}$$

where μ and the λ_j are Lagrange multipliers. Choosing μ in such a way that Eq. (C2) is satisfied and introducing the abbreviations

$$f_{j,n}(\mathbf{p}) = \frac{\partial f_j(\mathbf{p})}{\partial p_n} \tag{C5}$$

$$u_n = \sum_j \lambda_j f_{j,n}(\overline{\mathbf{p}}) \tag{C6}$$

yields

$$P(\mathbf{p}) = W(\mathbf{p}) \exp(-\sum u_n p_n)/Z(\mathbf{u}) \tag{C7}$$

with

$$Z(\mathbf{u}) = \int W(\mathbf{p}) \exp(-\sum u_n p_n)\,d\mathbf{p} . \tag{C8}$$

The expectation values $\overline{p_n}$ can be represented in terms of $Z(\mathbf{u})$,

$$\overline{p_n} = -\frac{\partial}{\partial u_n} \ln Z(\mathbf{u}) . \tag{C9}$$

An expression for the entropy H is obtained by substituting Eq. (C7) into Eq. (C1). This leads to

$$H = \ln Z(\mathbf{u}) + \sum_n u_n \overline{p_n} . \tag{C10}$$

It is easy to show that the functional

$$\phi = \ln Z(\mathbf{u}) + \sum_n u_n p_n + \sum_j \lambda_j [\overline{f_j} - f_j(\mathbf{p})] , \tag{C11}$$

obtained by adding the constraints multiplied with Lagrange multipliers to the entropy, is stationary with regard to a variation of u_n or p_n if they satisfy the relationships

$$u_n = \sum_j \lambda_j f_{j,n}(\mathbf{p}) \tag{C12}$$

$$\overline{p_n} = - \frac{\partial}{\partial u_n} \ln Z(\mathbf{u}) . \tag{C13}$$

These relationships define p_n and u_n as functions of $\boldsymbol{\lambda}$. If the p_n also satisfy Eq. (C3), they are equivalent to Eqs. (C6) and (C9). Furthermore,

$$\frac{\partial \phi}{\partial \lambda_k} = \sum_n \left[\frac{\partial}{\partial u_n} \ln Z(\mathbf{u}) + p_n \right] \frac{\partial u_n}{\partial \lambda_k} + \overline{f_k} - f_k(\mathbf{p})$$

$$+ \sum_m \sum_n \left[u_n - \sum_j \lambda_j f_{j,n}(\mathbf{p}) \right] \frac{\partial p_n}{\partial u_m} \frac{\partial u_m}{\partial \lambda_k} = 0 , \tag{C14}$$

if Eqs. (C11) and (C12) hold and if conditions (C3) are satisfied. The Hessian

$$\frac{\partial^2 \phi}{\partial \lambda_j \partial \lambda_k} = \sum_n \sum_m f_{j,n}(\mathbf{p}) \frac{\partial p_n}{\partial u_m} \frac{\partial u_m}{\partial \lambda_k}$$

$$= \sum_n \sum_m \frac{\partial u_n}{\partial \lambda_j} \left[\frac{\partial p_n}{\partial u_m} - \sum_\nu \sum_\mu \frac{\partial p_\nu}{\partial u_n} \sum_\ell \lambda_\ell \frac{\partial^2 f_\ell(\mathbf{p})}{\partial p_\nu \partial p_\mu} \frac{\partial p_\mu}{\partial u_m} \right] \frac{\partial u_m}{\partial \lambda_k} . \tag{C15}$$

Since $\partial p_n / \partial u_m$ is negative definite, a sufficient condition for the Hessian to be negative definite is that

$$\sum_\ell \lambda_\ell \frac{\partial^2 f_\ell(\mathbf{p})}{\partial p_\nu \partial p_\mu} \tag{C16}$$

be nonnegative definite. If this is the case, the Lagrange multipliers can be computed by minimizing $-\phi$ subject to conditions (C12) and (C13). Once the λ_j are known, the expectation values p_n are also known since Eqs. (C9) and (C13) are equivalent if $\mathbf{p}$ satisfies Eq. (C3).

For the case discussed in the main body of the paper, $J = 3$ and

$$\begin{aligned} f_1(\mathbf{p}) &= p_1 + p_2 + p_3 + p_4 \\ f_2(\mathbf{p}) &= p_1 - p_3 - 9p_4 \\ f_3(\mathbf{p}) &= -\sum p_n \log p_n \end{aligned} \tag{C17}$$

$$\overline{f}_1 = 1, \quad \overline{f}_2 = \overline{r} = 0.1, \quad \overline{f}_3 = 1.5 \tag{C18}$$

$$W(\mathbf{p}) = 1 \tag{C19}$$

Obviously, for positive λ_j, Eq. (C15) is negative definite and the λ_j can be computed by minimizing the functional $-\phi$ subject to Eqs. (C12) and (C13). The resulting p_n are shown in Eqs. (19).

Appendix D. Evaluation of an Integral

To compute the integral

$$I = \int_0^r \cdots \int_0^r \exp\left[-\sum_n u_n p_n\right] \delta\left[p - \sum_n p_n\right] dp_1 \ldots dp_N \tag{D1}$$

it is convenient to substitute the integral representation of the δ-function (Roos, 1969, p. 83)

$$\delta(x) = \frac{1}{2\pi} \int_{-\infty}^{\infty} \exp(i\omega x)\, d\omega \tag{D2}$$

into Eq. (D1) and change the order of integration. Then

$$\begin{aligned} I &= \frac{1}{2\pi} \int_{-\infty}^{\infty} \exp(i\omega p) \int_0^r \cdots \int_0^r \exp\left[-\sum_n (u_n + i\omega) p_n\right] dp_1 \ldots dp_N\, d\omega \\ &= \frac{1}{2\pi} \int_{-\infty}^{\infty} \prod_n \left[\frac{1 - \exp[-(u_n + i\omega) r]}{u_n + i\omega}\right] \exp(i\omega p)\, d\omega\,. \end{aligned} \tag{D3}$$

The remaining integral over ω can be solved by contour integration. To this aim we cast the product in the integrand into the form

$$\sum_{m=0}^{N} q_m \exp(-imr\omega) \sum_{n} \frac{a_n}{u_n + i\omega} . \qquad \text{(D4)}$$

The q_m are the coefficients of the polynomial

$$q(x) = \prod_{n} [1 - \exp(-u_n r)x] = \sum_{m=0}^{N} q_m x^m \qquad \text{(D5)}$$

and

$$a_n = \prod_{m \neq n} \left[\frac{1}{u_m - u_n}\right] . \qquad \text{(D6)}$$

The product in Eq. (D6) extends from 1 to N with the exception of $m = n$. Since the integral

$$\frac{1}{2\pi} \int_{-\infty}^{\infty} \frac{\exp[i\omega(p - mr)]}{u_n + i\omega} d\omega = \begin{cases} \text{sign}[u_n(p-mr)], & \text{if } (p-mr)u_n > 0 \\ (1/2)\,\text{sign}(u_n), & \text{if } (p-mr) = 0 \\ 0, & \text{if } (p-mr)u_n < 0 \end{cases} \qquad \text{(D7)}$$

substitution of Eq. (D4) into Eq. (D3) leads to

$$I = \sum_{n} a_n \sum_{\{m\}} q_m \exp[-u_n(p - mr)] , \qquad \text{(D8)}$$

where $\{m\}$ denotes all m in $0 \leq m \leq N$ that satisfy

$$u_n(p - mr) > 0 . \qquad \text{(D9)}$$

If $r > p$ and $u_n < 0$, representation (D7) simplifies to

$$I = \sum_{n} a_n \exp(-u_n p) . \qquad \text{(D10)}$$

Acknowledgment

I am indebted to Texaco USA for the permission to present this paper.

References

Jaynes, E. T. (1968) Prior probabilities, IEEE Trans. Syst. Sci. Cybern. **SSC-4**, 227.

Rietsch, E. (1977) The maximum entropy approach to inverse problems, J. Geophys. **42**, 489.

Roos, B. W. (1969) Analytic Functions and Distributions in Physics and Engineering (New York: Wiley).

ALGORITHMS AND APPLICATIONS

John Skilling[1] and S. F. Gull[2]

[1]Department of Applied Mathematics and Theoretical Physics, Silver Street, Cambridge, England

[2]Mullard Radio Astronomy Observatory, Cavendish Laboratory, Madingley Road, Cambridge, England

Maximum entropy, using the Shannon/Jaynes form $-\Sigma$ p logp, is an enormously powerful tool for reconstructing positive, additive images from a wide variety of types of data. The alternative form Σ log f, due to Burg, is shown to be inappropriate for image reconstruction in general, including radio astronomy, and also for the reconstruction of the profiles of power spectra.

An efficient and robust algorithm for maximizing the Shannon/Jaynes entropy subject to observational constraints is developed and described. Its most spectacular applications have been in radio astronomical interferometry, where it deals routinely with images of up to a million or more pixels, and with dynamic ranges well in excess of 10,000. The technique has also been applied to very long baseline interferometry, to x-ray and gamma-ray astronomy, to optical deconvolutions in astronomy and elsewhere, to tomographic reconstruction of plasma beams, to medical x-ray and radionuclide tomography, to optical reconstructions from sparse and blurred data, etc. Examples of typical applications are presented, concluding with a state-of-the-art million-pixel radio map of the supernova remnant Cassiopeia A.

C. Ray Smith and W. T. Grandy, Jr. (eds.),
Maximum-Entropy and Bayesian Methods in Inverse Problems, 83–132.

1. Introduction

This paper reviews the work of the 'Cambridge group' in maximum entropy. In chronological order of involvement in the work presented, this group consists of S. F. Gull (Cavendish Laboratory), G. J. Daniell (now at the Department of Physics, Southampton University), J. Skilling (Department of Applied Mathematics), R. K. Bryan (now at European Molecular Biology Laboratory, Heidelberg), M. C. Kemp (Department of Applied Mathematics), and M. T. Brown (Cavendish Laboratory). The review falls into three parts: why we use $-\Sigma$ p log p, how we program it, and how we apply it.

The first part describes theoretical work by Gull, Skilling, and Daniell. It starts (Section 2) by giving what they believe to be the correct basis of the maximum entropy method, following Jaynes (1968). For many types of positive image, this leads (Section 3) unequivocally to the form $-\Sigma$ p log p whenever one is interested in the configurational shape of the image. Because it is the analogy with time series that originally led to the alternative Σ log p form of entropy (Burg, 1967, 1972), Section 4 gives a short derivation of the entropy of a time series in terms of its spectral density. Whilst this result confirms and extends the Burg result, the analysis in Section 5 shows that it should not blindly be applied to the estimation of power spectrum profiles. Indeed, it is argued that $-\Sigma$ p log p is even here the more appropriate form for reconstructing power spectrum profiles from incomplete autocorrelation or other data. The Burg form Σ log p is a measure of the entropy of the probability distribution function (p.d.f) of the time series, whereas spectrum estimation is usually an image reconstruction problem in which the entropy ought to be that of the spectrum profile itself.

Turning in Section 6 to radio astronomy, the Burg form here measures the entropy of the random electric field vector on the ground (which defines the flexibility allowed to the electric field) instead of that of the pattern of radio emission from the sky. Section 7 then examines the reconciliation model of Kikuchi and Soffer (1977). Their entropy, which reduces to either of the above forms in appropriate limits, measures the entropy of a set of photons in a box. Again, this has nothing to do with the configurational entropy of an image of the sky although it would be appropriate if one were intending to predict occupation numbers of photons in a box.

The conclusion is firm. Whether it be for spectral analysis of time series, radio astronomy, or optical or x-ray astronomy, or for reconstruction of any other positive, additive image, there remains only one contender for the configurational entropy. It is $-\Sigma$ p log p. Shannon was right.

The second part reviews some of the algorithms that have been programmed, and describes the powerful numerical technique that we have adopted for general use. This is primarily the work of Skilling and Bryan.

Since entropy is intrinsically nonlinear, its numerical maximization is a constrained nonlinear optimization problem. The number of degrees of freedom in the problem is the number of cells (N) in the image, which can be up to a million or more. A successful algorithm must be able to cope with this and ought to be able to deal successfully with the wide variety of types of data arising from different experiments. It follows immediately from the size of the problem that direct matrix (N^2) operations are prohibited. It follows from the variety of types of data that only a simple and clear algorithm is likely to be adequately general. It is also desirable on efficiency grounds to reduce as far as possible the number of numerical transformations between image space and data space, since these tend to dominate the computing time. An algorithm with these properties is presented here.

In Section 8, a suitable form of the maximum entropy criterion is set up, and a survey of algorithms follows. Sections 9 and 10 discuss potentially promising algorithms that nevertheless prove to be insufficiently powerful. Sections 11, 12, and 13 develop a sequence of algorithms that, though they fail, nevertheless contain the basic ideas for a successful technique, and give the rationale for its structure. This successful technique is presented in Section 14 (which sets up the image-space structures needed) and in Section 15 (which gives the procedure for control of these structures).

Finally, Section 16 quotes the operational conclusion that the successful algorithm usually needs about 100 transformations between image space and data space to perform a correct maximum entropy reconstruction from typical experimental data. It is difficult to foresee any very substantial improvement on this figure.

The third part reviews a selection of our applications of maximum entropy and displays examples. Section 17 deals with optical imagery, in which the problem is usually one of straightforward deconvolution of blurred photographs or similar data sets (with suitable treatment of the photograph edges). Maximum entropy has been compared with various linear deconvolution algorithms by Burch (1980), and his simulations are reproduced. These clearly show how maximum entropy automatically gives the correct compromise between noise suppression and the superresolution that is attainable with good signal-to-noise. In Section 18, practical examples of deconvolutions from terrestrial and astronomical photographs are displayed. A further application is to sparsely sampled data, which cannot be naturally handled by linear deconvolution algorithms. Section 19 deals with related problems in x-ray and gamma-ray astronomy. Formally, these are similar to optical deconvolutions, usually with the extra difficulty of small numbers of detected photons.

Tomographic reconstructions are dealt with in Section 20. A common type of tomographic data is x-ray scans for medical diagnosis, and a high-resolution maximum entropy reconstruction of a section through a skull is shown. With poorer data, the quality of the reconstruction is impaired, and investigations are reported of the minimum number of scanning directions that may realistically be needed. Medical scanning by radionuclide emission is another tomographic reconstruction problem, investigated by Kemp. Yet another occurs in plasma diagnostics, where one wishes to determine the precise shape of an optically thin plasma beam.

Section 21 deals with radio astronomy, starting with the aperture synthesis work of Gull and Daniell (1978), which demonstrated that maximum entropy could be used routinely at respectably high resolutions. Finally,

Gull and Brown have applied maximum entropy to recalibrate aperture synthesis data in a "bootstrap" approach to image reconstruction, using the positivity constraint induced by the entropy to refine their knowledge of poorly known instrumental parameters. Similar ideas have been used in very-long-baseline interferometry.

Part One. Entropy and Images

2. Basis of Maximum Entropy Method

The entropy S of a probability distribution is (minus) a measure of its information content, defined by Shannon (1948) as

$$S = -\sum p_i \log p_i \tag{1}$$

for a discrete set of probabilities $\{p_i\}$, $\Sigma\, p_i = 1$. This is, apart from the arbitrary base of the logarithm, a unique measure satisfying the axioms one assigns to additive information (see, for example, Ash, 1965).

An alternative derivation of S starts with a large number N of equivalent random experiments, each of which can have any of m possible outcomes $i = 1,2,\ldots,m$. Let n_i be the number of occurrences of outcome "i" in the sequence ($\Sigma\, n_i = N$). In experiments of this type, the probability p_i of outcome "i" is taken to be the expectation of n_i/N, and one needs to consider the relative likelihood of the set of numbers $\{n_i\}$. If the experiments are truly random (as performed by the traditional team of monkeys), then the number of occurrences of $\{n_i\}$ will become proportional to the degeneracy $N!/\Pi\, n_i!$. In the limit $N \to \infty$, the logarithm of the degeneracy becomes NS, where S is the entropy defined above.

Jaynes (1968) was the first to use S as a tool for assigning probability distributions in the light of certain types of constraint, such as an ensemble average $\langle r \rangle = \Sigma\, r_i p_i$. He suggests that it is wise to choose the probability distribution in such a way as to keep as open a mind as possible concerning other quantities. To this end, he chooses that particular probability distribution consistent with the constraint(s) having maximum entropy S. Specifically, for the above example, maximizing $S = -\Sigma\, p_i \log p_i$ under $\langle r \rangle = \Sigma\, r_i p_i$ yields $p_i = \exp(-\lambda r_i)/Z(\lambda)$, where λ is the appropriate Lagrange multiplier (chosen to fit the constraint) and Z is its partition function.

3. Entropy of a Positive Additive Image

We now apply these ideas of information content to image formation. Many different types of image satisfy axioms of positivity and additivity. Such images can be directly visual, such as two-dimensional light distributions, or can be three-dimensional densities of (say) electrons in free space, or one-dimensional sequences of numbers in a computer memory or on magnetic tape, or even quantities such as absorptivity. A common feature of these types of image is that they can be displayed by electrons in a cathode ray tube, by grains of silver in a photographic plate, by holes in computer cards, or abstractly as sequences of numbers. The configurational structure (that is, apart from normalization) of each of these images satisfies what are, in fact, the Kolmogorov axioms of probability theory. If the image is represented as a sequence of positive numbers f_i ($i = 1,2,\ldots,m$) with corresponding proportions $p_i = f_i/\Sigma f_i$, then the relevant relationships are

(a) $p_i \geq 0$ (positivity)

(b) $\Sigma p_i = 1$

(c) $p_{i \cup j \cup \ldots} = p_i + p_j + \ldots$ (additivity) $(i \neq j \neq \ldots)$.

Accordingly, we can apply the whole of probability calculus to them, in particular defining their information content or entropy $S = -\Sigma p_i \log p_i$.

Reconstruction of an image, according to Jaynes, then consists of selecting that sequence $\{f_i\}$ that has maximum entropy consistent with whatever observational constraints apply.

We must stress that it is the image constructor himself who stipulates that the quantity he chooses to display has the properties of positivity and additivity that underlie the identification with probability. He may be guided by the physics of the observed system insofar as it is natural to consider free electrons, photons, etc., as having these properties. For other types of image, such as electrons in partly filled conduction bands, maps of Stokes' polarization parameters I,Q,U,V, or electromagnetic photon spectra, the identification with probability may be less straightforward, nonexistent, or nonunique—respectively.

What the identification does <u>not</u> depend on is the form of the constraints. This has been a source of great confusion. It does not depend on whether the object is viewed with fermions or bosons. It does not depend on whether the object is viewed directly or via its Fourier or Abel transform. It depends solely on the nature of the image being presented. We return to this point later.

4. Entropy of a Time Series

A different form of entropy has been derived for a time series (Burg, 1967, 1972; Edward and Fitelson, 1973; Ulrych and Bishop, 1975; Ulrych and Clayton, 1976), and has been applied—we believe inappropriately—to image formation in radio astronomy and elsewhere.

Some confusion has arisen concerning the derivation and meaning of the entropy, so for completeness we now give a derivation somewhat similar to that provided by Ulrych and Bishop (1975). We consider the case of a time series of some real random variable $x(t)$ with time-independent statistics, sampled at uniform (integer) values of t, and we wish to determine the entropy associated with the joint p.d.f. $p(\mathbf{x})$. Now, the entropy of a discrete probability distribution p_i is undoubtedly $S = -\Sigma p_i \log p_i$ as above.

Shannon and Weaver (1949) and many other authors subsequently generalize this to

$$S = -\int p(\mathbf{x}) \log p(\mathbf{x})\, d\mathbf{x} \tag{2}$$

for the continuous case.

Even this innocuous step has pitfalls. As pointed out by Jaynes (1968), there is a hidden assumption that the feasible values of $\mathbf{x}$ are uniformly distributed within some finite domain, and indeed one can see immediately that Eq. (2) is dimensionally incorrect. Jaynes gives the correct form

$$S = -\int p(\mathbf{x}) \log \frac{p(\mathbf{x})}{m(\mathbf{x})}\, d\mathbf{x}\ , \tag{3}$$

where $m(\mathbf{x})$ is the measure of the feasible values of $\mathbf{x}$. Fortunately, most time series (such as samples of position, velocity, or voltage) involve quantities for which the measure $m(\mathbf{x})$ should be taken to be uniform (Jeffreys, 1939; Jaynes, 1968), so use of the form (2) introduces only an unimportant additive constant to the entropy.

One now assumes that detailed knowledge of the previous samples is lost, and only some of the autocorrelation coefficients are retained. We may avoid much unpleasant algebra by restricting ourselves to the heuristically sufficient case of a periodic time series (with very large period N) in which the autocorrelation coefficients

$$\langle A_j \rangle = \langle N^{-1} \sum_{m=0}^{N-1} x_m x_{m+j} \rangle \tag{4}$$

diminish to zero as j approaches N/2. Angle brackets refer to ensemble averages, and we have introduced the summation into the definition of $\langle A_j \rangle$ to circumvent difficulties associated with stationarity, which are not relevant to the present problem.

To determine the p.d.f. $p(\mathbf{x} | \langle A_j \rangle)$ we use the technique of maximum entropy referred to in Section 2. Thus we maximize the integral (3) with $m(\mathbf{x})$ constant (equal to V^{-1}) over whatever large hypervolume V is available to $\mathbf{x}$, subject to

$$N^{-1} \sum_{m=0}^{N-1} \int p(\mathbf{x})\, x_m x_{m+j}\, d\mathbf{x} = \langle A_j \rangle \tag{5}$$

for j within some observation set Ω, and to

$$\int p(\mathbf{x}) d\mathbf{x} = 1\ . \tag{6}$$

Introducing Lagrange multipliers λ_j and μ, we obtain the solution

$$p(\mathbf{x}) = \exp\left[-\mu - \sum_{j\in\Omega}\sum_{m=0}^{N-1} \lambda_j \frac{x_m x_{m+j}}{N}\right] .$$

To diagonalize the quadratic form in x, it is convenient to Fourier transform via

$$\tilde{\xi}_k \equiv (F\xi)_k = N^{-1/2} \sum_{j=0}^{N-1} \xi_j \exp(-2\pi ijk/N)$$

to obtain

$$p(\mathbf{x}) = \exp\left[-\mu - N^{-1/2} \sum_{k=0}^{N-1} \tilde{\lambda}_k |\tilde{x}_k|^2\right]$$

where

$$\tilde{\lambda}_k = N^{-1/2} \sum_{j\in\Omega} \lambda_j \exp(-2\pi ijk/N) .$$

The normalization integral (6) gives

$$e^{-\mu}(N^{1/2}\pi)^{N/2} \prod_k \tilde{\lambda}_k^{-1/2} = 1$$

while the constraint integrals (5) give

$$\sum_k \int \exp(-2\pi ijk/N) \, |\tilde{x}_k|^2 \, p(\tilde{\mathbf{x}}) \, d\tilde{\mathbf{x}} = N\langle A_j\rangle ,$$

and hence the Lagrange multipliers λ_j, $j \in \Omega$, are defined for measured $\langle A_j\rangle$ by

$$(F^{-1}\tilde{\lambda}^{-1})_j = \frac{1}{2} \langle A_j\rangle . \tag{7}$$

The unmeasured $\langle A_j\rangle$ are obtained from Eq. (7) with $\lambda_j = 0$, $j \notin \Omega$. This is precisely the prescription for determining the Lagrange multipliers and unmeasured autocorrelation coefficients given by Burg.

Having determined the complete p.d.f. $p(\mathbf{x})$ of the time series, we may now evaluate its entropy (3) as

$$S = \frac{1}{2} \sum_{k=0}^{N-1} \log\langle W_k \rangle + \frac{N}{2} \log(2\pi e) + \log m \,, \tag{8}$$

where

$$\langle W_k \rangle = N^{-1} \sum_{j=0}^{N-1} \langle A_j \rangle \exp(2\pi ijk/N)$$

is the power spectrum.

The corresponding p.d.f., being maximally noncommittal about unmeasured parameters of the time series, is clearly to be preferred over all others. Accordingly, it is appropriate for the prediction problem, in which values $x_1, x_2, \ldots, x_{r-1}$, together with autocorrelation data, are to be used to predict the next value x_r. Likewise, it is appropriate for filling in gaps in a time series.

The entropy expression in Eq. (8) is the same, to within an additive constant, as that given by Shannon and Weaver (1949) for the increase of entropy $\delta S = +\Sigma \log W_k$ of a signal passed through an amplifier of known power response W_k. Because of this, it has conventionally been stated that the entropy of the time series must also be $S = +\Sigma \log W_k$, even when not all the W_k are measured. This form of entropy is subject to Jaynes' criticism of being dimensionally incorrect. The more direct derivation given here determines the additive constants needed to overcome this objection.

A further advantage of our derivation is that it makes no assumptions concerning the Gaussian (or other) nature of the statistics of the series. In particular, it is clear how to modify the above derivation to incorporate more general constraints such as higher order correlation coefficients. Whilst the incorporation of (say) fourth-order correlation information would in general be algebraically difficult, it is clear that there would be modifying constants in the entropy. Thus, there is nothing fundamental about the expression (8) for the entropy: it is merely an accident of one's having been exclusively concerned with measuring autocorrelation coefficients.

As our example, we consider a simple system of one degree of freedom x. The standard approach would be to measure the zeroth (only) autocorrelation coefficient, namely the variance $\langle x^2 \rangle = \sigma^2$. The maximum entropy algorithm yields the p.d.f.

$$p(x) = (2\pi)^{-1/2} \sigma^{-1} \exp\left[-\frac{x^2}{2\sigma^2}\right]$$

as expected. Its entropy is $S = 1.4189 + \log \sigma + \log m$. On the other hand, were $\langle x^4 \rangle = a^4$ to be measured instead, the p.d.f. would be

$$p(x) = \frac{2^{-3/2}}{(1/4)!} a^{-1} \exp\left[-\frac{x^4}{4a^4}\right],$$

giving $S = 1.3872 + \log \sigma + \log m$, where $\sigma = 0.8222a$ is the standard deviation of x. This is less than the variance-constrained entropy, because all distributions of variance σ^2 other than the Gaussian necessarily have lower entropy. We can safely conclude that there is nothing at all fundamental about Eq. (8). Such general validity as it does have is merely a reflection of the fact that exp(S) is a measure of the (hyper-)volume of $\mathbf{x}$ significantly occupied by $p(\mathbf{x})$.

5. Application to Spectral Analysis of a Time Series

Burg and authors following him use these results to construct spectra from sparse autocorrelation data. Their arguments do not run completely parallel with our derivation. They say that $\Sigma \log\langle W_k \rangle$ is the entropy of the time series and then proceed to maximize this form subject to the available constraints. The consequent maximization leads to results identical to those of Section 4. We have shown above, however, that $\Sigma \log\langle W_k \rangle$ is an absolute measure of the entropy of a p.d.f. of a time series only if the available constraints refer exclusively to the autocorrelation function. An equivalent statement is that the statistics of the time series are Gaussian [Eq. (5)]. Given this Gaussian assumption, the arguments of Burg and his followers are correct, because the entropy we must maximize is then adequately represented by $\Sigma \log\langle W_k \rangle$.

The claim is also made that this formula gives maximally noncommittal results about the spectrum itself. In one sense, this is right. Clearly this spectrum is maximally noncommittal about the samples of the time series, in that it allows their values to have maximum flexibility. It is easy to show that individual realizations of W have, in turn, an exponential p.d.f. $p(W_k) = \exp(-W_k/\langle W_k \rangle)$ with coefficients $\langle W_k \rangle$ given by the Burg formula, and to this extent the coefficients are optimal predictors of W_k. To give maximum freedom of choice concerning the samples x_j or indeed the spectral components W_k, one clearly needs the noisiest, most intense spectrum, as Eq. (7) confirms. If one is interested in the actual numerical values of individual x_j or individual W_k, the Burg algorithm is entirely appropriate. It allows each W_k (or x_j) to have as large a range as possible and if, perhaps because of the particular j for which $\langle A_j \rangle$ was measured, some particular W_k is very poorly determined, the Burg algorithm will return a correspondingly large value.

On the other hand, for many problems one is more interested in the shape of the spectrum as a whole—for example, if one wishes to display a line spectrum. In this case, one requires the configurational information (for example, the number of spectral lines) that is present. To guide us in this, we note that power spectra are positive and additive with uniform intrinsic measure. These are precisely the probability axioms that we have seen are widely applicable to images. Indeed, we suggest that a power spectrum is an image and should be displayed accordingly. The most noncommittal, or least informative, image is that which maximizes

$$S = -\sum p_k \log p_k , \qquad p_k = \frac{\langle W_k \rangle}{\sum \langle W_\ell \rangle} .$$

To conclude, if one is interested in predicting values in a time series with given autocorrelation data, it is most noncommittal to use the p.d.f. that has the Burg spectrum. On the other hand, if one is interesed in the shape of the spectrum as a whole, the Burg spectrum is not appropriate, and one should apply the idea of entropy directly to the spectrum itself.

6. Application to Radio Astronomy

Burg's results have been widely applied to the problem of image formation in radio astronomy (Ables, 1974; Ponsonby, 1973; Newman, 1977; Wernecke, 1977; Wernecke and d'Addario, 1977; and elsewhere). The electric field on the ground represents a two-dimensional space series whose power spectrum is the variable of interest, namely the angular distribution of radio flux across the sky, and the above authors have argued that one should draw an analogy between a time series and the electric field pattern. This argument is brought forward because it is often the autocorrelation of this electric field pattern that is observed (by radio interferometers), so the Burg algorithm for autocorrelation data appears appropriate. Accordingly, they maximize the form $S = \Sigma \log f_j$ for the entropy of the radio sky, where f_j represents the intensity from cell j of the sky.

The following are the major difficulties with this argument.

1. This form is the entropy of the two-dimensional space series. It would be appropriate to use this were one to wish to predict further samples of the electric field vector on the ground, or indeed individual values of f_j. As was the case for spectral analysis of time series, it is not appropriate for reconstructing the image representing the shape of the radio sky.

2. The entropy of the radio sky cannot depend on whether one observes with an interferometer or with a single dish. Equivalently, the form of the entropy cannot depend on the form of the available constraints.

Furthermore, it is often argued that the physics of the observing process should influence the image processor who produces maps of the radio sky. We believe that, on the contrary, accidents of observational technique should influence one's ideas of the shapes and structures of radio sources as little as possible. The techniques should be, and are, considered when formulating the observational constraints. There is no additional place for them in the entropy.

7. The "Reconciliation Model" of Kikuchi and Soffer

An entirely different derivation, actually of both formulas for the entropy, has been given by Kikuchi and Soffer (1977) and cited by Frieden and Wells (1978).

Kikuchi and Soffer consider the generation of an image from a map of photon fluxes. They ask the question "how probable is it that a given set of photon occupation numbers $\{n_i\}$ is observed in object space?" They very properly derive the formulas

$$S \equiv \log Q = \text{constant} - N \sum p_i \log p_i \qquad (9a)$$

and

$$S \equiv \log Q = \text{constant} + (z-1) \sum \log p_i \qquad (9b)$$

for the (relative) intrinsic probability Q of observing a pattern $\{n_i\}$ in the receiving cells, given that $N = \Sigma n_i$ photons are observed in all. In these formulas, z is the number of degrees of freedom of the photons, obtained by quantum statistics, and $p_i = n_i/N$ is the proportion in cell i. Equation (9a) holds in the nondegenerate limit $n_i \ll z$, and Eq. (9b) holds in the degenerate limit $n_i \gg z$. The two cases are separated according to $n_i/z = B_i(\nu)c^2/2h\nu^3$, where ν is the radiation frequency, B is the radiation intensity, and h is Planck's constant. These formulas are correct: Kikuchi and Soffer have determined the thermodynamic entropy of photons in a box.

However, they proceed to claim that $-\Sigma\, p_i \log p_i$ is the appropriate (configurational) entropy for optical astronomy where (by and large) $n_i \ll z$, whereas $\Sigma \log p_i$ is the appropriate form for radio astronomy where (by and large) $n_i \gg z$. This is misleading.

If one is interested in predicting the occupation numbers of photons in a box, one should indeed maximize the entropy in the form given by Kikuchi and Soffer, subject to any available constraints. This will give the most noncommittal estimate of the occupation numbers. For image processing, we maintain that one is simply not interested in this, and we reject the analogy on the same grounds that we rejected application of the Burg algorithm. It is ludicrous to suppose that our prejudices concerning the shape and structure of an astronomical object should depend on whether it is viewed in the optical or the radio bands.

We return to $-\Sigma\, p_i \log p_i$ as the only viable form of entropy for image reconstruction.

Part Two. Programming Maximum Entropy

8. Maximum Entropy Using Experimental Data

Pertaining to the numbers f_j representing the image to be reconstructed are observational data D_k related in some way, often linearly, to the image, and subject to some form of noise. A single statistic $C(f_1,f_2,\ldots,f_n)$ is set up to measure the misfit between the actual (noisy) data D_k and the data F_k that would be observed (in the absence of noise) if the observed quantity were correctly represented by the particular numbers f. Statistical analysis then indicates some upper bound C_{aim} to the values that C can plausibly take. The condition $C \leq C_{aim}$ then defines the set of feasible images f, which pass the given statistical test for consistency with the actual data. Using a single statistic is better than attempting to fit each separate datum, both because it is simpler and because it can tolerate occasional errors of several standard deviations.

A convenient choice for C is the chi-square (χ^2) statistic (Ables, 1974; Gull and Daniell, 1978; and others)

$$C(f) = \chi^2 \equiv \sum_k (F_k - D_k)^2/\sigma_k^2 , \tag{10}$$

which allows for noise through the variance σ_k^2 of datum k. The largest acceptable value C_{aim} is then given as a percentage point of the chi-square distribution, closely related to the number N of independent observations, and being about $(N + 3.29\sqrt{N})$ for 99% confidence. However, other statistical tests are possible (Bryan and Skilling, 1980) and may be preferable in certain circumstances.

The strict maximum entropy criterion requires one to select that particular feasible image that has the greatest entropy. Formally, the criterion is to maximize S subject to $C \leq C_{aim}$. Except in the trivial case where the global maximum of S (flat image) lies within the feasible set, the constrained maximum of S will lie at an extremal of $S-\lambda C$ for a suitable Lagrange multiplier λ (Fig. 1). From this, f can be determined via

$$\log \mathrm{DEF} - \log f_j = \lambda (\Sigma f) \frac{\partial C}{\partial f_j} . \tag{11}$$

where

$$\mathrm{DEF} = \exp (\Sigma p_i \log f_i) . \tag{12}$$

The DEF is a weighted mean of the f_j that can be interpreted as the default value to which f_j will tend if there are no data pertaining to cell j ($\partial C/\partial f_j = 0$).

In many applications, interferometry being typical, the total intensity Σf has a status different from individual pixel values, as indeed it does in

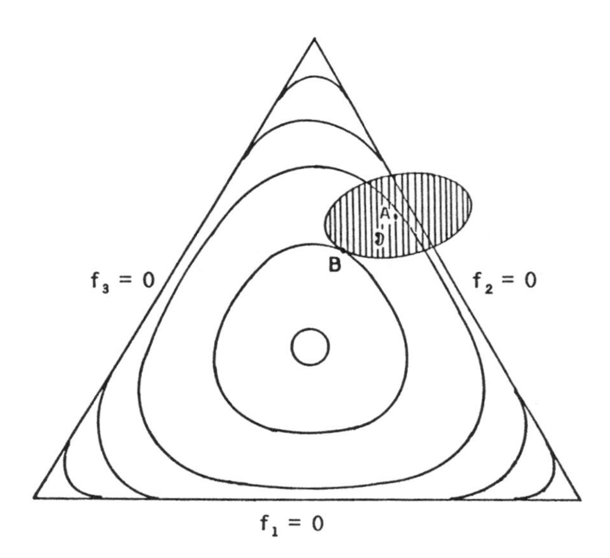

Figure 1. The S criterion and the χ^2 statistic in f space for a three-cell map normalized to Σ f = 1. S surfaces are convex and χ^2 surfaces are ellipsoids. A is the image that fits the data exactly. B is the maximum entropy image.

the entropy itself. The maximum entropy criterion may then be modified to maximize S subject to $C = C_{aim}$ and to some constraint on Σ f. This will be found at an extremal of $S - \mu\Sigma f - \lambda C$, where μ is an extra multiplier that has the operational effect of changing DEF. One can either choose μ, and hence DEF, to fit some given value of Σ f or, more simply, use DEF itself as a user-defined parameter. Formally, this is equivalent to modifying the entropy to

$$S = - \sum f_j \left[\log \frac{f_j}{DEF} - 1 \right] \tag{13}$$

whose derivatives are $\partial S/\partial f_i = \log DEF - \log f_i$ and $\partial^2 S/\partial f_i \partial f_j = -\delta_{ij}/f_i$. It is this latter form (13) of S that is used in this paper, though the algorithms that are presented can be modified fairly easily to cope with the strict form of S, or indeed with other modifications of it.

9. The 'Integral Equation'

Gull and Daniell (1978) attempted to maximize $Q = S - \lambda C$ at fixed λ by solving the extremal equation in the iterative form

$$f_j^{n+1} = \text{DEF} \exp\left[-\lambda \frac{\partial C(f^n)}{\partial f_j}\right] \quad (14)$$

where a superscript n denotes the nth iterate. This procedure had the attractive feature that successive iterates were all automatically positive, owing to the exponential function. Also, the algorithm allowed high values of f to develop in relatively few iterations, again because of the exponential. This was of considerable importance in Gull and Daniell's astronomical applications because the dynamic range of their images of f was often on the order of 1000.

Unfortunately, the exponential also introduced instability into the iteration, and they had to smooth successive iterates by setting

$$f_j^{n+1} = (1-p)f_j^n + p\,\text{DEF} \exp\left[-\lambda \frac{\partial C(f^n)}{\partial f_j}\right]$$

where p could be as small as 1%. Even then, the behavior of the algorithm was erratic and unstable, especially at high values of λ.

There did, of course, remain the difficulty of determining λ. Gull and Daniell started with $\lambda = 0$, whose solution was simply the flat image f_j = DEF, and then gently increased it, using as each initial iterate of f the end result of the iteration for the previous value of λ. As λ increased, as fast as possible to save computer time, more and more weight was given to the data as opposed to the entropy, and C decreased until, it was hoped, its correct value C_{aim} was reached.

It is straightforward to demonstrate that, at the maxima of $Q(\lambda)$, C is a monotonically decreasing function of λ for any linear experiment. However, numerical instability often set in before the correct value of C was reached, and the proportion p often had to be reduced so severely that the algorithm effectively stopped. This instability has also been noted by Willingale (1979). Nevertheless, the work of Gull and Daniell was of crucial importance in demonstrating the possibility of computing high-resolution maximum entropy images for a variety of different experiments.

10. Steepest Ascents

Another method (whose initial formulation can very soon be eliminated) is to maximize $Q = S - \lambda C$ by steepest ascents, using

$$f_j^{n+1} = f_j^n + \mu \frac{\partial Q(f^n)}{\partial f_j} \quad (15)$$

for suitable μ. The catastrophic disadvantage of this method is that in almost every case, whenever μ is sufficiently large to enable high values f_j of the image to develop significantly, there are also cells with negative $\partial Q/\partial f_j$ at which f_j becomes significantly negative. The entropy is not defined when $f_j < 0$, so an artificial prohibition on nonpositive f_j has to be introduced. Even so, ∇Q on the next iterate is dominated by those cells at which f_j became small, and the algorithm spends all its time inefficiently attempting to adjust these small values, instead of developing the high values in which one is usually more interested.

There is a variant of the steepest ascents that keeps f positive by working with log f instead of f. The variant uses

$$\log f_j^{n+1} = \log f_j^n + \mu \frac{\partial Q(f^n)}{\partial \log f_j}$$

or equivalently

$$f_j^{n+1} = f_j^n \exp\left[\mu f_j^n \frac{\partial Q(f^n)}{\partial f_j}\right],$$

but the extra factor f_j in the exponential makes this algorithm less stable than the integral equation.

11. Conjugate Gradients

The standard way of improving the steepest ascent algorithm is to use the conjugate gradient technique (Fletcher and Reeves, 1964) or a variant of it (for example, Polak, 1971). Instead of using ∇Q itself as a direction in which to look for a maximum of Q, one uses only that part ($\mathbf{e}^n$) of ∇Q which is conjugate to one or more previous directions ($\mathbf{e}^{n-1}$), defined by $\mathbf{e}^n\mathbf{e}^{n-1}:\nabla\nabla Q = 0$. In the terminology of this paper, the technique seeks a maximum of Q over the points $\mathbf{f}^n + x\mathbf{e}^n$ along a search direction $\mathbf{e}^n$, where

$$\mathbf{e}^n = -\nabla Q + \beta \mathbf{e}^{n-1}$$

with $\beta = |\nabla Q^{(n)}|^2 |\nabla Q^{(n-1)}|^{-2}$ or $\beta = \nabla Q^{(n)} \cdot [\nabla Q^{(n)} - \nabla Q^{(n-1)}] \, |\nabla Q^{(n-1)}|^{-2}$. The coefficient β in the incremental direction is derived on the assumption that Q is exactly quadratic in f. However, Q in maximum entropy is highly nonquadratic, and it may not be sensible to assume that second-derivative information can be carried forward for several iterates.

Nevertheless, as noted by Wernecke and d'Addario (1977) for a similar problem, conjugate gradients afford a considerable improvement over steepest ascents, although the algorithm remains plagued by negative values of f and still concentrates too much on small values. As specified here, it fails to solve realistically large maximum entropy problems, but a robust algorithm can eventually be developed from it.

12. Search Directions for the Unconstrained Problem (Fixed λ)

The conjugate gradient technique attempts to build up information about the n x n Hessian matrix $\nabla\nabla Q$ by using successive vectors ∇Q. It then uses a specific linear combination of the various ∇Q as a search direction along which Q is maximized by either a fixed coefficient β or an exact line search.

However, in the maximum entropy problem the main computational cost lies in generating the successive vectors ∇Q, each of which requires a transformation into data space and back to calculate ∇C. Scalar products between them are much quicker to compute. Accordingly, one can gain considerable extra flexibility at negligible extra computational cost by constructing, not merely one line along which to search, but rather a full subspace spanned by several vectors ∇Q.

Let the vectors $\mathbf{e}_1, \mathbf{e}_2, \ldots, \mathbf{e}_r$ ($r < 10$) be these base vectors. Then, within the subspace so spanned, one may construct a quadratic model

$$\tilde{Q}(x) = Q_0 + Q_\mu x^\mu + \frac{1}{2} H_{\mu\nu} x^\mu x^\nu \tag{16}$$

for the value of Q at increment $\delta\mathbf{f} = x^\mu \mathbf{e}_\mu$. The components of the model are $Q_\mu = \mathbf{e}_\mu \cdot \nabla Q$ and $H_{\mu\nu} = \mathbf{e}_\mu \mathbf{e}_\nu : \nabla\nabla Q$. Then $\tilde{Q}$ is maximized at

$$x^\mu = -(H_{\mu\nu})^{-1} Q_\nu , \tag{17}$$

the evaluation of which involves the trivial task of solving r simultaneous equations. The resulting value of $\tilde{Q}$, moreover, is greater than could have been obtained directly from conjugate gradients, because of the extra flexibility afforded by using a subspace.

As suggested above, $H_{\mu\nu}$ is obtainable from the current and the previous r-1 evaluates of ∇Q. However, this presupposes that Q remains close to quadratic even when f is incremented r-1 times. That is unlikely: indeed one would then be tempted to argue that the increments must be at least r-1 times too short. It is better to evaluate the search directions at the present position f as

$$\mathbf{e}_1 = \nabla Q , \quad \mathbf{e}_2 = \nabla\nabla Q \cdot \nabla Q, \ldots, \quad \mathbf{e}_r = (\nabla\nabla Q)^{r-1} \cdot (\nabla Q) .$$

Admittedly this involves throwing away information from previous iterates, but that information relates to rapidly changing second derivatives and would be unreliable.

Even so, some limit must be placed on the difference between successive iterates, as the quadratic model will still be inaccurate at large distances. In practice one can maximize $\tilde{Q}(x)$ subject to $|\delta f|^2 \leq \ell_0^2$ for some suitable ℓ_0. Suitably protected against negative f_j, this is a somewhat more promising algorithm.

13. Search Directions for the Constrained Problem

One difficulty with maximizing Q is that of λ, which still has to be iterated to fit $C = C_{aim}$. This double iteration is clumsy. However, using different values of λ involves using different proportions of S and C in Q, which suggests using two models in the subspace, one for S and the other for C, and attempting somehow to solve the actual problem of maximizing S subject to $C = C_{aim}$ without using λ explicitly. The subspace itself would be constructed from

(i) 2 directions, ∇S and ∇C,
(ii) 4 directions, $\nabla\nabla S$ and $\nabla\nabla C$ operating on (i),
(iii) 8 directions, $\nabla\nabla S$ and $\nabla\nabla C$ operating on (ii),

and so on. At depth r in this scheme there are now $2(2^r-1)$ search directions, so the models soon become unwieldy, though the scheme is potentially more powerful than merely maximizing at fixed λ.

It is now convenient to investigate the distance limit more closely. So far, the main disadvantage of the search direction algorithms has been their tendency to allow negative values of f. A distance limit $\Sigma\ (\delta f_i)^2 \geq \ell_0^2$ alleviates this but at the cost of drastically slowing the attainment of high values. However, the distance limit can be modified to overcome this defect. Logarithmic modification $\Sigma\ (\delta f_i/f_i)^2 \leq \ell_0^2$ is too severe on low values, and the intermediate form $\Sigma\ (\delta f_i)^2/f_i < \ell_0^2$ is a good operational compromise. It discriminates in favor of allowing high values to change more than low ones, but not excessively so. The actual value of ℓ_0^2 should be of the order of Σf on dimensional grounds, and values around $(0.1)\Sigma f$ or $(0.2)\Sigma f$ are useful in practice.

Now using a distance in this form is equivalent to putting a metric

$$g_{ij} = \frac{1}{f^i} \quad (i = j) \qquad \text{and} \qquad g_{ij} = 0 \quad (i \neq j) \tag{18}$$

onto image space (note covariant and contravariant indices). But this is just $-\nabla\nabla S$ (Bryan, 1980). This metric is far simpler and more convenient (and hence more powerful?) than the Hessian metric $g_{ij} = \partial^2 Q/\partial f^i \partial f^j$ normally used (Sargent, 1974) in variable-metric nonlinear optimization problems.

14. Entropy Metric

Using $-\nabla\nabla S$ as the metric is the single most important key to the development of a robust algorithm. With a non-Cartesian metric, the gradient directions $\nabla S = \partial S/\partial f^i$ and $\nabla C = \partial C/\partial f^i$ appear initially in covariant form. In order to increment f^i, their indices must be raised by g^{ij}, giving $f^i \partial S/\partial f^i$ and $f^i \partial C/\partial f^i$ as basic contravariant search directions. Furthermore, the second derivative matrices $\nabla\nabla S = \partial^2 S/\partial f^i \partial f^j$ and $\nabla\nabla C = \partial^2 C/\partial f^i \partial f^j$ must likewise be premultiplied by g^{ij} if they are to map contravariant vectors onto contravariant vectors. But $\nabla\nabla S = -g_{ij}$, so $\nabla\nabla S$ becomes minus the identity operator, and nothing new is obtained by operating with it. The family of search directions reduces to (schematically)

$$
\begin{array}{ll}
\text{(i)} & f(\nabla S)\ ,\ f(\nabla C) \\
\text{(ii)} & f(\nabla\nabla C)f(\nabla S)\ ,\ f(\nabla\nabla C)f(\nabla C) \\
\text{(iii)} & f(\nabla\nabla C)f(\nabla\nabla C)f(\nabla S)\ ,\ f(\nabla\nabla C)f(\nabla\nabla C)f(\nabla C)
\end{array}
\tag{19}
$$

and so on. Depth r is attained with just 2r search directions. Incidentally, the initial directions (i) alone subsume the "integral equation" approach, since the latter becomes

$$\delta f_j \equiv f_j^{n+1} - f_j^n \simeq f_j^n \frac{\partial Q}{\partial f_j}$$

near a maximum entropy map, so that it uses a specific multiple (unity) of a specific linear combination $(1:\lambda)$ of $f\nabla S$ and $f\nabla C$.

Furthermore, the factors of f^i in the search directions discriminate in favor of high values, and this helps to keep all the values positive. In fact it is rare for any cell to be sent negative when these search directions are properly controlled. Protection against negative values is still needed, but it does not slow the algorithm and is no longer a source of difficulty.

The family of directions is sufficiently powerful that the first four enable most practical problems to be solved. Indeed even this level of complexity is usually unnecessary, as the third and fourth directions can normally be replaced by a single difference combination, giving just three search directions:

$$
\begin{aligned}
\mathbf{e}_1 &= f\nabla S \\
\mathbf{e}_2 &= f\nabla C \\
\mathbf{e}_3 &= |\nabla S|^{-1} f(\nabla\nabla C)f(\nabla S) - |\nabla C|^{-1} f(\nabla\nabla C)f(\nabla C)\ .
\end{aligned}
\tag{20}
$$

Here the entropy metric is used to define the lengths

$$|\nabla S| = \left[\sum f^i \left[\frac{\partial S}{\partial f^i}\right]^2\right]^{1/2}, \qquad |\nabla C| = \left[\sum f^i \left[\frac{\partial C}{\partial f^i}\right]^2\right]^{1/2}.$$

With these three (or four or more) search directions, quadratic models for S and C

$$
\begin{aligned}
\tilde{S}(x) &= S_0 + S_\mu x^\mu - (1/2) g_{\mu\nu} x^\mu x^\nu\ , \\
\tilde{C}(x) &= C_0 + C_\mu x^\mu + (1/2) M_{\mu\nu} x^\mu x^\nu\ ,
\end{aligned}
\tag{21}
$$

where

$$S_\mu = \mathbf{e}_\mu \cdot \nabla S, \qquad g_{\mu\nu} = \mathbf{e}_\mu \cdot \mathbf{e}_\nu,$$
$$C_\mu = \mathbf{e}_\mu \cdot \nabla C, \qquad M_{\mu\nu} = \mathbf{e}_\mu \mathbf{e}_\nu : \nabla\nabla C \tag{22}$$

are constructed in the subspace

$$\mathbf{f} = \mathbf{f}_0 + x^\mu \mathbf{e}_\mu \equiv \mathbf{f}_0 + \delta\mathbf{f} \tag{23}$$

within which the squared length of the increment $\delta\mathbf{f}$ is

$$\ell^2 = g_{\mu\nu} x^\mu x^\nu . \tag{24}$$

Control of the algorithm now passes into the subspace, in order to determine suitable coefficients x^μ for the search directions. A variety of control procedures can be adopted, but the following technique has been found to be the most robust and powerful.

15. Control Procedures

15.1 Diagonalization in the subspace

This preliminary step simplifies the algebra. First, the base vectors $\mathbf{e}_\mu$ are normalized by scaling the model parameters, and the metric tensor $g_{\mu\nu}$ is diagonalized. The algorithm can now be protected against linear dependence of the search directions. Such dependence shows up as one or more unusually small eigenvalues of $g_{\mu\nu}$. Components of the model along the corresponding eigenvector(s) may reflect rounding errors rather than true structure, and such eigenvectors are discarded, reducing the subspace to that part spanned by eigenvectors having significant eigenvalues.

With the surviving eigenvectors rescaled to make the metric Cartesian, the distinction between covariant and contravariant indices disappears. Further simplification is effected by diagonalizing the revised form of $M_{\mu\nu}$ to give

$$\tilde{S}(x) = S_0 + S_\mu x_\mu - (1/2) x_\mu x_\mu,$$
$$\tilde{C}(x) = C_0 + C_\mu x_\mu + (1/2)\gamma_\mu x_\mu x_\mu, \tag{25}$$
$$\ell^2 = x_\mu x_\mu$$

where the γ_μ are the eigenvalues of $M_{\mu\nu}$, and all symbols are defined with respect to the new base vectors.

Apart from the protection against linear dependence, the above procedure is merely the simultaneous diagonalization of $g_{\mu\nu}$ and $M_{\mu\nu}$. Much hangs on accurate diagonalization in the subspace (especially for badly conditioned problems), and it is essential to diagonalize accurately.

15.2 Basic Control

The aim of the control procedure is to maximize $\tilde{S}$ over $\tilde{C} = C_{aim}$ subject to a distance constraint $\ell^2 \leq \ell_0^2$ ($\simeq 0.1\Sigma f$ or $0.2\Sigma f$). Unfortunately, this may be impossible. For very many applications, C is a convex (elliptical) function of f, for which all eigenvalues γ_μ are positive. There is then a minimum value

$$\tilde{C}_{min} = C_0 - (1/2)\gamma_\mu^{-1} C_\mu C_\mu$$

that $\tilde{C}$ can attain in the subspace. Clearly one should not attempt to aim below $\tilde{C}_{min}$, regardless of the value of C_{aim}. In fact, even attempting to reach values as low as $\tilde{C}_{min}$ is inappropriate since the resulting x is then determined purely by the structure of $\tilde{C}$ and not at all by $\tilde{S}$. It is better to set the more modest aim

$$\tilde{C}_{aim} = \max\left[\frac{2}{3}\tilde{C}_{min} + \frac{1}{3}C_0, C_{aim}\right], \tag{26}$$

which is always accessible.

The various maxima of $\tilde{S}$ over different values of $\tilde{C}$ may be parametrized by the Lagrange multiplier α in $\tilde{Q} = \alpha\tilde{S} - \tilde{C}$ (redefining Q using α instead of λ). Maximizing $\tilde{Q}$ yields

$$x_\mu = \frac{\alpha S_\mu - C_\mu}{\gamma_\mu + \alpha} \tag{27}$$

in which α is chosen to fit $\tilde{C} = \tilde{C}_{aim}$. The required range for α is the positive range (that is, assigning positive weight to the entropy)

$$\alpha_{min} < \alpha < \infty \tag{28}$$

where α_{min} is either 0 (for positive definite $\nabla\nabla C$), at which $\tilde{C}$ takes its minimum value $\tilde{C}_{min}$, or $\max(-\gamma_\mu)$ (for nonpositive $\nabla\nabla C$), at which the increment x_μ diverges. The upper limit $\alpha = \infty$ corresponds to global maximization of $\tilde{S}$ irrespective of $\tilde{C}$. Fortunately, the value of $\tilde{C}(x)$ increases monotonically in α, so a simple chop suffices to iterate toward $\tilde{C} = \tilde{C}_{aim}$.

The resulting x may, however, be too large to satisfy the distance constraint. To protect against this, the chop in α is redirected toward $\tilde{C} = C_0$ whenever α gives an increment with too large ℓ^2. The rationale for this form of protection is that the iterates f are all expected to lie reasonably close to the maxima of S over <u>some</u> value of C, so maximizing $\tilde{S}$ over the <u>existing</u> value C_0 is likely to give a closer iterate than attempting to reach a <u>different</u> value $\tilde{C}_{aim}$.

The α-chop normally behaves as in Fig. 2a or b. In any case it must always give a result in the range $\tilde{C}_{aim} \leq \tilde{C} \leq C_0$. However, the distance ℓ is

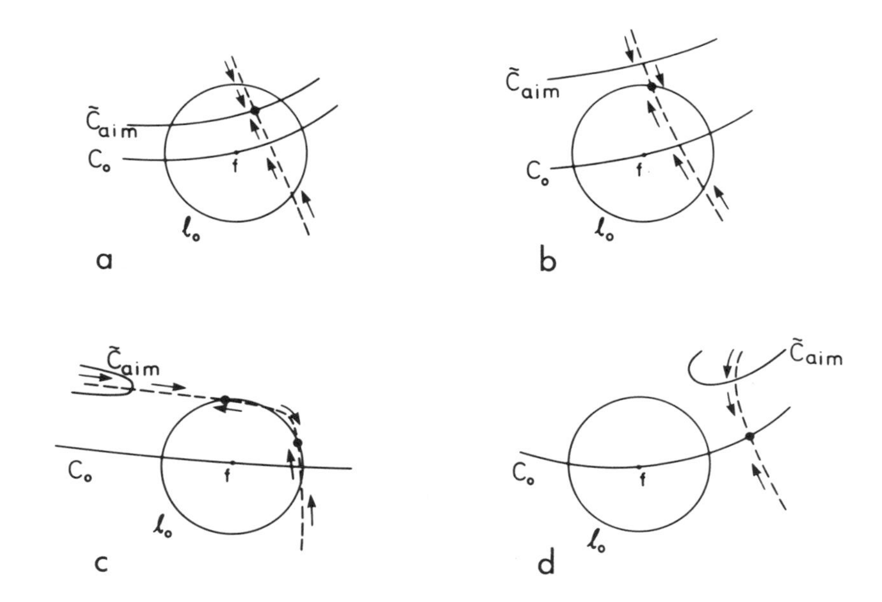

Figure 2. Operation of α-chop in the subspace. The maximum entropy trajectory, parametrized by α, is shown dashed. The circle centered on the current image f marks the maximum allowed distance ℓ_0. Results x are shown as filled circles (•). (a) Unique iterate $\tilde{C} = \tilde{C}_{aim}$, $\ell < \ell_0$. (b) Unique iterate $\tilde{C} > \tilde{C}_{aim}$, $\ell = \ell_0$. (c) Ambiguous iterate $\tilde{C} > \tilde{C}_{aim}$, $\ell = \ell_0$. (d) Too distant iterate $\tilde{C} = C_0$, $\ell > \ell_0$.

not monotonic in α, and this can lead to an ambiguity in the result of the chop (Fig. 2c), the progress of the chop depending on the particular values of α actually used. However, the ambiguity is harmless in that either answer for x gives a useful iterate. More seriously, the algorithm may be unable to find any sufficiently close value of x (Fig. 2d), especially if the current image f is far from a maximum entropy image. This case needs further protection.

15.3 Distance Penalty

If the α-chop cannot find a sufficiently close value of x, the distance constraint must be introduced explicitly into the maximization via a second Lagrange multiplier P, giving

$$\tilde{Q} = \alpha\tilde{S} - \tilde{C} - P\ell^2, \qquad P = \text{distance penalty} \geq 0.$$

This is maximized at

$$x_\mu = \frac{\alpha S_\mu - C_\mu}{P + \gamma_\mu + \alpha} . \tag{29}$$

Thus P can be interpreted as an increase of each eigenvalue γ_μ of C, giving a revised form

$$\tilde{C}_p(x) = C_0 + C_\mu x_\mu + (1/2)(P+\gamma_\mu)x_\mu x_\mu \tag{30}$$

that takes larger values than $\tilde{C}$ itself. $\tilde{C}_p$ is also more convex than $\tilde{C}$, and in the limit of large P its contours become almost spherical, so that the maximization of $\tilde{S}$ becomes better conditioned. An alternative interpretation of P, in which P/α increases the curvature of $\tilde{S}$, is not better conditioned, and the corresponding maximization algorithm is noticeably less efficient.

With a distance penalty invoked, α is chopped toward $\tilde{C} = \tilde{C}_{aim}$ as required, but the chop is redirected toward $\tilde{C}_p = C_0$ whenever the distance is too large. Redirecting on $\tilde{C}_p$ rather than on $\tilde{C}$ helps the algorithm because $\tilde{C}$ itself is always less than $\tilde{C}_p$ and hence will make useful progress toward $\tilde{C}_{aim}$ however the chop proceeds. For sufficiently large penalty P, the α-chop must be able to reach a result satisfying

$$\tilde{C}_{aim} \leq \tilde{C} < \tilde{C}_p \leq C_0 ,$$

and the smallest such P is used to give the final result x. A flowchart of this algorithm is shown in Fig. 3.

All that remains is to increment f by the multiples of x of the search directions, whilst protecting against stray nonpositive values. The algorithm is complete.

16. Summary of Recommended Algorithm

There are thus three main ingredients in the maximum entropy algorithm recommended here. They are—

1. The use of a subspace of search directions,
2. The entropy metric,
3. Controlling the algorithm directly on C, and not on λ.

The resulting program has proved highly robust and powerful. It and earlier versions have been used successfully in a wide variety of applications. Its operation is always checked by displaying the value of

$$\mathrm{TEST} = \frac{1}{2}\left[\frac{\nabla S}{|\nabla S|} - \frac{\nabla C}{|\nabla C|}\right]^2 . \tag{31}$$

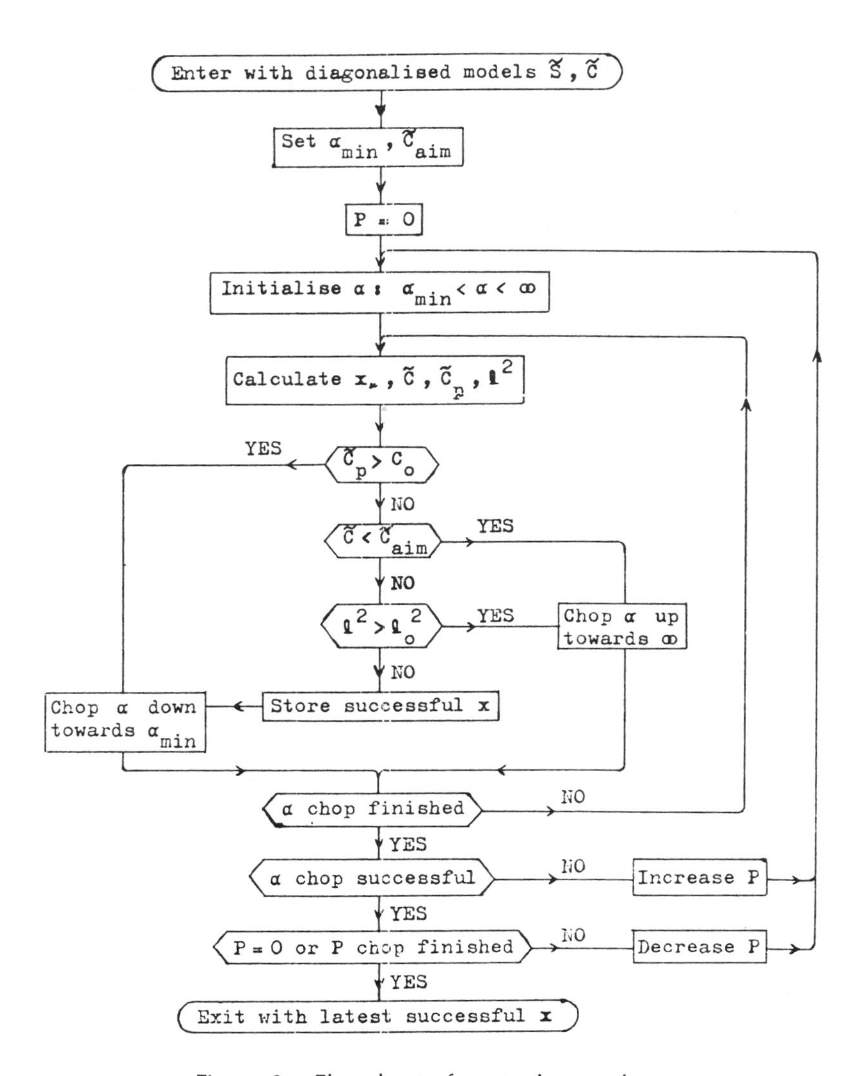

Figure 3. Flowchart of control procedure.

This measures the degree of nonparallelism between ∇S and ∇C, which is zero for a true maximum entropy image. Usually there is no difficulty in reaching TEXT < 0.01 or so at the correct value of C, which demonstrates that the correct, unique maximum entropy reconstruction has been attained.

Fortran implementations of the algorithm routinely perform maximum entropy calculations on arrays up to 1024×1024 in size, and there seems no bar in principle to still larger sizes. Almost regardless of size, the algorithm takes something like 20 iterations to reconstruct an image from an experiment with signal:noise of about 100:1. Each iteration involves six transformations between image space and data space, so the maximum entropy reconstruction is about 100 times slower than linear reconstructions. It does of course have the great advantages of noise suppression and positivity.

Part Three. Applications

17. Optical Imagery: Theory and Simulations

17.1 Deconvolution analysis

In these applications, one attempts to reconstruct an original object from blurred and noisy data, typically from a digitized photograph. In keeping with the previous notation, let

f_i = intensity of image in cell i, to be determined
D_i = intensity of data (photograph) in cell i
σ_i = standard deviation of datum i due to noise

where i is a two-dimensional index ranging over the N cells of the image. Let the blurring function of the camera (or other instrument) be

$$b_j = \text{intensity of p.s.f. in cell } j,$$

centered on the zero-offset position $j = 0$ and normalized to $\Sigma\, b = 1$. It is assumed that the p.s.f. is spatially invariant, as this greatly facilitates the computation.

Then the expected data that ought to be produced by the image f are

$$F_i = \sum b_j f_{i-j} = (f*b)_i , \tag{32}$$

where * denotes convolution.

Normally one uses $\chi^2 = \Sigma(F_i - D_i)^2/\sigma_i^2$ as the statistic, and seeks to maximize S subject to χ^2 taking some value close to N.

Finding χ^2 and its derivatives clearly involves convolution operations, whose computation is best performed by the fast Fourier transform (FFT) method.

17.2 Simulations

Burch (1980) has compared maximum entropy with four linear restoration methods, namely the inverse, Wiener, approximate Wiener, and constrained least-squares filters. He set up a simulated object consisting of five isolated points on a 64 × 64 grid, and blurred them with a Gaussian p.s.f. 12 pixels wide (full width, half power). He then added various amounts of noise to generate four different data sets, as shown in Fig. 4. His linear restorations are shown in Figs. 5, 6, 7, and 8, where restorations A through D correspond to degraded images A through D of Fig. 4. All of the restorations give oscillatory solutions with extensive regions of negative intensity, even though it is known a priori that the undegraded image is everywhere positive. Burch's maximum entropy reconstructions are shown in Fig. 9. Because of the positivity constraint, these reconstructions are not oscillatory and therefore have a very low noise level, and there is no spurious detail. In short, they speak for themselves, though in his paper Burch proceeds to analyze the reconstructions quantitatively.

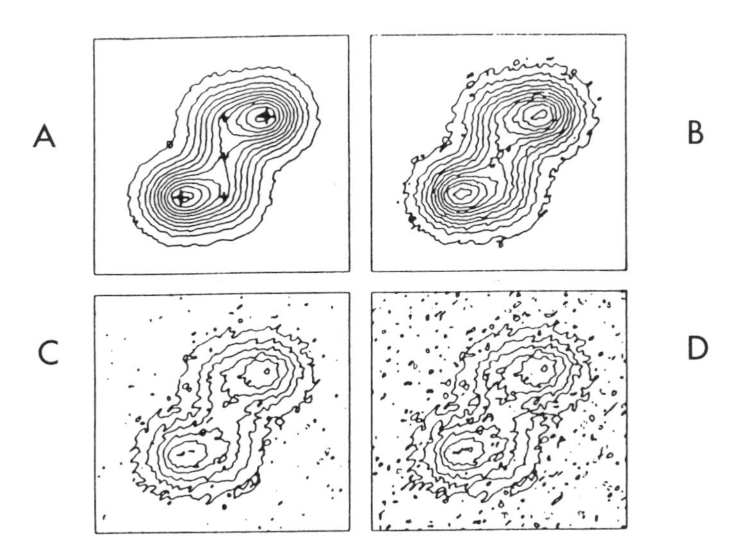

Figure 4. Contour plots of four degraded images used to test the methods for image restoration. The undegraded image is represented by the crosses on image A. This image was then degraded using a Gaussian blurring function, and differing amounts of noise were added to obtain the four images. Solid contours in A and B are at 5, 15, 25, ..., 95% of the peak intensity on the blurred image. Solid contours in C and D are at 10, 30, 50, 70, 90%. Regions of negative intensity are shown by the dotted contours. (Burch, 1980.)

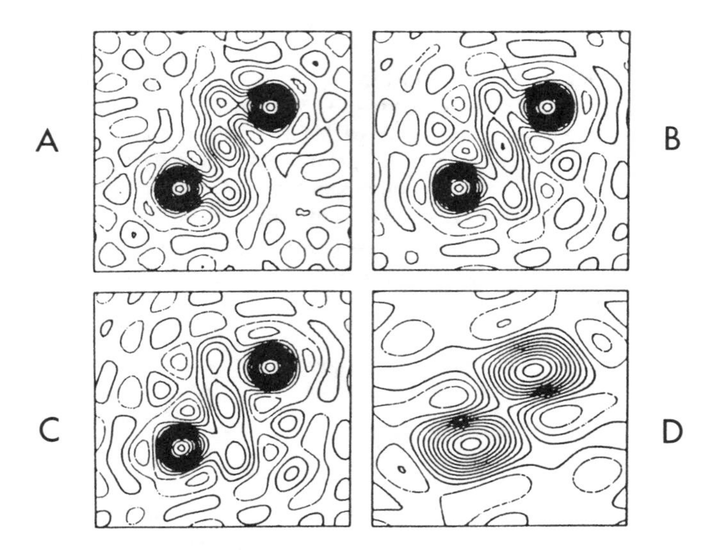

Figure 5. Contour plots of restorations obtained with inverse filters. Solid contours are at 5, 15, 25, ..., 95% of the peak intensity on each image. Regions of negative intensity are shown by the dotted contours. (Burch, 1980.)

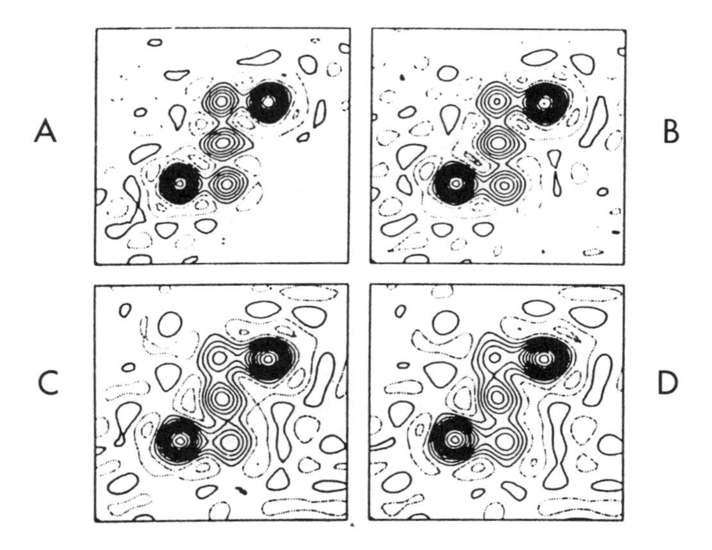

Figure 6. Contour plots of the restorations obtained with Wiener filters. Contours as in Fig. 5. (Burch, 1980.)

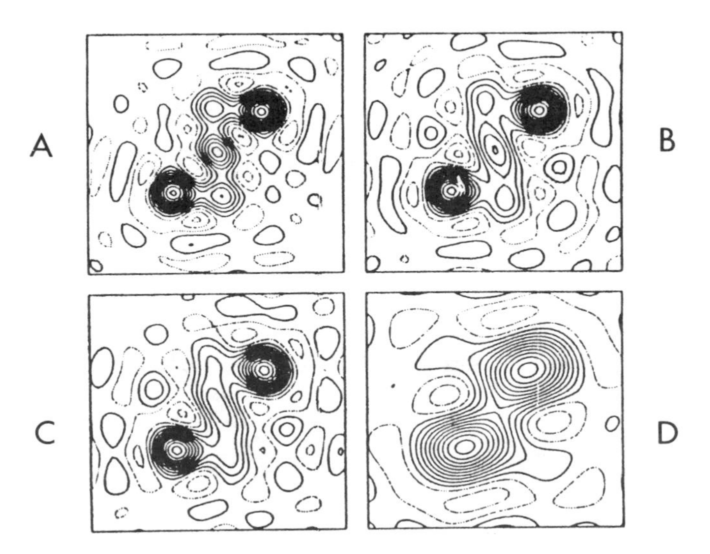

Figure 7. Contour plots of the restorations obtained with approximate Wiener filters. Contours as in Fig. 5. (Burch, 1980.)

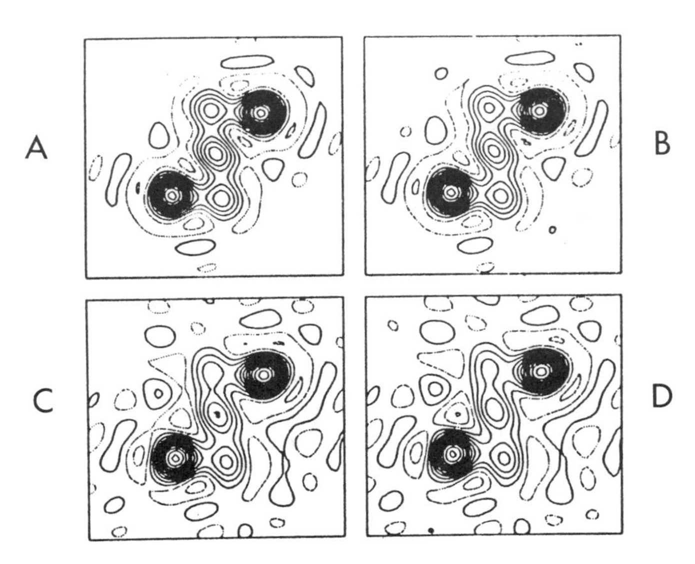

Figure 8. Contour plots of the restorations obtained by the method of constrained least squares. Contours as in Fig. 5. (Burch, 1980.)

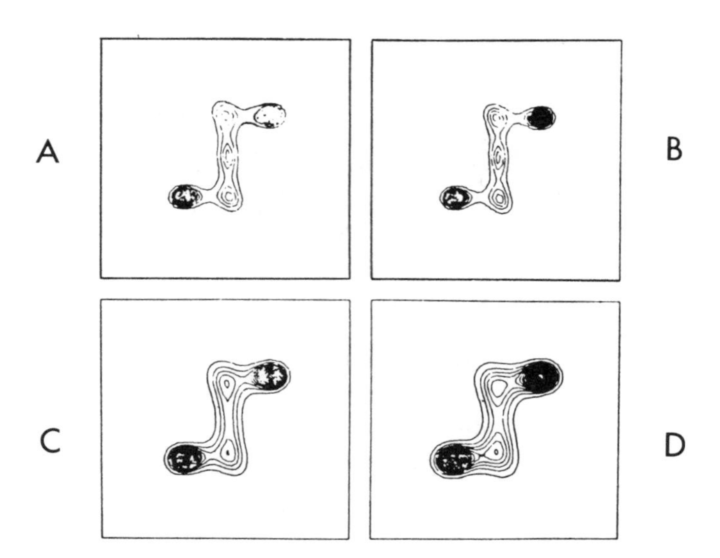

Figure 9. Contour plots of the restorations obtained with maximum entropy. Contour levels are 2, 7, 12, 17, 22, 27, ..., 92, 97% of the peak intensity on each image. (Burch, 1980.)

17.3 Edging a nonperiodic photograph

In any deconvolution, special attention must be paid to the edges of the picture. When transform techniques are used, the problem is exacerbated because it is implicitly assumed that the picture is wraparound periodic, with the top continuing directly from the bottom, and the left side from the right side. In practice, in any actual photograph D, there are almost always inevitably sharp discontinuities between the top and the bottom and between left and right. These can manifestly not be reproduced in any convolved trial data set $F = f*b$ since, whatever trial picture f is postulated, some wraparound continuity will appear after it is convolved with the point spread function b. It becomes impossible to deconvolve the sharp discontinuities actually present in the data, and in attempting to do so many algorithms produce "ringing" for several beamwidths around the edge of the picture.

This can be avoided by replacing the actual data near the edges by data blurred once more by the point spread function. Thus, mock data are constructed as

$$\alpha D + (1-\alpha)(D*b) , \tag{33}$$

where $\alpha = 1$ within the interior of the picture and $\alpha = 0$ on the edges. The rate of rolloff of α toward the edge is governed by the autocorrelation of the point spread function, and is thus automatically controlled by the p.s.f. width. Specifically,

$$\alpha(x,y) = 1 - b_1(x)b_2(y) \tag{34}$$

where $b_1(x)$ is the autocorrelation of the y-projection $\int b(x,y)dy$ and $b_2(y)$ is the autocorrelation of the x-projection $\int b(x,y)dx$.

This technique allows the algorithm to deconvolve the interior of the photograph correctly. However, within about one p.s.f.-width of the edge, the algorithm will be deconvolving $D*b$ by the p.s.f., and will thus tend to return to the original data, albeit with some suppression of noise.

18. Practical Optical Imagery

18.1 Standard maximum entropy deconvolutions

Figure 10, supplied by A. F. Lehar, shows a practical example of maximum entropy being used to reconstruct a blurred 128 × 128 pixel photograph of a car number-plate. This software is being used routinely by the UK police for both 128 × 128 and 512 × 512 images. A full 128 × 128 deconvolution can be performed in about 1 minute, and 512 × 512 in under an hour.

Figure 10. Maximum entropy in optical imagery: reconstruction of a car number-plate. (a) Blurred photograph. (b) 128 × 128 pixel reconstruction, assuming signal/noise to be 60. (Courtesy of Maximum Entropy Data Consultants.)

Figure 11 shows an example of the reconstruction of part of the text of "Hamlet" digitized on a 512 × 512 grid, and deconvolved by S. F. Burch at the UK Atomic Energy Authority's Harwell Laboratory. Again the original data set was an ordinary photograph. Occasional specks of dust and other imperfections on individual pixels confused the reconstruction somewhat because the algorithm tried to deconvolve them also. Burch removed this problem by switching off the contribution to χ^2 from those data pixels whose residuals became more than 7σ, since it was very plausible that such pixels were in error.

a

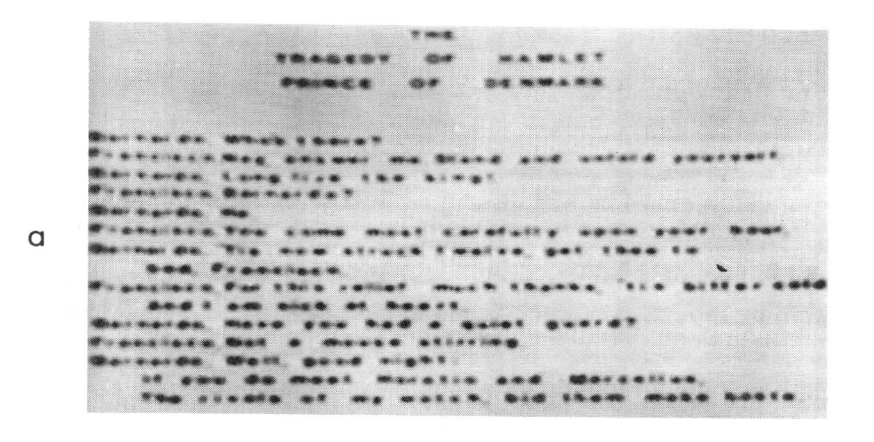

b

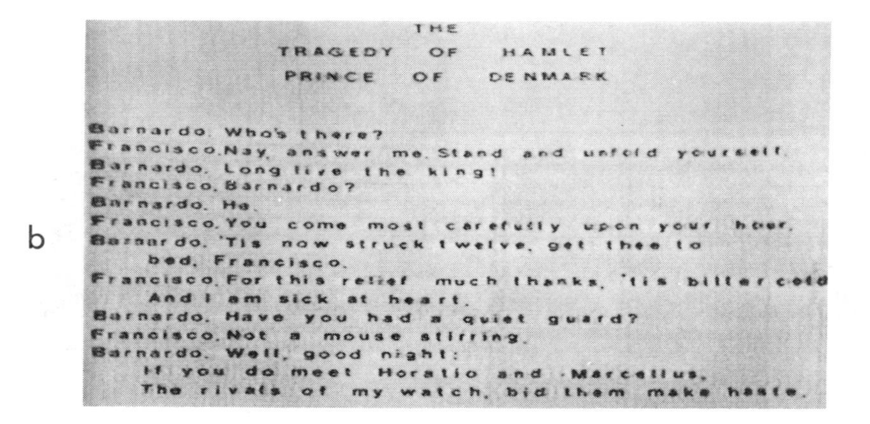

Figure 11. (a) Blurred photograph of part of text from "Hamlet." (b) 512 × 512 pixel reconstruction, with signal/noise of 100. (Burch, Gull, and Skilling, in preparation.)

An example from optical astronomy is the 128 × 128 picture (Fig. 12) of the galaxy M87 taken with the 200-inch telescope (J. Lorre) and deconvolved with maximum entropy by Bryan and Skilling (1980). The extra detail in the jet is real, as has been shown by more sophisticated (CCD)

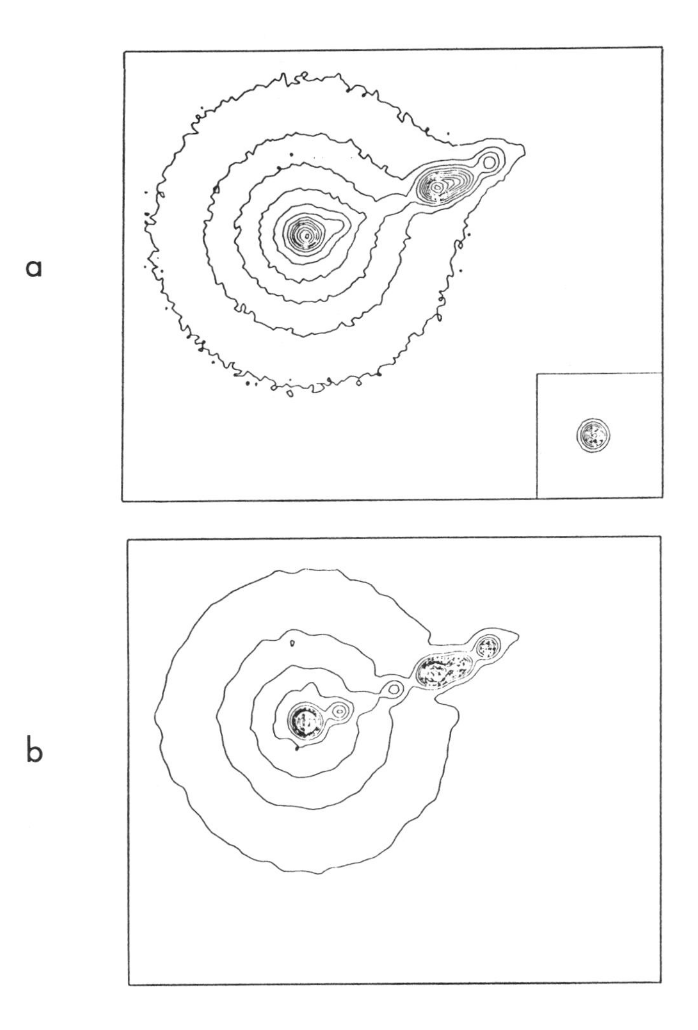

Figure 12. (a) Contour map of photograph of galaxy M87. Contour levels are 1, 2, 3, ..., 8, 9, 10, 12, 14, 16%, ... Inset: Contour map of telescope point spread function, to same scale. Contour levels are 10, 20, 30, ..., 90% of central maximum. (b) Maximum entropy reconstruction. Contour levels as in (a). (Bryan and Skilling, 1980.)

detecting equipment (Fielden, private communication). Interestingly, the χ^2 statistic was insufficiently powerful for this particular problem. The difficulty arose because the brightest parts of the image also had the largest errors σ, that being an unavoidable property of photographic plates. Accordingly, χ^2 was relatively insensitive to precisely those parts of the image in which one was most interested. The corresponding pixels all had markedly negative residuals, and the overall histogram of the residuals was very significantly skewed. A more sophisticated test, in which the residuals were forced to have the correct (Gaussian) histogram to within the expected noise, was used instead. Not only did this fit the histogram's second moment (χ^2), but all the other moments were automatically fitted as well. The price paid for using the nonquadratic statistic was a factor of about 2 in the computation time.

Figure 13, reproduced from Daniell and Gull (1980), shows 128 × 128 deconvolutions of "Susie" (supplied by Professor J. Harris), using the χ^2 statistic. For these reconstructions, the data set was the original photograph of "Susie," blurred by being convolved with a disc of 6-pixel radius, and with added noise. Figure 13b has a signal/noise ratio of 10 at the brightest part of the image, whereas Fig. 13d is of higher quality, with maximum signal/noise of 1000. The reconstructions show clearly the inevitable tradeoff between noise suppression and resolution. Figure 13c shows good noise suppression, but there was no increase in resolution. Figure 13e, having better signal/noise, was able to offer significant improvement in resolution as well as better discrimination of intensity levels. It is a noteworthy feature of the maximum entropy technique that the possible improvements in resolution are automatically determined by the confidence that can be placed in the degraded image.

18.2 Sparse data

Nothing in the algorithm requires all the σ to be the same. Indeed, a large proportion of the pixels can be entirely unmeasured, so that the data become sparsely distributed. Gull has demonstrated that maximum entropy still succeeds in reconstructing the image as long as more than about one pixel is measured in each p.s.f. area of the data plane (Fig. 14). Thus, Fig. 14b is a reconstruction from Fig. 14a, which sampled about 50 pixels per p.s.f. area, whereas Fig. 14d was reconstructed from Fig. 14c, which sampled only about 5 random pixels per p.s.f. area. If fewer pixels than this are measured, areas of the reconstruction become decoupled from the data and are pulled by the entropy toward a featureless average value. If this process is taken too far, the reconstructions become somewhat surreal. Thus, Fig. 14f is the reconstruction from Fig. 14e, which is about at the reconstruction limit of one sampled pixel per p.s.f. area.

Linear algorithms such as Wiener filters or constrained least-squares simply cannot cope with sparse data. They depend fundamentally on Fourier transforming the data, and such transforms cannot be performed on sparse data without also featuring the sampling pattern.

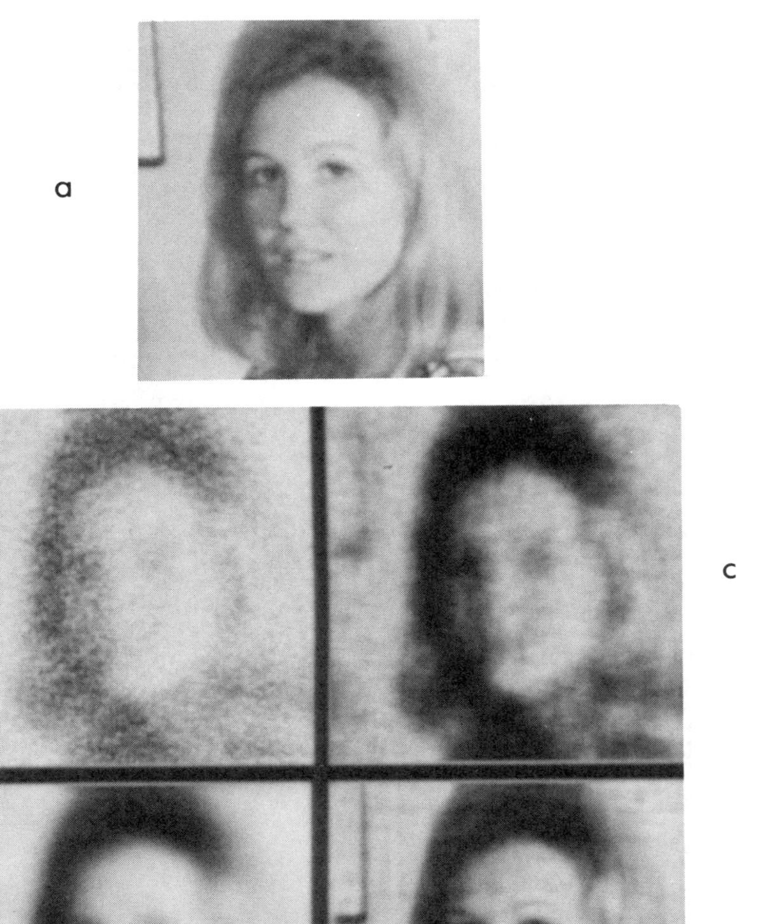

Figure 13. (a) Original test photograph of "Susie." (b) Degraded image: p.s.f. is disc of 6-pixel radius; maximum signal/noise is 10. (c) Restored image, showing noise suppression. (d) Degraded image: p.s.f is disc of 6-pixel radius; maximum signal/noise is 1000. (e) Restored image, showing improvement in resolution and dynamic range as well as noise suppression. (Daniell and Gull, 1980.)

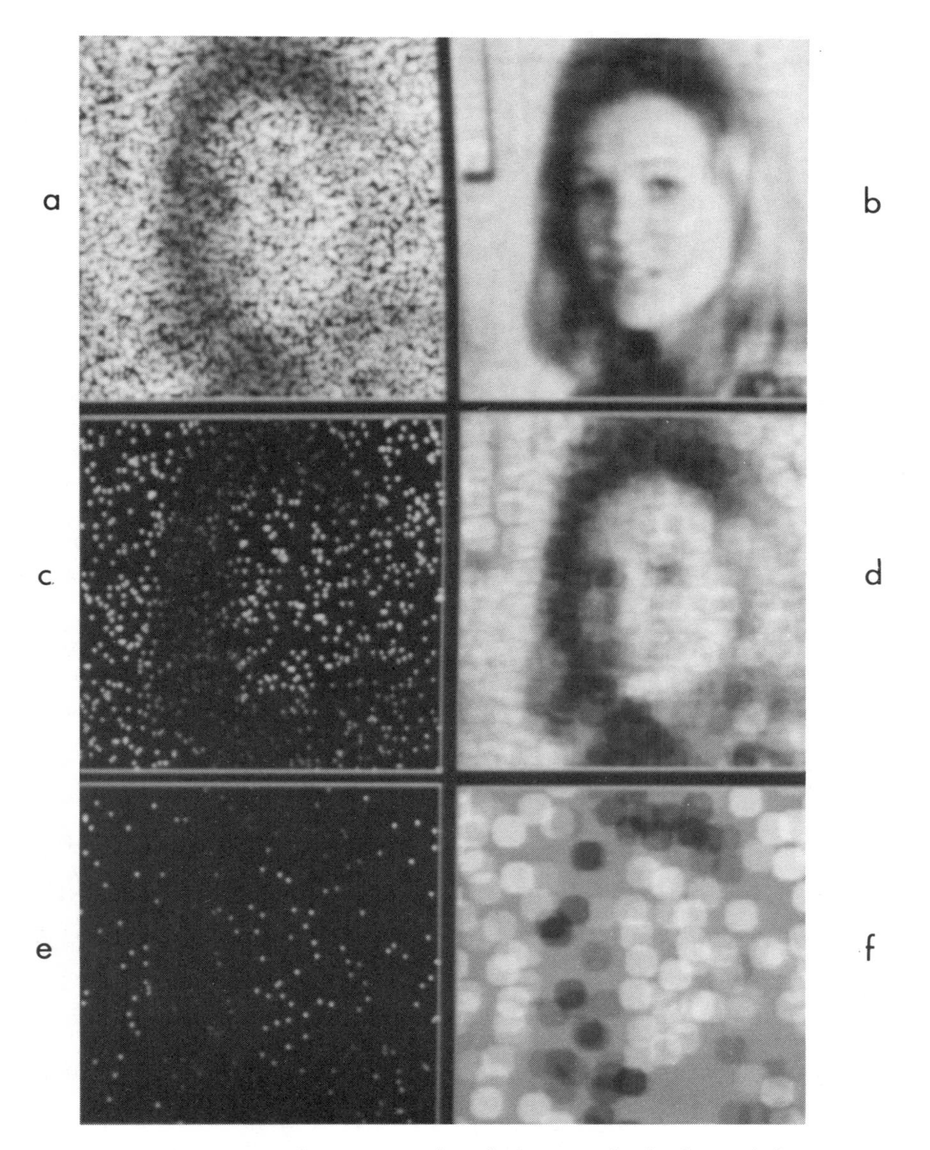

Figure 14. Restorations from Fig. 13d with data randomly discarded. (a) 50% of data discarded. (b) Reconstruction showing interpolation across p.s.f. (c) 95% of data discarded. (d) Reconstruction. (e) 99% of data discarded. (f) Reconstruction showing barely adequate sampling. (Gull.)

19. X-Ray and γ-Ray Astronomy

In x-ray and γ-ray astronomy, the observed data represent an (astronomical) object convolved with a known p.s.f. Often, the observing satellite is pointed in only a few specific directions, corresponding to sparse data in the optical case. A good example of this (Fig. 15) is x-ray data of the Cassiopeia A supernova remant taken with the Copernicus satellite (Fabian, Zarnecki, and Culhane, 1973) and deconvolved with maximum entropy by Gull and Daniell (1978). Again the maximum entropy reconstruction can be confirmed by comparing with other observations: It is virtually identical to a radio map at the same resolution (but see Fig. 26 for a state-of-the-art radio image). It is also superior to that produced by the algebraic reconstruction technique (ART) (Stevens and Garmire, 1973).

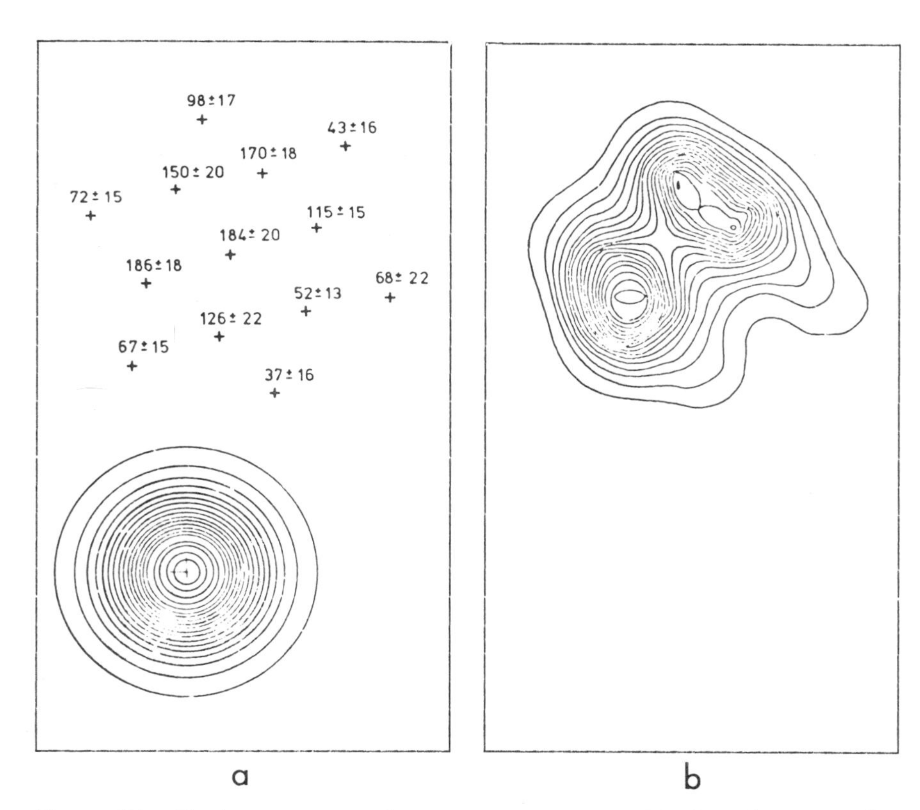

Figure 15. Maximum entropy in x-ray astronomy. (a) Counting rates for various positions near Cassiopeia A. Also shown are the experimental errors and a contour map of the p.s.f. (b) Maximum entropy reconstruction of (a). (Gull and Daniell, 1978.)

A similar reconstruction, of 3C273 (Fig. 16), has been performed by Willingale (1981) using the 'integral equation' algorithm.

Gamma-ray astronomy has the additional complication that very few photons are observed—too few for any χ^2 statistic to be applicable. Skilling, Strong, and Bennett (1979) maximized the entropy over constant likelihood

$$L = \text{Prob}\,(\text{data } D \mid \text{image } f), \tag{35}$$

this being the appropriate analog of χ^2, and fixed the required value of L by using a Kolmogorov-Smirnov nonparametric test for consistency between actual data D and simulated data F. They used about 1000 photons observed with the COS-B satellite (Scarsi et al., 1977) to obtain the maximum entropy map (Fig. 17) of the Crab Nebula region.

20. Tomography

In tomography one attempts to reconstruct a two-dimensional slice of a three-dimensional object from a finite number of projections within the plane of the slice. Thus the data represent line integrals across a two-dimensional image, with or without absorption, blurred transversely by the p.s.f. of the detector and corrupted by the inevitable noise. Although the transformation from object to signal remains linear,

$$F_k = \sum R_{kj} f_j \tag{36}$$

where R_{kj} = response of detector k to pixel j, it may be difficult or impossible to use fast algorithms to compute it. The simplest inversion technique is back projection, in which one approximates R(inverse) by R(transpose) and reconstructs according to

$$f_j(\text{back projection}) = \sum D_k R_{kj} .$$

This gives rise to a p.s.f. that behaves as 1/r and causes severe blurring of the reconstruction. Conventional reconstruction methods deconvolve this p.s.f. using a linear filter that is usually implemented as a data-space filter prior to back projection (Shepp and Logan, 1974).

20.1 X-ray transmission

Because of the computational burdens, much of the maximum entropy work has been done on 64 × 64 grids. However, Gull has reconstructed a 'phantom' brain (a skull filled with Perspex) on a 512 × 512 grid from 250 projections at each of 240 angles of view, using x-ray transmission data

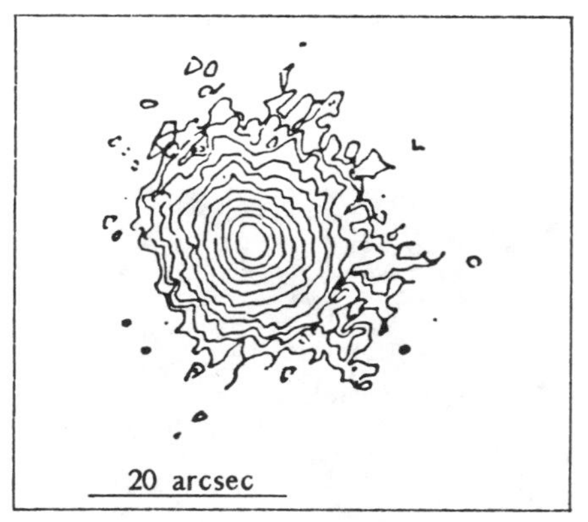

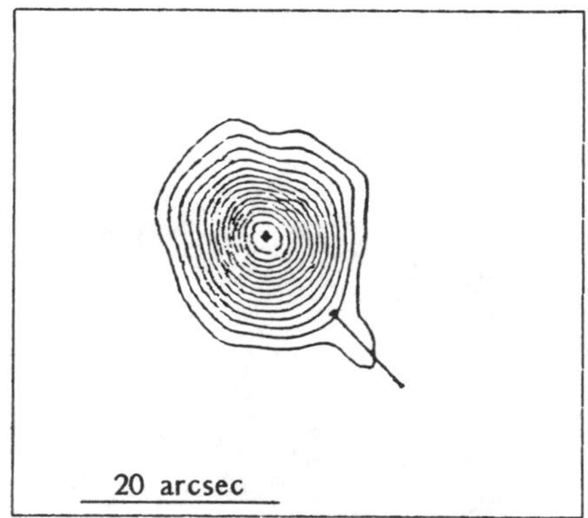

Figure 16. Maximum entropy in x-ray astronomy. (a) Contour plot of the raw count data (60,000 photons) from 3C273 binned in 4 x 4 arcsec pixels. Contour interval is logarithmic. (b) Contour plot of the maximum entropy deconvolution. The optical jet is shown as a bar; this region contains only 0.15% of the total x-ray flux. (Willingale, 1981.)

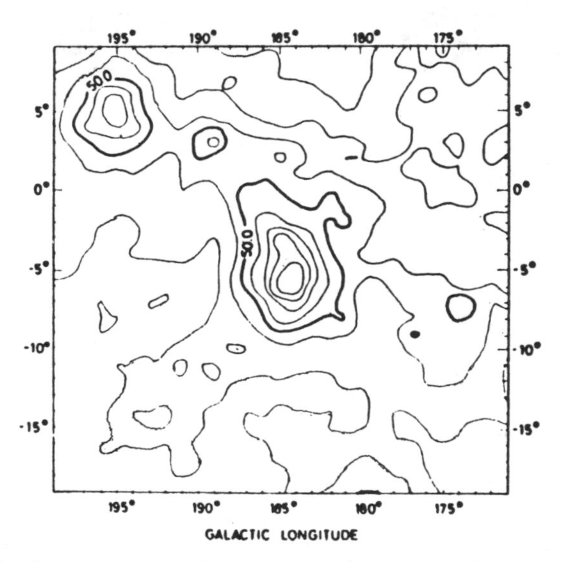

Figure 17. Maximum entropy in γ-ray astronomy. Reconstruction from about 1000 photons observed by COS-B in the galactic anticenter region centered on the Crab Nebula. Contour levels are 10, 20, 30, ..., 90% of maximum flux. (Skilling, Strong, and Bennett, 1979.)

from an EMI brain scanner (Fig. 18). The precision and clarity of this reconstruction clearly demonstrate the power of maximum entropy.

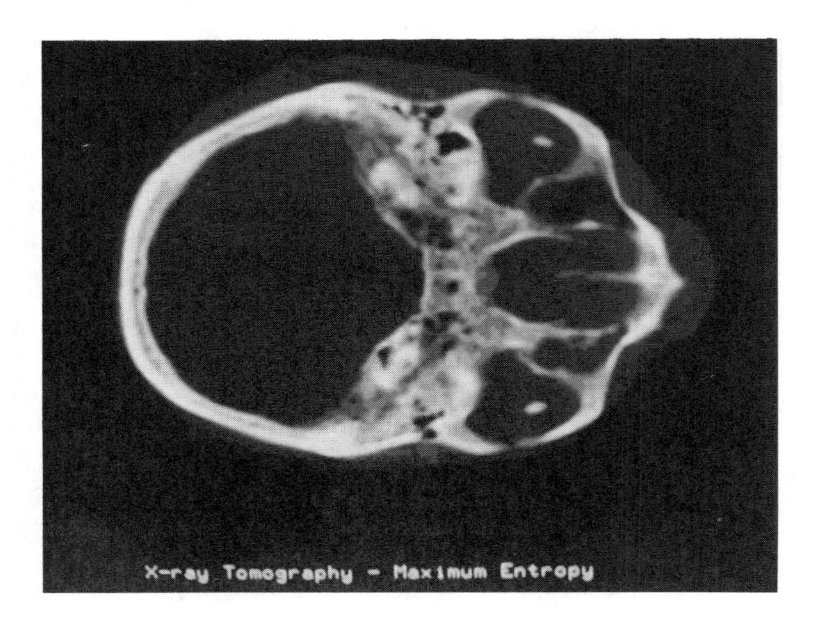

Figure 18. Maximum entropy in transmission tomography. Reconstruction on a 512 x 512 grid from x-ray transmission data taken with an EMI brain scanner (Gull).

Simulations of tomographic experiments have also been performed to compare maximum entropy with other techniques, and to determine the minimum number of projections needed for realistic reconstructions. Kemp (unpublished) has reconstructed a "disc plus ring" object on a 64 x 64 grid from as few as five viewing angles, each orientation having 40 parallel detectors spaced across the object (Fig. 19). Although the signal-to-noise ratio was given the relatively high value of 300, the reconstruction is decidedly imperfect, showing that five viewing angles are simply too few for good results for this object. A very similar reconstruction of the same simulated object using a maximum entropy technique was performed earlier by Minerbo (1979), although his algorithm required the detectors for each viewing angle to be parallel and nonoverlapping, and did not test properly whether the final reconstruction actually fitted the data to within the given noise. Kemp's implementation was based directly on the general transformation (36) and suffers from none of these limitations.

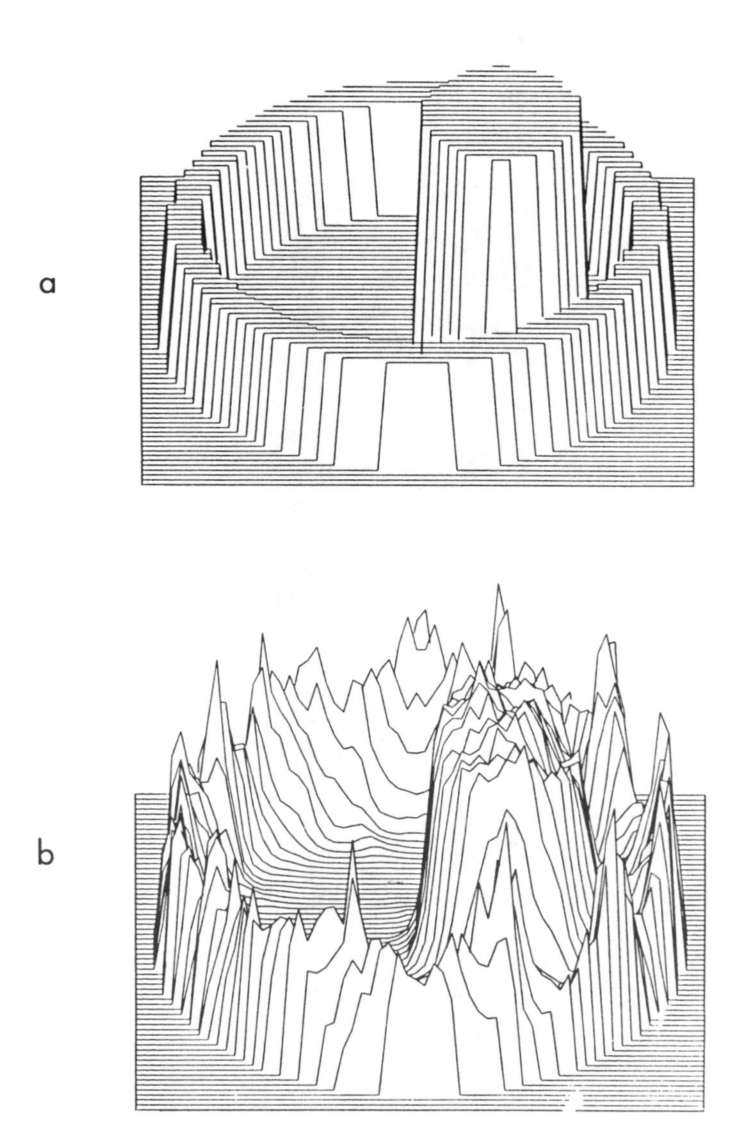

Figure 19. Tomographic simulation of a "disc plus ring" using only five projection angles. (a) Original object. (b) Maximum entropy reconstruction using 40 detectors for each angle and good signal/noise (300). (Kemp, 1980.)

20.2 Radionuclide emission

More realistically for medical applications, Kemp (1980) also reconstructed slices through a "phantom" liver containing various lesions and doped with γ-emitting radioisotope (Fig. 20). His data were taken with a GE rotating gamma camera provided by the Nuclear Medicine Departments

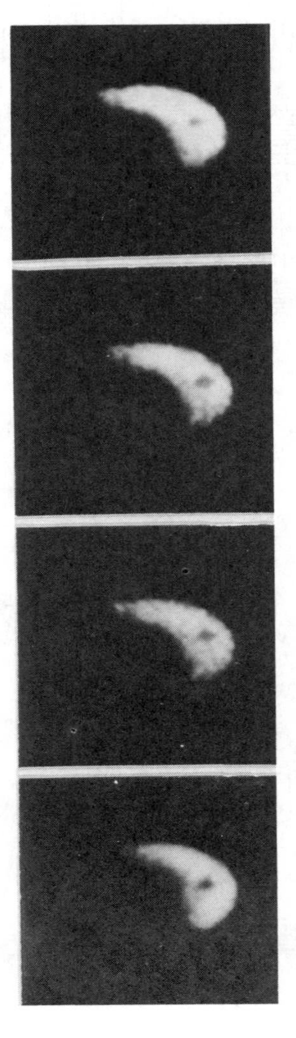

Figure 20. Maximum entropy in emission tomography. Reconstructions of a slice of a "phantom" liver, doped with radioisotope and observed with a GE rotating gamma camera. (Kemp, 1980.)

of Guy's and the Middlesex Hospitals, London, and different subsets of the data were used in his reconstruction. Absorption of γ-photons by the body surrounding the liver is a severe problem in this situation since the full "optical depth" of the body is about 6, and small errors in the assumed absorptivity have large effects on the reconstruction. Kemp was nevertheless able to show that maximum entropy needs to use only perhaps a quarter of the viewing angles needed by conventional algorithms in order to obtain comparable reconstructions. The hope is that use of maximum entropy can allow either diminished dosages to the patient or increased image resolution, or a tradeoff between the two.

Kemp also compared maximum entropy with filtered back projection for a simple object viewed from a few angles (Fig. 21). The artifacts along the viewing directions that are introduced by filtered back projection are

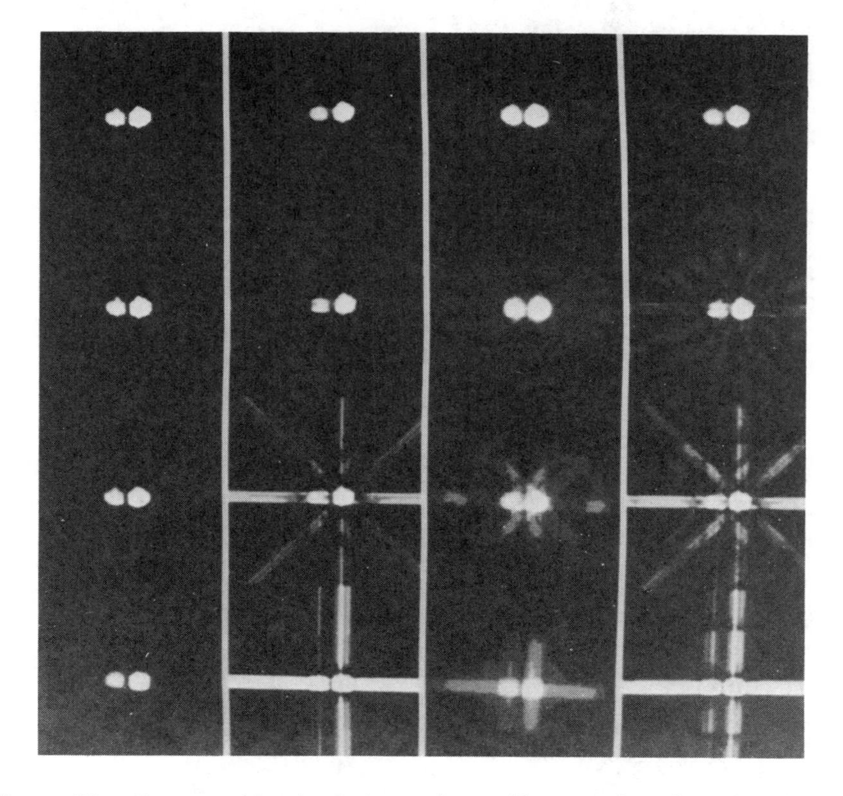

Figure 21. Tomographic simulations of two "hotspots" with and without a background, reconstructed from small numbers of projection angles. Columns 1 and 3 show maximum entropy reconstructions. Columns 2 and 4 show the corresponding filtered back-projection results, with regions of negative intensity suppressed. (Kemp, 1980.)

sharply reduced by maximum entropy. Reconstructions from a limited range of viewing angles are also shown (Fig. 22). This case is very difficult to treat using conventional methods, as large sectors of the Fourier transform of the object are completely unmeasured.

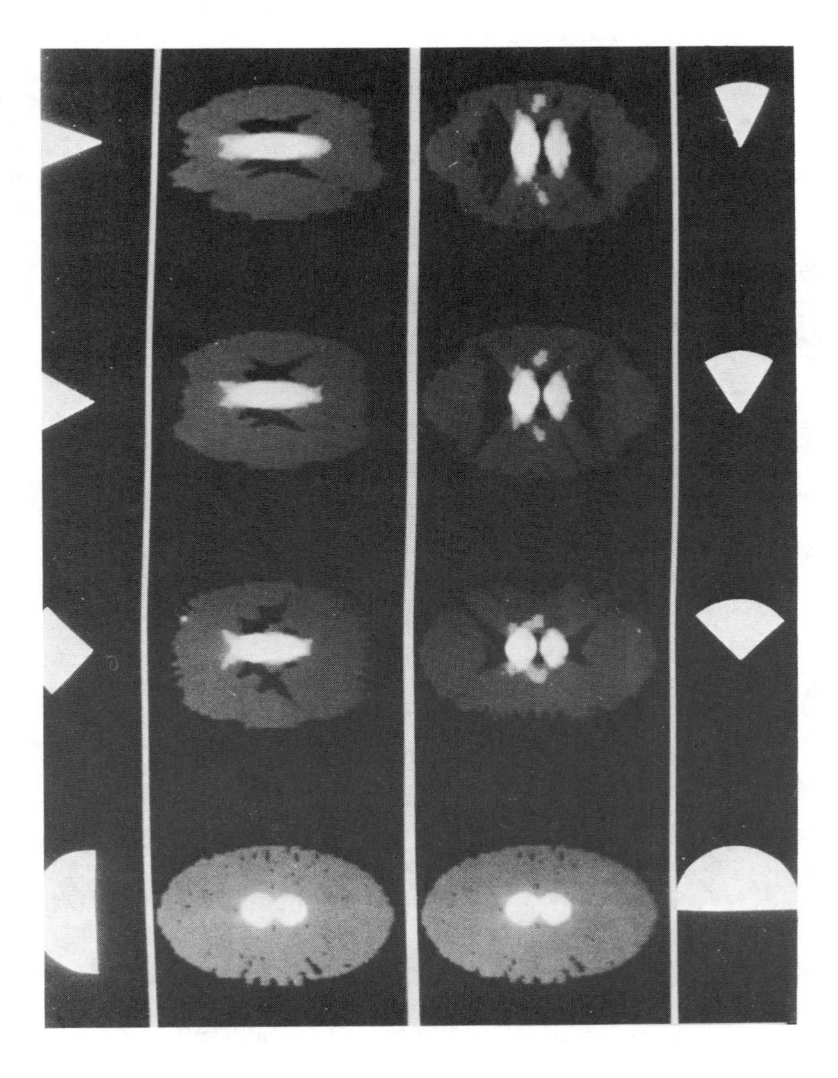

Figure 22. Tomographic simulations of two "hotspots" with a background. Maximum entropy reconstructions from limited angular ranges. (Kemp, 1980.)

20.3 Plasma diagnostics

Tomographic reconstruction has also been applied in plasma physics, to determine the cross-section of a neutral beam from the beam's transverse Hα emission. Maximum entropy software has been installed for G. A. Cottrell at Culham Laboratory to determine the precise shape of the injection beam for the Joint European Torus. With about 25 megajoules in the beam it is important to know exactly where the edges of the beam are, lest it inadvertently be allowed to touch the container walls. Figure 23 is a simulation of what will be a typical run using 12 pinhole scanners around the periphery of the beam, each having 60 photon detectors. In the reconstruction, even the outer contour (1% of maximum) is recovered well. This example shows clearly the effect of the positivity constraint inherent in maximum entropy.

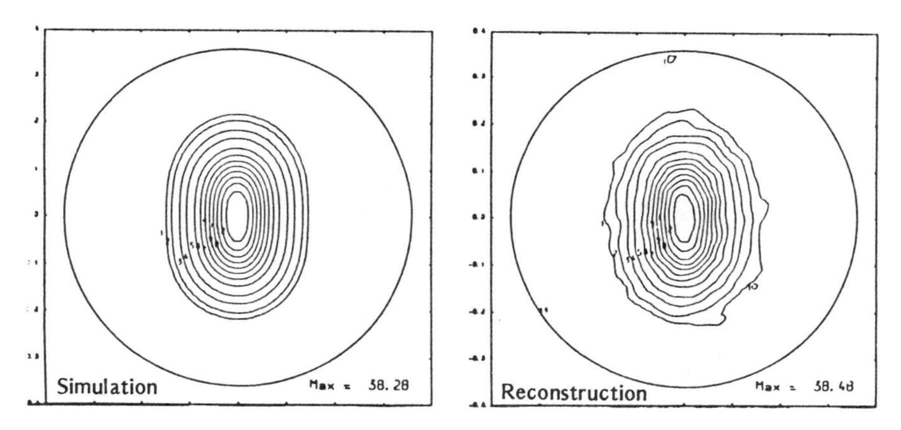

Figure 23. Maximum entropy in plasma emission tomography. Simulation and reconstruction of the 5-megawatt neutral injection beam for the Joint European Torus, observed in transverse Hα emission by 12 pinhole cameras each having 60 detectors. Contour levels are 1, 2, 5, 10, 20, 30, 40, ..., 90% of maximum. (Courtesy Culham Laboratory and G. A. Cottrell.)

21. Radio Astronomy

Maximum entropy can be used to improve the resolution of single-dish observations, just as for optical deconvolution. The more interesting application, though, is to interferometer data. Here the data are Fourier components

$$D_k = F_k + \text{noise}, \qquad F_k = f_j \exp(2\pi ijk/N) \qquad (37)$$

of the radio sky, in both amplitude and phase. The order k of the Fourier component corresponds to the spatial separation of the radio antennas.

21.1 Aperture Synthesis

By varying the antenna spacing, one can observe all the components out to some maximum value. This is the technique known as "aperture synthesis." Direct inversion of the data by the inverse Fourier transform $f_j = \Sigma D_k \exp(-2\pi ijk/N)/N$ yields reconstructions with inadmissibly high sidelobes produced by the sharp cutoff of k at the maximum antenna spacing. These can be alleviated by various filters ("aperture weighting") at the expense of resolution. They can be reduced further by algorithms such as CLEAN (Schwartz, 1978), which fits a sequence of point sources to the data (and can thus be inefficient at reconstructing large faint areas of the sky).

Maximum entropy gives still better results. Figure 24, from Gull and Daniell (1978) shows a conventional, aperture-weighted reconstruction of the radio galaxy 3C31, together with the central region of the maximum entropy reconstruction. The dynamic range of the map has improved from 50:1 to 1000:1. All negative sidelobes have been removed (because they are obviously impossible) and positive ones are greatly reduced, and the resolution has increased by a factor of about two. Gull and Daniell also give an example of large areas of low surface brightness being correctly reconstructed without difficulty.

21.2 Automatic Calibration by Maximum Entropy*

Recent years have seen the construction of larger interferometers and more sensitive receivers, so that now the quality of images produced is more often limited by systematic errors than by purer considerations of resolution or random noise. Especially at high frequency ($\gtrsim$15 GHz), the gain of radio receivers can change with time, leading to errors of typically a few percent in amplitude and, more seriously, of significant fractions of a radian in phase. The weather is also responsible for substantial phase errors in high frequency observations. The worst effect occurs when a warm or cold front passes over the telescope and introduces a phase gradient. The resulting aperture-synthesis artifacts take the form of lines radiating from point sources. These effects can be adequately modeled by a uniform gradient in phase over the telescope—yet another unknown parameter in the reconstruction.

The idea behind automatic calibration is, first, to reconstruct a maximum entropy map from the observed data, with all preliminary corrections applied to the best of one's ability. This is the most uniform, positive map consistent with the initially calibrated data. The nonlinear positivity constraint imposed by the entropy then allows the systematic phase errors to be retrieved in a way that would be impossible for a linear algorithm. Model data are constructed from the maximum entropy map and compared with the observed data via the orthodox χ^2 statistic. Because of the positivity

---

*This section reports recent unpublished work by Gull and Brown.

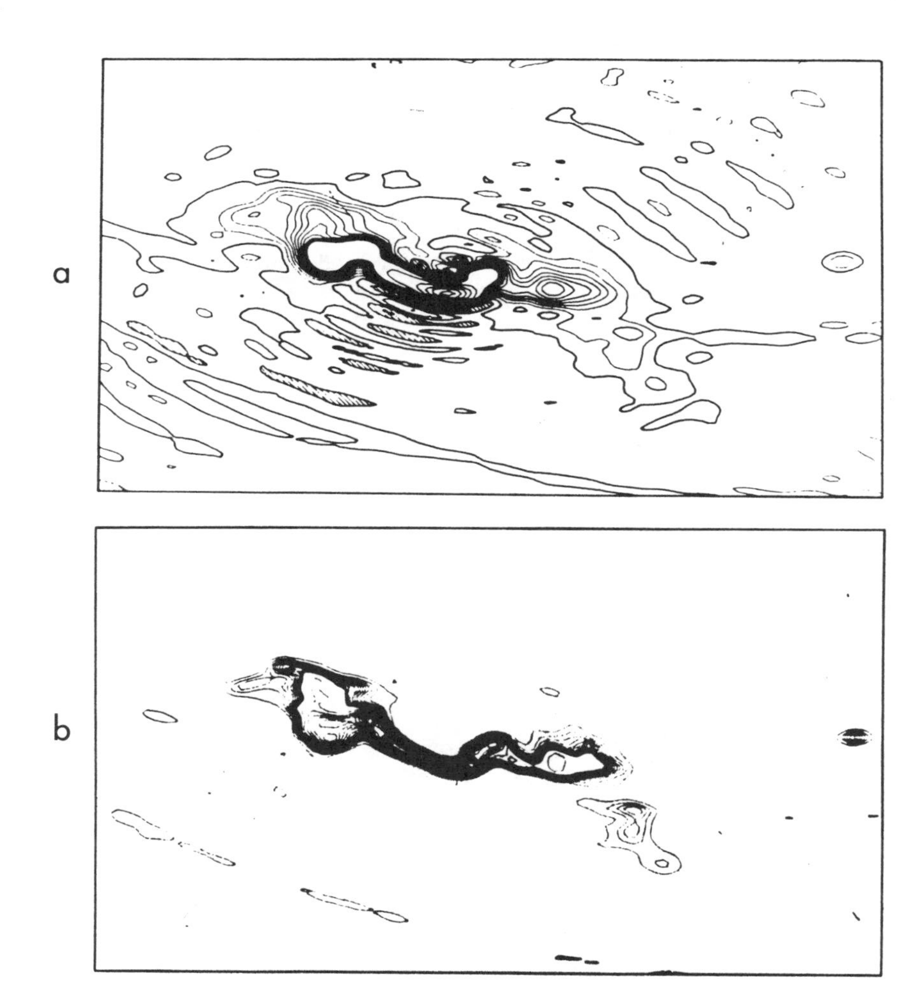

Figure 24. Maximum entropy in aperture-synthesis interferometry. (a) Aperture-weighted radio image of the galaxy 3C31. The contour interval is increased by a factor of 10 after the first 10 contours. Negative regions are shown hatched. (b) Reconstruction of central region of 3C31. Contour interval is increased by a factor of 10 after each set of 10 contours. (Gull and Daniell, 1978.)

constraint, this statistic will not be optimized over the calibration parameters: it will almost certainly become easier to fit the data with a positive map if the calibration parameters are adjusted somewhat. Accordingly, the uncertain antenna phases and other parameters are readjusted to minimize the misfit of the observations to the model. This makes the recorrected data more consistent with a positive map, which is a very plausible requirement to impose. With the data recalibrated, a new maximum entropy map is produced. In principle, this procedure could be iterated indefinitely, but in practice the scheme is stable (at least for phase errors less than $\pi/2$), and a single recalibration suffices.

Figure 25 shows the results of autocalibration applied to 30-GHz observations of W3(OH) made with the Cambridge 5-km telescope (Scott, 1981). The corrections applied were 5% to 10% in amplitude and up to 40 degrees in phase. Autocalibration of phase was also used to reconstruct the million-pixel (1024 x 1024) state-of-the-art map of Cassiopeia A shown in Fig. 26.

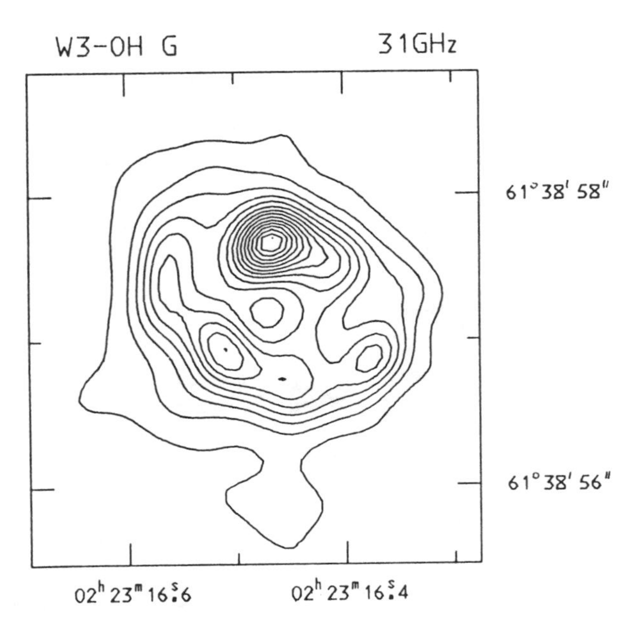

Figure 25. Autocalibrated maximum entropy radio map of W3(OH). The lowest contour corresponds to a brightness temperature of 165 K and the contour interval is 330 K. Open triangles denote positions of OH masers. (Scott, 1981.)

21.3 Very-Long-Baseline Interferometry

In very-long-baseline radio interferometry, the constituent receiving antennas are thousands of kilometers apart, and the baselines between them are not known to within many wavelengths. This means that the antenna phases are completely unknown and also change on a time scale of a few minutes. In effect, although the n antennas can be used to measure

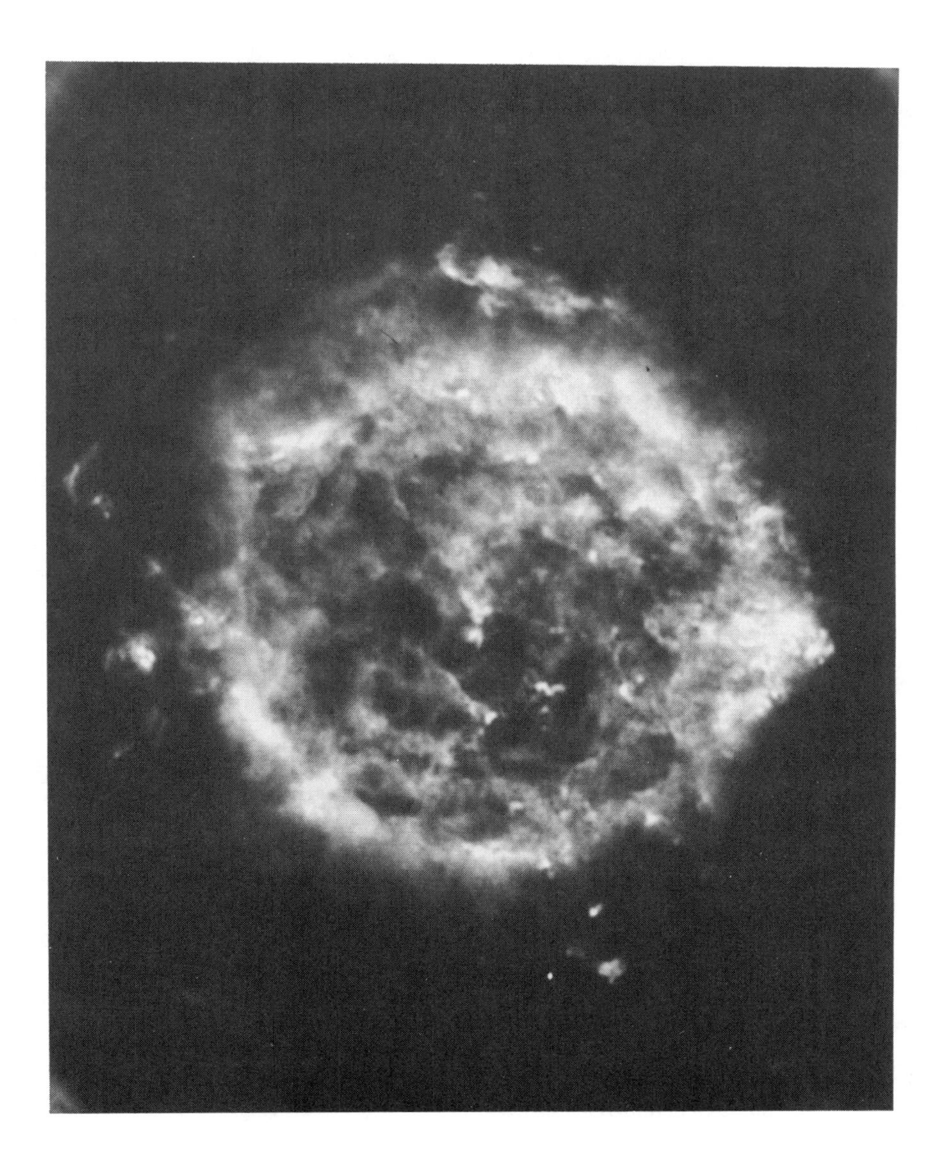

Figure 26. State-of-the-art radio map. Million-pixel maximum entropy reconstruction of the supernova remnant Cassiopeia A, observed with the 5-km aperture-synthesis radio telescope at Cambridge. Maximum entropy autocalibration and reconstruction by Gull and Brown.

n(n-1)/2 Fourier components, there are (n-1) undetermined relative phases. This problem is a relatively severe form of the phase autocalibration described above. Gull (unpublished) has used this technique on 3C345 data provided by the Cal Tech group to obtain the reconstruction shown in Fig. 27.

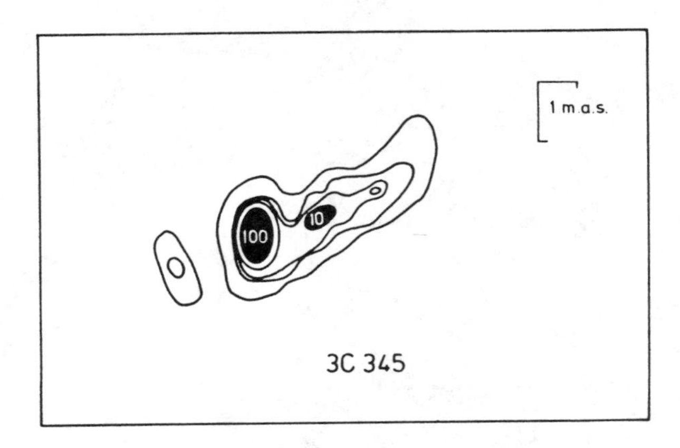

Figure 27. Maximum entropy in very-long-baseline interferometry. Reconstruction of 3C345. Note the 1 milliarcsec (m.a.s.) scale.

Conclusions

Maximum entropy has proved to be a formidably powerful tool for reconstruction of positive, linear images from a wide variety of types of experimental data. It is also a practical tool in that a robust, general algorithm has been developed that copes routinely with images of high resolution (up to a million pixels at present, almost certainly more in the future) and high dynamic range. Indeed, the superiority over other algorithms is most clearly demonstrated in just such high-contrast images.

The theoretical underpinning of maximum entropy appears to be disarmingly simple. In many problems, one asks for the shape of the object as a whole. Then the safest, least committal reconstruction is that image that has least information about the shape that one seeks. The natural measure of this information is (minus) the Shannon/Jaynes entropy $-\Sigma\, p \log p$, leading immediately and directly to the maximum entropy algorithm. There are, of course, many other questions one might ask about an image. One might seek unbiased estimates of the total flux from part or all of the image; maximum entropy, which imposes a deliberate bias towards uniformity, would be inappropriate. In time-series analysis, one might seek maximally noncommittal estimates of unmeasured samples; this leads to the Burg algorithm for the spectrum of the series. The list of alternative questions is endless.

The Shannon/Jaynes entropy deals directly with the simplest, most straightforward parameter one is likely to seek—namely what the image itself looks like. From this derives the generality and power that surely destine maximum entropy to be one of the standard tools of scientific analysis for the coming century.

References

Ables, J. G. (1974) Astron. Astrophys. Suppl. **15**, 383.
Ash, R. B. (1965) Information Theory (New York: Interscience).
Bryan, R. K. (1980) PhD Thesis, University of Cambridge.
Bryan, R. K., and J. Skilling (1980) Mon. Not. R. Astron. Soc. **191**, 69.
Burch, S. F. (1980) UKAEA, Harwell Report No. AERE-R-9671.
Burg, J. P. (1967) Paper presented at 37th meeting of Society of Exploration Geophysicists, Oklahoma City.
Burg, J. P. (1972) Geophysics **37**, 375.
Daniell, G. J., and S. F. Gull (1980) IEE Proc. (E) **5**, 170.
Edward, J. A., and M. M. Fitelson (1973) IEEE Trans. Inf. Theory **IT-19**, 232.
Fabian, A. C., R. Willingale, J. P. Pye, S. S. Murray, and G. Fabbiano (1980) Mon. Not. R. Astron. Soc. **193**, 175.
Fabian, A. C., J. C. Zarnecki, and J. L. Culhane (1973) Nature (London) Phys. Sci. **242**, 18-20.
Fletcher, R., and C. M. Reeves (1964) Comput. J. **7**, 149.
Frieden, B. R. (1972) J. Opt. Soc. Am. **62**, 511.
Frieden, B. R., and J. J. Burke (1972) J. Opt. Soc. Am. **62**, 1202.
Frieden, B. R., and D. C. Wells (1978) J. Opt. Soc. Am. **68**, 93.
Gull, S. F., and G. J. Daniell (1978) Nature **272**, 686.
Jaynes, E. T. (1968) IEEE Trans. Syst. Sci. Cybern. **SSC-4**, 227.
Jeffreys, H. (1939) Theory of Probability (Oxford: Clarendon).
Kemp, M. C. (1980) International Symposium on Radionuclide Imaging, **IAEA-SM-247**, 128, Heidelberg.
Kikuchi, R., and B. H. Soffer (1977) J. Opt. Soc. Am. **67**, 1656.
Minerbo, G. (1979) Comp. Graphics Image Processing **10**, 48.
Newman, W. I. (1977) Astron. Astrophys. **54**, 369.
Polak, E. (1971) Computational Methods in Optimization (New York: Academic Press).
Ponsonby, J. E. B. (1973) Mon. Not. R. Astron. Soc. **163**, 369.
Sargent, R. W. H. (1974) in P. E. Gill and W. Murray, eds., Numerical Methods for Constrained Optimization (New York: Academic Press).
Scarsi et al. (1977) Proc. 12th ESLAB Symposium, ESA-124, p. 3.
Schooneveld, C. van (1978) Proc. IAU Colloq. **49**, 197 (Dordrecht: D. Reidel).
Schwarz, U. J. (1978) Astron. Astrophys. **65**, 345.
Scott, P. F. (1981) Mon. Not. R. Astron. Soc. **194**, 23.
Shannon, C. E. (1948) Bell Syst. Tech. J. **27**, 379, 623.
Shannon, C. E., and W. Weaver (1949) The Mathematical Theory of Communication (Urbana, Ill.: Univ. Illinois).

Shepp, L. A., and B. F. Logan (1974) IEEE Trans. Nucl. Sci. **NS-21**, 21.
Skilling, J., A. W. Strong, and K. Bennett (1979) Mon. Not. R. Astron. Soc. **187**, 145.
Stevens, J. C., and G. P. Garmire (1973) Astrophys. J. **180**, L19.
Ulrych, T. J., and T. M. Bishop (1975) Rev. Geophys. Space Phys. **13**, 183.
Ulrych, T. J., and R. W. Clayton (1976) Phys. Earth Planet. Inter. **12**, 188.
Wernecke, S. J. (1977) Radio Sci. **12**, 831.
Wernecke, S. J., and L. R. d'Addario (1977) IEEE Trans. **C-26**, 351.
Willingale, R. (1979) PhD Thesis, University of Leicester.
Willingale, R. (1981) Mon. Not. R. Astron. Soc. **194**, 359.

ESTIMATING OCCURRENCE LAWS WITH MAXIMUM PROBABILITY, AND THE TRANSITION TO ENTROPIC ESTIMATORS

B. Roy Frieden

Optical Sciences Center, University of Arizona, Tucson, Arizona 85721

C. Ray Smith and W. T. Grandy, Jr. (eds.),
Maximum-Entropy and Bayesian Methods in Inverse Problems, 133–169.

1. Introduction

Suppose that events ξ_n, n = 1,...,N occur, where each ξ_n is one of a set of possibilities x_n, n = 1,...,M. For example, a particular set of $\{\xi_n\}$, the "sample"

$$x_8x_2x_5x_9x_9x_1x_8x_3x_5x_9x_9x_6x_1x_5x_2x_3x_1x_7x_4x_3$$

might occur. In the cases we shall consider, this is either a time sequence of events or a space sequence.

For the given sample, N = 20 and M = 9 (the largest subscript possible). Note that the occurrence rate of event x_1 is 3/20, that of x_2 is 2/20, etc. Call these occurrence rates in general p_n, n = 1,...,M. Suppose that the observer must somehow estimate the occurrence rates without observing the sample. This is the problem addressed in this paper. The information he has at hand about the unknown $\{p_n\}$ is data, such as moments, and "prior knowledge," which is all possible information about the $\{p_n\}$ aside from the data.

For example, suppose the sample is a page of print. Then the event sequence $\{\xi_n\}$ is a space sequence of letters, and $\{p_n\}$ are their unknown occurrence rates on the page. The prior knowledge could be information about the probable language in which the sample is written. Obviously this will have strong bearing upon the $\{p_n\}$ to expect prior to seeing the data.

Or, the sample may be an optical object. Then the event sequence $\{\xi_n\}$ is a sequence in time of photon radiation positions $\{x_n\}$. Here the prior knowledge could be a known probability law on the expected $\{p_n\}$ that defines a "class" of objects. These points will be enlarged upon below.

We shall suppose that the actual data are insufficient in number to form a unique estimate of the $\{p_n\}$. In this case there is a redundancy of solution. One method of lifting the redundancy of solution is to form an estimate $\{\hat{p}_n\}$ obeying

$$-\sum \hat{p}_n \ln \hat{p}_n = \text{maximum}\,, \tag{1}$$

maximum entropy. This is Jaynes' answer [1] to the problem. We shall show that this estimator follows from a straightforward maximum probability approach, but only under certain conditions of "prior knowledge" (prior to our observing the data). The same approach gives rise to estimation principles other than Eq. (1) as well, depending upon the form of prior knowledge at hand. This will become abundantly clear. In summary, these are:

(a) Jaynes' principle, Eq. (1), when the prior knowledge is strong enough to imply that the $\{p_n\}$ are close to a uniform probability law.

(b) A generalization of Jaynes' principle called "minimum relative entropy" or the "Kullbach-Leibler norm" [2], when it is known that the $\{p_n\}$ are close to some other definite probability law $\{Q_n\}$,

$$-\sum_n \hat{p}_n \ln \frac{\hat{p}_n}{Q_n} = \text{maximum} . \tag{2}$$

(c) No principle at all, when the prior knowledge is so vague that all laws $\{p_n\}$ are equally probable prior to our seeing the data [3].

(d) A principle

$$-\sum_n \hat{p}_n \ln \hat{p}_n = \text{minimum} \tag{3}$$

of minimum entropy, when the prior knowledge is in the form of negative information, that is, some probability law $\{R_n\}$ is known to not be a candidate solution. [Note that Eq. (3) is the exact opposite of Jaynes' Eq. (1).] Minimum entropy also arises under other conditions, for optical objects.

(e) A "generalized" Burg [4] principle [see Eq. (9) below]

$$\sum_n M_n \ln \hat{p}_n = \text{maximum} , \tag{4}$$

when an effort is made to estimate the prior probability law by observing M_n occurrences of event x_n, $n = 1,\ldots,M$, prior to the N events forming the data, where $M_n \ll Np_n$. The latter denotes the case of sparse prior knowledge. In the opposite situation $M_n \gg Np_n$ of strong prior knowledge, Jaynes' principle [1] follows once more.

(f) A principle

$$\sum_n (N\hat{p}_n + r_n) \ln(N\hat{p}_n + r_n) - \sum_n \hat{p}_n \ln \hat{p}_n = \text{maximum} , \tag{5}$$

when the prior knowledge is that the $\{\hat{p}_n\}$ are close to a law $\{q_n\}$, which is unknown except for following a known probability law $p(q_1)\ldots p(q_M)$. Each $p(q_n) = Cq_n^{r_n}$ for $0 \leq q_n \leq 1$, zero otherwise, with C a normalization constant and $r_n \geq 0$. This is a case where the most likely value of each q_n is 1, the least likely 0, and the power r_n dictates how dominant the value $q_n = 1$ is over all lesser values. In the particular case $N\hat{p}_n \gg r_n$, for example, when $r_n = 1$, principle (5) collapses to

$$\sum_n r_n \ln \hat{p}_n = \text{maximum} , \tag{6}$$

which is again a weighted Burg type. For the opposite case, $N\hat{p}_n \ll r_n$, the prior law $p(q_1)\ldots p(q_M)$ approaches a delta function about the one law $q_n = 1/M$, so Jaynes' law (1) again results.

(g) A principle [5]

$$-\sum_n (\hat{p}_n - a)\ln(\hat{p}_n - a) - \sum_n (z - \hat{p}_n)\,\ell(z - \hat{p}_n) = \text{maximum}\,, \qquad (7)$$

where $\hat{p}_n$ represents the photographic density at pixel position n in a photo. The photo has been prefogged to level a before the picture of an unknown scene is taken. Value z is the known upper level of (equivalent) density in the scene. Aside from the prior knowledge of a and z, it is assumed additionally that the unknown scene is close to being uniformly gray.

(h) A principle

$$-\sum_n m_n \ln m_n + \sum_n (m_n + z - 1)\ln(m_n + z - 1) = \text{maximum}\,,$$

$$m_n = N\hat{p}_n\,, \qquad (8)$$

where $\hat{p}_n$ represents the intensity of light in a pixel located at point $\mathbf{x}_n$ in an electromagnetic scene (of any wavelength). Parameter z is the number of quantum degrees of freedom for a photon in a resolution cell of the object. This relationship goes to Eq. (1) when m_n/z is small, or to Burg's estimator

$$\sum_n \ln \hat{p}_n = \text{maximum} \qquad (9)$$

when m_n/z is large. Prior knowledge that the unknown scene is close to uniform gray must also be the case, for result (8) to hold.

Principles (a) through (f) become much easier to understand when they are applied to specific cases. Two cases are summarized in Table 1.

For the first case, we return to the page of print consisting of N letters. Here, a statistical event x_n is the occurrence of the nth letter (for example, event x_1 is the occurrence of letter a, x_2 denotes b, etc.). Without seeing the page, we want to estimate the relative occurrences $\{p_n\}$ of the letters, given some indirect data about them (such as moments). We may also have "prior knowledge," which is information about the unknowns that is exterior to the data. A good estimation scheme uses all the information at hand, that is, all the data and all the prior knowledge. For example, if the language in use is known to be, or not to be, Russian, this information should be built into the estimate of the letter occurrences. The stronger and more restrictive the prior knowledge, the better will be the estimate of the occurrences.

Table 1. Maximum Probability (M.A.P.) Estimation Applied to Two Cases

M.A.P. estimator	Page of print	Picture (pointlike pixels)
(a) Maximum entropy, Eq. (1)	The language is known; each letter occurs with equal frequency (a most unusual language)	The object picture is known to be gray
(b) Minimum cross-entropy, Eq. (2)	The language is known; letter frequencies are unequal	The picture is known to be the Mona Lisa
(c) No principle at all	The language is unknown. The diversity of languages on Earth is such that, considering all languages, the relative occurrence of a is as likely to be 0.1 as 0.2 or 0.3, etc. through 1; similarly for b, etc. (Note that this is not the same as saying that a probably occurs as often as b, etc.; the latter is much stronger prior information)	The class of objects covers the Universe
(d) Minimum entropy, Eq. (3)	It is known that the language is not Russian	The picture is known to not be the Mona Lisa
(e) Generalized Burg, Eq. (4)	Before the data were received, a sample of the (unknown) language that formed the page was examined for its occurrence numbers $\{M_n\}$ of individual letters	One picture from the same class as the unknown was examined for its intensities $\{M_n\}$ The unknown is assumed to approximate these
(f) Close to generalized Burg (for lack of a better name)	The relative occurrence of a is more likely to be 1 than 0; similarly for the other letters. However, by normalization if a has relative occurrence near 1, the occurrences of the other letters must be near 0. Hence the printed page consists either of nearly all a's, or nearly all b's, or etc. The effect is stronger as the $r_n \to \infty$	The object picture tends to contain but a few very bright points, with the rest very subdued

The second case consists of a sharp object picture composed of photographic grains. Suppose these to be white grains on a background that is initially black as in a positive print. Each pixel contains a relative number p_n of grains that are to be estimated from indirect evidence (say, a blurred image of the picture). Here a statistical event x_n denotes the location of a given grain at the instant it develops, and the event sequence occurs in time. We assume also, for simplicity, that each pixel is truly a point, having therefore no degeneracy of position for its grains within. Later, we will consider the more realistic case of pixel "cells" of finite volume. The summary that follows holds only for the "point pixel" case.

The prior knowledge now takes the general form $p(q_1,\ldots,q_M)$ of a probability law on prior object intensities $\{q_n\}$, something that can be known if the class of objects is known.

For these two cases, various kinds of prior knowledge scenarios lead to the corresponding maximum-probability (M.A.P.) estimators in Table 1.

In the following sections we shall derive these results, along with others. We proceed from a general estimation principle, that of maximum probability, to specific cases of "prior knowledge," which make the principle take on its different forms.

2. Prior Knowledge Scenarios

A particular page of print

alt gcorn fat ikek jynj ahf figj hdksl eiei ghuj h eirh . . .

(it may be nonsense) exists, whose actual letter occurrences $\{p_n\}$ (normalized to unity) must be estimated. These represent unknown probabilities, as well, which define the writing style of the author. There are N letters on the page, and there are M different kinds of letters; for example, M = 26 for English. The page may not be observed; otherwise there is no estimation problem.

Two types of information are given about the page: direct data consisting of some moments of $\{p_n\}$; and indirect data, called "prior knowledge," regarding the expected values of $\{p_n\}$ prior to our observing the data. Call these $\{q_n\}$. These prejudices about $\{q_n\}$ may be expressed in the form of an assertion

$$p(q_1,\ldots,q_M)\,, \tag{10}$$

where

$$\sum_{n=1}^{M} q_n = 1\,,$$

as to the joint probability law for the expected occurrences. A statement $p(q_1,\ldots,q_M)$ is a statement about the probable language(s) on the unseen page.

Highly related to this letter occurrence problem is the optical restoration problem. Here the sample is a sharp object picture. It manifests itself during image formation as an unknown succession of photons radiating in time from resolution cell locations, such as

$$8\ 2\ 5\ 9\ 9\ 1\ 8\ 3\ 5\ 9\ 9\ 6\ 1\ 5\ 2\ 3\ 1\ 7\ 4\ 3\ . \tag{11}$$

The data image is invariant to the temporal order of its photons. Hence, we want to estimate the $\{p_n\}$, which are here photon densities at the cells, irrespective of the order in which they radiate from the cells. For example, the relative occurrence p_8 represents the number of grain events "cell 8" (see list above) divided by the total number (20 in the list), or 2/20, regardless of the placement order of "cell 8" in the list. This order-independence for photons corresponds to the order-independence for letters in the page-of-print problem.

It is also possible to model an object as successively developed photographic grains if the object is indeed a photo. Then a grain location, as in example (11), is analogous to a letter occurrence. The two problems are mathematically equivalent, provided there is no additional degeneracy within each resolution cell. In actuality, there will be, and we treat this in a later section. For now, we consider the simplest object case, that of temporal degeneracy alone, so as to enable treatment of the letter occurrence problem and the object estimation problem "in parallel," by the same development. The word "grain" shall refer generically to either a photographic grain or a photon. Later sections will distinguish between these two entities.

For convenience the grains are assumed to occur in a positive print, so that they are white against an otherwise black background. Then relative occurrence p_n represents the relative "whiteness" or brightness of the nth cell.

The direct information (the data) given about the $\{p_n\}$ is its sampled image. However, there is assumed to be not enough data, or they are not known with sufficient accuracy, to foster a unique solution to the problem. The indirect information (prior knowledge) given about the $\{p_n\}$ is a given probability law for the expected values of $\{p_n\}$, called the $\{q_n\}$, before the data are seen. This is a statement that the unknown object $\{p_n\}$ was selected from a "class" of objects $\{q_n\}$ having a joint probability $p(q_1,\ldots,q_M)$. The narrower the class, the higher the state of prior knowledge, and the more confined is $p(q_1,\ldots,q_M)$ made about a particular set $\{Q_n\}$ of the possible $\{q_n\}$. This is clarified next, by various examples.

Very Strong Prior Knowledge

Suppose the observer states that

$$p(q_1, \ldots, q_M) = \prod_{n=1}^{M} \delta(q_n - Q_n) . \tag{12}$$

Then regarding the printed-page problem, he is declaring the language to be known. Note that even for this strong case of prior knowledge the problem is not solved: the $\{Q_n\}$ merely represent average use of the language—whereas every writer has his own style, hence his own $\{p_n\}$ which, in general, is not identical with overall language use.

Regarding the optical object problem, an assertion, Eq. (12), on prior knowledge is a statement that a unique object $\{Q_n\}$ is expected. This would happen, for example, if the observer strongly expects the Mona Lisa, or a close version, to be the unknown object.

Very Weak Prior Knowledge

At the other extreme, if [3]

$$p(q_1, \ldots, q_M) = \prod_{n=1}^{M} \text{Rect}\,(q_n - 1/2),$$

$$\text{Rect}(x) \equiv \begin{cases} 1 \text{ for } |x| \leq 1/2 \\ 0 \text{ for } |x| > 1/2 \end{cases}, \tag{13}$$

then all values of each q_n on the interval (0,1) are declared to be equally probable. In effect, the observer admits that he has no prior knowledge of the language or of the object. All are equally probable. He will process the data with complete equanimity.

Negative Prior Knowledge

At still another extreme, the observer has only negative or indirect prior knowledge of the unknown. He knows, for example, that the language is not Russian (or related languages; see below). Let $\{R_n\}$ represent the letter occurrences overall for Russian. Aside from this knowledge, all other $\{q_n\}$ are equally probable. Hence, the overall probability law p is a combination of Eqs. (12) and (13),

$$p(q_1, \ldots, q_M) = (1 - \varepsilon)^{-1} \left[\prod_{n=1}^{M} \text{Rect}(q_n - 1/2) - \varepsilon \prod_{n=1}^{M} \delta(q_n - R_n) \right] . \tag{14}$$

Quantity ε is a small constant, measuring the amount of spread in language about Russian that must also be known to not be present. This might include, for example, related Slavic languages; but to keep things simple, we call the entire group "Russian." The use of Eq. (14) will not depend upon the size of ε, merely upon its not vanishing.

For an optical object, prior knowledge (14) would arise when the object is, for example, known to not be the Mona Lisa or a picture closely resembling it, but may otherwise be any picture with equal likelihood. The situation could be generalized rather easily. If it is known that the picture is not $\{R_n\}$ and also not $\{S_n\}$, etc., the delta function product in Eq. (14) is merely replaced by a sum of such terms. In this way negative prior knowledge of the most general kind may be described.

Power-Law Prior

Lying about "halfway" between complete determinism, Eq. (12), and complete randomness, Eq. (13), is the situation of power-law probabilities

$$p(q_1,\ldots,q_M) = B \prod_{n=1}^{M} q_n^{r_n}, \qquad r_n \geq 0, \ \ B \text{ constant}. \tag{15}$$

For $r_n = 0$, $n = 1,\ldots,M$, we are back to complete randomness, Eq. (13). For all $r_n = 1$, we have a product of triangle functions. For any set of positive $\{r_n\}$, Eq. (15) states that high letter occurrences are more probable (before the data are seen) than are low letter occurrences. But since the $\{q_n\}$ must obey normalization, the effect will be to have a few q's large, with all the rest much smaller. In other words, only a few letters will dominate in use. The effect becomes more dominant as the $\{r_n\}$ increase, until for the $\{r_n\}$ extremely large some one letter will dominate almost completely. The page of print would be expected to consist nearly entirely of a's, or of b's, etc.

The class of optical objects described by Eq. (15) would tend to consist of a relatively small number of randomly placed bright cells, the rest being dim. The higher the $\{r_n\}$, the more pronounced would be this effect, to the limit of one point containing all the energy of the picture.

Empirical Prior

Finally, consider an empirical determination of the prior. This is the situation where an attempt is made to define $p(q_1,\ldots,q_M)$ experimentally, in observation of a sample that is distinct from the unknown. A different page of print from the unknown, or a different object from the unknown, is observed. It consists of L letters or L grains. The page or the object is from the same "class" as the unknown, so the two should have a similar statistic $p(q_1,\ldots,q_M)$. The observed occurrences of letters or grains are

$\{M_n\}$. What is the inferred probability law $p(q_1,\ldots,q_M)$, contingent upon observation of occurrences $\{M_n\}$?

As will be shown [see Eq. (56)], it obeys

$$p(q_1,\ldots,q_M) = \frac{(L+M-1)!}{(M-1)!} \frac{1}{M_1!\ldots M_M!} q_1^{M_1}\ldots q_M^{M_M} . \tag{16}$$

3. The General Principle

Consider the list of 20 events in example (11). Event "1" occurs 3 times, "2" occurs twice, "3" occurs 3 times, etc. Let us denote these absolute occurrences as $\{m_n\}$, so that $m_1 = 3$, etc. Suppose, at first, that the prior information consists of specific values $\{Q_n\}$ for the expected relative occurrences. Then by the independence of event "1" from "2" and "3," etc., the probability of the occurrence numbers $\{m_n\}$ in example (11) is a simple product of probabilities

$$P(m_1,\ldots,m_{10}) \equiv P(3,2,3,\ldots,0) = Q_1^3\, Q_2^2\, Q_3^3 \ldots Q_{10}^0 . \tag{17}$$

Now the data are moments of the $\{m_n\}$ or an image of the $\{m_n\}$, etc. Being purely a function of the $\{m_n\}$, the data are blind to the particular order in which the events took place. Likewise, so is $P(m_1,\ldots,m_{10})$. In fact, the event $(m_1,\ldots,m_{10}) = (3,2,3,\ldots,0)$ occurs for every reordering of the events in sample (11). Let there be W such reorderings in all. W will in general depend upon the $\{m_n\}$, as discussed below. Then since each reordering has the same probability, Eq. (17), the total probability of the occurrence numbers in sample (11) obeys

$$P(3,2,3,\ldots,0) = W \cdot Q_1^3 Q_2^2 Q_3^3 \ldots Q_{10}^0 , \tag{18}$$

or more generally

$$P(m_1,\ldots,m_M) = W(m_1,\ldots,m_M) Q_1^{m_1} Q_2^{m_2} \ldots Q_M^{m_M} , \tag{19}$$

where the dependence $W(m_1,\ldots,m_M)$ must be found. As will be seen, this quantity depends upon the physical model followed by the individual events.

The maximum probable set of occurrences $(\hat{m}_1,\ldots,\hat{m}_M)$ consists of those that maximize Eq. (19). Once they are found, we may use the law of large numbers

$$\hat{m}_n = N\hat{p}_n \tag{20}$$

to estimate the relative occurrences $\{\hat{p}_n\}$. (This assumes that N is large.) However, this approach is a bit simplistic, as discussed next.

The probability law (19) states that if the $\{Q_n\}$ are the true probabilities of events $\{x_n\}$, then each possible set of occurrences $(m_1,\ldots,m_M)$ of the events obeys probability P as given. When we use the principle that $P(m_1,\ldots,m_M)$ is a maximum, we are saying that the actual occurrences are also the most probable set. But these are only most probable with respect to the assumed probabilities $\{Q_n\}$. Ideally, we would like them instead to be most probable with respect to the true probabilities $\{p_n\}$. However, as these are the very unknowns of the problem, it is impossible to accomplish this goal. Hence, as a more modest goal, in the approach (19) we settle for a realistic appraisal $\{Q_n\}$ of what the $\{p_n\}$ are.

The next sobering step in the overall approach is to admit of unknown, and hence random, departure from a definite pre-judgment $\{Q_n\}$. In the letter occurrence problem, this is the admission that any one of many languages might be present, each according to its own probability. This is reasonable. Before seeing the data, how could one claim to definitely know the language in use? One would have to be told such information, as additional prior knowledge.

In the optical object problem, the unknown is admitted to be randomly selected from a "class" of objects, each of which occurs with a certain probability. Again this is quite realistic. Before seeing its blurred image, how could we justify the claim that it is close to the Mona Lisa, or close to uniform grayness? Therefore, a joint probability law $p(q_1,\ldots,q_M)$ is now admitted as governing the probable set of relative occurrences $\{p_n\}$ prior to our seeing the data. This constitutes the prior knowledge we previously spoke of.

Use of such a law in the maximum probability approach is straightforward. It is a simple generalization of Eq. (19), to

$$
\begin{aligned}
P(m_1,\ldots,m_M) &= W(m_1,\ldots,m_M) \\
&\quad\times \int\cdots\int q_1^{m_1}\cdots q_M^{m_M}\, p(q_1,\ldots,q_M)\, dq_1\cdots dq_M \\
&= \text{maximum}\,, \qquad\qquad (21)
\end{aligned}
$$

where the q_n must be normalized to unity. (Recall that Eq. (19) actually represents $P(m_1,\ldots,m_M \mid Q_1,\ldots,Q_M)$. Next, use the "total probability" law to get Eq. (21).) Equations (20) and (21) form the general estimation principle to be used. This principle was previously derived for a specific case $W(m_1,\ldots,m_M)$.

Particular cases of prior knowledge, as reflected in functions p and W, will cause it to take on different forms. These will be examined in turn. The integral in Eq. (21) is, by the way, the Dirichlet integral, in the particular case where $p(q_1,\ldots,q_M)$ is constant.

4. Estimation Principles Under Different Conditions of Prior Knowledge

For simplicity, we shall first consider problems such as the page-of-print problem and the object-grain problem as previously defined. Objects having degeneracy within their resolution cells will be considered later. These include a more realistic model for the photographic object and photon-radiating objects, for which other types of degeneracy creep in.

Coefficient W represents the number of ways that m_1 letters x_1, m_2 letters x_2, etc., can arise on a printed page, independent of order. As an example, if the alphabet consisted of $M = 2$ letters a and b, if $N = 3$, and if $W(2,1)$ is required, we see that the event (2,1) can occur independently as aab, aba, or baa. Hence, $W(2,1) = 3$. In general, $W(m_1,\ldots m_M)$ follows a multinomial law

$$W(m_1,\ldots,m_M) = N!/m_1!\ldots m_M! \,. \tag{22}$$

This law would also be followed for the grain object case: For an $N = 3$-grain object consisting of $M = 2$ resolution cells, the picture (2,1) describing 2 grains in cell 1, and 1 in cell 2, can develop in time as cell events 112, or 121, or 211, so $W(2,1) = 3$ once again.

We now use Eq. (22) and various forms of prior knowledge $p(q_1,\ldots,q_M)$ to form various estimators from Eq. (21).

Use of Very Strong Prior Knowledge

The simplest case to consider is Eq. (12), where the language is known or the class of objects is extremely narrow. Substituting Eqs. (12) and (22) into principle (21), we obtain

$$P = (N!/m_1!\ldots m_M!)Q_1^{m_1}\ldots Q_M^{m_M} \,, \tag{23}$$

after use of the sifting property of the delta functions. Since P is to be maximized, ln P will be minimized as well. Taking the natural logarithm of Eq. (23) and using Stirling's approximation

$$\ln m! \simeq m \ln m \tag{24}$$

yields a principle

$$N \ln N + \sum_{n=1}^{M} m_n \ln(Q_n/m_n) = \text{maximum} \,. \tag{25}$$

If we finally use Eq. (20) to bring in the unknowns $\{\hat{p}_n\}$, and drop all additive and multiplicative constants that do not affect the maximum, we obtain

$$\sum_{n=1}^{M} \hat{p}_n \ln(Q_n/\hat{p}_n) = \text{maximum} . \tag{26}$$

The normalization of $\{p_n\}$ was also used. This is the Kullbach-Leibler norm of estimation. The left-hand side is also regarded sometimes as a form of information measure since it measures the degree to which an initial assessment $\{p_n\}$ of occurrence rate changes to values $\{Q_n\}$, after data or some other "information" is acquired about the process. This norm has been the subject of much recent investigation [2].

Principle (26) is correct as it stands only when data are lacking. However, the user always knows that the $\{\hat{p}_n\}$ obey normalization; he also may know that $\{\hat{p}_n\}$ has certain moments or, more generally, obeys certain data equalities

$$\sum_{n=1}^{M} \hat{p}_n f_{mn} = F_m , \qquad m = 1,\ldots,K . \tag{27}$$

This equality information may be built into the principle by the standard use of Lagrange multipliers, generally one multipler to a constraint. Hence, the total principle (26) now takes the form

$$\sum_{n}^{M} \hat{p}_n \ln(Q_n/\hat{p}_n) + \lambda \left[\sum_{n}^{M} \hat{p}_n - 1 \right] + \sum_{m}^{K} \mu_m \left[\sum_{n} \hat{p}_n f_{mn} - F_m \right]$$

$$= \text{maximum} . \tag{28}$$

The solution is found by setting $\partial/\partial\hat{p}_n = 0$, $n = 1,\ldots,M$, which yields

$$\hat{p}_n = Q_n e^{-1+\lambda+\Sigma\mu_m f_{mn}} . \tag{29}$$

This may be called a "generalized Boltzmann" probability estimate, in analogy with that gas statistic. The unknowns $\lambda,\{\mu_m\}$ must be found by substituting the form (29) for $\hat{p}_n$ back into the K data equations (27) and the normalization equation. This has a unique solution.

It is interesting to establish the solution when no data are at hand, so that all $\mu_m = 0$. Then solution (29) must only satisfy normalization. The result is simply

$$\hat{p}_n = Q_n . \tag{30}$$

For this reason, the $\{Q_n\}$ are called "bias values" for the unknowns.

Jaynes' Limiting Case

The principle (26) depends explicitly upon the choice of $\{Q_n\}$. What might be called a maximally noncommital case is

$$Q_n = \text{constant} = \frac{1}{M} . \tag{31}$$

That is, for the page-of-print problem the alphabet has equal probabilities for all letters. Or, for the object problem, a grain development event is as likely to occur in one cell as in any other.

The state (31) is Jaynes' state of "maximum prior ignorance." Certainly as regards the occurrence of a single letter or a single object grain, Eq. (31) admits of the most randomness. However, it predicts something very definite about the ensemble of the N events, when data are not given. It states that the object will tend to be uniformly gray, or that the page will contain the various letters in about equal amounts. This is basically why MacQueen and Marschak [3] objected to such a definition of maximum prior ignorance. In their view, maximum prior ignorance should mean that even an ensemble of events $\{x_n\}$ should not have a unique, expected histogram shape $\{q_n\}$. Before the image data are seen, no unique prediction should be possible about the individual intensities across the picture.

On the other hand, if the observer is not in a state of maximum prior ignorance, as when he knows from past observation that intensities should lie between certain bounds or that they in general tend to obey a certain probability law, he should use that information. Indeed, this and not Eq. (31) is the more usual state of prior knowledge. It is reflected in the chosen probability law $p(q_1,\ldots,q_M)$ of principle (21).

Returning to Jaynes' case, the use of probabilities (31) in Eq. (26) produces

$$-\sum_{n=1}^{M} \hat{p}_n \ln \hat{p}_n = \text{maximum} , \tag{32}$$

or maximum entropy. Hence, maximum entropy is optimum in a probabilistic sense, but only under very specific conditions. Prior knowledge must be very high. Hence, maximum entropy is not a panacea.

Use of Very Weak Prior Knowledge

We return now to the MacQueen-Marschak definition of maximum prior ignorance. This admits that all relative occurrences $\{q_n\}$ prior to one's seeing the data are equally probable. The probability law (13) defines this situation. Using Eq. (13) in the general principle (21)-(22) produces [3]

$$P(m_1,\ldots,m_M) = \text{constant} , \tag{33}$$

which means that all sets of occurrences $\{m_n\}$ or relative occurrences $\{p_n\}$ are equally likely. There is no estimation principle at all!

This result makes sense. Since all relative occurrences $\{q_n\}$ are equally likely a priori, there is no preferred set of absolute occurrences $\{m_n\}$ in the outcomes. Since the $\{q_n\}$ represent a prior estimate of the optical object, if all $\{q_n\}$ are equally likely to occur there is no a priori preferred object. A gray one is as likely as the Mona Lisa, prior to one's seeing the image.

Use of Negative Prior Knowledge

A form of prior knowledge that is slightly less vague than the preceding is where every object or language $\{q_n\}$ is equally likely, except for a very narrow class which it is known may not be present. For example, it is known only that the object is not the Mona Lisa or very similar pictures; or the language is not Russian or similar (say, Slavic) languages. This is obviously very weak information, and yet it does lead to an estimation principle.

The probability law describing this situation is Eq. (14), where $\{R_n\}$ is the picture or language known not to be present. Note that the delta function product consists of a narrow band of $\{q_n\}$ centered on $\{R_n\}$, which eliminates these events from the preceding uniform probability law. It puts a "notch" in each of the rectangle functions. Note that the notch has the correct area (number of missing events) integrated over $q_1,\ldots,q_M$ in the vicinity of $q_n = R_n$ to integrate $p(q_1,\ldots,q_M)$ out to 0 for just these events. Note also that $p(q_1,\ldots,q_M)$ is used only as a function to be integrated, in the principle (21), so it does not matter that its value is $-\infty$ at $q_n = R_n$ rather than the correct value 0.

When this "negative prior" (14) is substituted into principle (21), there results

$$P = \text{constant} - \varepsilon(1-\varepsilon)^{-1} \frac{N!}{m_1!\ldots m_M!} R_1^{m_1}\ldots R_M^{m_M}$$

$$= \text{maximum} . \tag{34}$$

This expression greatly simplifies. Because of the minus sign, P is a maximum when the second term is a minimum. Then setting ln P = minimum, as in the derivation of result (26), we get

$$\sum_{n=1}^{M} \hat{p}_n \ln (R_n/\hat{p}_n) = \text{minimum} . \tag{35}$$

This is minimum Kullbach-Leibler information, for general $\{R_n\}$, or minimum entropy in Jaynes' case of all $\{R_n\}$ equal! This is about as far from a principle of maximum entropy as one can go—showing once again that the form of an optimum estimation principle depends strongly upon the kind of prior knowledge one has at hand.

A Coin-Flip Problem

To see what kind of solution $\{\hat{p}_n\}$ to expect from principle (35), we consider a binary case $M = 2$. Then Eq. (35) becomes

$$-\hat{p}_1 \ln \hat{p}_1 - (1-\hat{p}_1)\ln(1-\hat{p}_1) + \hat{p}_1 \ln R_1 + (1-\hat{p}_1)\ln(1-R_1)$$

$$= \text{minimum}\,, \tag{36}$$

where we have used normalization $\hat{p}_1 + \hat{p}_2 = 1$ and $R_1 + R_2 = 1$. It is easy to solve Eq. (36) on the computer. Assigning some value to R_1 (prior knowledge), the form (36) was evaluated at a fine subdivision of $\{\hat{p}_1\}$ on the interval (0,1) and the minimum was selected. Results are as follows:

$$\begin{aligned} &\text{If } R_1 < 0.5,\ \hat{p}_1 = 1.0 \\ &\text{If } R_1 > 0.5,\ \hat{p}_1 = 0.0 \\ &\text{If } R_1 = 0.5,\ \hat{p}_1 = \text{either } 1.0 \text{ or } 0.0 \text{ (two equal minima)} \end{aligned} \tag{37}$$

To understand this surprising result, consider the following version of Bayes' primordial problem. A bag contains coins with biases p_1 between 0 and 1. Every bias is represented by the same number of coins, except for bias values R_1 lying between 0.2 and 0.3 (say) which are missing. A coin is randomly chosen from the bag and flipped $N = 10$ times, with m_1 heads occurring. What is the most likely value of m_1, or correspondingly of $p_1 = m_1/N$? According to results (37), it is value $m_1 = 10$, or $p_1 = 1$. All heads!

This actually makes sense. The key is to understand that for each possible bias p_1 selected, the number m_1 of occurrences can vary from 0 to N since

$$P(m_1\,|\,p_1) = \frac{N!}{m_1!(N-m_1)!}\, p_1^{m_1}(1-p_1)^{N-m_1}\,. \tag{38}$$

For example, even if the probability p_1 of a head is as low as 0.1, there is a finite probability that all heads, $m_1 = N$, will occur (see the curve labeled 0.1 in Fig. 1). Then $P(m_1)$ over all possible coins from the bag obeys the total probability law

$$P(m_1) = \frac{N!}{m_1!(N-m_1)!}\int_0^1 dp_1 p_1^{m_1}(1-p_1)^{N-m_1}\,\text{prob}(p_1)\,. \tag{39}$$

(This is the same as Eq. (21) for the case at hand.) Here, $\text{prob}(p_1)$ is the probability law for the various biases. If all biases are known to be present in equal amounts in the bag, $\text{prob}(p_1)$ is 1 over the integration limits (0,1). Then the total P is the superposition of all the curves in Fig. 1, where for clarity only discrete values of p_1 are used. This superposition is a constant,

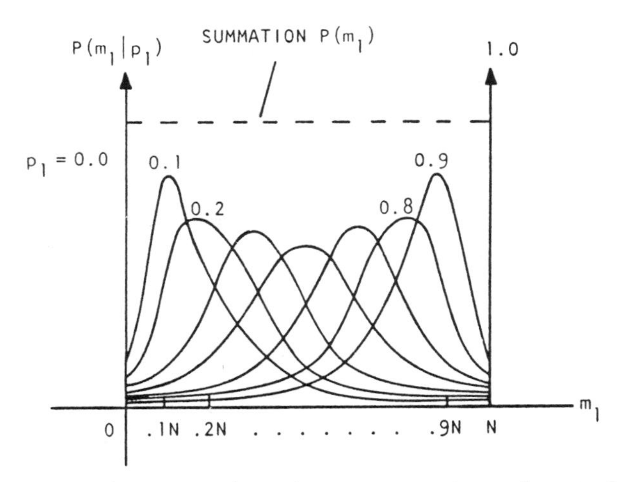

Figure 1. Binomial laws $P(m_1 | p_1)$ for various values of p_1 (solid curves), along with their summation (dashed).

which evaluation of the integral in Eq. (39) bears out, specifically because all the tails of the various curves of Fig. 1 add up to the same amount at each point m_1.

This allows us to understand, now, what happens if a curve $P(m_1 | R_1)$ is removed from Fig. 1 (the bias $p_1 = R_1$ is known not to occur). For example, let $R_1 = 0.2$. If that curve were removed from the figure, how would the previously constant $P(m_1)$ be affected? Since the curve $P(m_1 | 0.2)$ contributed most heavily toward the total $P(m_1)$ at value $m_1 = 0.2N$ (see the curve), removal of that curve would depress P most strongly at that m_1. Hence, value 0.2N would be the least probable value of m_1 to expect now after N coin flips. But more important, curve $P(m_1 | 0.2)$ also has its smallest contribution to P at value $m_1 = N$ (where its tail is the lowest). Therefore, P now remains highest at value $m_1 = N$. This corresponds to a relative occurrence for heads of $p_1 = N/N = 1$. Hence, a situation $\hat{p}_1 = 1$ of all heads is maximally probable when the bias $R_1 = 0.2$ is excluded from the realm of possibility.

More generally, the same reasoning shows that any value of $R_1 < 0.5$ would cause $P(m_1)$ to be maximum for $m_1 = N$, so that $\hat{p}_1 = 1$. This accounts for the first item in results (37).

Similarly, any value of $R_1 > 0.5$ would least depress the P value at $m_1 = 0$, so now $\hat{p}_1 = 0$. This accounts for the second item in results (37).

Finally, in the special case of $R_1 = 0.5$ (the missing coins are "fair" coins), the resulting $P(m_1)$ is maximum at both $m_1 = 0$ and $m_1 = 1$, so $\hat{p}_1 = 0$ or 1. Either all heads or all tails are most likely to occur, and with equal probability.

Monte Carlo Test

These effects were tested out by computer simulation. One million sequences of $N = 10$ flip events were generated. Each event consisted of a randomly chosen bias p_1, followed by outcome "head" or "tail."

Specifically, biases were chosen from a uniformly-random number generator on interval (0,1), excluding interval 0.2 to 0.3 (a case $R_1 \simeq 0.25$). After each bias p_1 was chosen, the coin was "flipped" by selecting another random number between 0 and 1. If the number was less than p_1, the verdict was "head"; if greater than p_1, the verdict was "tail." For each sequence of such $N = 10$ coin selection flips, the resulting m_1 (number of heads) was placed in a histogram $P(m_1/10)$, $m_1 = 0,1,\ldots,$or 10. After the number of occurrences at each m_1 was divided by 10^6, to effect normalization, the $P(m_1/10)$ of Fig. 2 resulted. The curve is indeed maximum at $q_1 = m_1/10 = 1$, verifying the theory.

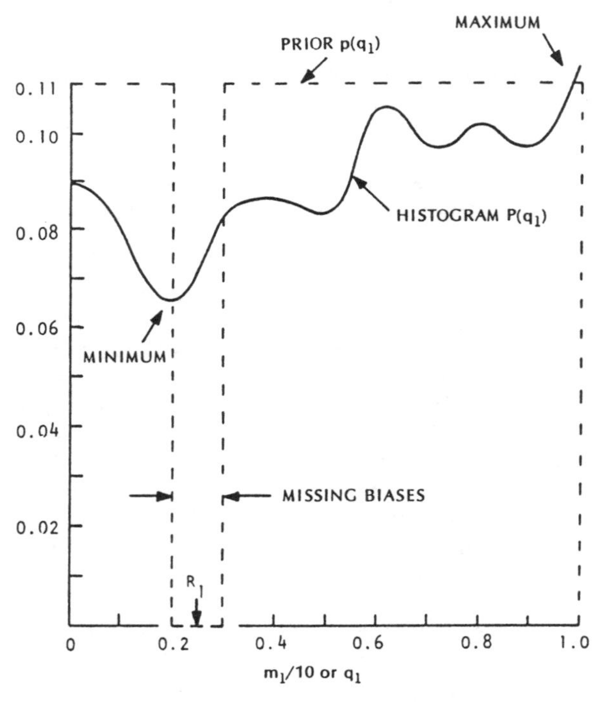

Figure 2. Histogram of the relative occurrence of rates $q_1 = m_1/10$ after one million experiments. Each experiment consisted of a randomly chosen bias for a coin and then a flip outcome for the coin, repeated 10 times. Here m_1 is the number of resulting head outcomes, and q_1 is the relative number of times head appears. Most of the erratic behavior of the histogram is due to the finite number of data that formed it.

It is interesting that, in this binary case, the answer p_1 does not depend upon the actual value of R_1 but merely upon which interval (0,0.5) or (0.5,1) it lies within. For the page-of-print problem, this is like stating that the estimated occurrences $\{\hat{p}_n\}$ should depend only upon knowing that some language is not present, but not upon knowing which language in particular. This is a strange effect, if true. It is precisely the opposite tendency of a maximum Kullback-Leibler solution, Eq. (30), where in the absence of data the estimated $\{\hat{p}_n\}$ exactly equal the letter occurrences $\{Q_n\}$ of the language known to be present.

But it should be remembered that this analysis has been carried through only for a dataless case. In the presence of data constraints, the particular values of the missing biases $\{R_n\}$ may yet play a more direct role in defining the $\{\hat{p}_n\}$. This is as yet unknown.

Use of Power-Law Prior Knowledge

The power-law case, Eq. (15), was intermediate between very weak and very strong prior knowledge. It is unusual (unique?) in permitting analytic integration of principle (21). The integral is Dirichlet's, with result

$$P(m_1,\ldots,m_M) = B\,\frac{(m_1+r_1)!\ldots(m_M+r_M)!}{m_1!\ldots m_M!}\,, \qquad B \text{ constant}\,. \tag{40}$$

Again setting $\ln P$ = maximum and using Stirling's approximation, Eq. (24), the estimation principle becomes

$$\sum_{n=1}^{M}(m_n+r_n)\ln(m_n+r_n) - \sum_{n=1}^{M} m_n \ln m_n = \text{maximum}\,,$$
$$m_n \equiv N\hat{p}_n\,. \tag{41}$$

The second sum by itself would represent maximum entropy. Hence the net principle is maximum entropy, reduced by the entropy of $\{(N\hat{p}_n + r_n)\}$.

Note that if all $r_n \to 0$, corresponding to uniform randomness (15) for the $\{q_n\}$, the MacQueen-Marschak result of no principle at all follows once again.

In the case of given data $\{F_m\}$, principle (41) is amended by Lagrange multipliers as in Eq. (28). Then the solution $\{\hat{p}_n\}$ may be found by setting $\partial/\partial p_n = 0$. This is

$$\hat{p}_n = \frac{r_n}{\exp\left(-\lambda + \sum_m \mu_m f_{mn}\right) - N}\,. \tag{42}$$

This may be called a 'Bose-type' probability estimate, for reasons given below.

In the case of no given data, we set all $\mu_m = 0$ in Eq. (42) and use normalization of the $\{\hat{p}_n\}$ to solve for λ. The result is a prior estimate

$$\hat{p}_n = \frac{r_n}{\sum_n r_n} . \tag{43}$$

Comparing this result with the dataless solution (30), we see that the $\{r_n\}$ act like bias values for the solution. This is somewhat unexpected since the $\{r_n\}$ are merely exponents in probability law (15) and not probabilities per se.

It is interesting to consider a 'gas law' problem. This is where data $\{F_m\}$ are in the form of moments

$$\sum_n \hat{p}(E_n)E_n^m = F_m \tag{44}$$

on energy levels $\{E_n\}$ for the gas particles. For the case M = 1 of a known average energy F_1, solution (42) becomes

$$\hat{p}(E_n) = \frac{r_n}{\exp(-\lambda - \mu_1 E_n) - N} . \tag{45}$$

(Parameters λ and μ_1 must satisfy normalization and F_1.) This is in the form of a Bose statistic on energy

$$\hat{p}(E_n) = \frac{g_n}{z^{-1}\exp(\beta E_n) - 1} , \tag{46}$$

if we identify

$$g_n = r_n/N , \quad z = Ne^{\lambda}, \text{ and } \beta = -\mu_1 . \tag{47}$$

N is now the total number of particles in the Bose gas.

We found before [see Eq. (29)] that maximum entropy (32) leads to derivation of the Boltzmann gas law, under the same data conditions just considered. Hence, we may conclude that the general maximum probability principle (21) will imply, in turn, all of the famous gas laws, depending on the form of prior knowledge injected into it.

It is interesting to consider limiting cases of principle (41) according to whether $r_n \ll m_n$ or $r_n \gg m_n$. In the first case, note that

$$\begin{aligned}(m_n + r_n)\ln(m_n + r_n) &= (m_n + r_n)\ln m_n(1 + r_n/m_n) \\ &\simeq (m_n + r_n)(\ln m_n + r_n/m_n) \\ &\to (m_n + r_n)\ln m_n \ ,\end{aligned} \tag{48}$$

so principle (41) becomes

$$\sum_n r_n \ln \hat{p}_n = \text{maximum} \tag{49}$$

after an irrelevant constant is discarded. This principle is Burg's [4], if all $r_n = 1$, a triangular prior law (15). Otherwise it might be called a "generalized Burg" estimator. Note that the higher an r_n is in Eq. (49), the more weight is given to maximizing its $\ln \hat{p}_n$ contribution to the sum. A high r_n in law (15) implies more extreme "all-or-nothing" behavior for the occurrence p_n. Hence, such a p_n will tend to be unpredictable. The maximum probability principle knows this through input parameter r_n, and forces it toward an average level $1/M$ through the action of principle (49). In this manner, it tries to offset the wild expected behavior of such a p_n.

Next, consider the opposite limit, $r_n \gg m_n$. Then

$$\begin{aligned}(m_n + r_n)\ln(m_n + r_n) &= (m_n + r_n)\ln r_n(1 + m_n/r_n) \\ &\simeq (m_n + r_n)\ln r_n \ .\end{aligned} \tag{50}$$

Using this in principle (41) and ignoring irrelevant constants produces a principle

$$\sum_n r_n \ln(r_n/\hat{p}_n) = \text{maximum} \ . \tag{51}$$

We see from comparison with Eq. (26) that this is a Kullbach-Leibler estimator, with each r_n mimicking a prior occurrence rate Q_n. Also comparing this with estimator (49), we observe that the dependence upon unknowns $\hat{p}_n$ is completely different for the two cases. Prior knowledge, this time of the relative size of each m_n and r_n, is again critical.

Use of Empirical Prior Knowledge*

The most direct form of knowledge about the priors $\{q_n\}$ is data $\{M_n\}$ consisting of the observed occurrences in a separate sample (distinct from the data). Hence, in the page-of-print problem, a different page than the unknown is actually observed for its letter occurrences $\{M_n\}$. Or, an object from the same class as the unknown is observed for its intensity profile $\{M_n\}$. How may such information be incorporated as prior knowledge? The answer is, in the form of a prior probability law $p(q_1,\ldots,q_M|M_1,\ldots,M_M)$, as shown next.

Bayes' rule states that the required law may be formed as

$$p(q_1,\ldots,q_M|M_1,\ldots,M_M) = P(M_1,\ldots,M_M|q_1,\ldots,q_M)\,\frac{p_0(q_1,\ldots,q_M)}{P(M_1,\ldots,M_M)}\,. \qquad (52)$$

Consider the right-hand quantities. By Eq. (23),

$$P(M_1,\ldots,M_M\,|\,q_1,\ldots,q_M) = \frac{L!}{M_1!\ldots M_M!}\,q_1^{M_1}\ldots q_M^{M_M}\,,$$

$$L \equiv \sum_n M_n\,. \qquad (53)$$

Next, consider probability $p_0(q_1,\ldots,q_M)$. This probability represents an assumption on the occurrences of the $\{q_n\}$, prior to observation of the empirical $\{M_n\}$ (it is a "pre-prior," in the spirit of Ref. [3]). Prior to these observations, it is reasonable to assign all possible sequences $(q_1,\ldots,q_M)$ the same probability of occurrence. This was the MacQueen-Marschak definition of maximum prior ignorance, Eq. (13),

$$p_0(q_1,\ldots,q_m) = \prod_{n=1}^{M} \mathrm{Rect}(q_n - 1/2)\,. \qquad (54)$$

It previously led to no estimation principle at all, Eq. (33). But that was in the absence of empirical data $\{M_n\}$. Results will be different now.

The denominator $P(M_1,\ldots,M_M)$ in Eq. (52) is the integral of the numerator. Using Eqs. (53) and (54), the integral is Dirichlet's, with result

$$P(M_1,\ldots,M_M) = \frac{L!(M-1)!}{(L+M-1)!}\,. \qquad (55)$$

---

*This general approach is due to MacQueen and Marschak [3].

This groundwork laid, Eq. (52) becomes

$$p(q_1,\ldots,q_M \mid M_1,\ldots,M_M) \equiv p(q_1,\ldots,q_m)$$
$$= \frac{(L+M-1)!}{(M-1)!} \frac{1}{M_1!\ldots M_M!} q_1^{M_1}\ldots q_M^{M_M} . \tag{56}$$

In the limit $M_m \to 0$, all m, of no empirical observation, this becomes unity. We would once again have the MacQueen-Marschak situation (13) of maximum ignorance for the prior $p(q_1,\ldots,q_M)$, and consequently no estimation principle, Eq. (33). However, the observation of any empirical data $\{M_n\}$ at all will give a definite shape to the prior law (52), that is, a definite bias. Then its use in the general principle (21) will give a definite probability law $p(m_1,\ldots,m_M)$ and estimator $\{\hat{p}_n\}$. In summary, the indefinite result (33) will follow only when there is a complete lack of empirical observation $\{M_n\}$. The existence of finite observations $\{M_n\}$ makes the difference, as shown next.

Having derived the prior probability law (56) for these conditions, we now want to substitute it into the general principle (21) to find the corresponding estimator. This becomes a Dirichlet integral, with result

$$p(m_1,\ldots,m_M) = C \frac{(M_1+m_1)!\ldots(M_M+m_M)!}{m_1!\ldots m_M!},$$
$$C \text{ constant}, \quad m_n \equiv N\hat{p}_n . \tag{57}$$

Again setting $\ln P$ = maximum and using Stirling's approximation (24), we have an estimation principle

$$\sum_n (M_n + N\hat{p}_n) \ln (M_n + N\hat{p}_n) - N \sum_n \hat{p}_n \ln \hat{p}_n = \text{maximum} . \tag{58}$$

This is of the same mathematical form as Eq. (41) for power-law prior knowledge. Hence, its solution follows the same form as Eq. (42), with each r_n replaced by M_n. Also, the Bose result (45) would again follow when the principle (58) is applied to gas particles whose mean energy is constrained.

We now notice two interesting limits of principle (58). First, suppose a meager amount of empirical knowledge to be present, because

$$M_n \ll N\hat{p}_n . \tag{59}$$

This corresponds to the limiting case (48) of power-law knowledge. Hence, the result

$$\sum M_n \ln \hat{p}_n = \text{maximum} \tag{60}$$

is analogous with Eq. (49). This is an empirically weighted Burg estimator. If all occurrences $\{M_n\}$ observed are equal, it is exactly Burg's estimator. Notice, however, the general tendency to weight more heavily in the sum those $\ln \hat{p}_n$ values whose occurrences are high in the empirical observations. In effect, the events whose occurrences are more strongly observed empirically will later be smoothed more by the estimator. Or, in reverse, events whose occurrences are not observed at all cannot be controlled in a future estimate that is based upon the presence of observation. This is a consistent result.

Next, consider the case where a large amount of empirical knowledge is present, that is,

$$M_n \gg N \hat{p}_n . \tag{61}$$

This corresponds to power-law case (50), with resulting estimator (51). The result here is

$$\sum_n M_n \ln (M_n/\hat{p}_n) = \text{maximum} . \tag{62}$$

Comparing this with estimator (26), we observe it to be an empirical version of the Kullbach-Leibler estimator. That is, the presumed form $\{Q_n\}$ for the unknown $\{p_n\}$ is merely replaced by the empirically observed values $\{M_n\}$. This is a satisfying result, quite in the spirit of the overall empirical approach of this section. One cannot help but think that the law of large numbers is coming into play. Because the number of observations M_n is now so high, it would indeed make sense to regard each ratio M_n/L as a very good approximation to prior Q_n which, since it is known with high precision, ought to obey a Kullbach-Leibler estimator [see Eqs. (23) through (26)].

Gas Law Cases

Let us apply the principle (58) to the case of gas particles whose mean energy is observed. According to solution (42), the estimate would obey

$$\hat{p}_n \equiv \hat{p}(E_n) = \frac{M_n}{\exp(-\lambda - \mu_1 E_n) - N} . \tag{63}$$

This result leads to the following paradoxical situation:

Suppose that a Boltzmann gas, obeying

$$p(E_n) = F_1^{-1} \exp(-E_n/F_1) , \tag{64}$$

is observed for its occurrences $\{M_n\}$ of energy levels $\{E_n\}$. F_1 is the known, mean energy, Eq. (44). Then a second sample of the Boltzmann gas has its mean energy measured. What is the maximum probable estimate $p(E_n)$ of the distribution of energy levels in the second sample? The answer is Eq. (63), a Bose gas statistic, and not the correct verdict, Eq. (64). How can this be?

The key lies in understanding that but a finite number L of observations are made in the first sample. Even though these were samples from a Boltzmann law, they do not uniquely imply a Boltzmann law, owing to their finiteness. In fact, they could have arisen from a whole range of probability laws, as given by Eq. (56). Then, when Eq. (56) is combined with the one piece of data on the mean energy F_1, the maximum probable estimate of $p(E_n)$ happens to be a Bose law. Hence, here the estimated law and the true (Boltzmann) law do not coincide.

If, on the other hand, L is very large, in fact so large that condition (61) is obeyed, then the estimation principle becomes Eq. (62). Its solution is Eq. (29) with $Q_m = M_n$, or

$$\hat{p}_n = M_n \exp(-1 + \lambda + \mu_1 E_n) \tag{65}$$

for a single datum on average energy. This is of the Boltzmann form, provided the $\{M_n\}$ are proportional to Boltzmann occurrences (64). They would be, since L is presumed very large and the gas truly Boltzmann. Hence, when the prior data are strongly Boltzmann, they overwhelm the Bose tendency of the estimate, and imply a probability law that is correct.

Let us place this less-than-ideal behavior in the proper perspective. We shall compare it with the use of maximum entropy, or its generalization, maximum Kullbach-Leibler, under similar conditions. This had the solution (29), which might be called "biased Boltzmann" (factors Q_n are biases). In particular, if only a constraint on average energy is known, it becomes

$$\hat{p}_n = Q_n \exp(-1 + \lambda + \mu_1 E_n) \,. \tag{66}$$

What if, now, the unknown gas is actually of the Bose type? If the user takes Jaynes' point of view of maximum ignorance, namely that $Q_n = 1/M$, then the estimate (66) incorrectly predicts a Boltzmann distribution for the gas!

What we have shown, then, is that the maximum probability approach can err, shown previously, and Jaynes' approach can err. However, Jaynes' approach will always give the same form of answer—Eq. (29) in general, or Eq. (66) for gas-law cases—whereas the maximum probability approach has the flexibility to give answers that depend upon the state of prior knowledge of the observer. As a consequence, it will be "right" more often than will Jaynes' approach (this is after all what "maximum probability" means).

More Realistic Photographic Object Model

We have previously assumed the object to consist of ideal pixels having no spatial extent. This model made the object-estimation mathematically identical with the page-of-print problem. Now we shall more realistically allow the pixels to be resolution cells of finite volume. In this case there are in general z locations within each cell for receiving photographic grains (see Fig. 3). Each location either contains a grain or doesn't, so the maximum brightness possible at a cell is z (the grains are "white," corresponding to a positive print). It is possible that the minimum brightness at a cell is also known; let this be a. Of course value a = 0 may always be used if a minimum is not known. This is minimal lower-bound knowledge. These assumptions comprise a "model" for the object, as in Fig. 3.

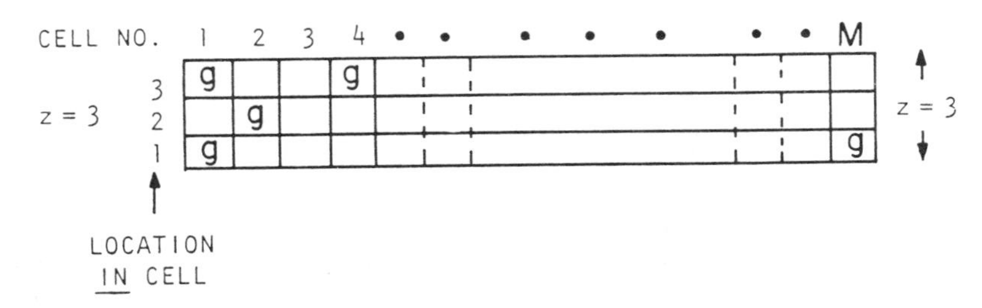

Figure 3. Photographic object model consisting of z grain locations per cell. One line of cells is shown, with z = 3. Grains are indicated by g. The particular object $\{m_n\} = (2,1,0,1,\ldots,1)$ is shown.

This more realistic model will give rise to a new degeneracy factor $W_{photo}(m_1,\ldots,m_M)$, as follows. The degeneracy W arises because the cell occurrences $(m_1,\ldots,m_M)$ for the grains can occur in many different time orders (as previously considered alone), and now also at many different locations within the cells. To satisfy lower bound a, let a fixed locations within each cell have grains. This could occur as a prefogging of the film. In any cell, this leaves $(m_n - a)$ grains to be distributed among $(z - a)$ locations.

W is then the number of ways $(m_1 - a)$ indistinguishable grains can be distributed over $(z - a)$ locations in cell 1, and independently $(m_1 - 1)$ may be distributed over $(z - a)$ locations in cell 2, and . . . , and $(m_M - a)$ may be distributed over $(z - a)$ locations in cell M—times the number of ways N! that any one sequence of the N grains may be reordered in sequence. This suggests that

$$W_{photo} = W_1 W_2 \ldots W_M N! , \tag{67}$$

where each W_n is the degeneracy of its cell. W_n is found as follows:

The grain occurrences within any one cell define a Bernoulli sequence of events "yes" or "no." Then its degeneracy W_n obeys a binomial law, and Eq. (67) becomes

$$W_{photo} = C \begin{bmatrix} z - a \\ m_1 - a \end{bmatrix} \cdots \begin{bmatrix} z - a \\ m_M - a \end{bmatrix} N! , \qquad (68)$$

where notation

$$\begin{bmatrix} k \\ j \end{bmatrix} \equiv \frac{k!}{j!(k-j)!}$$

defines the binomial coefficient. Also, C is a normalization constant that is independent of the $\{m_n\}$.

Combining Eqs. (67) and (68) shows that

$$W_{photo} = \frac{CN!}{(m_1 - a)! \cdots (m_M - a)!} \frac{[(z-a)!]^M}{(z - m_1)! \cdots (z - m_M)!} . \qquad (69a)$$

For a = 0, W is then the product of the old degeneracy, Eq. (22), times a new factor having to do with the finite size z of each cell. This factor is much greater than 1, so W_{photo} is much greater than W of Eq. (22), which is reasonable considering the added freedom due to finite cell size.

Estimation Laws for Photographic Objects

The general estimation principle is obtained by combining Eqs. (21) and (69a):

$$P(m_1, \ldots, m_M) = W_{photo} \int \cdots \int q_1^{m_1 - a} \cdots q_M^{m_M - a}$$

$$\times \, p(q_1, \ldots, q_M) \, dq_1 \ldots dq_M , \qquad (69b)$$

where $\Sigma q_n = 1$ and powers $(m_n - a)$ now replace m_n since only $(m_n - a)$ grains are being distributed. This principle will take on different forms, depending on the form of prior knowledge $p(q_1, \ldots, q_m)$ at hand, exactly as was carried through before for the point-pixel object case. For the sake of brevity, we shall be terse here, emphasizing the results.

Case of Very Strong Prior Knowledge

Again considering the case of prior probability law (12), Eqs. (69a) and (69b) yield

$$P = \frac{CN!}{(m_1 - a)! \cdots (m_M - a)!} \frac{[(z-a)!]^M}{(z - m_1) \cdots (z - m_M)} Q_1^{m_1 - a} \cdots Q_M^{m_M - a} . \qquad (70)$$

Setting $\ln P$ = maximum, and ignoring irrelevant constants, we obtain

$$\sum_n (m_n - a) \ln[Q_n/(m_n - a)] - \sum_n (z - m_n) \ln(z - m_n) = \text{maximum}, \tag{71}$$

as the estimation law for the $\{\hat{p}_n\}$; here, $m_n \equiv N\hat{p}_n$. The first sum is close to the Kullback-Leibler form (26) for point-pixels, and equals it if $a = 0$. The second sum is an extra term due to finite size z of the cells. It has the form of an entropy of the $(z - m_n)$ "missing events" in the cells.

In Jaynes' situation of $Q_n = 1/M$, principle (71) becomes

$$-\sum_n (m_n - a) \ln(m_n - a) - \sum_n (z - m_n) \ln(z - m_n) = \text{maximum} \tag{72}$$

since Σm_n is constrained to be N. This represents the entropy of the number $(m_n - a)$ of events above fog level $\underline{a}$, plus the entropy of the missing events. It has been derived before [5].

Case of Very Weak Prior Knowledge

In the case of the MacQueen-Marschak definition (13) of maximum prior ignorance, the principle (69a,b) becomes

$$P = \frac{K}{(z - m_1)! \ldots (z - m_M)!} = \text{maximum}, \quad K \text{ constant.} \tag{73}$$

The integral in Eq. (69b) now cancels the first factor in Eq. (69a). Setting $\ln P = \max$ and ignoring irrelevant constants, we obtain

$$-\sum_n (z - m_n) \ln(z - m_n) = \text{maximum}, \quad m_n \equiv N\hat{p}_n, \tag{74}$$

as the estimation principle. This is an interesting result. For the point-object, this weak state of prior knowledge led to no principle at all! In fact, the model at hand of finite resolution cells has made the difference: it represents strong enough prior knowledge to imply an estimation principle even under condition (13) of maximum prior ignorance.

It is interesting to compare Eq. (74) with Jaynes' maximum entropy principle (1). Whereas Jaynes' principle uses the entropy of the $\{\hat{p}_n\}$, Eq. (74) uses the entropy of $\{(z - N\hat{p}_n)\}$, the complementary occurrences to Jaynes'. These $\{(z - N\hat{p}_n)\}$ are the entropy of the unoccupied sites in the pixel cells.

In the "gas law" case of a single constraint on average energy, principle (74) has a solution

$$\hat{p}_n \equiv \hat{p}(E_n) = \frac{z}{N} - Ae^{-BE_n}, \qquad \text{A,B constants.} \tag{75}$$

This form implies that E_n is bounded from above. If it were not, $\hat{p}(E_n)$ could not be normalized. The upper bound might correspond to all z cell sites being filled.

Case of Empirical Prior Knowledge

Once again, assume that the observer has empirically observed the grain counts $\{M_n\}$ across an object that is of the same class as the unknown. These are finite data, where

$$\sum_n M_n = L \tag{76}$$

is any size relative to the total number of grain events N in the unknown object. How will knowledge of these $\{M_n\}$ affect the general estimation principle (69a,b)?

As in the derivation of Eq. (56), knowledge of the $\{M_n\}$ implies knowledge of a prior probability $p(q_1, \ldots, q_M)$, now of the form

$$p(q_1, \ldots, q_M) = Cq_1^{M_1 - a} \ldots q_M^{M_M - a}, \qquad \text{C constant.} \tag{77}$$

Substitution into principle (69a,b) gives

$$P = \frac{D(M_1 + m_1 - 2a)! \ldots (M_M + m_m - 2a)!}{(m_1 - a)! \ldots (m_M - a)!(z - m_1)! \ldots (z - m_M)!} = \text{maximum}, \tag{78}$$

D constant. [Compare with principle (57).]

Taking the natural logarithm yields the estimation principle

$$\sum_n (M_n + m_n - 2a) \ln(M_n + m_n - 2a) - \sum_n (m_n - a) \ln(m_n - a)$$

$$- \sum_n (z - m_n) \ln(z - m_n) = \text{maximum}, \qquad m_n \equiv N\hat{p}_n. \tag{79}$$

The interested reader can take the limits of this expression for $M_n \gg N\hat{p}_n$, that is, strong empirical data, and for $M_n \ll N\hat{p}_n$, weak empirical data. Results analogous to Eqs. (59) through (62) will result.

Optical Objects

An optical object consists of radiating photons from cell positions $\{x_n\}$. Here an unknown time sequence of photon positions $\{x_n\}$ occurs, as in

$$x_8x_2x_5x_9x_9x_1x_8x_3x_5x_9x_9x_6x_8x_5x_2x_3x_3x_7x_4x_3 \ . \tag{80}$$

The problem is again to determine the occurrence rates $\{p_n\}$ for the positions, without knowing the sequence. Instead, data in the form of sampled values of the image are given. These data are blind to the particular order, as in example (80), in which the positions occurred. However, because of the quantum nature of photons, some of the time sequences of type (80) are physically indistinguishable. Therefore, they cannot be counted toward the degeneracy factor for the process.

Instead, there is additional degeneracy due to other causes defining the physical situation. These factors cause each resolution cell to now contain z states or degrees of freedom in which its m_n photons may reside. (Note the similarity to the photographic grain model preceding.) Factors influencing z are the exposure time t and resolution cell area a, among others. Specifically, as shown by Kikuchi and Soffer [6],

$$z = \left[\frac{t}{\tau}\right]\left[\frac{A}{\sigma}\right], \tag{81}$$

where t is exposure time, τ is the temporal coherence time $1/\Delta\nu$ for the radiation of bandwidth $\Delta\nu$, A is the area of each detector aperture in the image plane, and σ is the coherence area subtended by the resolution cell at the image plane,

$$\sigma \simeq \frac{R^2\lambda^2}{a} \ . \tag{82}$$

R is the distance between object and image planes, and λ is the wavelength of the light.

Photons are Bose particles, so more than one photon can exist in each degree of freedom. For example, see Fig. 4, which shows an optical object $\{m_n\} = (3,3,1,\ldots,1)$. This object can occur in many different ways, according to how the three photons are distributed in cell 1, the three in cell 2, etc.

The degeneracy due to finite z at each cell is the Bose-Einstein expression [6]

$$W_n = \frac{(m_n + z - 1)!}{m_n!(z - 1)!} \ . \tag{83}$$

This is the number of ways the m_n photons can be distributed over z states, where the photons are indistinguishable and any number may occupy a given state.

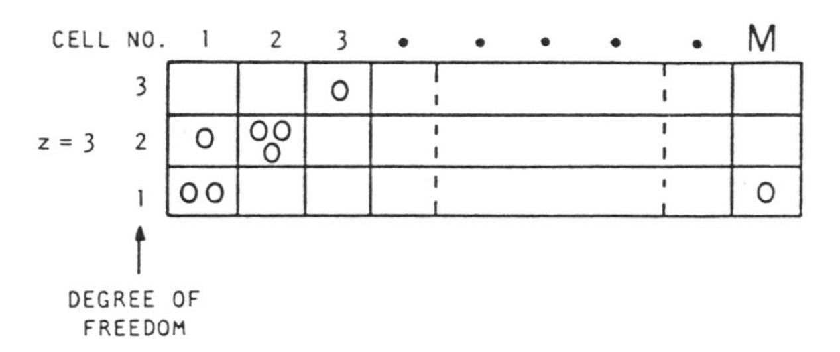

Figure 4. Optical object consisting of z degrees of freedom per resolution cell. One line of cells is shown, with z = 3. Photons are indicated by o. The particular object $m_n = (3,3,1,\ldots,1)$ is shown.

Since the cells are independent of one another, the new W, call it W_{opt}, obeys

$$W_{opt} = \prod_{n=1}^{M} \frac{(m_n + z - 1)!}{m_n!(z - 1)!} . \tag{84}$$

This degeneracy factor has two interesting limits. If z = 1, this means that there is but one degree of freedom per resolution cell. But since photons also are indistinguishable, there is now but one physically distinguishable arrangement of photons that form an object $\{m_n\}$. Physically, this situation occurs if, for example, the detection time t is a fraction of the coherence time, so that different time orders of arrival [as in example (80)] cannot be distinguished. The object is then completely nondegenerate, so its probability of occurrence now depends only upon the form of prior $p(q_1,\ldots,q_M)$.

If, on the other hand, $z \gg 1$, factor $(m_n + z - 1)!$ in Eq. (84) about cancels factor $(z-1)!$, so

$$W_{opt} \propto \prod_{n=1}^{M} (m_n!)^{-1} . \tag{85}$$

Here, W_{opt} is now of the form (22), as if photons were discrete entities like letters or grains. This is called the "particle limit" of optical degeneracy. In the particle limit, it is reasonable to expect optical objects to behave like the point-pixel objects previously considered. This suspicion will be borne out at Eq. (90).

Estimation Laws for Optical Objects

The general estimation principle is obtained by combining Eqs. (21) and (85):

$$P(m_1,\ldots,m_M) = W_{opt} \int \cdots \int q_1^{m_1} \cdots q_M^{m_M}\, p(q_1,\ldots,q_M)\, dq_1 \cdots dq_M$$

$$= \text{maximum}\,, \qquad (86)$$

where $\Sigma q_n = 1$. This principle will take on different forms, depending on the form of prior knowledge $p(q_1,\ldots,q_M)$ at hand about the object $(q_1,\ldots,q_M)$. (As before, ln P will be taken, and Stirling's approximation will be used.)

Very Strong Prior Knowledge

This is the case of prior probability law (12). Then Eqs. (84) and (86) yield

$$\sum_n m_n \ln(Q_n/m_n) + \sum_n (m_n + z - 1) \ln(m_n + z - 1) = \text{maximum}\,,$$

$$m_n \equiv N\hat{p}_n\,, \qquad (87)$$

as the estimation principle for object $\{p_n\}$. This resembles the corresponding estimation law (71) for photographic objects. The first term resembles the Kullbach-Leibler norm.

Jaynes' Limiting Case

When all the $\{Q_n\}$ are assumed equal, principle (87) becomes

$$-\sum_n m_n \ln m_n + \sum_n (m_n + z - 1) \ln(m_n + z - 1) = \text{maximum}\,,$$

$$m_n \equiv N\hat{p}_n\,. \qquad (88)$$

(Again we have used the normalization of Σm_n to ignore that term.) Note that when $z = 1$, or the object is completely nondegenerate, there is no estimation principle! All objects are then equally probable. Contrary to intuition, there is no bias toward the equal $\{Q_n\}$.

It is also interesting to examine principle (88) under the extreme conditions $m_n \ll z$ or $m_n \gg z$. Kikuchi and Soffer [6] had previously found that in the first limit Jaynes' entropy form (1) resulted, whereas in the second limit Burg's entropy form (9) resulted. However, these authors neglected the role of the prior $p(q_1,\ldots,q_M)$ in their analysis.

Taking the limit $m_n \ll z$ in Eq. (88), we note that

$$(m_n + z - 1) \ln(m_n + z - 1) = (m_n + z - 1) \ln(z - 1) \left[\frac{1 + m_n}{z - 1}\right]$$

$$\simeq (m_n + z - 1) \ln(z - 1) , \tag{89}$$

so Eq. (88) becomes

$$-\sum_n m_n \ln m_n = \text{maximum} . \tag{90}$$

(We again ignored a term in Σm_n since it is a constant.) Hence, in this 'dim' object case the estimator is again Jaynes'. The proviso is still a situation of very strong prior knowledge, all $Q_n = 1/M$, however. This was Kikuchi and Soffer's result [6] as well.

We also could have predicted that result (90) would follow from the assumption of z very large (as here assumed), since we found at Eq. (85) that photons behave like discrete particles in this limit. And discrete events were previously found to obey maximum entropy under the same prior information as here assumed; see Eq. (32).

Taking the opposite limit $m_n \gg z$ in Eq. (88), we have

$$(m_n + z - 1) \ln(m_n + z - 1) = (m_n + z - 1) \ln m_n \left[1 + \frac{z - 1}{m_n}\right]$$

$$\simeq (m_n + z - 1) \ln m_n . \tag{91}$$

Used in Eq. (88), this becomes

$$\sum_n \ln m_n = \text{maximum} , \tag{92}$$

after an irrelevant constant multiplier is ignored. This is Burg's estimator (9).

These are important results, since many optical objects fall into the bright-or-dim-object category [6]. For example, planetary astronomical objects have a very small m_n/z ratio in the visible and infrared regions. And bright radio astronomical objects often have a very large m_n/z ratio.

These results are limited in scope to a class of objects, Eq. (12), consisting of one uniformly gray one, $Q_n = 1/M$. This would be the case only if it were known that the object of interest is nearly featureless.

But these observations perhaps explain the relative success of maximum entropy in estimating some astronomical images. It has long been known by workers in the field that maximum entropy does a fine job of estimating objects consisting of isolated impulses against a uniform (or nearly uniform) background. These objects are indeed nearly featureless. Now we see that the approach is in fact optimum (ML sense) in these cases, provided the impulses are not too numerous and the overall object is not too bright.

A typical maximum entropy estimate of an astronomical object of this type is shown in Fig. 5. Most of the object is slowly varying background, as required. The object estimate on the right shows much more structure than do the image data on the left. In particular, note the spiral arms near the center of the object estimate. These are probably real.

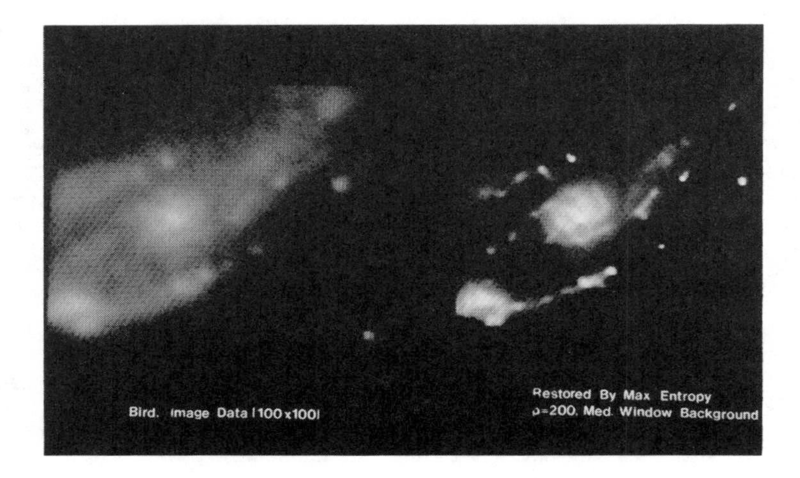

Figure 5. Use of the maximum entropy estimator upon an astronomical image. On the left is a portion of the "Bird" galaxy, an image blurred by atmospheric turbulence and taken by R. Lynds of Kitt Peak National Observatory. On the right is its estimated object, using a maximum entropy restoring program of the author. Note the emergence of resolved stars and galactic arms in the estimate.

Very Weak Prior Knowledge

With a prior probability law of the form (13), principle (86) becomes

$$\sum_n (m_n + z - 1) \ln(m_n + z - 1) = \text{maximum}, \quad m_n \equiv N\hat{p}_n . \tag{93}$$

We have used the fact that the integral in Eq. (86) becomes the Dirichlet integral, with value proportional to $m_1!m_2!\ldots m_M!$ Note the interesting case of $z = 1$. In this case of zero degeneracy [see discussion following Eq. (84)], the estimation principle (93) becomes one of minimum entropy, exactly the opposite of Jaynes' (1). The solution $\{\hat{p}_n\}$ would then tend toward a situation of no photons in most resolution cells, with nearly all packed into a few. This is a "bumpy" tendency, as opposed to the smoothness enforced by maximum entropy.

At this point it is necessary to discuss why the optical case should lead to a definite estimation principle (93), when under the same conditions the point-pixel case leads to no estimation principle. The reason is that the point-pixel model is the simplest possible model for an object consisting of points. As such, it represents a state of minimal prior knowledge. It lacks the complexity of the optical (or photographic) model, which after all has a parameter z representing prior knowledge about the object. The larger z is, the more complex the object is permitted to be: note from Eq. (84) that the object degeneracy W_{opt} grows with z. The opposite limit of $z = 1$, defining the nondegenerate object, defines the smallest degree of prior knowledge about the object. Hence, in summary, because they require for their description a higher degree of prior knowledge about the object, the photographic and optical models lead to estimation principles even when there is very weak prior knowledge $p(q_1,\ldots,q_m)$. Any model other than the point-pixel model ought to as well.

Empirical Prior Knowledge

Once again, assume that the observer has empirically observed an object $\{M_n\}$ which is of the same class as the unknown. He wants to use this information to bias the estimate. Accordingly, probability law (16) is used in estimator (86). The result is now a principle

$$\sum_n (m_n + z - 1)\ln(m_n + z - 1) + \sum_n (m_n + M_n)\ln(m_n + M_n) - \sum_n m_n \ln m_n = \text{maximum}, \qquad m_n \equiv N\hat{p}_n. \tag{94}$$

A state of minimal empirical knowledge would now be represented by the situation $M_n = 0$, $n = 1,\ldots,M$, no observed photons in the empirical object. Then Eq. (94) goes over into Eq. (93), as is consistent. If, further, $z = 1$, estimation principle (94) becomes minimum entropy once again.

A weighted-Burg type estimator (60) previously resulted, in the case of a point-pixel object and where empirical knowledge $\{M_n\}$ was meager, that is, when $M_n \ll m_n$. Using this condition in principle (94) now produces an estimator

$$\sum_n (m_n + z - 1)\ \ln(m_n + z - 1) + \sum_n M_n \ln m_n = \text{maximum}\,. \qquad (95)$$

Only the second sum is now weighted-Burg type, for a photon object.

An "empirical Kullbach-Leibler estimator" (62) previously resulted, in the point-pixel object case when empirical knowledge $\{M_n\}$ was strong, that is, when $M_n >> m_n$. Under this condition, principle (94) becomes

$$\sum_n (m_n + z - 1)\ \ln(m_n + z - 1) + \sum_n m_n \ln(M_n/m_n) = \text{maximum}\,. \qquad (96)$$

The second sum is of the Kullbach-Leibler form, as was sought, but the first term is additional. In the nondegenerate limit $z = 1$ this result goes into a linear estimator

$$\sum_n m_n \ln M_n = \text{maximum}\,, \qquad m_n \equiv N\hat{p}_n\,. \qquad (97)$$

Note how the number M_n of empirically observed photons in a cell n directly weights its later estimate $\hat{p}_n$. This makes sense since with a nondegenerate object the only statistical evidence an estimate $\hat{p}_n$ can be based upon is its prior observation M_n.

5. Summary

We have seen how the MacQueen-Marschak approach, suitably generalized, permits maximum probable (M.A.P.) estimates to be made for occurrence frequencies. Particular cases were treated: letter frequencies on a page, photographic grain densities across a scene, and photon densities across an object. It was shown that the estimation principle to use depends drastically upon the kind of prior knowledge the user has about the unknown occurrence frequencies. Under certain conditions the principle becomes Kullbach-Leibler, or Jaynes' maximum entropy; for other conditions it becomes that of minimum entropy, or no principle at all! Or it becomes a weighted-Burg type of principle, or a sum of entropylike forms. These results were summarized at the beginning of this work.

In the particular problem of optical estimation, it was found that the M.A.P. estimation principle is not Jaynes' maximum entropy form, but rather the form (84),(86) based on quantum considerations [6]. In particular cases defined by the state of prior knowledge, this goes into forms resembling the Kullbach-Leibler norm, Jaynes' entropy, Burg's entropy, minimum entropy, a linear estimator, or no estimator at all. Once again, prior knowledge plays a key role in forming the estimator.

In the particular case of nearly featureless objects, the estimator goes over into Jaynes' expression in the dim-object limit and into Burg's in the bright-object limit. This perhaps explains the recent success of maximum entropy estimation and Burg estimation upon certain astronomical images.

Since this chapter was written, some new work on the subject has appeared [7,8]. In Ref. [7], the approach parallels that of this chapter, with particular emphasis upon image restoration and particle size distributions. In Ref. [8], application is made to the estimation of emission spectra.

Acknowledgment

I would like to thank Bernard Soffer of Hughes Research Laboratories for referring me to the important work of MacQueen and Marschak [3], and for some important conversations on the overall problem.

References

1. E. T. Jaynes (1968) IEEE Trans. Syst. Sci. Cybern. **SSC-4**, 227.
2. J. E. Shore (1981) IEEE Trans. Acoust. Speech Signal Process. **ASSP-29**, 230.
3. J. MacQueen and J. Marschak (1975) Proc. Nat. Acad. Sci. USA **72**, 3819.
4. J. P. Burg (1967) Maximum entropy spectral analysis, paper presented at the 37th Annual Meeting, Society of Exploration Geophysicists, Oklahoma City.
5. B. R. Frieden (1973) IEEE Trans. Inf. Theory **IT-19**, 118.
6. R. Kikuchi and B. H. Soffer (1977) J. Opt. Soc. Am. **67**, 1656.
7. B. R. Frieden (1983) J. Opt. Soc. Am. **73**, 927.
8. B. R. Frieden (1984) Chapter 8 in P. A. Jansson, ed., Deconvolution, with Applications in Spectroscopy (New York: Academic Press).

FUNDAMENTALS OF SEISMIC EXPLORATION

Enders A. Robinson

University of Tulsa, Tulsa, Oklahoma 74104

The problem of the geophysicist is to determine the structure of the interior of the earth from data obtained at the ground surface. Ultimately the problem is to find a method of processing a whole seismogram that will give structure, composition, and source parameters. Such a problem is an inverse problem. The forward problem is one in which a model of the earth structure and seismic source is used to give properties of seismic motion. When the solution of the forward problem is known, one method of inversion is an iteration based on trial and error. The parameters of the model are readjusted according to some criterion until some satisfying agreement between the data and the computed wave motion is discovered. This iterative approach to the inverse problem has proven to be successful in many applications. Today, far more sophisticated inversion methods are being developed for seismic data. In this paper, we describe the present state of the art in which spectral methods in the form of lattice methods in general, and maximum entropy spectral analysis in particular, play the key role. An understanding of these spectral methods will provide a firm basis for appreciating the exciting new inversion methods that the future holds.

This paper treats the one-dimensional (1-D) case of a horizontally layered earth with seismic raypaths only in the vertical direction. This model exhibits a lattice structure that corresponds to the lattice methods of spectral estimation. It is shown that the lattice structure is mathematically equivalent to the structure of the Lorentz transformation of the special theory of relativity.

A horizontally stratified half-space bounded by a perfect reflector at the top gives rise to a seismogram that, when completed by the direct downgoing pulse at zero time and by symmetry about the origin for negative time, produces an autocorrelation function. If this autocorrelation is convolved with the corresponding prediction error operators of increasing length, we obtain a "gapped function," which deviates more and more from the perfect symmetry exhibited by the autocorrelation. This gapped function consists of the downgoing and upgoing waveforms at the top of each layer. The gap separates the two waveforms, and the gap width increases as deeper and deeper layers are reached. In particular, the width of the

C. Ray Smith and W. T. Grandy, Jr. (eds.),
Maximum-Entropy and Bayesian Methods in Inverse Problems, 171–210.

gap is a measure of the entropy of the seismogram at a given depth level—the deeper we go into the subsurface, the higher the entropy of the corresponding gapped function. We explore the nature of the gapped function as it relates to the Toeplitz recursion generating the prediction error operators, and we derive the synthetic seismogram in terms of wave motion measured in units proportional to the square root of energy. We obtain an explicit relationship between the partial autocorrelation function on the one hand, and the reflection coefficient sequence on the other.

The maximum entropy spectral estimate is directly related to the reflection coefficient sequence characterizing a given subsurface model, and these considerations, in turn, impinge on the philosophy of deconvolution operator design. Finally, we investigate both the physical and the mathematical foundations of seismic deconvolution, and we attempt to establish the implications of the success of this approach on spectral analysis.

1. Introduction

The following passage from Wiener's 1933 book The Fourier Integral indicates the philosophy of Wiener's thinking and his great personal appeal:

> Physically speaking, this is the total energy of that portion of the oscillation lying within the interval in question. As this determines the energy-distribution of the spectrum, we may briefly call it the "spectrum." The author sees no compelling reason to avoid a physical terminology in pure mathematics when a mathematical concept corresponds closely to a concept already familiar in physics. When a new term is to be invented to describe an idea new to the pure mathematician, it is by all means better to avoid needless duplication, and to choose the designation already current. The "spectrum" of this book merely amounts to rendering precise the notion familiar to the physicist, and may as well be known by the same name.

The limelight in geophysical exploration is on inverse methods. However, when one looks closer, he sees that these methods are underlain by the associated spectral methods. The spectral approach can be used in conjunction with inversion theory to obtain solutions which give us both physical insight and practical exploration tools. The spectral approach, as exemplified by the one-dimensional lattice (that is, layered earth) methods, provides practical schemes for seismic data processing. These spectral methods work and can be used routinely to solve exploration problems successfully in a large number of cases, even though the complete mathematical inverse solution is yet beyond our reach.

The mathematical structure of the one-dimensional synthetic seismogram for an elastic, horizontally stratified medium furnishes the theoretical framework from which many vital aspects of modern deconvolution technology continue to spring. We consider a model of a stratified earth, which consists of a set of stacked layers, and the free-surface seismogram. We may characterize the layered medium in two mutually complementary ways. On the one hand, we can express the symmetrized free-surface seismogram, or autocorrelation function, in terms of the reflection coefficients. This is the "direct" problem. On the other hand, we may determine the reflection coefficients corresponding to a given autocorrelation function. This is the "inverse problem." In a sense, the reflection coefficients describe the physical aspects of the layered medium problem, while an associated partial autocorrelation function provides the bridge enabling us to study this same layered medium from the vantage point of statistical communication theory. It is the central purpose of the present paper to explore such complementary aspects of the stratified earth model, and to establish what new insights might be gained. Our approach is based on a study of the change in entropy associated with the seismogram as it is recorded at deeper and deeper levels within the subsurface. We shall thus be able to establish a firm connection between the physical concept of the reflection coefficient on the one hand, and the statistical concept of the partial autocorrelation function on the other.

2. Entropy

The power spectrum is the Fourier transform of the autocorrelation. The estimation of the power spectrum of a stationary time series from partial knowledge of its autocorrelation function is a classical problem to which much attention has been given over the years. Almost all of this work is based on the use of window functions, whose properties can be analyzed by coventional Fourier methods. However, the window function approach can produce power spectral estimates that are negative in certain frequency ranges (despite the fact that power is intrinsically positive). Also in the window approach, the known autocorrelation values are modified by a weighting function, and the spectral estimate agrees with these weighted values instead of with the known autocorrelation values. Because of such shortcomings, Burg (1967) in his pioneering work introduced a new philosophy in spectral analysis based on general variational principles, and, in particular, the maximum entropy method that we will now discuss.

Given a limited set of autocorrelation coefficients together with the fact that a power spectrum $\Phi(\omega)$ must be nonnegative, we know that there are generally infinitely many power spectra in agreement with this information. Thus additional information is required, and a reasonable goal is to find a single function $\Phi(\omega)$ that is representative of the class of all possible spectra. To resolve this problem, some choice has to be made, and Burg made use of the concept of entropy. Maximum entropy spectral analysis is based on choosing that spectrum which corresponds to the most random or the most unpredictable time series whose autocorrelation coincides with the given set of values. This concept of maximum entropy is the same as that used in both statistical mechanics and information theory and, as we will see, represents the most noncommittal assumption possible with regard to the unknown values of the autocorrelation function.

The autocorrelation of a (discrete) time function represents a beautiful mathematical structure. The autocorrelation r_t as a function of time t is symmetric, $r_t = r_{-t}$ (for all integers t), and preserves all the spectral order of the time series. The autocorrelation is a positive definite function, and its largest value in magnitude, namely r_0, occurs as its central value.

Wherever there is order in nature there is a natural phenomenon, called entropy, that tends to break down this order to produce chaos. For an autocorrelation function this destructive entropy is represented by the prediction error operator. The prediction error operator is one-sided (or causal). In fact, we should speak of a whole sequence of prediction error operators, for there is one for each value of n, where n = 0,1,2,.... The effect of each of these destructive operators grows worse and worse for larger and larger values of n. In the end, as n becomes infinite, literally the whole positive-time side of the autocorrelation has been annihilated, and the negative-time side is a distorted image of its former self. As we know, an annihilated or zero correlation means chaos, and entropy in the guise of the prediction error operator has had its way. Entropy dislikes any connections or bridges, and when it meets a smooth function like the autocorrelation it wants to disconnect it to form a gap. This disfigured or gapped function lacks the symmetry of the autocorrelation. Let us look at its properties.

We start with the prediction error operator of order n, which we denote by $a_0, a_1, \ldots, a_n$. When this operator acts on the autocorrelation it produces the gapped function

$$g_t = \sum_{s=0}^{n} a_s r_{t-s},$$

where the gap occurs between g_0 and g_{n+1}; that is, $g_t = 0$ for $t = 1,2, \ldots, n$ (see Fig. 1). Thus the prediction error operator has broken down the auto-

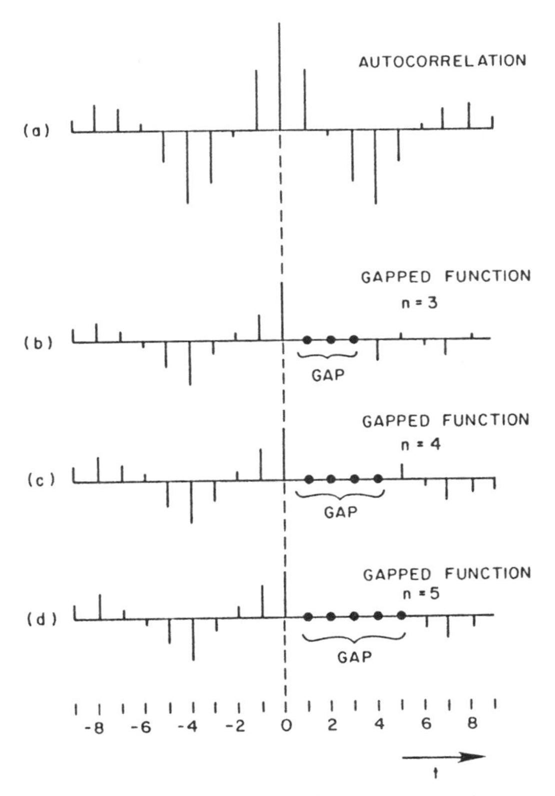

Figure 1. (a) Given autocorrelation function r_t; (b), (c), and (d) show gapped function g_t for n = 3, 4, and 5, respectively. The solid circles indicate the gap positions.

correlation function into two parts, and there is a gap of length n between them. The gap occurs from time 1 to time n, so the most important part of the positive-time side of the autocorrelation has been obliterated. It is customary to normalize prediction error operators by requiring that their leading coefficients be unity; thus, we let $a_0 = 1$. The remaining coefficients can be found by means of this gap requirement; that is, we find $a_1, a_2, \ldots, a_n$ from the values $r_0, r_1 = r_{-1}, \ldots, r_n = r_{-n}$ by solving the set of simultaneous equations

$$\sum_{s=0}^{n} a_s r_{t-s} = 0 \qquad \text{for } t = 1, 2, \ldots, n ,$$

which are called the normal equations. Because the autocorrelation coefficients form a Toeplitz matrix, they are called the Toeplitz normal equations.

Not satisfied with a gap of length n, we want a gap of length n+1. The Toeplitz recursion brings us from the prediction error operator $(a_0, a_1, \ldots, a_n)$, which produces a gap of length n, to the next prediction error operator $(a_0', a_1', \ldots, a_{n+1}')$, which produces a gap of length n+1. The essential property upon which the Toeplitz recursion depends is the fact that the distorted and gapped function g_t is not symmetric, as was the case for the autocorrelation function. Because g_t is not symmetric, g_t and its time reverse g_{-t} are linearly independent. As a result, we can form the new gapped function g_t' as a linear combination of g_t and g_{-t}. We write it as

$$g_t' = g_t - \gamma g_{-t} z^{n+1} ,$$

where γ is a constant to be determined and z^{n+1} represents the time shift required to line up the gap in g_{-t} with the gap in g_t. This equation can also be written as

$$g_t' = g_t - \gamma g_{-(t-n-1)} .$$

The gaps in the components of g_t' line up, so $g_t' = 0$ for $t = 1, 2, \ldots, n$; that is, the new gapped function g_t' necessarily has the same gap as the old gapped function g_t. However, we wish to increase the gap so as to include $t = n+1$, which means

$$g_{n+1}' = g_{n+1} - \gamma g_0 = 0 .$$

This equation allows us to solve for the constant γ:

$$\gamma = \frac{g_{n+1}}{g_0} .$$

Actually, we want γ in terms of the autocorrelation coefficients and the coefficients of the old prediction error operator. Because g_t is the convolution of a_t and r_t, we have

$$g_0 = a_0 r_0 + a_1 r_1 + \dots + a_n r_n$$

and

$$g_{n+1} = a_0 r_{n+1} + a_1 r_n + \dots + a_0 r_0 .$$

The value g_0 is called the (prediction-error) variance, and the value g_{n+1} is called the discrepancy. Hence, the coefficient γ is given by

$$\gamma = \frac{\text{discrepancy}}{\text{variance}} ,$$

and we have thus found γ. If we let Δ stand for the discrepancy and v for the variance, we have $\gamma = \Delta/v$.

Let us now find the new prediction error operator. We define the z-transforms of g_t' and g_t as

$$G'(z) = \sum_{t=-\infty}^{\infty} g_t' z^t , \qquad G(z) = \sum_{t=-\infty}^{\infty} g_t z^t$$

respectively. Then the z-transform of g_{-t} is

$$\sum_{t=-\infty}^{\infty} g_{-t} z^t = \sum_{s=-\infty}^{\infty} g_s z^{-s} = G(z^{-1}) .$$

Let us denote $G(z^{-1})$ by $\overline{G(z)}$ or simply by $\overline{G}$. Then the equation

$$g_t' = g_t - \gamma g_{-t} z^{n+1}$$

becomes

$$G' = G - \gamma \overline{G} z^{n+1} .$$

Because g_t is the convolution of a_t and r_t, the z-transform G is the multiplication of A and R, where A and R denote the z-transforms

$$A(z) = \sum_{t=0}^{n} a_t z^t \quad \text{and} \quad R(z) = \sum_{t=-\infty}^{\infty} r_t z^t$$

respectively. Because $G = AR$, we have $\overline{G} = \overline{A}\,\overline{R}$ where $\overline{A(z)} = A(z^{-1})$ and $\overline{R(z)} = R(z^{-1})$. Because r_t is symmetric, we have $R = \overline{R}$, so $\overline{G} = \overline{A}\,R$.

At this point, it is convenient to define the so-called reverse polynomial. The z-transform $A(z)$ is a polynomial of degree n with coefficients a_0, a_1, ..., a_n; i.e.,

$$A(z) = a_0 + a_1 z + \dots + a_n z^n .$$

The corresponding reverse polynomial, which we denote by $A^R(z)$, is the polynomial of the same degree but with coefficients in reverse order; i.e.,

$$A^R(z) = a_n + a_{n-1} z + \dots + a_0 z^n .$$

It is easy to see that a polynomial and its reverse are related by the equation

$$A(z) = z^n \overline{A^R(z)}$$

or equivalently,

$$A^R = z^n \overline{A} ,$$

which is

$$A^R(z) = z^n A(z^{-1}) .$$

Let us now return to our equation

$$G' = G - \gamma \overline{G} z^{n+1} ,$$

which we derived above. This equation becomes

$$\begin{aligned} A'R &= A R - \gamma \overline{A} \overline{R} z^{n+1} \\ &= A R - \gamma (\overline{A} z^n) z \overline{R} \\ &= A R - \gamma z A^R R \\ &= (A - \gamma z A^R) R . \end{aligned}$$

Thus we can obtain the polynomial A' of the new prediction error operator from the old prediction error operator A by the formula

$$A' = A - \gamma z A^R .$$

This process represents the Toeplitz recursion. In summary, the Toeplitz recursion says: First compute the ratio of discrepancy to variance:

$$\gamma = \frac{\Delta}{v}$$

where

$$\Delta = a_0 r_{n+1} + a_1 r_n + \ldots + a_n r_1$$

$$v = a_0 r_0 + a_1 r_1 + \ldots + a_n r_n$$

and then compute the new operator for step (n+1),

$$a'_0 = a_0 = 1$$

$$a'_1 = a_1 - \gamma a_n$$

$$a'_2 = a_2 - \gamma a_{n-1}$$

$$a'_3 = a_3 - \gamma a_{n-2}$$

$$\vdots$$

$$a'_{n-1} = a_{n-1} - \gamma a_2$$

$$a'_n = a_n - \gamma a_1$$

$$a'_{n+1} = -\gamma a_0 = -\gamma .$$

In actuality, one should not compute the variance by the formula given above, $v = a_0 r_0 + a_1 r_1 + \ldots + a_n r_n$, because this equation leads to unnecessary round-off errors. Instead, one should compute the variance by recursion. Thus, the new variance for the next step (that is, n+1) is

$$\begin{aligned} v' &= a'_0 r_0 + a'_1 r_1 + \ldots + a'_n r_n + a'_{n+1} r_{n+1} \\ &= a_0 r_0 + (a_1 - \gamma a_n) r_1 + \ldots + (a_n - \gamma a_1) r_n + (-\gamma a_0) r_{n+1} \\ &= (a_0 r_0 + a_1 r_1 + \ldots + a_n r_n) - \gamma (a_n r_1 + \ldots + a_1 r_n + a_0 r_{n+1}) \\ &= v - \gamma \Delta . \end{aligned}$$

Since $\Delta = v\gamma$, this equation becomes

$$v' = v(1-\gamma^2) ,$$

which is the required recursive formula for the variance. However, the new discrepancy still must be computed as above; that is, the new discrepancy (for the next step) will be

$$\Delta' = a_0' r_{n+2} + a_1' r_{n+1} + \dots + a_{n+1}' r_1 .$$

The new ratio will then be $\gamma' = \Delta'/v'$.

3. Lattice Model

One of the first scientists to treat a physical inverse problem was Lord Rayleigh, who in 1877 considered the problem of finding the density distribution of a string from knowledge of the vibrations. Marc Kac in 1966 aptly described this inverse problem in the title of his well-known lecture "Can one hear the shape of a drum?" Because seismic waves are sound waves in the earth, we can describe our inverse problem (that is, seismic exploration) as "Can we hear the shape of an oil field?"

The determination of the properties of the earth from waves that have been reflected from the earth is the classic problem of reflection seismology. As a first step in mathematical analysis, the problem is usually simplified by assuming that the earth's crust is made up of a sequence of sedimentary layers. The well known lattice model (Robinson, 1967) approximates the heterogeneous earth with a sequence of horizontal layers, each of which is homogeneous, isotropic, and nonabsorptive. This stratified model is subjected to vertically-traveling plane compressional waves, and thus it is a normal incidence model. It is assumed that the two-way travel time in each layer is the same and is equal to one time unit. In other words, the one-way travel time in each layer is taken to be one-half of the discrete unit of time. The upper half-space (the air) is called half-space 0, the first layer underneath is called layer 1, the next layer underneath is called layer 2, and so on. Interface 0 is the interface at the bottom of half-space 0, interface 1 is the interface at the bottom of layer 1, and so on. The stratified earth model can support traveling wave motion from the bottom to the top, which we call upgoing waves, and also traveling wave motion from the top to the bottom, which we call downgoing waves. At each interface between two adjacent layers, a traveling wave will be partially reflected and partially transmitted, the division of energy between the reflected and transmitted portions being governed by the reflection coefficient associated with that interface. Let c_n denote the reflection coefficient of interface n. The interpretation of this reflection coefficient is as follows.

We use the convention that all wave motion is measured in physical units that are proportional to the square root of energy. Thus, the square of the amplitude of a wave is in terms of energy. We define a "pulse" as a very narrow, spikelike waveform associated with a given discrete time instant, and we define the energy of the pulse as being equal to the square of its amplitude. If a downgoing pulse of unit amplitude is incident on interface n, then a reflected pulse and a transmitted pulse are produced (see Fig. 2a). Let the reflected pulse have amplitude c_n and the transmitted pulse have amplitude τ_n. The law of conservation of energy states that input energy is equal to output energy, so

$$1 = c_n^2 + \tau_n^2 .$$

Solving for τ_n, we have

$$\tau_n = \sqrt{1 - c_n^2} ,$$

where (by convention) we take the positive square root. The positive constant τ_n is called the (square root energy) transmission coefficient. In the case just given, we considered a downgoing incident unit pulse. If, instead, we consider an upgoing incident pulse striking interface n from the bottom, the reflection coefficient becomes $-c_n$ and the transmission coefficient remains the same, namely τ_n (see Fig. 2b).

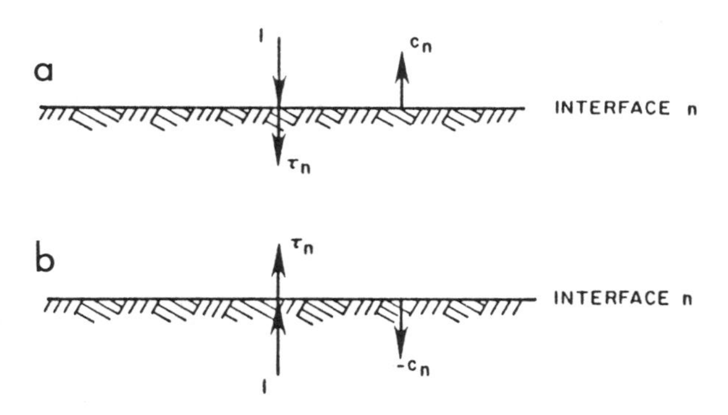

Figure 2. (a) Downgoing unit pulse incident on interface n from above. (b) Upgoing unit pulse incident on interface n from below.

As we have stated, the thickness of each layer was chosen so that it takes exactly one-half time unit for a wave to travel from one interface of the layer to the next. We make use of the unit delay operator z which, we recall, is defined as that operator that delays a time function by one time unit. It follows that $z^{1/2}$ is the operator that delays a time function by one-half time unit. Likewise, $z^{-1/2}$ is the operator that advances a time function by one-half time unit. In other words, $z^{1/2}$ delays and $z^{-1/2}$ advances a time function by an amount of time equal to the one-way travel time through the layer.

Let $d_n(t)$ and $u_n(t)$ be the downgoing and upgoing waves, respectively, at the top of layer n, and let $d'_n(t)$ and $u'_n(t)$ be the downgoing and upgoing waves, respectively, at the bottom of layer n (see Fig. 3). Because waves propagate without change of form within a given layer, the above waves satisfy

$$\begin{bmatrix} d_n'(t) \\ u_n'(t) \end{bmatrix} = \begin{bmatrix} z^{1/2} & 0 \\ 0 & z^{-1/2} \end{bmatrix} \begin{bmatrix} d_n(t) \\ u_n(t) \end{bmatrix} .$$

The waves at interface n are shown in Fig. 4. They can be related, as we shall now demonstrate. The wave $u_n'(t)$ consists of a part due to the reflection of $d_n'(t)$ and a part due to the transmission of $u_{n+1}(t)$; that is,

$$u_n'(t) = c_n d_n'(t) + \tau_n u_{n+1}(t) .$$

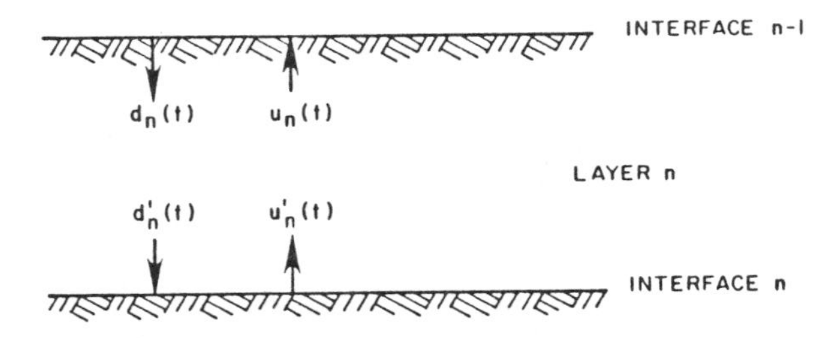

Figure 3. Upgoing and downgoing waves at top and bottom of layer n.

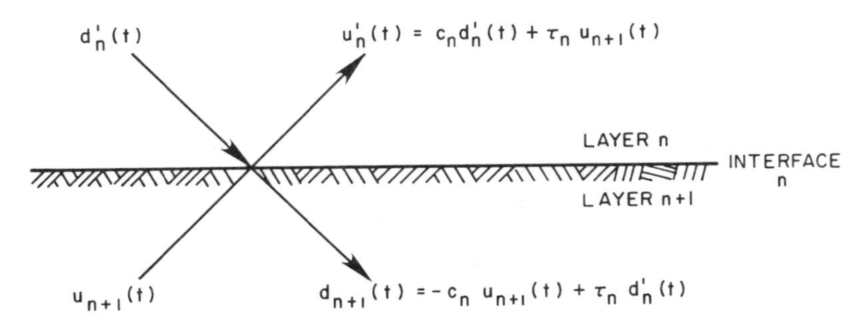

Figure 4. Wave motion relationships at interface n.

The wave $d_{n+1}(t)$ consists of a part due to the reflection of $u_{n+1}(t)$ and a part due to the transmission of $d_n'(t)$; that is,

$$d_{n+1}(t) = -c_n u_{n+1}(t) + \tau_n d_n'(t) .$$

If we solve these two equations for the unprimed quantities, we obtain

$$\begin{bmatrix} d_{n+1}(t) \\ u_{n+1}(t) \end{bmatrix} = \frac{1}{\tau_n} \begin{bmatrix} 1 & -c_n \\ -c_n & 1 \end{bmatrix} \begin{bmatrix} d_n'(t) \\ u_n'(t) \end{bmatrix} .$$

Combining this matrix equation with the one above, we obtain

$$\begin{bmatrix} d_{n+1}(t) \\ u_{n+1}(t) \end{bmatrix} = \frac{z^{-1/2}}{\tau_n} \begin{bmatrix} z & -c_n \\ -c_n z & 1 \end{bmatrix} \begin{bmatrix} d_n(t) \\ u_n(t) \end{bmatrix} .$$

By recursive use of this equation, we can relate the waves in any layer with those in any other layer, provided there are no sources or sinks in any intervening layer.

In particular, let us relate the waves in layer n+1 with those in layer 1. We have

$$\begin{bmatrix} d_{n+1}(t) \\ u_{n+1}(t) \end{bmatrix} = \frac{z^{-n/2}}{T_n} \begin{bmatrix} P_n^R & Q_n^R \\ Q_n & P_n \end{bmatrix} \begin{bmatrix} d_1(t) \\ u_1(t) \end{bmatrix} , \tag{1}$$

where T_n is defined as $\tau_n \tau_{n-1} \ldots t_1$, and where P_n and Q_n are defined by

$$\begin{bmatrix} P_n^R & Q_n^R \\ Q_n & P_n \end{bmatrix} = \begin{bmatrix} z & -c_n \\ -c_n z & 1 \end{bmatrix} \begin{bmatrix} z & -c_{n-1} \\ -c_{n-1} z & 1 \end{bmatrix} \cdots \begin{bmatrix} z & -c_1 \\ -c_1 z & 1 \end{bmatrix} , \tag{2}$$

whence we deduce that

$$\begin{bmatrix} P_n^R \\ Q_n \end{bmatrix} = \begin{bmatrix} z & -c_n \\ -c_n z & 1 \end{bmatrix} \begin{bmatrix} P_{n-1}^R \\ Q_{n-1} \end{bmatrix} . \tag{3}$$

We see that P_n and Q_n are <u>polynomials</u> and that

$$P_n^R(z) = z^n P_n(z^{-1}) \quad \text{and} \quad Q_n^R(z) = z^n Q_n(z^{-1})$$

are their <u>reverse polynomials</u>. The reader is referred to the next section (Lorentz transform) for more discussion of the model and these polynomials.

If we take the determinant of each side of matrix equation (2), we obtain

$$P_n P_n^R - Q_n Q_n^R = z^n(1-c_n^2)(1-c_{n-1}^2) \dots (1-c_1^2)$$

$$= z^n \tau_n^2 \tau_{n-1}^2 \dots \tau_1^2 = z^n T_n^2 . \qquad (4)$$

Let us now express the upgoing and downgoing waves in the layered medium as z-transforms. At the top of layer 1, the z-transform of $u_1(t)$ is

$$U_1(z) = u_1(1)\, z + u_1(2)\, z^2 + u_1(3)\, z^3 + \dots ,$$

and the z-transform of $d_1(t)$ is

$$D_1(z) = d_1(0) + d_1(1)\, z + d_1(2)\, z^2 + d_1(3)\, z^3 + \dots .$$

Next let us take the z-transform of the waves at depth. The z-transform of $d_{n+1}(t)$ is

$$D_{n+1}(z) = d_{n+1}\left(\frac{n}{2}\right) z^{n/2} + d_{n+1}\left(\frac{n}{2}+1\right) z^{(n/2)+1} + \dots$$

$$= T_n\, z^{n/2} + d_{n+1}\left(\frac{n}{2}+2\right) z^{(n/2)+2} + \dots \qquad (5)$$

and the z-transform of $u_{n+1}(t)$ is

$$U_{n+1}(z) = u_{n+1}\left(\frac{n}{2}+1\right) z^{(n/2)+1} + u_{n+1}\left(\frac{n}{2}+2\right) z^{(n/2)+2} + \dots$$

$$= T_n\, c_{n+1}\, z^{(n/2)+1} + u_{n+1}\left(\frac{n}{2}+2\right) z^{(n/2)+2} + \dots . \qquad (6)$$

In the above expressions for $D_{n+1}(z)$ and $U_{n+1}(z)$, we have made use of the relationships

$$d_{n+1}\left(\frac{n}{2}\right) = T_n , \qquad U_{n+1}\left(\frac{n}{2}+1\right) = c_{n+1}\, d_{n+1}\left(\frac{n}{2}\right) = c_{n+1}\, T_n .$$

We observe that $d_{n+1}(n/2)$ is simply the direct transmitted pulse entering layer n+1 from above, while

$$c_{n+1}\, d_{n+1}\left(\frac{n}{2}\right)$$

is the first pulse of the upgoing wave produced by an upward reflection at interface n+1.

We can now relate the above z-transforms by means of Eq. (1). Taking the z-transform of this matrix equation, we have

$$\begin{bmatrix} D_{n+1} \\ U_{n+1} \end{bmatrix} = \frac{z^{-n/2}}{T_n} \begin{bmatrix} P_n^R & Q_n^R \\ Q_n & P_n \end{bmatrix} \begin{bmatrix} D_1 \\ U_1 \end{bmatrix} . \tag{7}$$

Let us replace z by z^{-1} in Eq. (7). We obtain

$$\begin{bmatrix} \overline{D}_{n+1} \\ \overline{U}_{n+1} \end{bmatrix} = \frac{z^{n/2}}{T_n} \begin{bmatrix} z^{-n}P_n(z) & z^{-n}Q_n(z) \\ Q_n(z^{-1}) & P_n(z^{-1}) \end{bmatrix} \begin{bmatrix} \overline{D}_1 \\ \overline{U}_1 \end{bmatrix},$$

where $\overline{D}_1 = D_1(z^{-1})$, $\overline{U}_1 = U_1(z^{-1})$, and so on. Accordingly, we have

$$\begin{bmatrix} \overline{D}_{n+1} \\ \overline{U}_{n+1} \end{bmatrix} = \frac{z^{-n/2}}{T_n} \begin{bmatrix} P_n(z) & Q_n(z) \\ z^nQ_n(z^{-1}) & z^nP_n(z^{-1}) \end{bmatrix} \begin{bmatrix} \overline{D}_1 \\ \overline{U}_1 \end{bmatrix}$$

$$= \frac{z^{-n/2}}{T_n} \begin{bmatrix} P_n & Q_n \\ Q_n^R & P_n^R \end{bmatrix} \begin{bmatrix} \overline{D}_1 \\ \overline{U}_1 \end{bmatrix} ,$$

which can be written

$$\begin{bmatrix} \overline{U}_{n+1} \\ \overline{D}_{n+1} \end{bmatrix} = \frac{z^{-n/2}}{T_n} \begin{bmatrix} P_n^R & Q_n^R \\ Q_n & P_n \end{bmatrix} \begin{bmatrix} \overline{U}_1 \\ \overline{D}_1 \end{bmatrix} . \tag{8}$$

Combining matrix equations (7) and (8), we obtain

$$\begin{bmatrix} D_{n+1} & \overline{U}_{n+1} \\ U_{n+1} & \overline{D}_{n+1} \end{bmatrix} = \frac{z^{-n/2}}{T_n} \begin{bmatrix} P_n^R & Q_n^R \\ Q_n & P_n \end{bmatrix} \begin{bmatrix} D_1 & \overline{U}_1 \\ U_1 & \overline{D}_1 \end{bmatrix} . \tag{9}$$

Now we compute the determinant of each side of Eq. (9). For the left-hand side, we obtain immediately that the determinant is

$$D_{n+1}\,\overline{D}_{n+1} - U_{n+1}\,\overline{U}_{n+1} .$$

For the right-hand side, we have

$$\det \left[\frac{z^{-n/2}}{T_n} \begin{Bmatrix} P_n^R & Q_n^R \\ Q_n & P_n \end{Bmatrix} \right] \det \begin{bmatrix} D_1 & \overline{U}_1 \\ U_1 & \overline{D}_1 \end{bmatrix}$$

$$= \frac{z^{-n}}{T_n^2} (P_n^R P_n - Q_n^R Q_n)(D_1 \overline{D}_1 - U_1 \overline{U}_1) .$$

Here we use the relationship $\det(kC) = k^m \det C$, where k is a scalar factor multiplying a square matrix C of order m. But by Eq. (4) we have already shown that

$$P_n^R P_n - Q_n^R Q_n = z^n T_n^2 .$$

Therefore the operation of taking the determinant of each side of Eq. (9) gives

$$D_{n+1} \overline{D}_{n+1} - U_{n+1} \overline{U}_{n+1} = D_1 \overline{D}_1 - U_1 \overline{U}_1 . \tag{10}$$

This relationship states that the net downgoing energy in layer n+1 is equal to the net downgoing energy in layer 1.

We have thus described the model for a layered medium. Besides its interest in exploration geophysics, the layered-earth model is of interest in spectral estimation, for it is mathematically the same as a lattice model. These networks are useful in building models of many processes that occur in engineering practice, such as the acoustic tube model for digital speech processing. The lattice model provides methods for adaptive spectral estimation. We now want to discuss some of the properties of this model that have been useful in exploration geophysics.

4. Lorentz Transformation

When Maxwell derived the electromagnetic wave equation, it soon became known that it is not invariant under the Galilean transformation. However, it is invariant under the Lorentz transformation, and this observation was a key factor in Einstein's development of the special theory of relativity. The Lorentz transformation can be written as

$$D_2 = (1-c_1^2)^{-1/2} (D_1 - c_1 U_1)$$

$$U_2 = (1-c_1^2)^{-1/2} (-cD_1 + U_1)$$

where D_1 and U_1 are respectively the time and space coordinates of an event in frame 1, D_2 and U_2 are respectively the time and space coordinates of the event in frame 2, and c_1 (where $|c_1| < 1$) is the velocity (in natural units, such that the velocity of light is unity) between the two frames. The Lorentz transformation is a consequence of the invariance of the interval between events. By direct substitution, it can be shown that the coordinates of two events must satisfy the equation

$$D_2^2 - U_2^2 = D_1^2 - U_1^2$$

on transition from one frame of reference to the other.

We now want to find the relationship between the waves in the lattice model. Instead of the conventional treatment, we will try to put this relationship in a more general setting. We know that the waves in each layer obey their respective wave equation. Let $D_1(z)$ and $U_1(z)$ be respectively the z-transforms of the downgoing wave and the upgoing wave at the top of layer 1, and let $D_2(z)$ and $U_2(z)$ be the corresponding z-transforms for layer 2. We then say that <u>wave motion</u> must be related by the <u>Lorentz transformation</u>

$$D_2(z) = (1-c_1^2)^{-1/2} [z^{1/2}D_1(z) - c_1 z^{-1/2}U_1(z)]$$

$$U_1(z) = (1-c_1^2)^{-1/2} [-c_1 z^{1/2}D_1(z) + z^{-1/2}U_1(z)] .$$

The constant c_1 (where $|c_1| < 1$) is the reflection coefficient of the interface between the two layers. This Lorentz transformation is a consequence of the invariance of the net downgoing energy in the layers. By direct substitution, it can be shown that

$$D_2\overline{D}_2 - U_2\overline{U}_2 = D_1\overline{D}_1 - U_1\overline{U}_1 ,$$

where the bar indicates that z is to be replaced by z^{-1}; i.e., $\overline{D(z)} = D(z^{-1})$. This equation says that the <u>net downgoing energy</u> in each layer is the same. Because there is no absorption, this energy relationship is a physical fact implied by the model.

The Lorentz transformation between two adjacent layers can be written in matrix form as

$$\begin{bmatrix} D_{k+1} \\ U_{k+1} \end{bmatrix} = \frac{z^{-1/2}}{\tau_k} \begin{bmatrix} z & -c_k \\ -c_k z & 1 \end{bmatrix} \begin{bmatrix} D_k \\ U_k \end{bmatrix}$$

where we have used the symbol τ_k to denote the transmission coefficient $(1-c_k^2)^{1/2}$. Robinson (1967) defines the polynomials $P_k(z)$ and $Q_k(z)$ and the reverse polynomials $P_k^R(z)$ and $Q_k^R(z)$ (with superscript R for reverse) by Eq. (2), which is

$$\begin{bmatrix} P_k^R & Q_k^R \\ Q_k & P_k \end{bmatrix} = \begin{bmatrix} z & -c_k \\ -c_k z & 1 \end{bmatrix} \begin{bmatrix} z & -c_{k-1} \\ -c_{k-1} z & 1 \end{bmatrix} \cdots \begin{bmatrix} z & -c_1 \\ -c_1 z & 1 \end{bmatrix} .$$

By inspection, we can find the first and last coefficients of these polynomials. We have

$$P_k^R(z) = c_1 c_k z + \dots + z^k \qquad Q_k^R(z) = -c_k + \dots -c_1 z^{k-1}$$

$$Q_k(z) = -c_1 z + \dots -c_k z^k \qquad P_k(z) = 1 + \dots + c_1 c_k z^{k-1}$$

The polynomials for adjacent layers are related by

$$\begin{bmatrix} P_k^R & Q_k^R \\ Q_k & P_k \end{bmatrix} = \begin{bmatrix} z & -c_k \\ -c_k z & 1 \end{bmatrix} \begin{bmatrix} P_{k-1}^R & Q_{k-1}^R \\ Q_{k-1} & P_{k-1} \end{bmatrix} .$$

This equation gives the Robinson recursion (1967),

$$P_k = P_{k-1} - c_k z \, Q_{k-1}^R, \qquad Q_k = Q_{k-1} - c_k z \, P_{k-1}^R . \tag{11}$$

Let us now subtract the two recursion equations to obtain

$$P_k - Q_k = (P_{k-1} - Q_{k-1}) - c_k z (Q_{k-1}^R - P_{k-1}^R) .$$

Consider the free-surface case, that is, the case when the reflection coefficient of the surface of the ground has magnitude unity. Thus we let $|c_0| = 1$, where c_0 is the reflection coefficient of interface 0. For definiteness, let $c_0 = 1$. If we define the polynomial A_k as $A_k = P_k - c_0 Q_k = P_k - Q_k$, we obtain the Levinson recursion (1947),

$$A_k = A_{k-1} + c_k z \, A_{k-1}^R . \tag{12}$$

The polynomial of the second kind is defined as $B_k = P_k + c_0 Q_k = P_k + Q_k$, and we see that it satisfies the recursion

$$B_k = B_{k-1} - c_k \, z \, B_{k-1}^R .$$

The two recursions given in this section govern the structure and the analysis of layered earth (lattice) models, the A recursion for the free-surface case, and the P,Q recursion for the non-free-surface case.

5. Free-Surface Seismogram

The spectral methods used in geophysics are closely connected with the inverse problem. Basically, the spectral approach provides methods for the approximate solution of many simplified inverse problems. We are concerned in this paper with such specialized solutions and not with the general case. In this sense, this paper deals with the production-type data processing methods in everyday use in geophysical exploration, which serve not perfectly but well.

Since the energy crisis, a greatly increased research effort has been devoted to problems in the geophysical exploration for oil and natural gas. General inverse scattering methods from the mainstream of physics and applied mathematics have been introduced to the geophysical exploration industry. The pioneering work of Jack Cohen and Norman Bleistein has caused a revolution in the direction of research on the long-standing unsolved problems in the seismic exploration for oil. We do not treat these more powerful methods in this paper.

We now want to consider an idealized seismic experiment. The source is a downgoing unit impulse introduced at the top of layer 1 at time instant 0. This pulse proceeds downward, where it undergoes multiple reflections and refractions within the layered system. Some of the energy is returned to the top of layer 1, where it is recorded in the form of a seismic trace, which we denote by the sequence r_1, r_2, r_3, ..., where the subscript indicates the discrete time index.

There are two types of boundary conditions commonly imposed on the top interface (interface 0 with reflection coefficient c_0). One is the free-surface condition, which says that interface 0 (the air-earth interface) is a perfect reflector; that is, the free-surface condition is that $|c_0| = 1$. The free-surface condition approximately holds in the case of a marine seismogram taken in a very smooth, calm sea, so the surface of the water (interface 0) is virtually a perfect reflector. The other condition is the non-free-surface case. A typical land seismogram is generated by the lattice model with the non-free-surface condition $|c_0| < 1$.

In summary, an ideal marine seismogram is generated by the lattice model with the free surface condition $|c_0| = 1$. For the moment, suppose that $c_0 = 1$. Thus the reflection coefficient for upgoing waves is $-c_0$, which is -1, so an upcoming pulse $-r_n$ is reflected to the downgoing pulse r_n. The well known acoustic tube model for human speech may be described as a lattice model with the free-surface condition at the lips.

Kunetz (1964) gave the solution for the inversion of a free-surface reflection seismogram. This inversion method yields the reflection coefficient series c_1, c_2, c_3, ..., from which the impedance function of the earth as a function of depth may be readily computed. Robinson (1967) reformulated the Kunetz solution in terms of the Levinson recursion (1947) and gave a computer program to do both the forward process (generation of the synthetic seismogram) and the inverse process (inversion of the seismogram to obtain the reflection coefficients).

The celebrated inversion method of Gelfand and Levitan (1955) represents the solution of the inversion problem for non-free-surface reflection seismograms. This method has been in the mainstream of physics for many years and has been further developed and extended by many physicists and mathematicians. The discrete form of the Gelfand-Levitan equation is derived by Aki and Richards (1980) for the case of a finite inhomogeneous medium, that is, an inhomogeneous medium bounded by homogeneous media at both ends. Robinson (1982) treats the discrete Gelfand-Levitan equation and gives a derivation that holds for an unbounded inhomogeneous medium, and then discusses dynamic deconvolution, which is a means of solving the Gelfand-Levitan equation. Dynamic deconvolution makes use of the interactive recursion of the P and Q polynomials (Robinson, 1967). This recursion for the non-free-surface case represents the counterpart of the Levinson recursion for the free-surface case.

Let us excite our stratified system with a surface source. In particular, let us assume that a unit downgoing source pulse δ_t is set off at time zero just below the first interface (i.e., $\delta_t = 1$ for $t = 0$ and $\delta_t = 0$ for $t \neq 0$). We want to consider the case for which the surface (i.e., interface 0) is a perfect reflector with reflection coefficient $|c_0| = 1$; that is, either $c_0 = 1$ or $c_0 = -1$ (see Fig. 5). The case $c_0 = +1$ corresponds to the pressure (or stress) reflection coefficient $(Z_1-Z_0)/(Z_1+Z_0) \simeq 1$ since $Z_0 \simeq 0$, where Z_0 and Z_1 are the characteristic impedances of layers 0 and 1, respectively. The case $c_0 = -1$ corresponds to the particle displacement, particle velocity, or particle acceleration reflection coefficient $(Z_0-Z_1)/(Z_0+Z_1) \simeq -1$, since $Z_0 \simeq 0$. These situations occur in marine seismic exploration, for then the first layer is water, and the surface of the water tends to act as a perfect

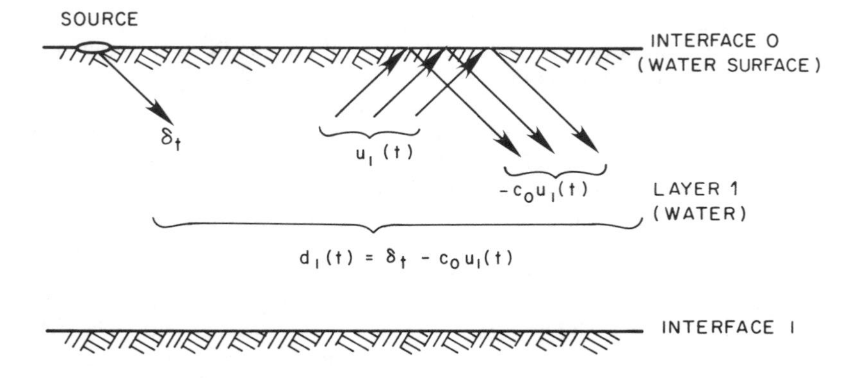

Figure 5. Wave motion relationships at the top of layer 1. (Note: The water layer may consist of more than one "mathematical" layer, where each "mathematical" layer has a thickness corresponding to a travel time of one-half time unit.)

reflector. As a result, the upgoing waveform $u_1(t)$ at the top of layer 1 is reflected at the surface to yield a downgoing waveform $-c_0u_1(t)$, which simply is $-u_1(t)$ if $c_0 = 1$ or is $u_1(t)$ if $c_0 = -1$. In other words, the upgoing waveform $u_1(t)$ is reflected perfectly into the downgoing waveform $-c_0u_1(t)$, where $|c_0| = 1$. The entire downgoing wave $d_1(t)$ at the top of layer 1 is the sum of the initial unit source pulse δ_t and the reflected waveform $-c_0u_1(t)$; that is,

$$d_1(t) = \delta_t - c_0u_1(t) . \tag{13}$$

The source pulse occurs at time $t = 0$. The first pulse in $u_1(t)$ travels down from the surface, is reflected from interface 1, and returns to the surface; the total elapsed time is the two-way travel time in layer 1, or one time unit. Thus, the upgoing wave $u_1(t)$ has its first nonzero value at $t = 1$, so $u_1(t)$ is the time series

$$u_1(1),\ u_1(2),\ u_1(3),\ \ldots ,$$

which has its first break (or arrival time) at time $t = 1$. The downgoing wave $d_1(t)$ is the time series

$$1,\ -c_0u_1(1),\ -c_0u_1(2),\ -c_0u_1(3),\ \ldots ,$$

which has its first break (or arrival time) at time $t = 0$. In other words, the first break of the upgoing wave $u_1(t)$ at the top of layer 1 occurs one whole time unit after the first break of the downgoing wave $d_1(t)$ at the top of the same layer (see Fig. 6). A little thought shows that this property is quite general, namely the first break of the upgoing wave $u_{n+1}(t)$ at the top of layer n+1 occurs one whole time unit after the first break of the downgoing wave $d_{n+1}(t)$ at the top of the same layer. In addition, we see that the

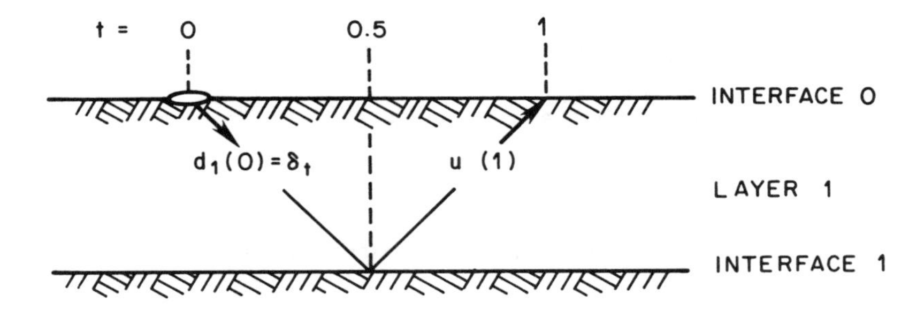

Figure 6. The first break of the upgoing wave at the top of layer 1 occurs one time unit after the first break of the downgoing wave at the top of this same layer.

first break of the downgoing wave $d_{n+1}(t)$ is produced by the unit source pulse traveling directly down through the n layers from the surface through interface n. The direct pulse is multiplied by the transmission coefficient for each of these n interfaces, and thus the amplitude of the direct pulse is $\tau_1 \tau_2 \cdots \tau_n$. The direct pulse also suffers a delay of one-half time unit as it travels through each of the n layers, so its travel time is n/2. Thus, the downgoing wave $d_{n+1}(t)$ at the top of layer n+1 can be written as

$$\underset{\substack{\uparrow \\ \text{direct pulse}}}{d_{n+1}(\tfrac{n}{2})}, \quad d_{n+1}(\tfrac{n}{2} + 1), \quad d_{n+1}(\tfrac{n}{2} + 2), \quad \cdots \quad ,$$

where the arrival time (the travel time of the direct pulse) is n/2. Therefore, the direct pulse $d_{n+1}(n/2)$ has amplitude $\tau_1 \tau_2 \cdots \tau_n = T_n$, and occurs at time n/2. As we have seen, the upgoing wave $u_{n+1}(t)$ arrives one time unit after the downgoing wave $d_{n+1}(t)$; hence, the upgoing wave $u_{n+1}(t)$ has the form

$$\underset{\substack{\uparrow \\ \text{first pulse}}}{u_{n+1}(\tfrac{n}{2} + 1)}, \quad u_{n+1}(\tfrac{n}{2} + 2), \quad \cdots ,$$

where the arrival time (the travel time of the first pulse) is n/2 + 1. In fact, the first pulse $u_{n+1}(n/2 + 1)$ of the upgoing wave is due solely to the reflection of the direct pulse $d_{n+1}(n/2)$ from interface n+1 (see Fig. 7). Hence, the first pulse $u_{n+1}(n/2 + 1)$ has amplitude

$$\tau_1 \tau_2 \cdots \tau_n c_{n+1} = T_n c_{n+1}$$

and occurs at time n/2 + 1.

Let us now convert to z-transforms. We first take the z-transforms of the waves at the surface. The z-transform of $u_1(t)$ is

$$U_1(z) = u_1(1)z + u_1(2)z^2 + u_1(3)z^3 + \cdots$$

and the z-transform of $d_1(t)$, as given by Eq. (13), is

$$D_1(z) = 1 - c_0 \, U_1(z) \, .$$

Adding $(-c_0)$ times matrix equation (7) to matrix equation (8), we obtain

$$\begin{bmatrix} -c_0 D_{n+1} + \bar{U}_{n+1} \\ -c_0 U_{n+1} + \bar{D}_{n+1} \end{bmatrix} = \frac{z^{-n/2}}{T_n} \begin{bmatrix} P_n^R & Q_n^R \\ Q_n & P_n \end{bmatrix} \begin{bmatrix} -c_0 D_1 + \bar{U}_1 \\ -c_0 U_1 + \bar{D}_1 \end{bmatrix} . \tag{14}$$

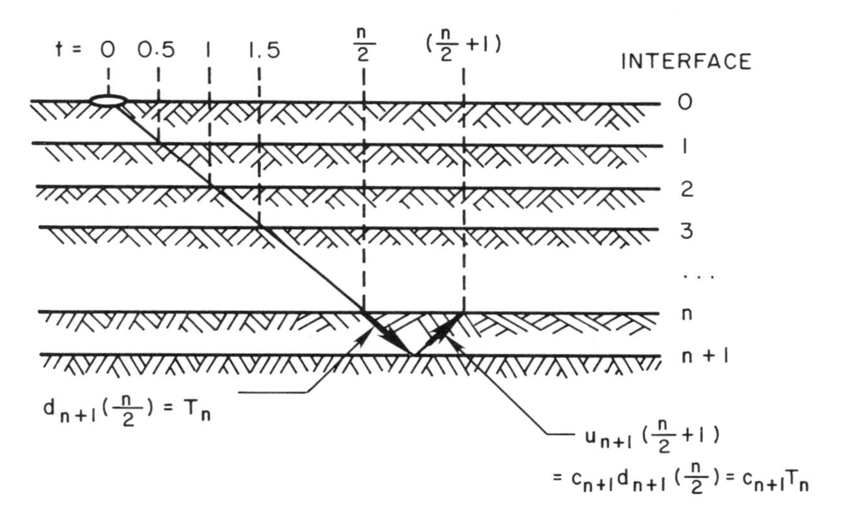

Figure 7. The direct downgoing pulse $d_{n+1}(n/2)$ at the top of layer n and its relationship to the first upgoing pulse $u_{n+1}(n/2 + 1)$ at the top of this layer. This first upgoing pulse is due solely to the reflection of the direct downgoing pulse from interface n+1.

Let us now return to the subject of net downgoing energy. Since n is arbitrary, Eq. (10) shows that the net downgoing energy in every layer is the same. In fact, if we go down to infinite depth ($n = \infty$), there is no upcoming wave (that is, $U_\infty = 0$). Therefore at infinite depth the net downgoing energy is $D_\infty \overline{D}_\infty$, which is in the form of a spectral function (Robinson and Treitel, 1978). In other words, $D_\infty \overline{D}_\infty$ is the z-transform of the autocorrelation of the downgoing wave $d_\infty(t)$ at infinite depth. Let us denote this autocorrelation function by the symbol r_t, which agrees with the notation already introduced. Thus, the autocorrelation r_t has the z-transform

$$R(z) = \sum_{t=-\infty}^{\infty} r_t z^t = D_\infty \overline{D}_\infty .$$

By Eq. (10), we therefore have

$$R = D_{n+1} \overline{D}_{n+1} - U_{n+1} \overline{U}_{n+1} = D_1\overline{D}_1 - U_1\overline{U}_1 .$$

Now we come to one of the most interesting results in this field, which was first given by Kunetz (1964). Equation (13) states that $D_1 = 1 - c_0 U_1$, so the net downgoing energy in the first layer is

$$D_1\overline{D}_1 - U_1\overline{U}_1 = (1-c_0U_1)(1-c_0\overline{U}_1) - U_1\overline{U}_1$$

$$= 1 - c_0U_1 - c_0\overline{U}_1 + (c_0^2-1)U_1\overline{U}_1$$

$$= 1 - c_0U_1 - c_0\overline{U}_1 ,$$

because $c_0^2 = 1$. The interesting result occurs by combining the above two equations to yield

$$R = 1 - c_0U_1 - c_0\overline{U}_1 \qquad (\text{when } |c_0|^2 = 1) .$$

This equation may be written as either

$$R = (1 - c_0U_1) - c_0\overline{U}_1 = D_1 - c_0\overline{U}_1 \tag{15}$$

or

$$R = (1 - c_0\overline{U}_1) - c_0U_1 = \overline{D}_1 - c_0U_1 . \tag{16}$$

The function R is the z-transform of the autocorrelation r_t. The center value $r_0 = 1$ represents the source, which is the first break of the downgoing wave. From Eqs. (15) and (16) we see that $R = \overline{R}$, which demonstrates the symmetry of the autocorrelation. Equation (16) may be interpreted as follows. Flip all downgoing waves around the time origin so that they run backward in time. With this convention, we can describe

$$\overline{D}_1 = 1 + d_1(1)z^{-1} + d_1(2)z^{-2} + \ldots$$

as the z-transform of the "flipped" downgoing wave

$$\begin{array}{ccccccc} \ldots, & d_1(2), & d_1(1), & 1, & 0, & 0, & \ldots . \\ & \uparrow & \uparrow & \uparrow & \uparrow & \uparrow & \\ & t=-2 & t=-1 & t=0 & t=1 & t=2 & \end{array}$$

The function $-c_0U_1$ is of course the z-transform of the component of the downgoing wave due to the reflection of the upgoing wave at the free surface, that is, the z-transform of

$$\begin{array}{ccccccc} \ldots, & 0, & 0, & 0, & -c_0u_1(1), & -c_0u_1(2), & \ldots , \\ & \uparrow & \uparrow & \uparrow & \uparrow & \uparrow & \\ & t=-2 & t=-1 & t=0 & t=1 & t=2 & \end{array}$$

or, equivalently, the z-transform of

$$\begin{array}{cccccc} \ldots, & 0, & 0, & 0, & d_1(1), & d_2(1), \ \ldots . \\ & \uparrow & \uparrow & \uparrow & \uparrow & \uparrow \\ & t=-2 & t=-1 & t=0 & t=1 & t=2 \end{array}$$

The function R is the z-transform of the autocorrelation

$$\begin{array}{cccccc} \ldots, & r_2, & r_1, & r_0=1, & r_1, & r_2, \ \ldots . \\ & \uparrow & \uparrow & \uparrow & \uparrow & \uparrow \\ & t=-2 & t=-1 & t=0 & t=1 & t=2 \end{array}$$

Hence, Eq. (16) says that

$$(\ldots, r_2, r_1, 1, r_1, r_2, \ldots)$$

$$= (\ldots, d_1(2), d_1(1), 1, 0, 0, \ldots) + (\ldots, 0, 0, 0, -c_0u_1(1), -c_0u_1(2),\ldots)$$

$$= (\ldots, d_1(2), d_1(1), 1, d_1(1), d_1(2), \ldots)$$

$$= (\ldots, -c_0u_1(2), -c_0u_1(1), 1, -c_0u_1(1), -c_0u_1(2), \ldots) ,$$

or, in words, that the autocorrelation function is equal to the downgoing wave symmetrized about the time origin.

Let us now see what the stratified earth does to this autocorrelation function, which we have defined to occur just below the surface. We shall see that as the waveforms travel deeper and deeper the destructive effect of entropy becomes greater and greater. Entropy, in the form of the earth's strata, tears our autocorrelation apart, inserting larger and larger gaps as we go deeper and deeper, until at last the entire right-hand side is destroyed, yielding nothing but chaos in the bowels of the earth. The autocorrelation is the symmetrized marine seismogram observed at the surface due to a surface source. We shall now follow the action of the waveforms as they travel down into the earth, and witness firsthand the destruction of the seismogram.

Let us use Eqs. (5) and (6) to form

$$-c_0U_{n+1} + \overline{D}_{n+1} = \ldots + d_{n+1}(\tfrac{n}{2} + 1)\ z^{-(n/2)-1} + T_n\ z^{-n/2}$$

$$- T_n\ c_0c_{n+1}\ z^{(n/2)+1} - c_0u_{n+1}\ (\tfrac{n}{2} + 2)\ z^{(n/2)+2} + \ldots . \qquad (17)$$

Thus, $-c_0U_{n+1} + \overline{D}_{n+1}$ is the z-transform of the waveform

$$\{\ldots, \underset{t=-(n/2)-1}{\underset{\uparrow}{d_{n+1}(\tfrac{n}{2}+1)}}, \underset{t=-(n/2)}{\underset{\uparrow}{T_n}}, \underset{\text{Gap of n time points}}{\underset{\uparrow}{\underbrace{0,\ldots,0}}},$$

$$\underset{t=(n/2)+1}{\underset{\uparrow}{-c_0T_nc_{n+1}u_{n+1}(\tfrac{n}{2}+1)}}, \underset{t=(n/2)+2}{\underset{\uparrow}{-c_0u_{n+1}(\tfrac{n}{2}+2)}}, \ldots\} .$$

Matrix equation (14) represents two scalar relationships, the second of which is

$$-c_0U_{n+1} + \overline{D}_{n+1} = \frac{z^{-n/2}}{T_n}[Q_n(-c_0D_1 + \overline{U}_1) + P_n(-c_0U_1 + \overline{D}_1)] . \quad (18)$$

By Eq. (15) we may write, bearing in mind that $c_0^2 = 1$,

$$-c_0D_1 + \overline{U}_1 = -c_0(D_1 - c_0\overline{U}_1) = -c_0R ,$$

while by Eq. (16) we have

$$-c_0U_1 + \overline{D}_1 = R .$$

Hence, Eq. (18) is

$$-c_0U_{n+1} + \overline{D}_{n+1} = \frac{z^{-n/2}}{T_n}[Q_n(-c_0R) + P_nR]$$

$$= \frac{z^{-n/2}}{T_n}(P_n - c_0Q_n)R . \quad (19)$$

Let us now define the operator A_n as

$$A_n = P_n - c_0Q_n , \quad (20)$$

where $|c_0| = 1$, and Eq. (19) becomes

$$A_n R = T_n z^{n/2}(-c_0U_{n+1} + \overline{D}_{n+1}) . \quad (21)$$

If we substitute Eq. (17) into Eq. (21), we finally obtain

$$A_n R = \ldots + T_n d_{n+1}(\tfrac{n}{2}+1)z^{-1} + T_n^2 - T_n^2 c_0 c_{n+1} z^{n+1}$$

$$- T_n c_0 u_{n+1}(\tfrac{n}{2}+2) z^{n+2} + \ldots .$$

The right-hand side of this equation is the z-transform of the time function

$$\{ \ldots, \underset{t=-1}{\underset{\uparrow}{T_n d_{n+1}(\tfrac{n}{2}+1)}}, \underset{t=0}{\underset{\uparrow}{T_n^2}}, \underset{\text{Gap of n time points}}{\underset{\uparrow}{\underbrace{0, \ldots, 0}}},$$

$$\underset{t=n+1}{\underset{\uparrow}{-T_n^2 c_0 c_{n+1}}}, \underset{t=n+2}{\underset{\uparrow}{-T_n c_0 u_{n+1}(\tfrac{n}{2}+2)}}, \ldots \} .$$

We now see that the operator A_n defined by Eq. (20), when operating on the autocorrelation R, produces a gap of length n.

We can, therefore, identify the above gapped time function as

$$(\ldots, g_{-1}, g_0, \underset{\text{Gap of n time points}}{\underset{\uparrow}{\underbrace{0, \ldots, 0}}}, g_{n+1}, g_{n+2}, \ldots) \quad ,$$

and A_n as the z-transform of the prediction error operator, which occurs in the Levinson recursion. Recall that the parameter γ was defined as the ratio of the discrepancy to the variance; from the above correspondence, we see that the parameter γ is in fact the negative of the product of the reflection coefficients c_0 and c_{n+1}; that is,

$$\gamma = \frac{g_{n+1}}{g_0} = \frac{-T_n^2 c_0 c_{n+1}}{T_n^2} = -c_0 c_{n+1} . \tag{22}$$

In conclusion, the operator with z-transform $A_n = P_n - c_0 Q_n$ with $|c_0| = 1$ is indeed the prediction error operator for the autocorrelation given by the symmetrized marine seismogram. This operator, when convolved with the autocorrelation r_t, produces a gapped function g_t. The left-hand side of the gapped function (including time 0) is equal to T_n times the flipped downgoing wave (measured from its first arrival). The gap occurs for the time points from 1 through n. The right-hand side of the gapped function starting at time n+1 is equal to $(-T_n c_0)$ times the upgoing wave (measured from its first arrival), where $|c_0| = 1$. As a result, we can write

$$G_n = A_n R , \tag{23}$$

where G_n is the z-transform of the gapped time function containing a gap of n time points.

6. Partial Autocorrelation

Norbert Wiener published in 1930 his classic paper "Generalized Harmonic Analysis," which he considered his finest work. In his introduction, he stated that he was motivated by the work of researchers in optics, especially that of Rayleigh and Schuster. However, Wiener demonstrated that the domain of generalized harmonic analysis was much broader than optics. Among Wiener's results were the writing down of the precise definitions of and the relationship between the autocovariance function and the power spectrum. The theorem that these two functions make up a Fourier transform pair is today known as the Wiener-Khintchine theorem. Whereas Khintchine worked with an existential approach (involving, for example, an ensemble of time series), Wiener adopted a constructive approach (involving a section of increasing length of a single time series).

In time series analysis the autocorrelation function is a well known concept. Because the partial autocorrelation function is less well known, we briefly review it here. Let x_t (where the time index t takes on integer values) be a stationary time series. The autocorrelation function is defined as

$$r_s = E(x_{t+s} x_t) ,$$

where E denotes the expectation operator and s is the time shift. For convenience, we normalize the autocorrelation function so that $r_0 = 1$. For each positive integer n, we can define a prediction error series ε_t and a hindsight error series η_t associated with this value of n. The <u>prediction error time series</u> is given by

$$\varepsilon_t = \sum_{s=0}^{n} a_s x_{t-s} ,$$

where the operator coefficients $a_0 = 1, a_1, a_2, \ldots, a_n$ are chosen to minimize $E\varepsilon_t^2$. If we go through this minimization procedure, we find that the required operator coefficients $1, a_1, \ldots, a_n$ can be found from the first part of the autocorrelation $1, r_1, \ldots, r_n$ by solving the Toeplitz normal equations

$$\sum_{s=0}^{n} a_s r_{t-s} = 0 \qquad \text{for } t = 1, 2, \ldots, n .$$

Thus, we see that the operator $1, a_1, \ldots, a_n$ found here is identical to the operator $1, a_1, \ldots, a_n$ given previously, which explains why we called it the prediction error operator. It also turns out from the minimization procedure that the minimum value of $E\varepsilon_t^2$ is

$$E\varepsilon_t^2 = a_0 r_0 + a_1 r_1 + \ldots + a_n r_n = g_0 \qquad (\text{where } a_0 = 1,\ r_0 = 1),$$

which explains why we called g_0 the (prediction error) variance.

Let us now define the hindsight error series η_t. The hindsight error time series for a given value of n is

$$\eta_t = \sum_{s=0}^{n} a_s\, x_{t+s} \qquad (\text{where } a_0 = 1).$$

If we minimize $E\eta_t^2$, we find that the required hindsight error operator coefficients $1, a_1, \ldots, a_n$ satisfy the same set of normal equations as for the prediction error operator, and the minimum value of $E\eta_t^2$ is also equal to g_0. Thus, the operator coefficients of the prediction error operator and the hindsight error operator are identical, and in addition

$$E\varepsilon_t^2 = E\eta_t^2 = g_0 .$$

The difference between the two operators is that the prediction error operator acts on past values of the time series, while the hindsight error operator acts on future values. That is,

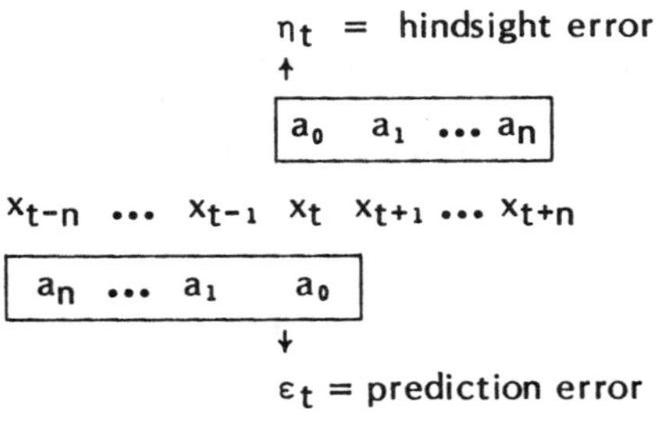

η_t = hindsight error

↑

a_0	a_1	$\ldots$	a_n

$x_{t-n} \;\ldots\; x_{t-1} \;\; x_t \;\; x_{t+1} \;\ldots\; x_{t+n}$

a_n	$\ldots$	a_1	a_0

↓

ε_t = prediction error

In brief, the prediction error ε_t is the error in forecasting x_t from the past values $x_{t-1}, \ldots, x_{t-n}$, while the hindsight error η_t is the error in "backcasting" x_t from the future values $x_{t+1}, \ldots, x_{t+n}$.

Let us now treat the same datum points and consider the corresponding prediction error and hindsight error:

$$\begin{array}{l}
\eta_{t-n-1} \\
\uparrow \\
\boxed{a_0 \quad a_1 \quad a_2 \quad \cdots \quad a_{n-1} \quad a_n}
\end{array}$$

$$x_{t-n-1} \quad x_{t-n} \quad x_{t-n+1} \quad \cdots \quad x_{t-2} \quad x_{t-1} \quad x_t$$

$$\begin{array}{r}
\boxed{a_n \quad a_{n-1} \quad \cdots \quad a_2 \quad a_1 \quad a_0} \\
\downarrow \\
\varepsilon_t
\end{array}$$

Here, the prediction error ε_t is the error in forecasting x_t from the past datum points $x_{t-n}, x_{t-n-1}, \ldots, x_{t-1}$. The hindsight error η_{t-n-1} is the error in backcasting x_{t-n-1} from the future datum points $x_{t-n}, x_{t-n-1}, \ldots, x_{t-1}$. We see that both sets of datum points are indeed the same, and represent all the data values between the point x_{t-n-1} that is being backcasted and the point x_t that is being forecasted. In other words, η_{t-n-1} represents the error in x_{t-n-1} and ε_t represents the error in x_t, after we have removed the effect of the common intervening variables $x_{t-n}, x_{t-n+1}, \ldots, x_{t-1}$. As a result, the correlation between η_{t-n-1} and ε_t represents the net correlation between x_{t-n-1} and x_t after the effects of all intervening variables are removed. This net correlation is called the "partial autocorrelation." The partial autocorrelation coefficient is defined in the form

$$(\text{Partial autocorrelation coefficient}) = \frac{E\varepsilon_t \eta_{t-n-1}}{\sqrt{E\varepsilon_t^2}\,\sqrt{E\eta_{t-n-1}^2}} .$$

To find an expression for this coefficient in terms of well known quantities, let us first consider the cross correlation between the prediction error series ε_t and the given time series x_t. This cross correlation is

$$E\varepsilon_t x_{t-s} = E\left(\sum_{j=0}^{n} a_j\, x_{t-j}\right) x_{t-s}$$

$$= \sum_{j=0}^{n} a_j E x_{t-j} x_{t-s} = \sum_{j=0}^{n} a_j r_{sj} = g_s .$$

In words, this cross correlation is the gapped function g_s defined previously. Likewise, we find that the cross correlation between the hindsight error series η_t and the given time series x_t is

$$E\eta_t x_{t-s} = \sum_{j=0}^{n} a_j r_{s+j} = g_{-s} \,,$$

since $r_{-s-j} = r_{s+j}$, because the autocorrelation function r_s is symmetric. In words, this cross correlation is the flipped gapped function g_{-s}.

Let us now find an expression for the covariance term in the numerator of the partial autocorrelation coefficient. We have

$$E\varepsilon_t \eta_{t-n-1} = E\varepsilon_t \sum_{s=0}^{n} a_s x_{t-n-1+s}$$

$$= \sum_{s=0}^{n} a_s E\varepsilon_t x_{t-n-1+s} = \sum_{s=0}^{n} a_s g_{n+1-s}$$

$$= g_{n+1} + a_1 g_n + a_2 g_{n-1} + \dots + a_n g_1 \,.$$

Because of the gap (that is, $g_1 = g_2 = \dots = g_n = 0$), this equation becomes

$$E\varepsilon_t \eta_{t-n-1} = g_{n+1} \,,$$

and we saw at the beginning of this section that

$$E\varepsilon_t^2 = E\eta_t^2 = g_0 \,.$$

Use of Eq. (22) now enables us to state that the partial autocorrelation coefficient between x_t and x_{t-n-1} is

$$(\text{Partial autocorrelation coefficient}) = \frac{g_{n+1}}{\sqrt{g_0}\sqrt{g_0}} = \frac{g_{n+1}}{g_0}$$

$$= \gamma = -c_0 c_{n+1} \,.$$

In other words, <u>the partial autocorrelation coefficient for two time series points separated by n+1 time units is $-c_0$ times the reflection coefficient c_{n+1}</u>. Therefore, the partial autocorrelation function is the set

$$1, -c_0c_1, -c_0c_2, -c_0c_3, \dots$$

in the marine seismic model. This is the desired relationship between the partial autocorrelation function and the reflection coefficients of a layered system in the marine case ($c_0^2 = 1$), that is, in the free-surface case.

7. Maximum Entropy

The fundamental assumption involved in maximum entropy spectral analysis is that the stationary process under consideration is the most random or the least predictable time series that is consistent with the given measurements. Specifically, the given measurements are the known autocorrelation coefficients, namely,

$$r_k = \frac{1}{2\pi} \int_{-\pi}^{\pi} \Phi(\omega) e^{i\omega k} d\omega \qquad (\text{for } |k| \leq n) . \tag{24}$$

In terms of information theory, we require that the entropy per sample of time series be a maximum. From the work of Shannon (1948) it follows that the entropy is proportional to the integral of the logarithm of the power spectrum; that is, the entropy is

$$\int_{-\pi}^{\pi} \log \Phi(\omega) d\omega . \tag{25}$$

Therefore the required maximum entropy power spectrum is that function $\Phi(\omega)$ that maximizes Eq. (25) under the constraints in Eqs. (24), as given by Burg (1967).

Before we discuss entropy, let us find an expression for the prediction error variance. As we have seen, the prediction error variance for the prediction error operator A_n is

$$g_0 = T_n^2 = \tau_1^2 \tau_2^2 \ldots \tau_n^2 = (1 - c_1^2)(1 - c_2^2) \ldots (1 - c_n^2) .$$

That is, the prediction error variance for the operator $1, a_1, \ldots, a_n$ is equal to the two-way transmission coefficient from the surface through layer n. If we let n tend to infinity, we see that the final prediction error variance for an infinitely long prediction error operator is σ^2, given by

$$\sigma^2 = T_\infty^2 = \prod_{j=1}^{\infty} (1 - c_j^2) .$$

Let us consider the class of all autocorrelation functions that have the same initial section

$$1, r_1, r_2, \ldots, r_n .$$

The tails of all members in this class will be different, but all the beginnings are identical. Because the prediction error operator is determined from normal equations that involve only this common initial section, all members of the class have the same prediction error operators for integers 1, 2, ..., n. As a result, all members of the class have partial autocorrelation functions all with the same beginning, namely

$$1, -c_0c_1, -c_0c_2, \ldots, -c_0c_n \qquad (|c_0| = 1) .$$

Thus, all members of the class have final prediction errors of the form

$$\sigma^2 = [(1-c_1^2)(1-c_2^2) \ldots (1-c_n^2)] [(1-c_{n+1}^2)(1-c_{n+2}^2) \ldots] ,$$

where the first factor is fixed, that is,

$$\sigma^2 = T_n^2 [(1-c_{n+1}^2)(1-c_{n+2}^2) \ldots] , \tag{26}$$

where

$$T_n^2 = (1-c_1^2)(1-c_2^2) \ldots (1-c_n^2) = \text{constant} .$$

Now the question is: Which member of the class has the largest final prediction error variance? The answer is easy, for it may be seen by inspection. Each factor $(1-c_{n+1}^2)$, $(1-c_{n+2}^2)$, ... is maximized if we let $c_{n+1} = c_{n+2} = \ldots = 0$, and therefore

$$\sigma_{max}^2 = T_n^2 .$$

Thus, the required member is the autocorrelation function for which all reflection coefficients from n+1 on are zero. Because this autocorrelation has the greatest final prediction error variance, it represents the time series with the smallest ultimate predictability in the class considered. Since a decrease in predictability means an increase in uncertainty or entropy, this particular autocorrelation with maximum σ^2 is called the maximum entropy autocorrelation, and its Fourier transform is called the maximum entropy spectrum.

What does the maximum entropy case mean in terms of our other quantities? Previously, we said that the new prediction error polynomial A' is obtained from the previous one, A, by the formula

$$A' = A - \gamma z A^* ,$$

where γ is given by $\gamma = -c_0c_{n+1}$, as was shown at the end of the previous section. Because all the reflection coefficients from c_{n+1} on are zero in the maximum entropy case, it follows that all the prediction error operators for n+1, n+2, ..., are the same as the prediction error operator for n. Thus, the final prediction error operator is of finite length and equal to $a_0 = 1$, a_1,

$a_2, \ldots, a_n$. This operator is determined from the normal equations with knowledge of the autocorrelation coefficients $r_0 = 1, r_1, r_2, \ldots, r_n$. Because the final prediction error operator annihilates the entire right-hand side of the autocorrelation (that is, the final prediction error operator results in $g_t = 0$ for $t = 1, 2, \ldots$), it follows that the finite operator $a_0 = 1, a_1, a_2, \ldots, a_n$ annihilates the right half of the autocorrelation. Let us now interpret this result.

The spectral information, or structure of a time series as preserved in its autocorrelation function, represents the negentropy (negative entropy) of the time series. As we have seen, the entropy as manifested by the prediction error operator attacks the autocorrelation function and breaks down its negentropy. The result is the gapped function g_t with a gap from $t = 1$ to $t = n$. The maximum entropy autocorrelation has by definition minimum negentropy, and the prediction error operator for integer n in this case not only produces a gap from 1 to n, but indeed produces a gap from 1 to infinity. In other words, the maximum entropy autocorrelation is that member of the class first to have its right-hand side completely annihilated by the sequence of prediction error operators.

Let us now look at a numerical illustration of these matters. We consider the model shown in Fig. 8, which consists of four layers of finite thickness, the two-way travel time in each such layer being one time unit. Layers 0 and 5 are the upper and lower half-space, respectively. The numerical values of the reflection coefficients are as indicated. Figure 9 shows the progressive destruction of the autocorrelation function as we descend from the top of layer 1 to the top of layer 5, the lower half-space.

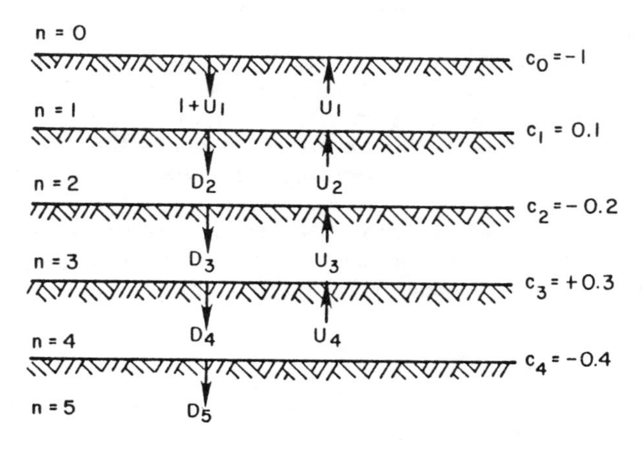

Figure 8. A model having four layers of finite thickness, with reflection coefficients c_i as shown.

At the top of layer 1, we have by Eqs. (5) and (16) that

$$R = \overline{D}_1 - c_0 U_1 = 1 - c_0 \overline{U}_1 - c_0 U_1 = 1 + \overline{U}_1 + U_1$$

since in the present example we have chosen $c_0 = -1$. The symmetric autocorrelation function R corresponding to our model is pictured as the plot for n = 1 in Fig. 9. The flipped downgoing wave at the top of layer 1 is shown to the left of the origin t = 0, and the upgoing wave at the top of this layer

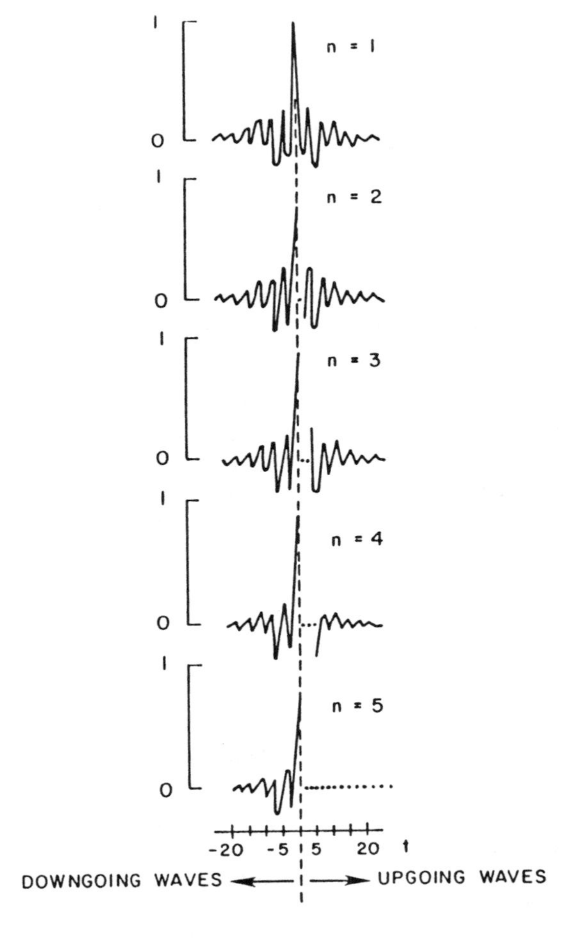

Figure 9. Progressive destruction of the autocorrelation function associated with the layered model of Fig. 8. The dots (•) indicate the gap positions.

is shown to the right of t = 0. Because $c_0 = -1$, we have here that

$$-c_0u_1(1) = u_1(1) = d_1(1),$$
$$-c_0u_1(2) = u_1(2) = d_1(2),$$
$$\vdots$$

and at the top of layer 1 the upgoing wave is identical to the downgoing wave. In fact, the autocorrelation R is the symmetrized marine seismogram.

To produce the gapped function G_n at the top of layer n, where n = 2, 3, 4, and 5, we use Eq. (23),

$$G_n = A_n R ,$$

where A_n is the prediction error operator

$$A_n = P_n - c_0Q_n = P_n + Q_n \qquad (\text{since } c_0 = -1)$$

[see Eq. (20)]. These gapped functions are shown in Fig. 9 for each value of the layer index, n. Note that the gap increases in size from one time point (n = 1) at the top of layer 2 to two time points (n = 2) at the top of layer 3 and so on until, at the top of the lower half-space (layer 5), the gap becomes infinitely long. At this point the entire right-hand side of the autocorrelation R has been annihilated, and we see that $A_4 = (1, a_1, \ldots, a_4)$ is the final prediction error operator. This operator is determined from the normal equations with knowledge of only the first n+1 = 4+1 = 5 autocorrelation coefficients, namely $r_0 = 1$, r_1, r_2, r_3, and r_4.

We next use Eq. (26) to obtain the prediction error variance, σ^2, for the present example:

$$\sigma^2 = T_4^2[(1-c_5^2)(1-c_6^2)\ldots] .$$

But since $c_5 = c_6 = 0$, the value of the final and maximum prediction error variance is

$$\sigma_{max}^2 = T_4^2 = (1-c_1^2)(1-c_2^2)(1-c_3^2)(1-c_4^2) = 0.72649 .$$

Thus, the autocorrelation $R = (r_0 = 1, r_1, r_2, r_3, r_4, \ldots)$ is the one for which all reflection coefficients from n+1 = 4+1 = 5 on are zero. Since this autocorrelation has the maximum final prediction error variance, it is in fact the maximum entropy autocorrelation, and its Fourier transform is the maximum entropy spectrum. The waveforms shown to the left of the origin t = 0 in Fig. 9 represent T_n times the flipped downgoing wave at the top of layers 2, 3, 4, and 5, as measured from their first arrivals. Similarly, the waveforms to the right of t = 0 represent $-T_n c_0$ times the upgoing wave at the tops of these same layers. There is, of course, no upgoing wave in the lower half space (layer 5) because all reflection coefficients for $n > 4$ vanish.

8. Concluding Remarks

Purely mathematical reasoning alone cannot yield physical insight, and if anything physical results from the mathematics, physical considerations must have been employed in the theory in the first place. The problem, therefore, is to find out where physics enters the theory. The first step in any appraisal is to separate the mathematical equations from the physical situation accompanying them. The mathematics has consequences and properties of its own, which are of purely algebraic or analytic significance; they remain true independently of any physical hypothesis; indeed, they may be as applicable to economics as they are to geophysics. When we look at deconvolution, we must therefore separate the physical hypothesis from the attendant mathematical approach. Deconvolution entails both of these aspects; in its application, one cannot have one without the other. Without going into the entire story here, let us merely say that one of the physical assumptions of deconvolution is the hypothesis of a stratified or layered system in the earth, which produces unwanted reverberations on the seismic records. Let us emphasize that this hypothesis is physical; it has been verified countless times by indirect as well as direct evidence, the latter provided by the drilling of wells for oil and gas. In fact, few hypotheses in science can be more physical to the human senses than this one.

Scientific questions can be resolved only by scientific criteria, that is, through the verification of theoretical models by means of experimental data. The layered medium that we have considered in this paper represents a theoretical model. In particular, when n layers overlie a homogeneous half space (that is, when $c_{n+1} = c_{n+2} = c_{n+3} = \ldots = 0$), we have seen that a finite-length prediction error operator suffices to break down the autocorrelation. This prediction error operator is the deconvolution operator. Such an operator is necessarily minimum-delay or, equivalently, minimum-phase-lag. The deconvolution operator characterizes the n-layer stratified system. It has been used for many years in the oil exploration industry in day-to-day operations. Because the prediction error is the essence of seismic deconvolution, the focus of attention is at the end of the observation time interval, so that new primary reflected events can be detected. This aspect explains the fact that deconvolution operators are one-sided; as a result, many of the computational formulas favor prediction as opposed to hindsight. Deconvolution was developed within the oil exploration industry, and it is only in the past few years that its value has been appreciated in other fields, such as speech and image processing. Apparently, this delayed interest is due to the fact that much traditional time series theory has been concerned with spectral analysis. Spectral analysis is at a premium in static problems, where the attention is equally distributed over the entire observation interval. In such cases, symmetry is then also at a premium.

To meet the demand for spectral estimators in other sciences, the deconvolution process can be related to the "maximum entropy spectral analysis" as originated by Burg (1967, 1975). The maximum entropy method (MEM) rests on a physical hypothesis as well as on mathematical considerations. The physical hypothesis, whether stated explicitly or not, is that the physical system is an n-layered stratified medium which, as we have seen, is indeed a maximum entropy system. The MEM assumption, however, should be used with considerable care in unknown physical situations not readily amenable to verification. Complete symmetry characterizes the mathematics of MEM. The essential feature of the Burg (1967) algorithm is to put the hindsight errors and the prediction errors on exactly equal footing. The deconvolution operator, being nonsymmetric, is used only as an intermediate result. The phase spectrum of the process, being antisymmetric, is ignored, and the amplitude spectrum, being symmetric, is kept as the end product; namely, the square of the amplitude spectrum is the maximum entropy power spectrum. The discarding of the phase spectrum, however, means that some of the mathematical implications of the physical hypothesis may be lost. We feel, therefore, that both the phase spectrum and the amplitude spectrum should be considered whenever the MEM approach is used.

The normal incidence synthetic seismogram constitutes a classic example of the fruitful symbiosis between a physical theory, namely the horizontally layered medium, and the mathematical consequences arising from the postulated physics. The entropy of the synthetic seismogram increases monotonically as the surface excitation penetrates to deeper and deeper strata. This increase in entropy manifests itself by the growing gap that the layering introduces into the originally symmetric autocorrelation function. We have obtained an explicit relationship between the partial autocorrelation function and the reflection coefficients, and have indicated how the behavior of these entities governs the behavior of the associated seismograms. Finally, we have tried to show how the above concepts affect the estimation of the maximum entropy spectrum and, perhaps most important, we have tried to establish why these concepts take on such practical significance when they are applied to the seismic deconvolution problem.

In the late 1950s a digital revolution occurred because of the introduction of transistors in the building of digital computers, which made possible reliable computers at much lower cost than previously. As a result, the seismic industry completely converted to digital technology in the early 1960s, a long 10 years after the first digital results were obtained. Since then, nearly every seismic record taken in the exploration of oil and natural gas has been digitally deconvolved and otherwise digitally processed by these methods. The final result of the digital processing of seismic data was the discovery of great oil fields that could not be found by analog methods. These include most of the offshore discoveries, as in the North Sea, Gulf of Mexico, and Persian Gulf, as well as great onshore discoveries in Alaska, Asia, Africa, Latin America, and the Middle East, made in the past 20 years. Today an oil company will deconvolve and process as many as a million seismic traces per day.

Whereas the digital revolution came first to the geophysical industry largely because of the tremendous accuracy and flexibility afforded by large digital computers, today we are in the midst of a universal digital revolution of epic proportions. Digital signal processing is a dynamic field that involves the exploration of new technology and the application of the techniques to new fields. The technology has advanced from discrete semiconductor components to very large-scale integrated circuits (VLSIs) with densities above 100,000 components per silicon chip. The availability of fast, low-cost microprocessors and custom high-density integrated circuits means that increasingly difficult and complex mathematical methods can be reduced to hardware devices as originally envisaged by Wiener, except the devices are digital instead of analog. For example, a custom VLSI implementation of linear predictive coding is now possible, requiring a small number of custom chips. Whereas originally digital methods were used, at great expense, only because the application demanded high flexibility and accuracy, we have now reached the point that anticipated long-term cost advantages have become a significant justification for the use of digital rather than analog methods.

In closing our discussion of one-dimensional inversion, let us mention the fixed entropy estimate (Robinson, 1978) of the seismic acoustic log. This scheme is an autocorrecting iterative method, based on a forward model, to compute the acoustic impedance of the earth from the reflection seismogram. This method was offered commercially by Digicon, Inc., in 1968-1969 under the trade name S.A.L. (for seismic acoustic log), and it has proved successful as a practical inversion scheme in seismic data processing centers.

References

Aki, K., and Richards, P. G. (1980) Quantitative Seismology, vol. II (San Francisco: W. H. Freeman).

Bardan, V. (1977) Comments on dynamic predictive deconvolution, Geophys. Prospect. **25**, 569-572.

Burg, J. P. (1967) Maximum entropy spectral analysis, Society of Exploration Geophysicists Meeting, Oklahoma City.

Burg, J. P. (1975) Maximum Entropy Spectral Analysis, Ph.D. thesis, Dept. of Geophysics, Stanford University, 123 pp.

Gelfand, I. M., and Levitan, B. M. (1955) On the determination of a differential equation by its spectral function, Am. Math. Soc. Trans. **1**, 253-304.

Kac, Marc (1966) Can one hear the shape of a drum? Am. Math. Mon. **73**, 1-23.

Kunetz, G. (1964) Généralisation des opérateurs d'antirésonance à un nombre quelconque de réflecteurs, Geophys. Prospect. **12**, 283-289.

Levinson, N. (1947) The Wiener RMS error criterion in filter design and prediction, J. Math. Phys. **25**, 261-278.

Rayleigh (1877) The Theory of Sound (New York: Dover, 1945, reprint of 2nd (1894) edition).

Robinson, E. A. (1967) Multichannel Time Series Analysis with Digital Computer Programs (Oakland, Calif.: Holden-Day), 298 pp. (revised edition, 1978).

Robinson, E. A. (1975) Dynamic predictive deconvolution, Geophys. Prospect. **23**, 779-797.

Robinson, E. A. (1978) Iterative identification of non-invertible autoregressive moving-average systems with seismic applications, Geoexploration **16**, 1-19.

Robinson, E. A. (1982) Spectral approach to geophysical inversion by Lorentz, Fourier, and Radon transforms, Proc. IEEE **70**, 1039.

Robinson, E. A., and Treitel, S. (1978) The fine structure of the normal incidence synthetic seismogram, Geophys. J. R. Astron. Soc. **53**, 289-310.

Shannon, C. E. (1948) A mathematical theory of communication, Bell Syst. Tech. J. **27**, 379 and 623.

Wiener, N. (1930) Generalized harmonic analysis, Acta Math. **55**, 117-258.

Wiener, N. (1933) The Fourier Integral and Certain of Its Applications (Cambridge: Cambridge University Press).

APPLIED SEISMOLOGY

Enders A. Robinson

University of Tulsa, Tulsa, Oklahoma 74104

Geophysical inversion seeks to determine the structure of the interior of the earth from data obtained at the surface. In reflection seismology, the problem is to find inverse methods that give structure, composition, and source parameters by processing the received seismograms. The pioneering work of Jack Cohen and Norman Bleistein on general inverse methods has caused a revolution in the direction of research on longstanding unsolved geophysical problems. This paper does not deal with such general methods but instead gives a survey of some production-type data processing methods in everyday use in geophysical exploration.

The fundamental digital processing techniques used to analyze seismic records taken in the exploration for petroleum and natural gas are described. These methods include the determination of the seismic velocity function for a prospect, static and dynamic corrections to the seismic traces, multiple coverage of a prospect and stacking, deconvolution, and migration. The concept of minimum-phase, or, what is the same thing, the concept of minimum-delay, is defined and several theorems on minimum-delay are given. The statistical model used in the method of the predictive deconvolution of seismic traces is presented, and it is shown that the method of predictive deconvolution is valid for removing either long-period or short-period reverberations even in the case of a non-minimum-delay source pulse.

The purpose of seismic migration is to find the coordinates of inaccessible subsurface points by use of seismic reflection measurements taken on the earth's surface. The three most popular conventional migration methods are the finite-difference method, the Kirchhoff method, and the frequency-wavenumber method. Each is based on a stratified model, that is, a two-dimensional model in which all the interfaces encountered by propagating waves must be horizontal. Such a model is called a 1-1/2 dimensional (1.5-D) model, because a true two-dimensional model would allow dipping interfaces. The 1.5-D model corresponds to a one-dimensional W.K.B. approximation to the wave equation. The surface seismic data can be downward continued (that is, backtraced) into the earth in conventional

C. Ray Smith and W. T. Grandy, Jr. (eds.),
Maximum-Entropy and Bayesian Methods in Inverse Problems, 211–242.

migration because the 1.5-D model has only flat horizontal interfaces, so no unknown dips are ever encountered. If in the real earth there is just one dipping interface, then all the conventional migration results below this interface are wrong. Thus in any practical situation, finite-difference migration, Kirchhoff migration, and frequency-wavenumber migration suffer serious specification errors. Larner depth migration, which in practice is replacing the 1.5-D methods, uses a true two-dimensional model, and, although it is more difficult to implement, it is correctly specified. As a justification for the use of 1.5-D wave equation migration on seismic data, in this paper we propose the random reflector-dip hypothesis. This statistical justification is the only reason that conventional migration methods give satisfactory results on actual seismic data.

1. Introduction

Deposits of petroleum and natural gas are located in reservoirs deep below the surface of the earth. To find these resources, geophysical methods of exploration are used, the seismic method being the most important. Exploration seismology is divided into the branches of reflection seismology and refraction seismology. Most of the exploration for petroleum and natural gas is done by reflection seismic methods.

Reflection seismology is a method of mapping the subsurface structure of the earth. An energy source is set off at or near the surface. The source may be a dynamite explosion, a weight-dropping machine, a vibration-inducing machine, or some other means of transmitting seismic energy into the earth. The seismic waves travel from the source to subsurface layers within the earth, from which they are reflected back to the surface. The returning seismic waves are then recorded by instruments. From knowledge of the arrival times of the seismic events that are reflected from the subsurface layers, the subsurface structure of the earth is then contoured. The earth's sedimentary layers are approximately horizontal, but they do have features such as anticlines, unconformities, and faults that can serve as traps for petroleum and natural gas.

The seismic method is based upon the propagation of elastic waves in the earth and their reflection and refraction due to changes in the earth's velocity-density distribution. Active sources of energy are required—dynamite, air guns, and chirp signal generators are the most widely used sources. Detection of the faint pressure or particle motion at or near the surface is achieved by use of sensitive pressure gauges or geophones. The received signal is amplified, recorded (usually by digital methods), processed (usually in a digital computer), and displayed in a form that is interpretable in terms of geologic structure, stratigraphy, and hydrocarbon content. Displays of reflection data look like slices through the earth. It is important to understand why such a simplistic viewpoint often fails.

The refraction method peels off the layers, divulging the gross features of the velocity distribution with depth.

The pictorial view of the reflection method shows wavefronts of seismic energy progressing outward from the source with reflections at discontinuities in the geologic section. The received energy is recorded at or near the surface on a spread of detectors located at various distances from the source. The arrival times of reflections are observable, and the distances traveled can be calculated provided the velocity distribution of the beds traveled through is known.

When the shot point is located in the middle of a spread of detectors, a "split-spread" record is produced. When the source is offset from the end of the spread, "end-on" records are produced; in marine work, this is done by using a towed cable containing a spread of detectors.

The first arrivals on each trace travel a refraction path from the source to the detector in the usual case on land where low-velocity weathered material overlies a higher velocity subweathering. In sand/shale sequences, an abrupt velocity change from the order of about 2,000 ft/s to about 6,000 ft/s occurs at the natural water table. The presence of limestones and other high velocity rocks near the surface will, of course, influence the ray paths of the first breaks. On marine data, the first breaks arrive directly through the water, depending upon the velocity of the bottom sediments and the water depth, until the critical distance for refractions from the water bottom is exceeded. Reflections can be identified by the in-phase lineups across the spread. A given reflection arrives progressively later as the offset (the source-to-detector separation) increases because the length of the travel path to a given reflector progressively increases with offset. The reflection lineups have somewhat erratic arrival times because of near-surface variations.

To map the subsurface, the geophysicist must convert the received seismic traces that record the events as a function of time into traces that record the events as a function of depth. In other words, a time function recorded at the surface must be transformed into a depth function. Unlike radio waves, seismic waves have a velocity that is very much dependent upon the medium. That is, the velocity of the seismic waves changes as the waves travel into the earth. Generally, the velocity increases with depth, although occasionally there may be layers in which a decrease of velocity occurs. For a given surface point, the velocity plotted as a function of depth is called the velocity function.

In all fields of geophysics, and in seismology in particular, much of the basic data collected is in the form of time series. The analysis of these time series provides methods through which physical understanding may be derived from the large mass of data. In some cases the data are uncomplicated and the physical information they contain may be abstracted by straightforward analytical methods. More frequently, however, the data are complicated, and refined computer analyses are necessary.

Any observed geophysical phenomenon is bound up with a large number of other phenomena that are not observed. For example, if we select any given geophysical process as the subject of study, we would find that there are central relationships that determine the basic features of the process, but also there are nonessential relationships that affect only some secondary features. In studying the process one must discover and take into account the essential features and at the same time disregard the unimportant details caused by subsidiary features. Thus it is not a geophysical phenomenon in all its complexity which is subject to analysis, but a simplified model whose behavior coincides basically with the behavior of the phenomenon except for details of a minor nature. The criterion for the correctness of a model is the agreement between theoretical results and practical data. Moreover, the model of a geophysical phenomenon should be constructed on the basis of making explicit its connections with related phenomena. The subdivision of factors into essential and nonessential ones, moreover, depends not only on the specific nature of the geophysical phenomenon itself, but also on the actual problem to be solved. Of course, it is often not possible to find such an ideal model in geophysics, so compromises between the ideal and the obtainable must be made.

In classical geophysics a great many models are known in which the behavior of the system is fully determined by initial-value and boundary-value conditions. Such is the case in the earth model first treated in a classic paper by Lamb (1904). This model is that of a perfectly elastic, homogeneous, isotropic medium bounded by a free plane surface. Lamb showed that a vertical or horizontal impulse along a line on the surface produces a P-pulse, an S-pulse, and a Rayleigh pulse, in that order. The physical laws that apply to models of this sort are known as dynamical laws. These laws are characteristic in cases where there is a unique specification of the consequences of a given cause.

In addition to models of geophysical phenomena which lead to the setting up of dynamical laws, other models lead to the formulation of laws of a different nature, namely statistical laws. To clarify this concept, let us consider as an example a model taken from geology. Time plays a peculiar role in geology. Whereas in most sciences time can be taken as an independent variable which is assumed to be known, the geologist sees time as a dependent variable. As a result the geologist is faced with problems unique to his science in his effort to measure time quantitatively in terms of events that have occurred over billions of years. Geologic time over these past eons can be defined only by observations and measurements taken on the earth, moon, and planets together with astronomical evidence. These measurements, whether they be geological, paleontological, geochemical, radiological, geophysical, or astronomical, are subject to imperfections. These errors lead to statistical fluctuations in the reported measurements of geologic time. As a result, geologic time as we know it is not a uniform variable but is a variable subject to chance effects; that is, geologic time is a random variable.

Let us consider another example, namely the problem of determining the depths of the stratigraphic layers in the earth by seismic prospecting for oil and natural gas. The deep sedimentary layers were laid down in geologic time in an unsystematic way, and thus the seismic events produced by these layers are unsystematic in space and time. If it were possible to have at our disposal unlimited computational means and extremely detailed and accurate data, a dynamic model could be constructed that would make use of the laws of wave propagation to give the seismic wave motion. However, the practical and theoretical difficulties connected with the solution of such a problem are virtually insurmountable. As an alternative we can construct a statistical model. Since the knowledge of the geophysicists working on the prospect is incomplete and imperfect, this partial knowledge can be treated from a statistical point of view. Thus the depths and reflectivities of the geologic beds can be treated as random variables. The use of the term "random variable" does not imply that the variables which represent depths and reflectivities are ones whose values are uncertain and can be determined in any given instance by a "chance" experiment. In other words, the seismic variables are not random in the sense of the frequency interpretation of probability, because such variables are fixed by the geologic structure. Instead the variables representing depths and reflectivities are similar to, say, a variable which represents the billionth digit in the expansion of $\pi = 3.1415926...$, which although unknown is a definite fixed number. A major justification of using such a statistical approach in geophysics is the fact that large amounts of data must be processed, and any data in large enough quantities take on a statistical character, even if each individual piece of data is of a deterministic nature. This approach would correspond to speaking of the probability distribution of the billionth digit in the expansion of π, and whether two successive digits in the expansion are uncorrelated. In the same manner we can speak of the probability distribution of a deep reflection coefficient, and whether two successive reflection coefficients from deep interfaces are uncorrelated.

Geophysical phenomena are usually so complex and so variable that methods that are strictly deterministic are often inadequate. In fact, methods that do not take into account the statistical character of the data have been known to produce useless and sometimes extremely misleading results. The situation in geophysics is complicated further by the enormity of the pertinent observational data. Although it might be tempting to perform careless analyses because of this large volume of data, this situation must be avoided. Adequate results in petroleum exploration can result only from extremely careful and tedious examination and analysis of the seismic data. In this paper the fundamental digital processing techniques used to analyze seismic records taken in the exploration for petroleum and natural gas are described. These methods include static and dynamic corrections to the seismic traces, multiple coverage of a prospect and stacking, determination of the seismic velocity function, deconvolution, and seismic migration.

2. Static and Dynamic Corrections

The direct problem in geophysics may be thought of as the determination of how seismic waves propagate on the basis of known make-up of the subsurface of the earth. The inverse problem is to determine the subsurface make-up on the basis of wave motion observed at the surface of the earth. The inverse problem is not unique to geophysics. A large part of our physical contact with our surroundings depends upon an intuitive solution of inverse problems. In many other real-world problems, we must also infer the size, shape, and texture of remote objects from the way they transmit, reflect, and scatter traveling waves. For example, in x-ray computerized tomography, it is necessary to combine x-ray scans taken at different angles to form a cross-sectional image which represents the internal details of the scanned structure. In nondestructive testing, a three-dimensional refractive-index field is reconstructed from holographic measurements taken at different angles. In electron microscopy, three-dimensional biological structures are deduced from two-dimensional electron micrographs taken at different tilt angles. Optimization techniques are being developed for digital image reconstruction from various types of projections.

It is desired with the reflection method to know the depths (or arrival times) vertically below a datum. This requires application of two corrections to the seismic field data: a static correction and a dynamic correction. Let us consider static and dynamic corrections to the seismic traces.

Because of lateral variations in the thicknesses of the near-surface layers, each trace is corrected by a time shift referred to as a static correction. The static correction is additively composed of a source correction and a receiver correction, and has the effect of placing the source and receiver on a fictitious horizontal datum plane. The static correction overcomes the effects of changes in elevation at source and receiver locations, changes in source and receiver depths, variations in velocity and thickness of the weathered zone on land, and variations of water depth and subbottom velocity on marine data. For a given source-receiver location, the static correction is a constant time shift applied to the corresponding seismic trace.

Suppose the source is at depth h below a surface point whose elevation is e_s, and the receiver and datum elevations are e_r and e_d. Suppose the source is beneath the weathering and in a sub-weathered zone with constant velocity v_1. The time correction T_{sd} to move the source to datum is

$$T_{sd} = [e_d - (e_s - h)]/v_1 .$$

The receiver at zero offset can be moved to the source depth by subtracting the "uphole time" T_u (the travel time from the source to the surface) from the record time. The receiver can then be moved to datum by the time correction T_{sd}. The time correction T_{rd} to move the zero-offset receiver to datum is then

$$T_{rd} = - T_u + T_{sd} .$$

The correction to move both source and zero-offset receiver to datum is the static correction T_s

$$T_s = 2T_{sd} - T_u .$$

The static correction applies to receivers close to the source location and is measurable at each source point. Receiver points intermediate to the source points usually have an additional correction due to changes in elevation or weathering thickness and velocity. First break arrivals are often used in determining corrections at points intermediate to the source points. Under the assumptions that the base of the weathering is flat and the weathering velocity v_0 is constant, elevation changes will result in an addition correction of $(e_s - e_r)/v_0$.

The near-surface velocity distribution is usually determined by an uphole survey, recording uphole times as a function of shot depth. From this, the depth of the base of the weathering and the average weathering and subweathering velocities are obtained.

Static corrections are always critical on land data, and sometimes critical on marine data, and constitute one of the most significant problems in reducing the data to a meaningful form for interpretation.

The dynamic correction converts each trace to the equivalent trace that would have been received if the source and receiver had been at the same lateral point, namely the point midway between the actual source and receiver locations. In this conversion we are referring to traces made up only of so-called primary reflections. According to ray theory, a primary reflection results from a ray path traveling down from the source to the reflecting horizon and then traveling directly upward to the receiver. Given the velocity function, this ray path can be computed by means of Snell's law. If the layers are horizontal, then all the reflection points (or depth points) are always directly beneath the midpoint between source and receiver. If the layers are dipping, then the depth points are offset from the midpoints. As a result, the dynamic corrections depend on both the velocity function and the dip of the reflecting beds. The component of the dynamic correction due to the separation of source and receiver is called the normal moveout correction, and the component due to dip is called the dip correction. The dynamic correction overcomes the effect of offset distance between source and receiver. Increasing the offset increases the arrival time of a reflection from a given layer due to increased distance traveled to the reflector and back to the surface. This geometric effect is called "normal movement," abbreviated NMO.

In the no-dip case with constant velocity v to the reflector, the reflection time at offset distance x is T_X and the reflection time at zero-offset is T_O. The direct arrival time at offset distance x is x/v. These three arrival times constitute the sides of a right triangle obtained by using the source point, the image point of the source, and the receiver point as the three vertices of the triangle. By the theorem of Pythagoras [see Fig. 1(a)],

$$T_X^2 = T_O^2 + (x/v)^2 .$$

The time-distance curve of T_X vs. x is a hyperbola with vertex at $T = T_O$ and $x = 0$. The asymptotes to the hyperbola are $T = \pm x/v$. The difference between T_X and T_O is the normal movement Δ given by

$$\Delta = \sqrt{T_O^2 + (x/v)^2} - T_O .$$

For a given reflector (a given T_O), the NMO increases as x increases. At a given offset distance, the NMO decreases as the reflector depth increases; that is, the correction on a given trace changes with record time—the correction is dynamic rather than static.

In the dipping case, with dip angle α, the hyperbola is shifted with its vertex located at $T = T_O \cos\alpha$ and $x = vT_O \sin\alpha$. The normal moveout is not symmetrical about $x = 0$ as in the no-dip case. [See Fig. 1(b).]

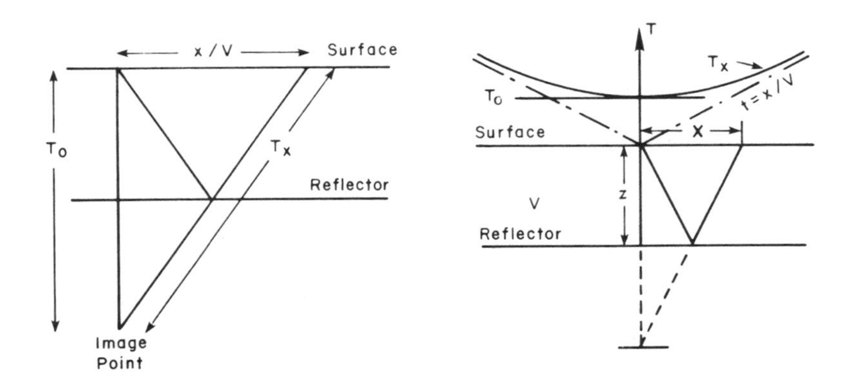

(a) NO DIP

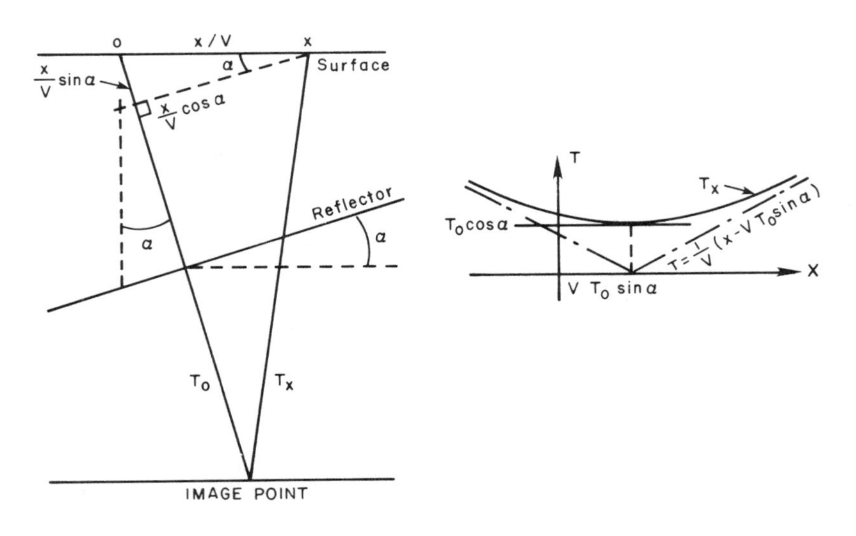

(b) DIPPING BEDS

Figure 1. Normal moveout correction on seismic data.

Parabolic approximations to the hyperbolic moveout expressions are obtained by substituting $T_O + \Delta$ for T_X, expanding the resulting $(T_O + \Delta)^2$ term, and neglecting Δ^2. In the no-dip case,

$$\Delta = \frac{1}{2T_O} \left(\frac{x}{v}\right)^2$$

and in the dipping case,

$$\Delta = \frac{1}{2T_O} \left(\frac{x}{v}\right)^2 - \frac{x}{v} \sin \alpha \ .$$

In summary, there are four important corrections that must be made to the recorded seismic traces: the source correction, the receiver correction, the normal moveout correction, and the dip correction. The source and receiver corrections are the static corrections, and the normal moveout and dip corrections are the dynamic corrections.

3. Multiple Coverage and Stacking

Each seismic trace is a time series made up of reflected events together with various interfering waves and noise. The desired reflected events are the primary reflections, that is, waves that travel down to a given reflector and then back up to the surface where they are recorded. An important type of undesirable interfering wave is the multiple reflection. A multiple reflection is a seismic event that has undergone more than one reflection between source and detector. As a seismic wave travels through the earth, it is split into a reflected and a transmitted wave at each acoustic discontinuity. Each of these waves upon encountering another discontinuity splits into a pair of waves, and so on ad infinitum. The presence of multiple reflections makes the identification of primary reflections difficult or impossible on the recorded seismic traces. Thus it is necessary to process the traces so as to attenuate the multiples as much as possible.

One way to make the static and dynamic corrections as well as to attenuate the multiples is through the use of redundancy. The engineering aspect of collecting seismic data is concerned with an acquisition scheme in which adequate redundancy is incorporated into the system in order to overcome the effects of the multiples. To achieve this goal, reflection seismic data are collected in a special way. For each point source, a spread (or array) of detectors is laid out. The source is activated and the traces are recorded. Then the entire configuration of source and detectors is moved laterally to a new surface location and the process is repeated. The configuration is moved in small enough increments that each depth point is covered several times, and thus multiple coverage of the depth point is obtained. For example, in sixfold coverage six seismic traces are recorded for each depth point. In twenty-four-fold coverage, twenty-four traces are recorded for each depth point. Thus the recording of seismic data by a multiple coverage scheme introduces a considerable amount of overlap or redundancy.

Consider seismic data collected in the following manner (Fig. 2). An in-line spread of n traces with equal spacing Δd is used with the source offset from the nearest trace by Δd. (Any arbitrary but fixed offset distance may be used.) After the first shot is taken, the entire system is moved by $\Delta s = \Delta d/2$ for the second shot, $2\Delta s$ for the third shot, etc. In the no-dip case, the following raypaths to a given interface have a common reflection point: trace 1 on record 1, trace 2 on record 2, ..., trace n on record n. The offset distances for these traces are Δd, $2\Delta d$, ..., $n\Delta d$. The reflection from this interface will have increasing normal moveout on the successive traces. The fold depends upon the number of traces and the ratio $\Delta s/\Delta d$. In the above case, the ratio $\Delta s/\Delta d = 1/2$ and the fold equals n. If the ratio is unity, then the fold equals n/2. The general relationship is that the fold equals $(n/2)/(\Delta s/\Delta d)$, which holds for any $\Delta s/\Delta d$ ratio for which the fold has an integer value less than or equal to n.

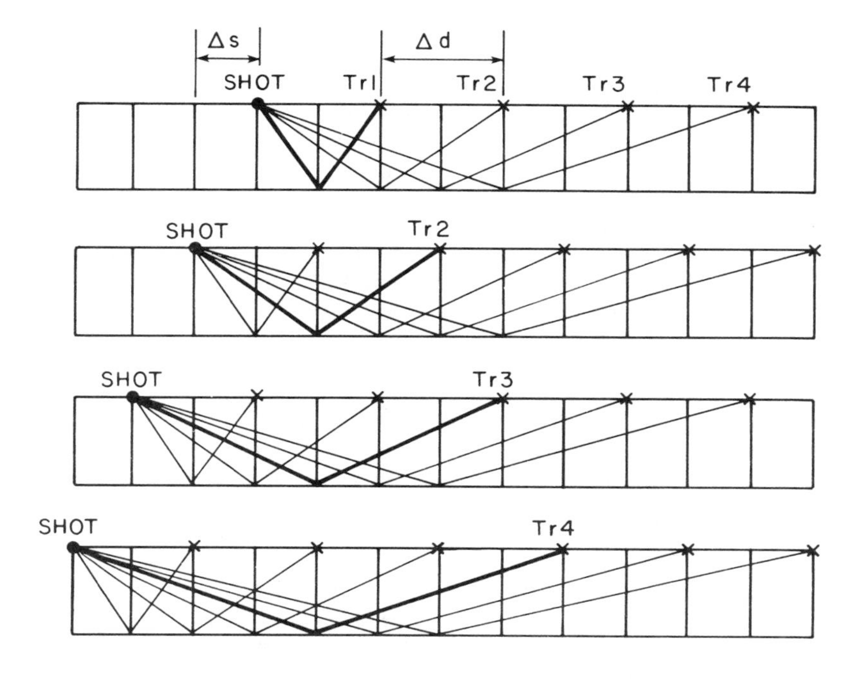

Figure 2. Field procedure for data collection; $\Delta s = (1/2)\Delta d$.

All the traces in the prospect can be sorted into various types of data sets. A data set in which all the traces have the same midpoint is called a common midpoint gather. Another type of data set is the common source gather, made up of all those traces with the same source. The common receiver gather is made up of all those traces with the same receiver. Finally the common offset gather is made up of all those traces with the same offset distance between source and receiver. The redundancies introduced by multiple coverage allow the geophysicist both to make the necessary static and dynamic corrections and to attenuate severely the unwanted interference such as multiples and noise.

For example, let us consider the normal-moveout component of the dynamic correction. Let us consider a common midpoint gather. Under the ceteris paribus (other things being equal) assumption, let us suppress for the time being all corrections except the normal moveout correction. The normal moveout correction would convert each primary-reflection trace in a common midpoint gather into the same equivalent trace, namely that primary-reflection trace that would have been received if source and receiver had been at the common midpoint in question. In other words, under the appropriate normal moveout correction, the primary reflections of all the traces in a common midpoint gather will be in phase, thus making the corrected traces coherent. Of course, the multiple reflections will not be in phase after the normal moveout correction, as the correction is the correct one only for primary events. Because we do not know the appropriate normal moveout correction beforehand, we must estimate it from the data. One way of estimation is the trial-and-error procedure; we make many trials and pick the one with the least error. More specifically, let us compute the coherency of the traces in a common midpoint gather after a given normal moveout correction. If we plot the coherency for many different corrections, the appropriate correction would be that one for which the coherency is a maximum, under the conditions of the ceteris paribus assumption. In practice, a preliminary normal moveout correction is often applied to the data using a preliminary velocity function. In such a case this discussion would then refer to a second-order correction, namely the residual normal moveout correction, which in effect represents a correction of the preliminary velocity function. The same ceteris paribus approach discussed here for normal moveout corrections can also be used to make appropriate source, receiver, and dip corrections in the different types of gathers of the traces. On the other hand, instead of making the corrections one at a time under ceteris paribus assumptions, we can use a simultaneous approach. Because of the complexity of the problem, iterative methods are more readily devised than direct methods.

We have seen that the redundancy of the seismic data allows us to make static and dynamic corrections. This same redundancy allows us to attenuate unwanted interference such as multiple reflections. As we have seen, the normal moveout correction puts all the primary reflections in phase on the traces in each common midpoint gather. Because the ray path of a multiple reflection is different from the ray path of a primary reflection with the same arrival time, the normal moveout correction does not put the multiple reflections in phase. Hence if we add together all the corrected traces in a common midpoint gather, the primary reflections will add in phase and the multiple reflections will add out of phase. The operation of adding the corrected traces in a common midpoint gather is called stacking; the result is an output trace that is called the stacked trace for that midpoint. The stacked trace preserves the primary reflections because they were added in phase, but the multiples are severely attenuated on the stacked trace because they were added out of phase. In summary, stacking is the process of obtaining one output trace for each midpoint such that the primary reflections are preserved and the multiple reflections as well as incoherent noise are destroyed.

Considering the complexity of multiple reflection generation, it may seem remarkable that the seismic reflection method based upon primary reflections works at all. The explanation seems to be that although multiples exist whenever there are primaries, their amplitudes relative to primaries are small in the case when the reflectivity function has the characteristics of white noise (that is, whenever the reflectivity function has the properties of a sequence of uncorrelated random variables). For a futher discussion of this case, see Robinson (1980). When this is not the case, multiple reflections are a serious problem. Most poor records probably are due to severe multiple interference. Multiple reflections have been recognized as a source of seismic noise since the beginnings of seismic prospecting. Geophysics, Volume 13, Number 1, 1948, was devoted to current knowledge of multiples, which consisted primarily of identification. Reproducible recording, which must necessarily precede techniques for multiple suppression, and the communication theory model of the seismic method, which gives deeper insight into multiple generation, were both yet to come.

Multiples have a wide range of coherence. Those easily identified on field data arise from multiple bounces between interfaces having large reflection coefficients. Their ray paths can be deduced from the observed primary arrival times. A multiple having twice the arrival time of a primary is called a W-type. One whose arrival time is the sum of two different primary arrival times is called a peg-leg multiple. Multiples such as the peg-leg and W-types having three bounces in the path are called first-order multiples. Second-order multiples have five bounces, third-order, seven bounces, etc. Because reflection coefficients have magnitude less than one, the multiple amplitudes tend to decrease as the order increases. Given an upper bound on subsurface reflection coefficients of 0.3 and a surface coefficient near unity, it is apparent that first-order multiples have an upper bound of $(0.3)^3 = 0.027$ if all bounces are internal (all within the subsurface) and $(0.3)^2 = 0.09$ if the surface is the middle bounce. The latter, called the first-order surface multiples and abbreviated FOSM, is the most significant subset of first-order multiples. With second-order multiples, the upper bounds are $(0.3)^3 = 0.027$ for surface multiples and $(0.3)^5 = 0.00243$ for internal multiples. Higher-order multiples have still smaller upper bounds. It is apparent from considering the upper bounds that only first-order surface multiples, and possibly first-order internal and second-order surface multiples, can compete with primary reflections in regard to amplitude and will be the only ones recognizable as coherent events, except in special cases. All other multiples merely contribute to the underlying random noise.

Stacking serves two purposes in regard to signal-to-noise improvement:

(1) In the usual case where the velocity increases with depth, a surface multiple arriving at the same time as the primary reflection will have greater normal moveout because it sees a lower average velocity than the primary. Hence, after normal moveout correcting of the primary, there will be a residual normal moveout on the multiple. Stacking the traces will enhance the primary because it is lined up by applying its normal moveout and will tend to suppress the multiple because it is out of phase due to the residual normal moveout (see Fig. 3).

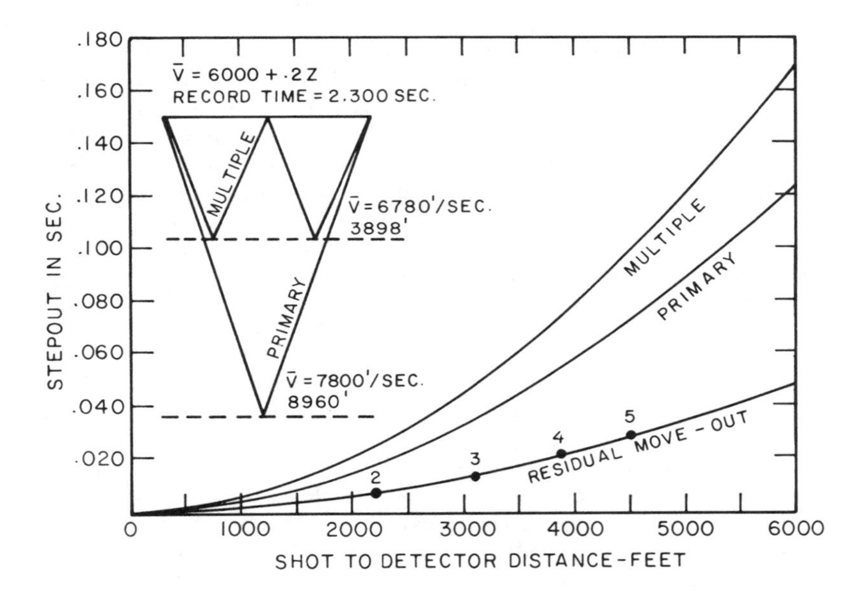

Figure 3. Normal moveout of primary and multiple that have the same arrival time (Mayne, 1962).

(2) Random noise present on the data will be suppressed compared to signal. The ratio of signal amplitude to random noise power increases as the square root of the fold, provided the reflection signal is identical on each trace and the random noises on the traces are mutually uncorrelated. Correlation of signal and noise causes the signal-to-noise ratio to be less than the square root of the fold.

Stacking of data collected at ever increasing separation of source and receiver about a common midpoint has proved to be an extremely powerful processing procedure in improving the signal-to-noise ratio and is especially useful in suppressing coherent multiples. Where the layering is horizontal, the common midpoint coincides with the projection of the common depth point to the surface. Where the layers are dipping, the stacked data do not have a common depth point, but rather have reflections from a subsurface segment whose length increases with dip. To enhance primaries and suppress multiples, it is essential to stack the data along the normal moveout curves appropriate for the primary reflection. In the usual case of increasing velocity with depth, the moveout at any fixed record time T_0 will be less for the primary reflection than for any coherent multiple arriving at that time. The larger the differential normal moveout between primary and multiple reflections, the easier it is to differentially suppress the multiples. Where velocity decreases with depth, it may be difficult to identify primaries and multiples and impossible to separate them by stacking.

Stacking is a multiple suppression technique with wide usage. Its effectiveness depends upon the differential movement between primary and multiple reflections that occurs in the usual case where velocity increases with depth. Then a primary event will see a larger average velocity than the multiple events at any given record time. Hence, the primaries have less moveout than the multiples. Dynamic time shifting the primaries to line them up for all record time at various source-detector separations will force the multiples to lie along their differential moveout curves. Then summing the set of traces will augment primary reflections because they are in phase and will suppress multiples because they are not in phase. Simple stacking (summing) is not the most effective use of the potential power of the pattern response of an n-fold stack. The pattern response has nonuniform spacing because the differential moveout curve is hyperbolic. Simple stacking applies uniform weighting; hence, the pattern is overweighted where the spacing is small. To make the pattern have uniform sensitivity, the weight should be divided by the spacing. This simple procedure will produce an improved stack.

4. Seismic Velocity Function

In reflection seismology there are two equally important variables: time of reflected events and velocity. From knowledge of these variables the depths to the reflecting horizons can be determined. Because there are important lateral changes in velocity, the velocity function will vary from one geographic location to another. As a result, a given velocity function cannot be assumed to be valid for an entire prospect, but must be continually corrected from place to place over the area of exploration.

One method of measuring the velocity function is to "shoot a well." The method uses a deep hole, namely an oil well. The velocity is determined by placing seismic detectors in the hole at various depths and then setting off a seismic source at or near the surface. Estimates of velocity involve measuring travel times of the seismic waves from the source to the detectors. To obtain vertical travel times, the measurements must be corrected for angularity and refraction effects because of the source deviation from vertical. Well shooting gives good estimates of the gross velocity layering.

The invention of the continuous velocity log (CVL) by Summers and Broding (1952) provided for the first time the fine detail of the velocity layering. The CVL tool consists of an acoustic transmitter and either one or two pressure detectors, with acoustic insulators separating the elements by known fixed distances. The source is usually an electromechanical device that produces acoustic pulses in the 10- to 20-kHz frequency range, at a repetition rate of about 20 pulses per second. As the tool is moved slowly up the hole, the set of travel-time measurements produces essentially a continuous measure of transit time either between transmitter and receiver or between the two receivers, depending upon the tool configuration.

The velocity obtained from the CVL measurements is the velocity averaged over the separation distance between elements. For good velocity estimates, the refraction effects must be compensated for. This is especially important with the one-receiver tool where the pulse refracts into the formation at the source and out of the formation at the receiver. With the two-receiver tool, the refraction effects are minimized because only differences in the refraction paths out of the formation at the two receivers must be considered. This has led to almost universal usage of the two-receiver system.

By integration of the transit time curve, a time-depth function can be constructed. Errors in the transit time measurements caused by formation damage during drilling, borehole cavities and other irregularities, low signal amplitude due to absorption, calibration inaccuracies, etc., will introduce errors in the time-depth computation. For this reason, the integrated CVL curve is usually tied to a well survey in order to produce a time-depth function that is accurate throughout the entire section of open hole that is being logged. The CVL cannot measure formation velocities that are less than the borehole fluid velocity, as may happen in gas sands.

Experience has shown that seismic velocity in Tertiary basins can often be approximated by a linear function of depth, $v = v_0 + az$, where $v_0 = 6000$ ft/s and $a = 0.6\ s^{-1}$ are reasonably good estimates in the absence of any measurements. This applies in the Texas-Louisiana Gulf Coast, the San Joaquin Valley in California, and the Maturin Basin of Venezuela, according to Slotnik (1959). Thus, at $z = 5000$ ft, $v = 9000$ ft/s, and at $z = 20{,}000$ ft, $v = 18{,}000$ ft/s. At greater depth, the linear relationship tends to break down.

More accurate detailed velocity information available from well shooting and acoustic velocity logs can be used to build wavefront charts, but this information holds only in the vicinity of the well and must be extrapolated with caution into the unknown away from the well.

Estimates of the velocity distribution with depth can be made from seismic surface measurements. In fact, it is usually necessary to estimate the velocity function by surface measurements since oil wells are available only in old prospects. The moveout is defined as the time difference of the same event recorded at two detectors separated by a given lateral distance in a reflection prospect. From measurements of moveouts at various times along the records, the velocity function can be estimated. In other words, the velocity function is estimated by considering the time differentials Δt of the same event received by a lateral array of detectors. Any such estimate always depends upon a ceteris paribus assumption.

Velocity estimations from surface seismic measurements using Δt analysis based upon normal moveout have been made for many years. The weaknesses in these techniques arise from the following: (1) small errors in Δt result in large errors in v and consequently in the calculated depth, and (2) straight line paths in a medium with uniform velocity are often assumed rather than using least-time paths in layered media that obey Snell's law. The introduction of stacking procedures gave rise to techniques of velocity estimation directly from data that are being collected in a routine fashion. This allows a dense collection of velocity estimation points. The widespread use of stacking has led to an abundance of velocity data that have proved useful for lithologic studies, in addition to its primary purpose of enhancing the stack to increase the primary-to-multiple ratio and to improve the signal-to-noise ratio.

Let us now consider the estimation of interval velocities from surface seismic measurements. A very significant breakthrough in precision of estimating interval velocities from surface measurements was made possible by use of the common midpoint gather. Under the assumption of horizontal layering (no dip), the least time path through a layered medium for small offsets leads to the concept of the RMS velocity, v_{rms}. The RMS velocity in the n-layer case is given by

$$v_{rms,n} = \left[\sum_{i=1}^{n} T_i v_i^2 \Big/ \sum_{i=1}^{n} T_i \right]^{1/2},$$

where T_i is the two-way travel time and v_i is the interval velocity in the ith layer. The importance of the RMS velocity, which is a mathematical quantity devoid of physical meaning, is that the interval velocity layering is measurable from the RMS velocity distribution. The interval velocity in the nth layer is obtained by subtracting the equation for the RMS velocity in the (n-1)-layer case from the RMS velocity in the n-layer case. The result is

$$T_n v_n^2 = v_{rms,n}^2 \sum_{i=1}^{n} T_i - v_{rms,n-1}^2 \sum_{i=1}^{n-1} T_i .$$

The interval velocity in the nth layer, v_n, is then given by the Dix equation (1955)

$$v_n = \left[(v_{rms,n}^2 \sum_{i=1}^{n} T_i - v_{rms,n-1}^2 \sum_{i=1}^{n-1} T_i) / T_n \right]^{1/2} .$$

All quantities on the right-hand side are measurable from the data; the RMS velocities are the slopes of reflection arrival times in (T_X^2, x^2) space and $\Sigma_i T_i$ are the arrival times of the events duly compensated for filter delay.

The RMS velocity differs from the average velocity, which is defined by

$$v_{ave,n} = \sum_{i=1}^{n} T_i v_i \Big/ \sum_{i=1}^{n} T_i \, .$$

The RMS velocity is always greater than or equal to the average velocity, as shown most clearly by Al-Chalabi (1974), who relates the two velocities through a quantity that he calls the heterogeneity factor H. The relationship is

$$v_{rms} = (1+H)^{1/2} v_{ave} \, ,$$

where the heterogeneity factor to the nth interface, H_n, is given by

$$H_n = \frac{1}{z_n^2} \sum_{i=1}^{n-1} z_i \sum_{j=i+1}^{n} z_j (v_i - v_j)^2 / v_i v_j \, .$$

The heterogeneity factor is nonnegative, and equals zero if and only if $v_i = v_j$ for all (i,j), that is, when the earth is homogeneous.

A vast literature has grown up around velocity estimation from common surface point data. The paper by Taner and Koehler (1969) is most instructive in the procedures and uses of what they call the velocity spectra. All traces with common point are gathered together and subjected to a velocity search procedure. At a fixed zero-offset time T_O, the set of n-traces with different offset distances x_i, i = 1,2,...,n, is swept with a collection of hyperbolas satisfying the equation $T_X^2 = T_O^2 + (x/v)^2$, where the collection is indexed on v. Summing the trace amplitudes along each hyperbola and squaring the result will give an amplitude-squared (energy) function of v. The velocity that produces the maximum energy will produce the best stack at time T_O, and this velocity is called the stacking velocity v_s at T_O. This procedure is repeated for a collection of zero-offset times, generally equally spaced in time at increments of t. The result is called the velocity spectrum at common surface point. It consists of a three-dimensional display of stacked energy as a contour in (v_s, T_O) space, or some equivalent display. A trace of the stacked energy maxima as a function of zero-offset time T_O is useful in identifying the coherent events.

Several schemes have been used to measure the stacked energy as a function of stacking velocity, including use of normalized and unnormalized correlation functions, and semblance, in addition to squaring the sum. These coherency measures are discussed by Neidell and Taner (1971). The velocity spectra examples from Taner and Koehler use semblance, which they describe as a normalized output/input energy ratio. Most investigators do not believe that there is significant difference between measures, and squaring the sum is often used because it uses the least computer time.

The collection of gathered traces corrected with the stacking velocity function obtained by connecting the stacked energy maxima associated with primary reflections is a useful display to determine the correctness of the stacking velocity function. Often a collection of gathers stacked with a set of constant velocities is used as the prime data from which the stacking velocity function is deduced.

The major applications of velocity spectra are to:

(1) Determine the velocity function for optimal stacking.
(2) Identify multiple reflections.
(3) Estimate interval velocities.
(4) Predict lithology, such as of shale and limestone masses, sand/shale ratios, and anomalous overpressured zones.

Much confusion has resulted in the use of stacking velocity to estimate interval velocities. Although the stacking velocity is related to Dix's RMS velocity, they are not the same entity. The RMS velocity is valid for small dip and short offset distances. The stacking velocity increases as the offset distance increases. The stacking velocity is always greater than or equal to the RMS velocity, even when there is no dip. As the dip increases, the difference between the stacking and RMS velocities increases. Because of these differences, the insertion of stacking velocities into the Dix equation for determining interval velocities always results in interval velocities that are too large. Again, Al-Chalabi (1974) has most clearly related stacking and RMS velocities, by determining the bias B in the estimate of RMS velocity from the stacking velocity, $B = v_s - v_{rms}$. He shows that the bias B is a nonnegative quantity that increases as the quantity

$$\sum_{k=1}^{n} F_k \, p^{2k}$$

increases. F_k is a complicated nonnegative function of the velocities and thicknesses of the layers that increases as the heterogeneity factor H increases, and p is the ray parameter. Normally the bias decreases with depth because the ray parameter decreases with depth and its contribution swamps out any possible increase due to F_k. However, if there is an increase in heterogeneity with depth, the increase in F_k may overcome the decrease due to the ray parameter. An example from the North Sea shows increase in bias with depth due to greatly increasing heterogeneity with depth. An example from Alaska shows large bias near the surface because of the permafrost layer.

5. Deconvolution

In terms of linear filter theory, the seismic signal process is the convolution of the seismic pulse with the earth's reflectivity. This process is a mixture of determinism and randomness, as the seismic pulse has a deterministic waveform whereas the reflectivity is a random function. This concept of the seismic signal process was developed by Robinson (1954, 1957). This work considers the reflectivity to be a set of weighted impulses whose weights and arrival times are mutually uncorrelated. The seismic pulse is deterministic because its characteristics obey basic physical laws. With the invention of the CVL in the 1950s, for the first time a detailed measurement of velocity versus depth was available. Combining this measurement with later developed continuous density measurements gives the acoustic impedance of earth in fine detail. From the acoustic impedance, the reflectivity is easily derived. It was then possible to compute synthesized seismograms from the CVL data that matched closely in many instances the seismic field data obtained near the well. At last an adequate model of the seismic data generation process was available.

The linear filter model of the seismic reflection process has the two characteristics common to all mathematical models: It is precisely defined mathematically, and it is only an approximation to reality. The primary purpose of a model is to answer some questions about the "thing" being modeled. Its usefulness lies in its success in giving reasonably correct answers. The answers may be quantitatively inaccurate but qualitatively useful. Even when the answers are partially wrong, the model is useful, as it provides a framework for describing deviations from ideality, and this may lead to a better model. Models are important from another standpoint—and this is very important in the case of the linear filter model of the seismic process—in that they force a certain precision of thought on the part of the problem formulator that had not been demanded by the relatively nebulous sort of thinking that preceded the model.

The seismic pulse b(t) is time-invariant in the model. However, in real life its waveform changes as it travels through the earth. This is called earth filtering. Earth filtering causes an increasing loss in amplitude due to absorption and scattering as the frequency increases. Earth filtering occurs simultaneously with the reflection process, and introduces time variation into the reflection process in the real earth.

Synthetic seismograms are often made under the assumption that the acoustic impedance is adequately characterized by the velocity distribution obtained from a CVL. The fact that synthetics made using this approximation are in many instances markedly similar to field records means that the linear filter model is realistic. The invention of the CVL made it possible for the first time to characterize the earth in all its complexity by a parameter that is directly related to reflections on seismic records.

The filter theory approach leads to models containing all multiples. The continuous velocity function from the CVL is sampled to give an "n" layered earth, and all reflections, primary and multiple, are obtained by solving the one-dimensional wave equation, taking into account the reflection and transmission effects at each boundary. The sampled reflectivity functions give correct primary reflections, provided the sampling is done in accordance with the time-domain sampling theorem.

It is convenient to consider ghosts and reverberations as distortions of the seismic pulse, and to represent them in the model as a filter h(t) in cascade with b(t). Convolution of b(t) and h(t) produces the distorted seismic pulse c(t) that enters the reflection path to produce the seismic data s(t) (see Fig. 4).

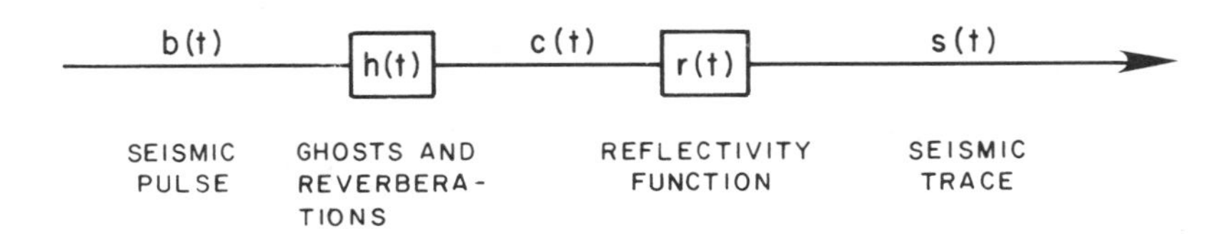

Figure 4. Seismic communication channel.

Ghosts at the source result when energy travels upward from a buried shot, is reflected downward by reflectors at or near the surface, and then follows the initial downgoing signal into the reflection path. Thus, the ghost energy produces ghost reflections that are superimposed upon the primary reflections. Ghosts are also produced when detectors are located beneath the surface. In the model, the additional filter in cascade with b(t) may include the additional absorption in the ghost path as well as the acoustic layering above the buried sources and detectors.

Water reverberations represent a special kind of multiple reflection, a short-period multiple that results from two unique factors: (1) a large reflection coefficient at the water-rock interface plus a nearly perfect reflector at the water surface, and (2) very low absorption in the water layer. As a result of these factors, multiple reflections within the water layer produce highly distorted seismograms that are difficult to interpret. Energy trapped in the near-surface layers keeps being reflected back and forth, and this energy becomes attached to the primary reflections as they travel through the near-surface layers. As a result, instead of having sharp, clear reflections with good time resolution, one has reflections followed by long reverberation trains. These trains overlap with the trains of succeeding primary reflections, and thus the whole seismic trace is given a ringing or sinusoidal character from which it is difficult or impossible to pick out the onset of the primary reflections. The solution of this problem consists of canceling the energy in each reverberation train but leaving intact the primary reflections, thus increasing the resolution of all the reflected events. This process is called deconvolution and is accomplished in the following way. If we consider the wavelet made up of a primary with its attached reverberation, we know from energy considerations that the wavelet is minimum-delay. Moreover, all such wavelets have approximately the same shape. Because the primaries result from geologic beds laid down with irregular thicknesses, the arrival times of the primaries are effectively random. Hence the autocorrelation function of the trace is the same as the autocorrelation function of the wavelet, and so from this autocorrelation function we can compute the required inverse (or deconvolution) operator. The application of this operator to the trace yields the deconvolved trace, namely, a trace where the reverberation components of the wavelets have been removed, thereby increasing the resolution of the primary reflections. The process of deconvolution can also be extended to remove long-period multiple reflections as well. In practice, deconvolution can be applied to the seismic traces either before or after stacking, depending upon cost and other considerations.

An important physical concept often encountered in the study of seismic waves is the concept of minimum-delay. For example, as we have seen in the foregoing discussion of deconvolution, a reverberatory wavelet is minimum-delay. Because the concept of minimum-delay is in fact identical to the concept of minimum-phase, let us first discuss minimum-phase.

The concept of minimum-phase shift was first given by Bode (1945), who worked in the domain of continuous time. Because we are interested primarily in digital computer applications where discrete time is used, we shall translate Bode's work into discrete time. His definition of minimum-phase, given on page 121 of his book, becomes:

> Definition. A stable one-sided time sequence is called minimum-phase if its z-transform has no zeros within the unit circle.

His main theorem, given on page 242 of his book, becomes:

> Theorem. A stable one-sided time sequence is minimum-phase if and only if the net phase-lag displacement between zero and π is zero. The net phase-lag displacement between zero and π of a stable one-sided time sequence of the non-minimum-phase type is always positive.

Bode's theorem thus states that a stable one-sided time sequence is minimum-phase if and only if the phase-lag at π equals the phase-lag at zero. Furthermore, the phase-lag at π is greater than the phase-lag at 0 for a non-minimum-phase, stable, one-sided time sequence.

Robinson (1962) extended these results in the following ways. The class of all stable one-sided sequences with a given amplitude spectrum is considered. Then each of the following conditions is necessary and sufficient that a member of the class be minimum-phase:

(a) The inverse is one-sided.
(b) The phase-lag is a minimum over the entire frequency range.
(c) The group-delay (the derivative of the phase-lag with respect to frequency) is a minimum over the entire frequency range.

(d) The partial energy (the partial sum of squares of the sequence up to time t) is a maximum for all positive times t.

These more precise results were possible because a stable one-sided time sequence was considered not by itself, but as a member of a class of other stable one-sided time sequences all with the same amplitude spectrum. Because all members of the class have the same total energy (due to the fact that all members have the same amplitude spectrum), and because the partial energy of the minimum-phase member exceeds the partial energy of a non-minimum-phase member, it follows that the energy of the minimum-phase member is more concentrated at the front of the sequence. Thus minimum-phase sequences are also called minimum-energy-delay sequences, or simply minimum-delay.

At this point one might ask the question: In terms of information transfer, what does the concept of minimum-delay mean? One approach is to consider the flow of information through a system. A system can destroy information in the sense that the flow of information out of the system is less than the flow of information into the system. In such a case there is a net loss of information; this lost information cannot be recovered. A more favorable situation is the case where a system does not destroy information but only delays it. In this case the total information that comes out of the system over all time is the same as the total information that went into the system, but at any given instant the system withholds information in the sense that information has gone in but has not yet come out. If we wait long enough, we can ultimately obtain this retained information at the system output terminal. The cost to us is delay; we must wait for information at the system's pleasure. The most favorable situation is the case where the system neither destroys nor delays information. In this case the total information that has come out of the system up to any given time instant is the same as the total information that has gone into the system up to the same time instant. Such a system does not withhold any information from us; it merely converts the information into a different form with loss of neither time nor content. Certainly this is the most efficient system; it is the minimum-delay system. The minimum-delay system is the one that produces optimum information transfer.

We now want to present the statistical time series model used in the method of the predictive deconvolution of seismic traces (Robinson, 1954) and show that the method of predictive deconvolution is valid for removing either long-period or short-period reverberations even in the case of a non-minimum-delay source pulse.

Let us now make use of the following statistical model. A system of surface layers (the upper one of which may be water) overlies deep reflecting horizons of random spacing and reflection characteristics. If a unit source spike is excited at the surface, the surface layers produce unwanted reverberatory wavelets that mask the desired reflection series from the deep horizons, from which the geologic structure can be contoured. This model has two characteristic features, namely (1) the statistical feature that the wanted primary events are represented by an uncorrelated random reflection series, and (2) the deterministic feature that the unwanted reverberations attached to the primary events have the same minimum-delay wavelet shape.

In practice the source is not a unit spike but is a mixed delay wavelet. The unwanted wavelet that masks the reflections is now the convolution of the source pulse with the reverberating wavelet. This more realistic model has (1) the statistical feature that the wanted primary events are represented by an uncorrelated random reflection series and (2) the deterministic feature that the unwanted wavelets attached to the primary events have the same mixed-delay wavelet shape, namely that given by the convolution of the minimum-delay reverberatory wavelet with the mixed-delay source pulse. The specification of the model is completed by the assumption that the seismic trace is equal to the convolution of the uncorrelated random reflection series with the unwanted wavelet.

The period of the reverberation is defined as the time interval between successive bursts of the reverberatory energy. For example, let us consider the reverberation produced by a water layer as the topmost layer, as in the case of offshore exploration. Let us neglect all multiples except those from the water bottom. For a pressure pulse, a hard bottom gives no phase shift whereas a soft bottom gives a 180° phase shift. Also, for a pressure pulse, the reflection from the water surface that transforms an upgoing wave into a downgoing wave gives a 180° phase shift. The surface reflection coefficient has magnitude one, whereas the sea bottom reflection coefficient has magnitude less than one. If we let c be the reflection coefficient of the sea bottom, then the reverberating motion due to a unit spike source will be in the form of a pulse train with amplitudes

$$1, -c, c^2, -c^3, \ldots,$$

where the time interval (or reverberatory period) between successive amplitudes is equal to the two-way travel time in the water. This reverberatory pulse train is minimum-delay.

First, let us consider the method of predictive deconvolution in the case of long-period reverberations. More particularly, let us consider the case in which the period of the reverberation is greater than the length of the source pulse. In this case, when we convolve the source pulse with the reverberation pulse train to obtain the so-called unwanted wavelet, there will be quiet intervals in the unwanted wavelet. These quiet intervals are due to the fact that the source pulse is not long enough to span the intervals between reverberatory bursts. As a result we can think of the unwanted wavelet as a pulse train of geometrically attenuated source pulses, where the source pulses in this train do not overlap but are well separated from each other. Under the uncorrelated random feature of the reflection coefficient series, the autocorrelation of the unwanted wavelet is approximately equal to the autocorrelation of the seismic trace. This last autocorrelation we can compute directly from the observed seismic trace, and thus we can obtain the needed autocorrelation of the unwanted wavelet. As we know, prediction operators depend only upon the autocorrelation function, and hence we can compute the prediction operator that predicts future values of the unwanted wavelet from its past values. More specifically, we compute that prediction operator with prediction distance equal to the reverberatory period. Let us now convolve this prediction operator with the unwanted wavelet, which we recall is a geometrically attenuated train of separated source pulses. With a prediction distance of one period, the prediction operator cannot predict at all the source pulse lying within the first period. However, the prediction operator can use the source pulse lying in the first period as the past values, and so the operator can perfectly predict the source pulse lying in the second period. By the same argument the prediction operator can predict every succeeding pulse in the pulse train. In summary, the prediction operator cannot predict at all the first pulse, but can predict all the succeeding pulses in the unwanted wavelet. Hence, the prediction error is equal to the source pulse lying in the first time interval. Instead of dealing with the prediction operator, we can deal with the corresponding prediction error operator. When the prediction error operator is convolved with the unwanted wavelet, the result is the prediction error, which we have seen is the source pulse. In other words, the prediction error operator compresses the unwanted wavelet into the source pulse. When applied to the seismic trace, the prediction error operator in effect replaces each unwanted wavelet with the source pulse. The result is the so-called deconvolved trace. Since the source pulse is much shorter than the unwanted wavelet, seismic resolution is achieved. We see that this result was obtained even under the assumption of a non-minimum-delay source pulse. That is, the method of predictive deconvolution removes the reverberating energy despite the fact that the source pulse is mixed-delay.

If, in fact, the finite-length source pulse is minimum-delay, the deconvolved trace represents the wanted reflection series filtered by the minimum-delay source pulse. Since a minimum-delay source pulse has a sharper leading edge than any of its mixed-delay counterparts, greater seismic resolution can be expected in the minimum-delay case. Of course, the ultimate example of a minimum-delay source pulse is a spike. In this case the deconvolved trace is the wanted reflection series.

We have shown that the method of predictive deconvolution is valid for removing long-period reverberations even in the case of a non-minimum-delay source pulse. Of course, greater resolution can be expected in the case of a minimum-delay source pulse. In either case, if by some ancillary means we can determine the actual shape of the source pulse, then an optimum least-squares shaping filter can be computed to transform the source pulses on the deconvolved trace into some desired shape that increases resolution. Such an optimum shaping filter can be designed whether or not the source pulse is minimum-delay.

Now let us consider the method of predictive deconvolution in the case of short-period reverberations. In this case, the source wavelet, again assumed to be of finite length, has length greater than the reverberation period. Therefore, in the unwanted wavelet, the tail of the first source wavelet will overlap the leading edge of the next source wavelet, and so on down the entire reverberatory sequence. Again we compute the prediction error operator with prediction distance equal to the period of reverberation. When this operator is applied to the unwanted wavelet, the prediction error is equal to this signal: The signal is equal to the source pulse up to a time equal to the reverberatory period, but after this time the signal has a distorted tail that can extend to infinity. This signal has better resolution than the unwanted wavelet, despite the distorted tail. In the case of a minimum-delay source wavelet, the distorted tail vanishes, and thus resolution is better. However, in any case, resolution can be improved by least-squares waveform shaping, provided the shape of the source pulse can be found by some auxiliary means.

In conclusion, the method of predictive deconvolution is valid for removing either long-period or short-period reverberations in the case of a minimum-delay or mixed-delay source pulse. The best seismic resolution is attained in the case of long-period reverberations and a minimum-delay source pulse, and the worst resolution is attained in the case of short-period reverberations and a mixed-delay source pulse. In any case, if the shape of the source is known, seismic resolution can be further improved by a least-squares waveform shaping filter.

6. Seismic Migration

After static and dynamic corrections are applied, the data are horizontally stacked and displayed as a record section in time as a function of surface location along a seismic line. The data received at each surface point are plotted vertically below that point. This display will be a true time-slice through the earth only if all layers have no dip and if no discontinuities or structures exist. This is not the usual case, however, and the time record section is a more complicated transformation of the subsurface layering. On the time section, faults produce diffractions, synclines produce multibranch reflections if their foci are beneath the surface, and anticlines appear broader than they actually are.

In order to display a true slice through earth, it is necessary to migrate the data to place reflections at the originating surfaces, to compress diffractions back to their originating points, and to move multibranches into their true structural positions. The example in Fig. 5 shows seismic data before and after migration. The latter is a good approximation of a time-slice through the earth. Converting the time scale to depth, by applying the velocity distribution to the time section, will give a good approximation to a depth slice through the earth.

Geophysicists are well aware that a seismic record section is not a "slice through the earth," but is instead a transformation of the three-dimensional earth's reflectivity into a plane where each recorded event is located vertically with respect to the source-receiver midpoint. Migration is the inverse transformation, carrying the plane into the true three-dimen-

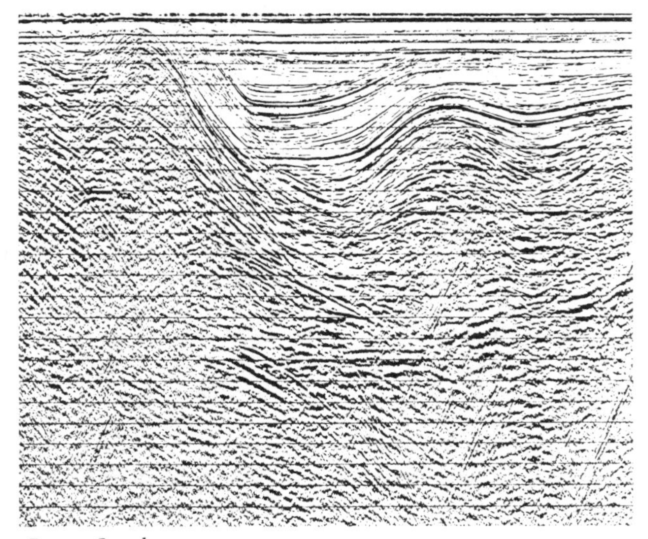

Brute Stack

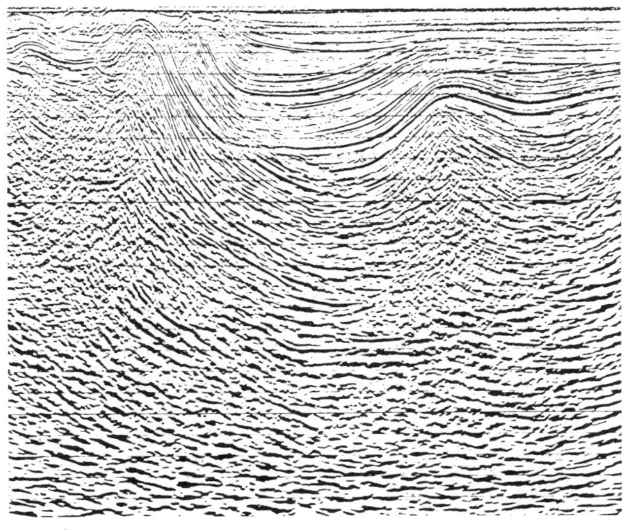

Final Section

Figure 5. An example of seismic data. Courtesy of Western Geophysical Company, Houston, Texas.

sional reflectivity. Steep dip, curved surfaces, buried foci, faults and other discontinuities—each contribute their unique characteristics to the seismic record section, and in complexly faulted and folded areas make interpretation of the geology from the seismic sections difficult unless the data are migrated.

The basic problem in exploration seismology is to find the coordinates of underground interfaces (Robinson, 1981). Let us simplify the problem as much as possible. We will consider the problem of finding the coordinates of a small body located at one point, which is customarily called a point diffractor. This problem is useful because a subsurface reflecting interface can be assumed to consist of a number of point diffractors spaced extremely close together. If we devise a method to locate one point diffractor, then we can use the same method to locate all of the point diffractors that make up the reflecting interfaces.

Although the methodology can be extended to three spatial dimensions, we consider only two spatial dimensions in this paper. The coordinate x represents the horizontal dimension and the coordinate z the vertical dimension. Let us choose the origin of coordinates (0,0) on the earth's surface, and let depth z be measured vertically down. A point diffractor is located underground, so its coordinates are (x_0,z_0) where $z_0 > 0$. A point on the surface of the earth has coordinates (x,0). In general, a point on any horizontal line above the point diffractor has coordinates (x,z), where $z < z_0$.

Now we make an important assumption commonly used in seismic analysis, the stratified earth assumption. This assumption is that the velocity of the rocks varies only in the vertical dimension; that is, the velocity v(z) is a function of z and not of x. Thus, on any horizontal line the velocity is the same at all points. Although, nominally, the stratified earth model has two dimensions (2-D), x and z, variation can take place in only one dimension, z. Consequently, a stratified model is called a 1-1/2 dimensional model (1.5-D model).

We further assume that a source of energy is located at each point diffractor. This source of energy is initiated as an impulse at time t = 0, and the seismic waves propagate outward from the point diffractor. This concept can be extended to the case of entire reflecting interfaces, and thus it is called the exploding reflector hypothesis. The idea of the exploding reflector hypothesis was given by Loewenthal et al. (1976). Instead of the sources being on the surface, each reflector is considered to be composed of a series of point sources. The magnitude of each source equals the value of the reflection coefficient at that point. All the sources are set off at zero time, and the resulting wave motion is recorded at the surface.

Under the foregoing assumptions, the waves are propagated as time increases, as follows. At time t = 0 all the reflectors explode, and thus they all instantly disappear. What is left is wave motion propagating out from where the reflectors were. This wave motion propagates in a medium specified by the velocity function v(z). This medium is stratified, because the velocity does not change along any horizontal line. Since the original reflecting horizons are gone, they cannot influence the propagating waves in any way. Instead, the stratified medium alone determines the raypaths as the waves find their way up to the surface. We have thus described the forward model, that is, the propagation model.

The inverse, or depropagation, model is the one that characterizes the data processing method of migration. At the surface of the earth, we have recorded the received waves. These recorded waves are received at each surface point (x,0) as a function of time, so we can represent them as u(x,0,t). This function plus the velocity function v(z) represents the given quantities.

The idea of migration is to backtrack down the raypaths until we hit a diffracting point, that is, one of the points that exploded. Because all points exploded at time zero, we keep reducing the time as we backtrack. When the time is reduced to zero, we know we are at a diffraction point. For example, if an event occurs at time equal to 3 seconds, we backtrack 3 seconds into the earth and stop. The stopping position is the position of the point diffractor. We know how to backtrack because the raypaths are those of the stratified velocity model, which we can determine from the given velocity function v(z). The raypaths have nothing to do with the unknown reflecting horizons, because by assumption these unknown interfaces exploded and disappeared at time $t = 0$. However, once we have found all the diffracting points, we assume they delineate the unknown interfaces. We have thus succeeded in migrating the data.

We have described the migration process in the case of a stratified earth (1.5-D model). Not only does it neglect all multiple paths, but its assumptions are self-contradictory unless all the geologic reflecting horizons are horizontal. Of course, in the case of horizontal interfaces, migration is not needed because all the least-time raypaths go straight up. If the interfaces are not horizontal, then the hypothetical raypaths of the model can deviate greatly from the raypaths in the earth, and the results of wave equation migration can be significantly in error.

Wave equation (1.5-D) migration cannot correctly locate any diffraction point that lies under a dipping interface. Let us explain. The image ray from the diffraction point is the least-time path, so it strikes the surface at right angles (see Fig. 6). The apex of the hyperbolic time-distance curve lies at this point of emergence of the image ray. The wave equation (1.5-D) migration model specifies that the interface is horizontal, and thus the image ray of the model is vertical. As a result, wave equation (1.5-D) migration locates the diffraction point directly under the apex of the time-distance curve. The actual diffraction point will never be at that point, unless the true interface is level. Thus there will always be a specification error between migrated point and true point; this error increases as the angle of dip of the interface increases. Thus the wave equation (1.5-D) migration method (whether implemented by finite differences, Kirchhoff, or frequency-wavenumber) cannot locate any points under a dipping interface; the answers it does yield are due to an incorrectly specified model, not to the geophysics. There is no geophysical basis for wave equation (1.5-D) migration; for dipping interfaces the wrong answer is always obtained. The

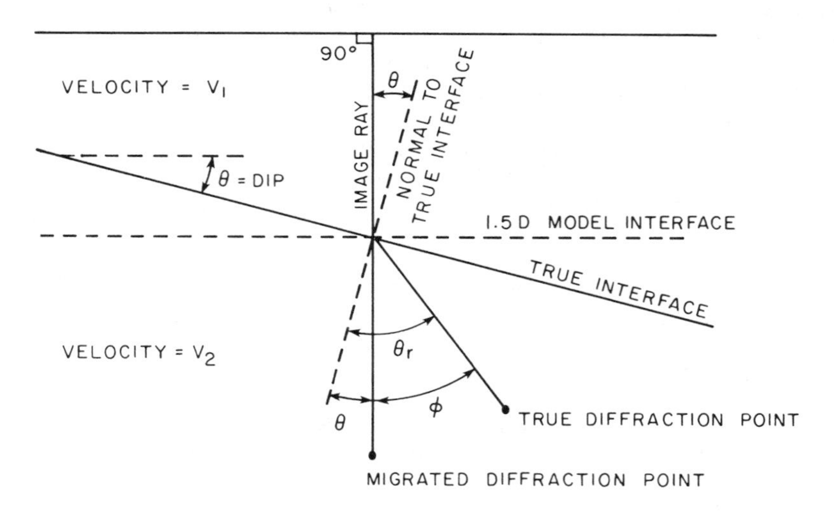

Figure 6. The specification error for 1.5-D migration (finite-difference, Kirchhoff, or f-k) is given by the angle $\phi = \theta_r - \theta$, where $\sin \theta_r = (v_2/v_1) \sin \theta$. Since θ is arbitrary, the specification error is arbitrary.

only justification for wave equation migration is a statistical one in which the effects of random dips of the various interfaces tend to cancel each other out. We are back therefore to the random reflection hypothesis of predictive deconvolution (Robinson, 1954). We hereby propose in this paper the random reflector-dip hypothesis as the justification for performing wave equation migration on seismic data. This statistical justification is the only reason that wave equation migration gives satisfactory results on actual seismic data. For good results, the dips must be small as well as random.

Specification error can be corrected only by throwing out the methods based upon the wrong model and starting over again. This is what depth migration (Larner, 1977) does. It was Larner (1977), Hubral (1977), and Hubral and Krey (1980) who identified the serious failings of the conventional methods. The Larner depth migration method is based on a full 2-D or 3-D model, and thus specification error is eliminated. Research in progress is now directed at reducing the implementation error for depth migration. Even though the implementation is more difficult for depth migration, the advantages of using a correct model greatly outweigh this factor. We cannot go into the Larner depth migration method here, but we do want to say that the work of Larner as well as of the other geophysicists, physicists, and mathematicians working on the geophysical inverse problem represents a conceptual revolution in geophysics, even as digital recording and computing resulted in a technical revolution in the past two decades. Whereas in the past the emphasis was on computer programs, such as finite-difference algorithms, today it is on geophysical models.

7. Concluding Remarks

Today, with the efficient and reliable equipment available, it is possible to collect large amounts of high-quality geophysical data. In the digital processing of the data much regard must be given to the underlying analytic bases of the methods. Emphasis must be placed on the design of geophysical surveys, and in particular on the fundamental geophysical models being used. Geophysicists make use of advanced solid-state electronics equipment to record the seismic data; the data are analyzed with large-scale digital computers and for some purposes with minicomputers. Not only are reflection times used to determine geologic structure, but velocity measurements are used to indicate rock types. Also amplitude anomalies, the so-called bright spots, on the records are used as possible indicators of natural gas. As a result of using more geophysical parameters, the geophysicist can obtain a better map of the subsurface structure than ever before. In the past 30 years geophysical science and technology have had an unparalleled development. Concurrent with this development, the application of these geophysical methods, more specifically the application of reflection seismology, has led to the discovery of a supply of petroleum and natural gas whose real value dwarfs all past wealth. Petroleum is indeed liquid gold, for a nation with an adequate supply of petroleum is truly wealthy. Nations have been transformed by the discovery of oil, and the Middle Eastern nations have wealth beyond the imagination. A provident nature has endowed the cradle of civilization with the great wealth of the world. As long as petroleum is available, our civilization is secure; we live in the age of petroleum. It is unfortunate that worldwide the amount of natural gas that is thrown away (flared) is almost equal in energy content to the amount of petroleum that is used. This natural gas is a by-product of the petroleum production. Let us hope that someday political barriers will disappear so that pipelines can be built and this waste will stop.

References

Al-Chalabi, M. (1974) An analysis of stacking, RMS, average, and interval velocities over a horizontally layered ground, Geophys. Prospect. **22**, 458-475.

Bode, H. W. (1945) Network Analysis and Feedback Amplifier Design (Princeton, N.J.: Van Nostrand).

Chun, J. H., and C. A. Jacewitz (1981) Fundamentals of frequency domain migration, Geophysics **46**, 717-733.

Dix, C. H. (1955) Seismic velocities from surface measurements, Geophysics **20**, 68-86.

Hubral, P. (1977) Time migration, some ray theoretical aspects, Geophys. Prospect. **25**, 728-745.

Hubral, P., and Th. Krey (1980) in K. Larner, ed., Interval Velocities from Seismic Reflection Time Measurements, S.E.G. Monograph, Society of Exploration Geophysicists, Tulsa.

Lamb, H. (1904) On the propagation of tremors over the surface of an elastic solid, Phil. Trans. R. Soc. London (A) **203**, 1-42.

Larner, K. (1977) Depth Migration, Western Geophysical Company, Houston, Texas.

Larner, K., L. Hatton, and B. Gibson (1981) Depth migration of imaged time sections, Geophysics **46**, 734-750.

Loewenthal, D., L. Lu, R. Roberson, and J. Sherwood (1976) The wave equation applied to migration, Geophys. Prospect. **24**, 380-399.

Mayne, W. H. (1962) Common reflection point horizontal stacking technique, Geophysics **27**, 927-938.

Neidell, N. S., and M. T. Taner (1971) Semblance and other coherency measures for multichannel data, Geophysics **36**, 482-497.

Robinson, E. A. (1954) Predictive decomposition of time series with application to seismic exploration, Ph.D. thesis, Dept. of Geology and Geophysics, MIT, 252 pp. (also published in Geophysics **32**, 418-484, 1967).

Robinson, E. A. (1957) Predictive decomposition of seismic traces, Geophysics **22**, 767-778.

Robinson, E. A. (1962) Random Wavelets and Cybernetic Systems (High Wycombe, England: Charles Griffin and Co.), 135 pp.

Robinson, E. A. (1967) Multichannel Time Series Analysis with Digital Computer Programs (Oakland, Calif.: Holden-Day), 328 pp.

Robinson, E. A. (1980) Physical Applications of Stationary Time Series (New York: Macmillan), 314 pp.

Robinson, E. A., and H. Wold (1963) Minimum delay structure of least-squares/eo ipso predicting systems for stationary stochastic processes, in M. Rosenblatt, ed., Time Series Analysis (New York: Wiley), pp. 192-196.

Silvia, M. T., and E. A. Robinson (1979) Deconvolution of Geophysical Time Series in the Exploration for Oil and Natural Gas (Amsterdam: Elsevier Scientific Publishing Co.), 274 pp.

Summers, G. C., and R. A. Broding (1952) Continuous velocity logging, Geophysics **17**, 598-614.

Taner, M. T., and F. Koehler (1969) Velocity spectra, digital computer derivation and application of velocity functions, Geophysics **34**, 859-881.

SPECTRAL ANALYSIS AND TIME SERIES MODELS: A GEOPHYSICAL PERSPECTIVE

Tad J. Ulrych

Department of Geophysics and Astronomy, University of British Columbia, Vancouver, B.C., Canada

C. Ray Smith and W. T. Grandy, Jr. (eds.),
Maximum-Entropy and Bayesian Methods in Inverse Problems, 243–272.

1. Introduction

I was both delighted and honored to be invited to participate in this workshop on maximum entropy. When Ray Smith conveyed to me the list of invited speakers, however, I found myself in somewhat of a quandary, namely, what should I talk about? John Burg is certainly more qualified to talk about maximum entropy spectral estimation than I am. John Skilling and Roy Frieden would cover two-dimensional applications. And of course Edwin Jaynes would talk about it all. After some reflection, I decided that a useful contribution to this workshop might be to link maximum entropy as formulated by Burg with autoregressive (AR) modeling in time series analysis and with the ubiquitous problem of deconvolution, particularly in the framework of geophysical data analysis. My motivation for doing this is to hopefully seduce a few of the participants in this workshop who are interested in entropy but unfamiliar with geophysics into devoting their many and varied talents to some of the important and unsolved problems in geophysical signal processing, which may, perhaps, be tackled by means of the principle of maximum entropy.

The relationship of maximum entropy to AR modeling has been explored in the literature in a number of publications (see for example Ulrych and Bishop, 1975; Makhoul, 1975; Ulrych and Ooe, 1979; Scargle, 1981; Kay and Marple, 1981). At the same time, the connection between the AR model and predictive deconvolution, or the unit delay Wiener spiking filter, is well recognized in the geophysical literature. My aim, then, in this paper is to bring these important fields together by exploring the intimate connecting links that exist between them.

To my mind, the relationship between spectral analysis, time series models, and geophysical signal analysis may be expressed in the flow diagram that is illustrated in Fig. 1. Although my colleague Don Russell remarked on seeing this figure that it reminded him of the wiring diagram to his washing machine, nevertheless I hope the diagram and the ensuing discussion will establish the relationships that exist and that are indeed important.

As an example, let me mention here one of the blocks in Fig. 1 that links many of the elements of the flow diagram. This is the famous Levinson recursion (Levinson, 1947). This algorithm for solving a set of simultaneous equations that have the Toeplitz form plays an integral part in the scheme of things. It links the maximum entropy method and the AR model by virtue of the fact that both the prediction error filter and the AR parameters are estimated using this recursion. The connection between the minimum phase assumption, so central in much of seismic processing, AR models, autoregressive-moving average (ARMA) models, and the physical layered earth model is also established, as I will show, by means of the Levinson recursion.

I do not pretend in this paper to undertake a comprehensive review of the many topics that are illustrated in Fig. 1. My aim, rather, is to present an overview with references to publications where the relevant details may be found.

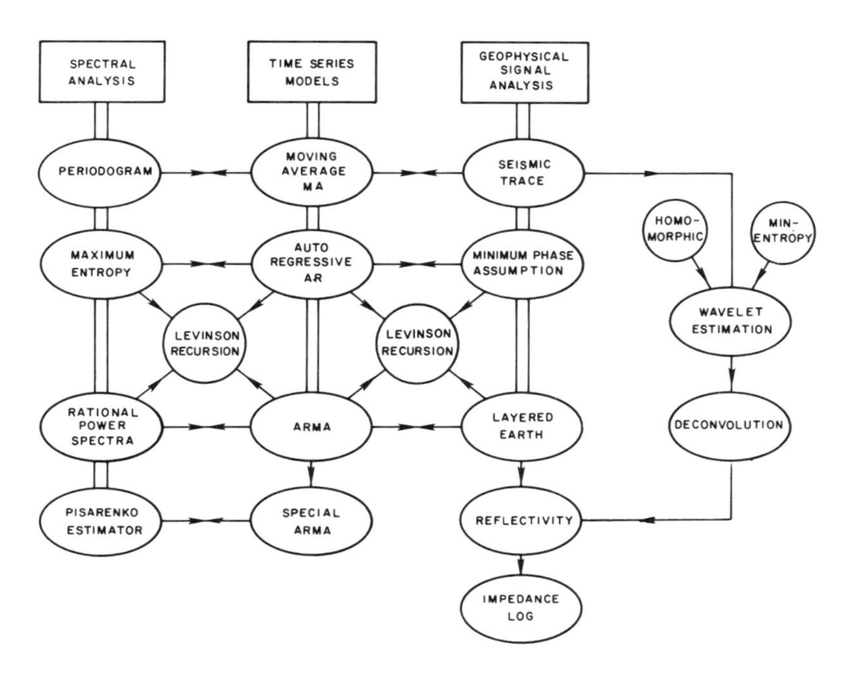

Figure 1. Flow diagram representation of the connections between spectral analysis, time series modeling, and geophysical data processing.

This paper is divided very broadly into two sections. The first deals with the links between spectral analysis and models. The second discusses the connections of models to certain aspects of geophysical data analysis. In particular in this section I deal with some of the latest work in the problem of the inversion of the acoustic impedance from surface seismic reflection records.

2. Models and Spectra

The starting point in this description of the interrelationships that I mentioned in the introduction is the moving average (MA) model, which the Wold decomposition theorem (Wold, 1938) tells us is the most general in the sense that any real-valued stationary process x_t allows the decomposition

$$x_t = u_t + v_t$$

where u_t is deterministic (i.e., $u_{t+\alpha}$ can be predicted for any α from its past values u_{t-1}, u_{t-2}, etc., without error) and v_t is nondeterministic. Further, the stochastic component v_t has the one-sided MA representation

$$v_t = \sum_{i=0}^{\infty} \psi_i \, q_{t-i}$$

where $\psi_0=1$ and q_t are independent identically distributed random variables with

$$E[q_t] = 0 \qquad \text{and} \qquad E[q_t q_s] = \sigma_q^2 \, \delta_{ts} \, .$$

In general we are interested in the case where u_t either does not exist or can be removed, and where the MA model is of finite length m+1, say. Then,

$$x_t = \sum_{i=0}^{m} b_i \, q_{t-i} \, . \tag{1}$$

A word here about the determination of the MA parameters b_i. The usual least squares approach leads to the normal equations

$$\mathbf{b} = \frac{1}{\sigma_q^2} E[x_t \mathbf{q}_t] \, ,$$

where

$$\mathbf{b}^T = (1, b_1, \dots, b_m)$$

and

$$\mathbf{q}_t^T = (q_t, q_{t-1}, \dots, q_{t-m}) \, .$$

Since $E[x_t \mathbf{q}_t]$ is not generally known, **b** must be determined from the covariance structure of the observations. This immediately leads to the concept of spectral factorization and minimum phase (or the equivalent minimum delay property in the time domain), which is one of the important links in the scheme shown in Fig. 1.

The spectral representation of the MA model depicted in Eq. (1) is simply represented in z transform notation. From Eq. (1), $X(z) = B(z)\,Q(z)$, where we define the z transform of a function f_t to be

$$F(z) = \sum_{n=0}^{\infty} f_n \, z^n \, .$$

Consequently

$$R_{xx}(z) = X(z)X(1/z) = R_{bb}(z) \, \sigma_q^2 \, . \tag{2}$$

Evaluating the power spectrum $R_{xx}(z)$ on the unit circle, we obtain

$$R_{xx}(\omega) = \sigma_q^2 \mid 1 + \sum_{k=1}^{m} b_k e^{-i\omega\Delta tk} \mid^2 , \tag{3}$$

which of course represents the periodogram method of spectral analysis. In obtaining Eq. (2) from Eq. (1), I have assumed an infinite realization of the process x_t. In general, of course, we have only a finite sample and consequently all the familiar issues that are involved with windowing, periodicity assumptions, and the like apply.

The finite MA model described in Eq. (1) is central in geophysical signal analysis, where I refer primarily to signals that arise in the seismic prospecting for oil and gas. The seismic trace is modeled as the convolution of a wavelet w_t that results from the propagation of elastic energy in the ground, with a random impulse series, q_t, which represents the impulse response of a layered earth.

Thus, the seismic trace x_t is given by

$$x_t = \sum_{i=0}^{m} w_i \, q_{t-i} + n_t , \tag{4}$$

where the term n_t represents an additive noise component.

One of the principal objectives of seismic data processing is to recover the series q_t from the observations x_t. Equation (4) shows the complexity of this task. We have in fact one equation and three unknowns. Clearly the approach to this problem of deconvolution must be by means of the introduction of constraints to the solution. I will have more to say about this when we consider the determination of the acoustic impedance as shown in Fig. 1. For the moment we consider the constraints most generally adopted in seismic deconvolution, those of the minimum delay nature of the wavelet and of randomness of the impulse series q_t. (An excellent discussion of these assumptions and of the ensuing techniques is given by Robinson and Treitel, 1981.)

As we have already seen in Eq. (2), the assumption of whiteness for q_t allows us to write

$$R_{xx}(z) = R_{ww}(z) \, \sigma_q^2 + \sigma_n^2 . \tag{5}$$

The estimation of the seismic wavelet that is required in the deconvolution must be obtained from the power spectrum (or the autocorrelation) of the observations. Since all phase is lost, we resort to the minimum phase theorem, which allows us to associate a phase spectrum with the amplitude spectrum determined from Eq. (5). Indeed, given the amplitude spectrum of the wavelet, $A(\omega)$, a unique minimum phase spectrum, $\theta(\omega)$, may be computed by first expanding $\log A(\omega)$ in a Fourier series

$$\log_e A(\omega) = \sum_{s=0}^{\infty} h_s \cos \omega s ,$$

where

$$h_s = \frac{2}{\pi} \int_0^{\pi} \log_e A(\omega) \cos \omega s \, d\omega ,$$

and then forming

$$\theta(\omega) = \sum_{s=1}^{\infty} h_s \sin \omega s .$$

All computations may be carried out using the FFT. Other approaches are possible and are described by Claerbout (1976).

The appeal of the minimum phase assumption lies not only in its mathematical simplicity. As Robinson (1964) has shown, there is justification from the physical point of view if we consider the fact that the minimum phase member of the family of wavelets with identical autocorrelation functions has the least energy delay. The assumption of minimum delay for the seismic wavelet w_t, or equivalently for the MA parameters b_t, allows a simple interpretation of the resulting model. Ignoring the noise term in Eq. (4) for the moment, we have in the z domain

$$X(z) = W(z) \, Q(z) .$$

Since W(z) is minimum phase by assumption, all zeros of W(z) lie outside the unit circle $|z| = 1$, and consequently $W^{-1}(z) = G(z)$ is a convergent, stable, series. In other words, the inverse of the wavelet is a realizable, one-sided, causal, and also minimum delay operator. We can now write

$$X(z) \, G(z) = Q(z) ,$$

and for $g_0 = 1$ the inverse transform yields

$$X(z)(1 + g_1 z + \ldots + g_p z^p + \ldots) = Q(z) .$$

The inverse operator is of course infinitely long, but for obvious reasons we will consider a finite operator of length p+1. If $a_i = -g_i$, we obtain

$$x_t = a_1 x_{t-1} + a_2 x_{t-2} + \ldots + a_p x_{t-p} + q_t , \qquad (6)$$

or

$$x_t = \mathbf{x}^T_{t-1} \, \mathbf{a} + q_t ,$$

which represents the well-known autoregressive time series model of order p. Here, $\mathbf{a}^T = (a_1, a_2, \ldots, a_p)$ is the prediction operator and $\mathbf{g}^T = (1, g_1, \ldots, g_p)$ is the prediction error operator (peo), which is the finite minimum delay inverse of the wavelet. The AR model leads to the method of predictive deconvolution (Silvia and Robinson, 1979) of seismic data. If the assumption that the wavelet is minimum delay is correct, then the innovation, q_t, of the AR model represents the desired impulse response of the layered earth. I should emphasize at this point that, whether the wavelet is in fact minimum delay or not, a minimum delay inverse can always be computed. Of course, if the assumption is incorrect, the deconvolved innovation will have little resemblance to the actual reflectivity series.

Let us now look for a moment at the spectral representation of the AR process. Since

$$X(z)\,G(z) = Q(z)$$

and

$$E[q_t q_s] = \sigma_q^2\,\delta_{ts}\,,$$

we obtain

$$|X(z)|^2 = \frac{\sigma_q^2}{|G(z)|^2}\,.$$

Evaluating this expression on the unit circle, we obtain the power spectrum

$$R_{xx}(\omega) = \frac{2\,\sigma_q^2}{\left|1 + \sum_{k=1}^{p} g_k \exp(-i2\pi fk)\right|^2}\,. \tag{7}$$

It is both interesting and important that this formulation is identical to the maximum entropy (ME) power spectrum that is obtained by maximizing the entropy, H_B, of a Gaussian process with spectral density S(f). That is, one maximizes

$$H_B = \frac{1}{4f_N}\int_{-f_N}^{f_N} \log_e S(f)\,df + \log_e(2\pi e)^{1/2}$$

subject to the constraint that S(f) be consistent with the known, or at least assumed known, autocovariances.

I will not dwell on the ME estimate here. John Burg's paper in this volume is the place to seek details. My task is to incorporate the ME approach into the scheme of Fig. 1. In so doing, I will mention those aspects of ME estimation that I consider particularly important.

Let us consider briefly the problem of parameter estimation. The equations for determining the vector **g** that result either from the ME formulation or from the ℓ_2 minimization of the quantity $E[q_t^2]$ in the AR representation are

$$\mathbf{R}\mathbf{g} = \sigma_q^2 \mathbf{i} \; . \tag{8}$$

Here, **R** is the (p+1) × (p+1) Toeplitz autocovariance matrix and $\mathbf{i}^T = (1,0,\ldots,0)$. The solution of these equations is implemented by means of the well-known Levinson recursion, which updates the peo of length n to length n+1. Given, for an example of n = 2, that the known autocovariances are r_i, i = 1,3, and that g_{12} and g_{22} have been previously computed, the Levinson scheme computes the updated operator coefficients g_{13}, g_{23}, and g_{33}. In this notation g_{ik} is the ith coefficient of the kth order operator.

The recursion for this small illustrative example may be written

$$\mathbf{R}_3 \begin{bmatrix} 1 \\ g_{13} \\ g_{23} \\ g_{33} \end{bmatrix} = \mathbf{R}_3 \left[\begin{bmatrix} 1 \\ g_{12} \\ g_{22} \\ 0 \end{bmatrix} + g_{33} \begin{bmatrix} 0 \\ g_{22} \\ g_{12} \\ 1 \end{bmatrix} \right] = \begin{bmatrix} P_2 \\ 0 \\ 0 \\ e_2 \end{bmatrix} + g_{33} \begin{bmatrix} e_2 \\ 0 \\ 0 \\ P_2 \end{bmatrix} = \begin{bmatrix} P_3 \\ 0 \\ 0 \\ 0 \end{bmatrix} \tag{9}$$

where $\mathbf{R}_3$ is the 4x4 Toeplitz autocovariance matrix, P_2 and P_3 are the estimates of σ_q^2 in Eq. (8) for the particular order, and e_2 is defined by the recursion. The really important thing about Eq. (9) is that, as can be seen, the recursion is written in terms of a forward and backward system of equations. This is the crux of the Levinson recursion and is at the heart of other approaches to AR parameter estimation. The importance of the Levinson recursion lies not only in the fact that it is a very efficient method of solving Toeplitz normal equations. There are many other aspects to this recursion, some of which I have attempted to illustrate in the flow diagram of Fig. 1. One of the most useful and fundamental ones is that the Levinson scheme guarantees that the peo is computed as a minimum delay operator. This, of course, is vital from the stability point of view of AR modeling and also from the requirement that the ME spectral estimator be nonnegative. The minimum delay nature of the peo may be established by use of Rouché's theorem, and since this theorem crops up in a number of places, particularly in establishing properties of the layered earth model, I will briefly present the argument here.

The general form of the recursion that is given in Eq. (9) for the specific example is

$$G_{m+1}(z) = G_m(z) + k_m z^{m+1} G_m(1/z) , \tag{10}$$

where k_m is the so-called reflection coefficient. Rouché's theorem states that if two polynomials $P(z)$ and $Q(z)$ satisfy $|P(z)| > |Q(z)|$ on $|z| = 1$, then $P(z)$ and $P(z) + Q(z)$ have the same number of zeros inside the unit circle. If we identify $G_m(z)$ with $P(z)$ and $k_m z^{m+1} G_m(1/z)$ with $Q(z)$, we conclude that, provided $|k_m| < 1$, $P(z) + G(z) = G_{m+1}(z)$ and $G_m(z)$ have an equal number of zeros in $|z| = 1$. A repeat of this argument leads to the conclusion that $G_{m+1}(z)$ and $G_m(z) = 1$ have the same number of zeros in $|z| = 1$ and hence $G_{m+1}(z)$ is minimum phase. That $|k_m| < 1$ follows from the iteration and is an expression of the physical bounds of the reflection coefficient of the real earth.

The contribution of John Burg lies not only in the development of the ME power spectral formulation, but also in the extension of the Levinson recursion in a manner that is consistent with the philosophy of the ME approach. Thus, while the normal Levinson scheme computes the reflection coefficients from an estimate of the autocovariance function, which is usually calculated assuming a zero extension of the data, Burg's estimate of k_m makes no such assumption. Indeed, in terms of the forward and backward errors of the mth order filter, $e^{+}_{t,m}$ and $e^{-}_{t,m}$, respectively,

$$k_m = \frac{2 \sum\limits_t e^{+}_{t,m} e^{-}_{t,m}}{\sum\limits_t [(e^{+}_{t,m})^2 + (e^{-}_{t,m})^2]} . \tag{11}$$

The elegance of Eq. (11) is that it guarantees that $k < 1$, and consequently coupled with Eq. (10) the Burg scheme provides a minimum delay peo without the restrictive assumptions about the data that are otherwise required. The Burg prediction error recursion may be written as

$$\begin{bmatrix} e^+ \\ e^- \end{bmatrix}_{m+1} = \begin{bmatrix} 1 & -k_m z \\ -k_m & z \end{bmatrix} \begin{bmatrix} e^+ \\ e^- \end{bmatrix}_m . \tag{12}$$

It is both interesting and important that this recursion also relates directly to the downward continuation of surface waveforms for the layered earth model. I will return to this subject later.

The importance of determining the correct length for the peo in ME spectral analysis has often been mentioned in the literature. A very successful approach when the data can be well approximated by an AR model is by means of the Akaike FPE and AIC criteria (Akaike, 1969, 1974). Basically, the rationale for the FPE (final prediction error) is as follows: Let x_t be a realization of the AR process for which the order is to be determined. We obtain $\hat{a}_{x,m}$, m = 1,2,...,M, from x_t by normal means. If y_t is another realization of this AR process, we can form

$$\hat{y}_t = \sum_{m=1}^{M} \hat{a}_{x,m} y_{t-m}$$

and determine

$$FPE = E[(y_t - \hat{y}_t)^2] .$$

We adopt as the order of the process, p, that M for which FPE is minimized. Akaike has shown that an efficient estimator is

$$(FPE)_M = \frac{N + (M + 1)}{N - (M + 1)} s_M^2 ,$$

where s_M^2 is the residual sum of squares.

The AIC criterion of Akaike is based on the Kullback-Leibler (1951) mean information measure. In case of normally distributed innovations,

$$AIC = N \log_e s_M^2 + 2M .$$

Again, we take as the correct order, p, that M which minimizes the AIC.

Another measure that has been found useful is the CAT criterion of Parzen (1974). This criterion, which is given by

$$CAT(M) = \frac{1}{N} \sum_{j=1}^{M} \hat{s}_j^{-2} - \hat{s}_M^{-2} ,$$

where

$$\hat{s}_j^2 = \left[\frac{N}{N-j}\right] s_j^2 ,$$

has been investigated by Landers and Lacoss (1977), who concluded that all three criteria estimated the order well, provided the process was in fact well approximated by an AR model.

The ME estimator has often been applied to the study of harmonic processes, particularly in the case when the length of the data series presents resolution difficulties. It is important to realize that although the AR representation models noiseless harmonic processes exactly, the addition of noise alters the AR model to an ARMA model [see Eq. (16) and after]. In this case the FPE, AIC, or CAT criterion cannot be used reliably, and Ulrych and Bishop (1975) have suggested the rule of thumb that the order p should be

$$\frac{1}{3} N < p < \frac{1}{2} N . \tag{13}$$

This rule has been borne out by the work of Berryman (1978). For high signal to noise ratios (SNR) the ME estimate is not particularly sensitive to the model order provided that Eq. (13) is adhered to. The situation for small SNR is not so simple and the choice of model order can have a significant effect (Kay, 1979). Since the model we are now dealing with is in fact an ARMA model, it may be accurately represented by an infinite order AR process. However, the problem with using high model orders when the SNR is low is that spurious peaks arise as a result of perturbations in the estimated noise poles which attempt to model the ideal, flat noise background. These issues are well discussed by Kay (1979). It is also important to remember that the ME estimate is an estimate of the spectral density (Lacoss, 1971; Burg, 1975). In other words, correct values of the power of the process require integration of the ME spectral estimate. John Burg has illustrated this very clearly during this workshop. This point is particularly important in low SNR cases when the positions of the harmonic poles with respect to the unit circle are determined inaccurately. The great importance of the ME estimator in spectral analysis is well demonstrated by the recent work of Currie (1981).

Adaptive methods of determining the AR parameters have extended the application of the ME method to nonstationary time series. These methods are also important in the deconvolution of time varying seismic wavelets. The approaches that have found application are the Kalman filter, recursive least squares (RLS), and the Widrow and Hoff LMS algorithm (Widrow and Hoff, 1960; Berkhout and Zaanen, 1976; Guarino, 1979; Willsky, 1979). Since the RLS and the Kalman formulations for the application I have mentioned are quite similar, the Kalman filter transition matrix being a unit matrix in the RLS formulation, I will only briefly consider the latter.

Define a data matrix $\mathbf{X}_n$ as

$$\mathbf{X}_n = \begin{bmatrix} x_1 & 0 & \cdot & \cdot & \cdot & 0 \\ x_2 & x_1 & \cdot & \cdot & \cdot & 0 \\ \cdot & \cdot & & & & \cdot \\ \cdot & \cdot & & & & \cdot \\ \cdot & \cdot & & & & \cdot \\ x_n & x_{n-1} & \cdot & \cdot & \cdot & x_{n-p+1} \end{bmatrix}$$

where x_i, $i = 1,\ldots,n$, are the data points.

With the definitions

$$\mathbf{a}^T = (a_1, a_2, \dots, a_p)$$

$$\mathbf{x}^T = (x_2, x_3, \dots, x_{n+1})$$

and

$$\mathbf{e}^T = (e_2, e_3, \dots, e_{n+1})$$

the AR system may be written

$$\mathbf{x} = \mathbf{X}_n \mathbf{a} + \mathbf{e} \; . \tag{14}$$

The least squares solution to Eq. (14) is

$$\mathbf{a} = [\mathbf{X}_n^T \mathbf{X}_n]^{-1} \mathbf{X}_n^T \mathbf{x} \; .$$

We wish to update the pth order filter at time step j, $\mathbf{a}_j$, to produce a new filter, $\mathbf{a}_{j+1}$, that depends on the new data point x_{j+1}. Since $[\mathbf{X}_n^T \mathbf{X}_n]$ is no longer Toeplitz, Levinson recursion cannot be used. A recursive approach is, however, possible and is applied in the following manner.

It may be shown (Berkhout and Zaanen, 1976) that if $\mathbf{Q}_j = [\mathbf{X}_j^T \mathbf{X}_j]^{-1}$ and $\mathbf{x}_j^T$ is the jth row of $\mathbf{X}_j$, then

$$\mathbf{a}_{j+1} = \mathbf{a}_j + \mathbf{Q}_{j+1} \mathbf{x}_{j+1} [x_{j+1} - \mathbf{x}_{j+1}^T \mathbf{a}_j]$$

and

$$\mathbf{Q}_{j+1} = \mathbf{Q}_j - \frac{\mathbf{Q}_j \mathbf{x}_{j+1}}{J+1} \mathbf{x}_{j+1}^T \mathbf{Q}_j$$

where $J = \mathbf{x}_{j+1}^T \mathbf{Q}_j \mathbf{x}_{j+1}$.

No matrix inversion is required in the computation of the updated filter. The RLS approach is equivalent to the covariance method of parameter estimation (Makhoul, 1975), for stationary inputs. Since, as can be seen from the structure of $\mathbf{x}_n$, the filter is updated from the whole past history of the input, it seems that some type of memory function is required.

The second approach that we have found very useful in a number of different contexts, including spectral analysis, strong motion simulation, and adaptive deconvolution, is the LMS algorithm due to Widrow and Hoff (1960). The LMS filter is adapted as

$$\mathbf{a}_{k+1} = \mathbf{a}_k + \eta \, \hat{\boldsymbol{\nabla}} [E(e_k^2)] \; .$$

Here, $\hat{\boldsymbol{\nabla}} [E(e_k^2)]$ is the estimate of the gradient of the mean squared error e_k^2 with respect to $\mathbf{a}_k$ and is computed simply as

$$\hat{\boldsymbol{\nabla}} [E(e_k^2)] = \boldsymbol{\nabla} (e_k^2) = -2 \, e_k \, \mathbf{x}_{k-1}^T$$

where $\mathbf{x}_{k-1}^T = (x_{k-1}, x_{k-2}, \dots, x_{k-p})$. Hence,

$$a_{k+1} = a_k + \mu e_k x_{k-1} , \tag{15}$$

where μ must be specified and is a tradeoff between adaptation speed and mean squared error. An example of the tracking ability of the LMS algorithm is shown in Fig. 2, where the input signal consists of two sinusoids with an SNR of 25 dB.

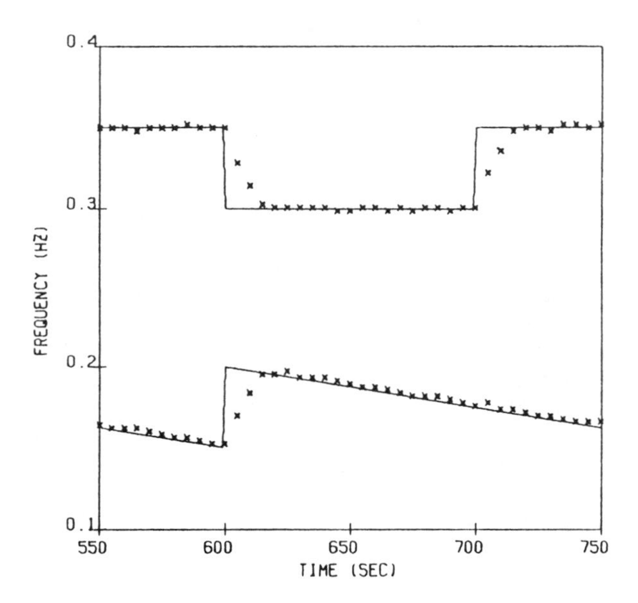

Figure 2. Behavior of the LMS algorithm for a sinusoidal input. Computed frequencies (xxx). Actual frequencies (—).

We now move to the next level of the flow chart diagram of Fig. 1 by the simple expedient of adding noise to the AR representation. As demonstrated by Pagano (1974), when observation noise is added to an AR series x_t, the resulting time series model requires a filter that has both poles and zeros in its transfer function. This fact may be simply demonstrated (Kay, 1979). The AR process

$$x_t = - \sum_{k=1}^{p} g_k x_{t-k} + e_t$$

is corrupted by noise to yield $y_t = x_t + n_t$, where $E[x_t n_t] = 0$. Forming the spectral density estimate of y_t, we obtain

$$R_{yy}(z) = Y(z)\,Y(z^{-1}) = \frac{\sigma_e^2}{G(z)G(z^{-1})} + \sigma_n^2$$

$$= \frac{\sigma_e^2 + \sigma_n^2\,G(z)G(z^{-1})}{G(z)G(z^{-1})} \;.$$

Identifying $\sigma_e^2 + \sigma_n^2 G(z)G(z^{-1})$ with $\sigma_w^2 B(z)B(z^{-1})$, we obtain an ARMA representation for y_t that is written in general form as

$$y_t + \sum_{i=1}^{p} g_i y_{t-i} = w_t + \sum_{j=1}^{q} b_j w_{t-j} \;; \qquad E[w_t w_s] = \sigma_w^2 \delta_{ts} \;. \tag{16}$$

It is important to remember in determining the ARMA coefficients that $G(z)$ must be invertible and consequently minimum phase.

From a spectral point of view the importance of the ARMA model is that, since both poles and zeros are modeled, the spectral profile may be approximated by a transfer function of low combined order. In other words, the ARMA representation is a parsimonious one from the point of view of the number of coefficients.

In general the ARMA coefficients are obtained by first solving for the vector of AR parameters, $\mathbf{a}^T = (a_1, a_2, \ldots, a_p)$, and then determining the MA parameters (the vector $\mathbf{b}^T = (b_1, b_2, \ldots, b_q)$) by spectral factorization. An estimate of $\mathbf{a}$, designated by $\hat{\mathbf{a}}$, for an ARMA (p,q) process may be determined, assuming p and q are known, from the normal equations

$$\begin{bmatrix} r(q) & r(q-1) & \cdots & r(q-p+1) \\ r(q+1) & r(q) & \cdots & r(q-p+2) \\ \vdots & \vdots & & \vdots \\ r(q+p-1) & r(q+p-2) & \cdots & r(q) \end{bmatrix} \begin{bmatrix} \hat{a}_1 \\ \hat{a}_2 \\ \vdots \\ \hat{a}_p \end{bmatrix} = \begin{bmatrix} r(q+1) \\ r(q+2) \\ \vdots \\ r(q+p) \end{bmatrix} .$$

These equations may be used since, for $k > q$, the autocovariances $r(k)$ depend only on the AR parameters. Notice that this system of equations is Toeplitz but not symmetric, and the denominator of the ARMA model is not guaranteed to be minimum phase.

An interesting approach to the problem of ARMA spectral computation is the iterative scheme of Treitel et al. (1977). This technique uses the Wiener least squares spiking and shaping filters to produce a denominator that is guaranteed to be minimum phase by virtue of the properties of the Levinson recursion.

One of the problems in ARMA modeling is, of course, the determination of the model order, (p,q). Several approaches have been proposed. One is the Akaike AIC criterion, which has found application in diverse fields, for example Ooe's work on the Chandler wobble (Ooe, 1978). The S array approach of Gray and Morgan (1978) shows promise in this regard. Another procedure that is useful has been suggested by Woodward and Gray (1979) and is based on the partial autocorrelation (pac) array approach of Box and Jenkins (1976). The pac function is composed of the reflection coefficients, k_m (or $-g_{mm}$), in the Levinson recursion. The generalized pac criterion establishes a tableau of pac values as a function of AR and MA order. Woodward and Gray have shown that for an ideal ARMA (p,q) process the pac coefficients have the following distinguishing properties. If we denote the kth prediction coefficient computed for AR order k and MA order j by a^j_{kk}, then

$$a^q_{pp} = a^{q+i}_{pp} = a_p , \qquad i = 1, 2, \ldots,$$

and

$$\begin{aligned} a^q_{kk} &\neq 0 \quad \text{for } k \leq p \\ &= 0 \quad \text{for } k > p . \end{aligned}$$

As Ulrych and Clayton (1976) have shown, a very interesting ARMA model results when one considers a time series composed of harmonics with random additive noise. Since a harmonic process may be represented by a second order AR model, we may write for p/2 harmonics

$$x_t = \sum_{k=1}^{p} a_k x_{t-k} .$$

With additive white noise, the combined process becomes

$$y_t = x_t + n_t = \sum_{k=1}^{p} a_k x_{t-k} + n_t .$$

Substituting

$$x_{t-k} = y_{t-k} - n_{t-k} ,$$

we obtain

$$y_t = \sum_{k=1}^{p} a_t y_{t-k} + n_t - \sum_{k=1}^{p} a_k n_{t-k} . \qquad (17)$$

This is an ARMA process that is special in the sense that the orders and parameters of the AR and MA components are the same. The vector $\mathbf{g}^T = (1, -a_1, -\ldots, -a_p)$ may be shown to be the solution of the eigenproblem

$$\mathbf{R}_{yy}\,\mathbf{g} = \sigma_n^2\,\mathbf{g} \tag{18}$$

where $\mathbf{R}_{yy}$ is the Toeplitz autocovariance matrix obtained from y_t. Further, the desired $\mathbf{g}$ is the eigenvector (normalized such that $g_0 = 1$) corresponding to the minimum eigenvalue. As shown by Robinson (1967), the zeros of $G(z)$, the z transform of the eigenvector, are all located on the unit circle. The positions of these zeros identify the frequencies of the harmonic process y_t. As pointed out by Ulrych and Clayton (1976), the method of computing $\mathbf{g}$ from Eq. (18) and determining the roots of $G(z)$ is identical to the spectral estimate suggested by Pisarenko (1973). The resolution properties of the estimator may be established by writing Eq. (18) in the form

$$\lim_{\gamma\to 1} (\mathbf{R}_{yy} - \gamma\,\sigma_n^2\,\mathbf{I})\,\mathbf{g} = (1-\gamma)\,P_n\mathbf{i}\;. \tag{19}$$

Equation (19) is now in the form of the normal equations for determining the peo for the ME estimate with the exception that a fraction γ of the noise variance is subtracted from the diagonal of the autocovariance matrix. The potential increase in resolution resulting from a decrease in the zero lag autocovariance is illustrated in Fig. 3.

The Pisarenko estimator is the most highly resolved estimator for harmonic processes. Its efficiency depends very critically on the SNR of the input. This follows immediately from the realization that the locations of the spectral lines are determined from the eigenvector corresponding to the minimum eigenvalue. As is well known, this eigenvector is most susceptible to noise.

The relationship between the AR/ME peo, which we write as $\tilde{\mathbf{g}}$ for this discussion, and the Pisarenko eigenvector $\mathbf{g}$ is very interesting. If we designate the eigenvectors and eigenvalues that are the solutions to Eq. (18) as $\mathbf{g}_k$ and λ_k, $k = 1,2,\ldots,p$, respectively, where $\lambda_p > \lambda_{p-1} > \ldots > \lambda_1$, we can write

$$\tilde{\mathbf{g}} = \sigma_n^2 \sum_{k=1}^{p} \frac{g_{k_1}\,\mathbf{g}_k}{\lambda_k}\;. \tag{20}$$

(Note that in this notation the Pisarenko $\mathbf{g}$ is $\mathbf{g}_1$.)

The intriguing thing about Eq. (20) is that the particular weighting of the eigenvectors expressed by the r.h.s. of Eq. (20) leads always to a minimum delay result. This follows from the fact that, as I have discussed previously, $\tilde{\mathbf{g}}$ is minimum delay. Treitel and Ulrych (1979) attempted to show the minimum delay property of $\tilde{\mathbf{g}}$ directly from the r.h.s. of Eq. (20) but, as

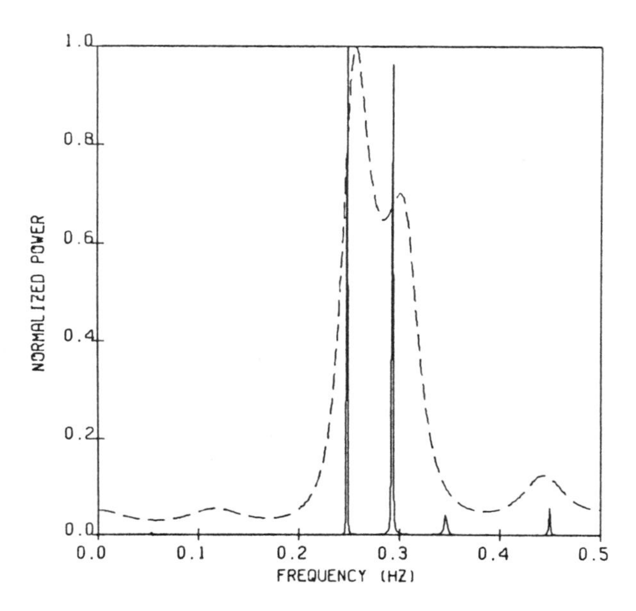

Figure 3. Increase in resolution resulting from a decrease in the zero lag autocovariance. Normal AR/ME estimate (---). Modified estimate (—).

Makhoul (1981) has pointed out, did so incorrectly. A correct proof using the eigenvalue-eigenvector decomposition might lead to new insight into the structure of Toeplitz matrices.

From a geophysical point of view, the ARMA representation is an extremely useful one and relates directly to the very general model of the layered earth. This model, which is extensively used in the exploration industry, considers the earth as made up of layers of different but constant properties. Each layer of constant two-way travel time contains an upgoing and downgoing elastic wave component that is a result of energy input at the surface of the layered earth. We will denote these components in the kth layer by U_k and D_k respectively, where U_k and D_k may be defined in terms of pressure, particle velocity, or admittance of the material. As several authors have shown (see Robinson and Treitel, 1981, and Claerbout, 1976, for details), a relationship exists between the U and D waves in the kth layer and those in the k+1th layer and is given by

$$\begin{bmatrix} U \\ D \end{bmatrix}_{k+1} = \frac{1}{t_k\sqrt{z}} \begin{bmatrix} 1 & c_k z \\ c_k & z \end{bmatrix} \begin{bmatrix} U \\ D \end{bmatrix}_k , \tag{21}$$

where the matrix is the layer matrix, c_k is the reflection coefficient between the kth and the k+1th layer and is related to the acoustic impedance, z is the unit delay operator, and t_k is the transmission coefficient across the layer.

Let us now recall the Burg scheme for updating the forward and backward errors that was given in Eq. (12) and is repeated here for convenience,

$$\begin{bmatrix} e^+ \\ e^- \end{bmatrix}_{m+1} = \begin{bmatrix} 1 & -k_m z \\ -k_m & z \end{bmatrix} \begin{bmatrix} e^+ \\ e^- \end{bmatrix}_m . \qquad (22)$$

A comparison of Eqs. (21) and (22) allows us to identify the pacs, k_m, with reflection coefficients in the layered earth model. We can make use of this identity by considering, for example, the reflection seismology geometry shown in Fig. 4 (Claerbout, 1976). The input energy is in the form of an impulse that is returned to the surface by the layers as -Q(z). Assuming the surface to be perfectly reflective, Q(z) is sent back into the earth, and E(z) escapes into the basement half space. Since we have assumed an input impulse, Q(z) is in effect the reflection seismogram that has been deconvolved for the source signature. An approximate expression for Q(z) in terms of C(z), the z transform of the primary reflection series (that is, the series of reflection coefficients from the layer boundaries), is

$$Q(z) \simeq \frac{C(z)}{A(z)} . \qquad (23)$$

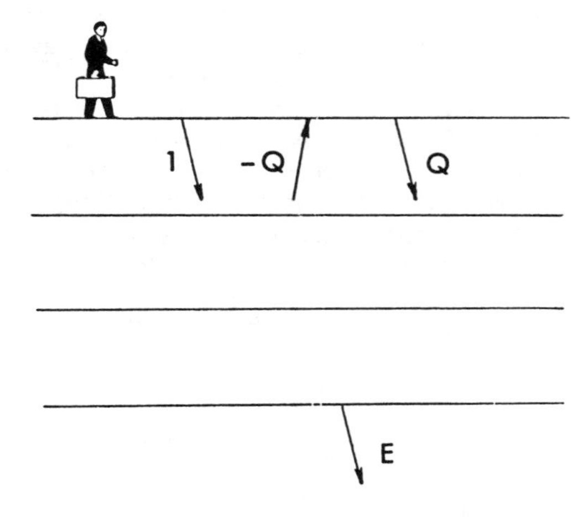

Figure 4. Geometry of the layered earth model for reflection seismology.

Equation (23) is an ARMA model where $A(z)$ is a minimum phase reverberation operator. In other words, the reflection seismogram may be modeled as an ARMA process. To obtain the series of interest, $C(z)$ from $Q(z)$, one applies the energy flux theorem to the model to yield the equation

$$1 + Q(z) + Q(1/z) = \frac{K}{G(z)G(1/z)}, \tag{24}$$

where K is a constant and $G(z)$ is another minimum phase operator. Identifying coefficients of equal powers of z, we obtain for an example of three layers (Claerbout, 1976)

$$\begin{bmatrix} q_0 & q_1 & \cdot & q_3 \\ q_1 & q_0 & \cdot & q_2 \\ \cdot & \cdot & \cdot & \cdot \\ \cdot & \cdot & \cdot & \cdot \\ q_4 & q_3 & \cdot & q_1 \end{bmatrix} \begin{bmatrix} 1 \\ g_1 \\ g_2 \\ -c_3 \end{bmatrix} = \begin{bmatrix} K \\ 0 \\ \cdot \\ \cdot \\ \cdot \end{bmatrix} .$$

Thus, the reflection coefficients may be derived from the observed, deconvolved surface record by means of the ubiquitous Levinson recursion. In this regard, an interesting connection of the layered earth model to the principles of maximum and minimum entropy has been recently established by Treitel and Robinson (1981). As Sven Treitel has pointed out, this simple model of the earth has still much to teach us.

Many of the links that I outlined in Fig. 1 are now complete. In the present section I assumed that the observed response, $Q(z)$, was deconvolved for the blurring effect of the seismic wavelet. In fact, the process of deconvolution is of central importance in geophysical signal analysis, particularly in the goal of determining the subsurface acoustic impedance from surface observations. In the next section I will trace out the remaining part of the flow chart of Fig. 1 with particular reference to some recent work by Oldenburg et al. (1983).

3. Deconvolution of Geophysical Data

Returning to the MA model of the seismic trace, we can state the problem of deconvolution in the following manner. Given the trace $x_t = w_t * q_t$, we wish to find a filter h_t such that $x_t * h_t = q_t$. Clearly this view leads naturally to the problem of estimating the seismic wavelet w_t. We have already mentioned the most common approach, which entails the minimum delay assumption for w_t and the whiteness assumption for q_t. These assumptions allow the modeling of the MA process as an AR or predictive process and lead to the widely applied technique of predictive deconvolution. This approach is quite general in that the predictive distance, α, can be varied. Thus, $\alpha = 1$ corresponds to spiking minimum delay deconvolution, whereas $\alpha > 1$ can be used to suppress short and long period multiples. In this regard, the LMS algorithm has been used by Griffiths et al. (1977) in an adaptive manner.

The assumption of minimum delay may be a very poor one. Figure 5 illustrates this clearly. Figure 5a shows a family of band-limited wavelets with the same amplitude spectrum but different phase spectra, ranging from the minimum delay member to the maximum delay member, each succeeding wavelet having two more zeros moved from outside to inside the unit circle. Figure 5b illustrates the deconvolution of these wavelets with the minimum delay member. As can be seen, any perturbation of the phase spectrum leads to a very poor deconvolution.

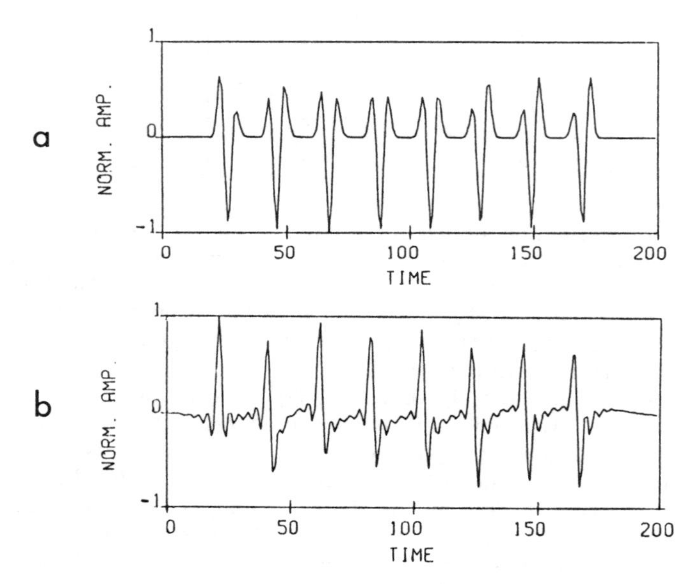

Figure 5. (a) A family of band-limited wavelets with the same autocovariance ranging from minimum to maximum delay. (b) Deconvolution of the suite of wavelets shown in (a) by the minimum delay member.

To overcome the severe limitations of the minimum delay assumption (and the whiteness assumption as well), a number of different wavelet estimation methods have been devised. An interesting approach applied to seismology by Ulrych (1971) and, in a modified form, by Tribolet (1979), is based on the work of Oppenheim et al. (1968) and Schafer (1969). The basic idea behind this technique, called homomorphic deconvolution, rests on a logarithmic transformation from the frequency to the cepstral domain.

Since

$$X(z) = W(z)\, Q(z) \; ,$$

we can take logarithms to obtain

$$\log X(z) = \log W(z) + \log Q(z) \; .$$

Evaluating the inverse z transform of log X(z) on $|z| = 1$, we obtain in the cepstral domain $\hat{x}_n = \hat{w}_n + \hat{q}_n$.

If the cepstral contributions of the wavelet and the spike series are distinct, $\hat{w}_n$ may be separated and $\hat{w}_t$ obtained by an inverse homomorphic transformation. In practice many difficulties are encountered, principally as a result of noise and ensuing problems in unwrapping the phase curve in the neighborhood of small values of the amplitude spectrum. The problem of noise is particularly severe in the homomorphic approach since it is not clear what happens to noise in the logarithmic computation. A useful approach is that of cepstral averaging, which has been proposed and used by several authors (see Lines and Ulrych, 1977, for a discussion and references).

A particularly interesting recent approach to wavelet estimation known as minimum entropy deconvolution (MED) has been proposed by Wiggins (1978) and has stimulated research into a class of wavelet estimators that lead to what has been termed parsimonious deconvolution. The MED approach seeks a measure of the simplicity or simple structure of the resulting deconvolved spike train. Like the homomorphic approach, MED does not require the normal constraining assumptions regarding either the wavelet or the spike train.

If f_t is the MED inverse filter, then

$$f_t * x_t = u_t \simeq q_t \; .$$

The criterion that Wiggins uses to define the simple structure of u_t is called the varimax and is defined by

$$V = \frac{\sum_t u_t^4}{\left(\sum_t u_t^2\right)^2} \; .$$

Solving for f_t by forming $(\partial V/\partial f_k) = 0$ results in the equations $\mathbf{Rf} = \mathbf{v}$, where $\mathbf{f}$ is the vector of filter coefficients,

$$v_k = \sum_t u_t^3\, x_{t-k} \; ,$$

and **R** is a Toeplitz matrix of weighted autocovariances of the input. Although these equations are highly nonlinear, they may be solved iteratively very efficiently by use of the Levinson recursion.

Extensions to this approach are based on the fact that the criterion V is, in effect, a measure of the kurtosis of the distribution of u_t (Claerbout, 1978; Ooe and Ulrych, 1979; Godfrey, 1979; Gray, 1979; Postic et al., 1980). Gray (1979) introduced a parsimonious criterion of the form

$$C = \frac{\frac{1}{n}\sum_t |u_t|^r}{\left(\frac{1}{n}\sum_t |u_t|^s\right)^{r/s}} .$$

An excellent discussion of various aspects of MED has recently been presented by Donoho (1981).

Since the problem of deconvolution is horribly nonunique, it is clear that some constraints must be introduced. MED introduces the constraint that the output be spiky and simple. In a recent paper, Oldenburg et al. (1981) have combined this constraint on the reflectivity series with the constraint that the wavelet be "smooth" in some sense by means of linear inverse theory. Their analysis first proceeds by obtaining an estimate, $\hat{q}_t$, of q_t by using MED, for example, then by solving the discretized equations

$$x(t_j) = x_j = \int_0^{t_w} w(t)\, g_j(t)\, dt + n_j$$

$$= (w,g_j) + n_j , \qquad j = 1,2,\ldots,N, \quad (25)$$

where t_w is the time domain duration of the wavelet, $g_j(t)$ are the kernel functions

$$g_j(t) = g(t,t_j) = \hat{q}(t-t_j) , \qquad 0 \leq t-t_j \leq t_w,$$

and (w,g_j) denotes the inner product of the two functions.

Oldenburg et al. (1981) use the method of spectral expansion (Parker, 1977) to solve Eq. (25). This approach is very versatile in that it can be applied to produce either "smallest" models by minimizing (w,w), flattest models by minimizing (w',w'), or smoothest models by minimizing (w'',w''). Further, the wavelet is expanded in terms of basis functions that depend on the eigenstructure of the covariance matrix of approximate reflectivity. It is known that basis functions that correspond to small eigenvalues are generally oscillatory in nature. Thus some desired smoothness may be achieved by discarding basis functions associated with small eigenvalues.

The problems of wavelet estimation and deconvolution are, to my mind, two quite distinct issues. The first entails the estimation of a band-limited (generally 8 to 40 Hz, say) time function. The second entails the recovery of a full-band reflectivity series from the band-limited observations given an approximate knowledge of the source signature.

The necessity of obtaining a full-band reflectivity function lies in the requirement that the subsurface impedance that is derived from it contain both the low and high frequency components that characterize the subsurface. That this is so has been clearly demonstrated by Oldenburg et al. (1983), who applied the linear inverse approach of Backus and Gilbert (1970) in a manner initiated by Oldenburg (1981) in a recent paper on deconvolution. As applied to deconvolution, the Backus-Gilbert (BG) approach is, essentially, the following.

Given the wavelet w_t, we search for an inverse filter, f_t, such that

$$w_t * f_t = \delta_t$$

where δ_t is the Dirac delta function. Since w_t is band-limited, recovery of a delta function in a linear manner is not possible, and we must be satisfied with a band-limited result; that is,

$$\begin{aligned} f_t * y_t &= f_t * w_t * q_t + f_t * n_t \\ &= a_t * q_t + f_t * n_t \\ &= \langle q_t \rangle + \delta\langle q_t \rangle \,. \end{aligned}$$

The inverse filter, f_t, is determined by minimizing an objective function that incorporates both the desire for deltaness and the desire to minimize the variance of $\delta\langle q_t \rangle$. Such an objective function is

$$\psi = \phi \cos\theta + \gamma \sin\theta$$

where $\phi = \sum_t (w_t * f_t - \delta_t)^2$ expresses the desired deltaness, $\gamma = \mathrm{var}[\delta\langle q_t \rangle]$, and θ is the well known BG tradeoff parameter, $0 \leq \theta \leq \pi/2$, chosen to trade off the desired resolution versus the output variance.

A fundamental role in this approach to deconvolution is played by the function a_t, which is the averaging or resolving kernel. In fact, BG show that, whereas the q_t obtained by methods such as the spectral technique, which can be called construction methods, are nonunique, the averages, $\langle q_t \rangle$, obtained by the BG appraisal approach (see Oldenburg, 1981) are unique in that all models that reproduce the data will generate the same $\langle q_t \rangle$. A view of the Wiener shaping filter through the BG formalism has been recently explored in detail by Lines and Treitel (1981).

We are now ready to take the final step in the flow chart of Fig. 1. The end result in the processing of seismic exploration data is to obtain a description of the subsurface. Such a description is contained in the acoustic impedance function, ξ_k, which is defined in terms of the layer properties of our layered earth model as $\xi_k = \rho_k v_k$ where ρ_k and v_k are the density and velocity respectively. The reflection coefficient at the base of the kth layer is defined in terms of the impedances by

$$q_k = \frac{\xi_{k+1} - \xi_k}{\xi_{k+1} + \xi_k} .$$

Rearranging terms, one obtains the impedance, ξ, in terms of the reflection coefficients

$$\xi_{k+1} = \xi_1 \prod_{k=1}^{N} \left[\frac{1 + q_k}{1 - q_k}\right] . \tag{26}$$

As we have seen previously, the recovery of q_k from the band-limited seismogram following standard procedures leads to averages of the reflectivity function, $\langle q_k \rangle$.

Oldenburg et al. (1983) have shown that, defining $\eta_t = \ln(\xi_t/\xi_0)$ (note that we have defined these functions as functions of time), the recovered impedance $\hat{\eta}_t$ is also obtained approximately in terms of averages as

$$\hat{\eta}_t \simeq a_t * \eta_t = \langle \eta_t \rangle .$$

It is clear, therefore, that low and high frequency information about the acoustic impedance is absent from the band-limited result. Since this information, and in particular the low-frequency part, is vital in the recovery of subsurface behavior, the full-band recovery of the reflectivity function is of central importance in seismic data processing. I will consider two approaches that have been recently investigated by Oldenburg et al. (1983) and that show considerable promise. An example of the application of the two techniques to a real data set will be shown at the end of this section.

The first approach is an AR spectral extension scheme (Lines and Clayton, 1977; Lines et al., 1980), and depends on the observation that, since the ideal reflectivity function consists of a series of spikes, $q_t = \sum_k q_k \, \delta(t-k)$, the Fourier transform of q_t, Q_f, is a sum of complex sinusoids:

$$Q_f = \sum_k q_k \exp(-2\pi i f t_k) .$$

As shown by Clayton and Ulrych (1977), a good model for Q_f may be obtained by using a complex AR process. Such a representation allows us to model Q_f beyond the limits of the wavelet band. In other words, the complex process Q_f is extrapolated outside the known band by means of a complex prediction filter.

The second approach adopted by Oldenburg et al. (1983) uses linear programming (LIP) to minimize the ℓ_1 norm of some objective function. The ℓ_1 norm of a vector $\mathbf{r}$ is defined as

$$||\mathbf{r}||_1 = \sum_{i=1}^{m} |r_i| .$$

The use of the ℓ_1 norm was suggested by Claerbout and Muir (1973) and was first applied to the problem of seismic deconvolution by Taylor et al. (1979). (See Taylor, 1981, for a review of the application of the ℓ_1 norm to deconvolution.) Important properties of the ℓ_1 norm as far as recovery of phase information is concerned are discussed by Scargle (1977, 1981). The 1981 paper is an excellent treatment of the modeling of astronometric time series.

The importance of the ℓ_1 norm in conjunction with linear programming (using the Simplex algorithm in this case) as applied to deconvolution can best be illustrated by a simple numerical example. We consider the underdetermined system of equations

$$\mathbf{A}\,\mathbf{x} = \mathbf{d}\,, \tag{27}$$

where

$$\mathbf{A} = \begin{bmatrix} 1 & 2 & 3 \\ 3 & 1 & -1 \end{bmatrix}$$

$$\mathbf{x}^T = (x_1, x_2, x_3)$$

and

$$\mathbf{d}^T = (2, 1)\,,$$

and we seek solutions for $\mathbf{x}^T$ using the ℓ_2 and ℓ_1 norms. The ℓ_2 solution, ${}^2\mathbf{s}$, is obtained by minimizing $\mathbf{x}^T\mathbf{x}$ subject to Eq. (27) as constraints and is given by

$${}^2\mathbf{s} = \mathbf{A}^T(\mathbf{A}\mathbf{A}^T)^{-1}\,\mathbf{d}\,.$$

For this example, ${}^2\mathbf{s}^T = (1/3, 1/3, 1/3)$. The ℓ_1 norm solution using the Simplex algorithm, ${}^1\mathbf{s}$, is obtained by minimizing

$$\sum_{i=1}^{3} |{}^1s_i| = \sum_{i=1}^{3} |a_i - b_i| , \qquad a_i > 0,\ b_i > 0 ,$$

subject to the constraints $\mathbf{Ax} = \mathbf{d}$. The solution that is obtained is ${}^1\mathbf{s}^T = (0,1,0)$. This result represents a general property of the LIP solution in that, in contrast to the ℓ_2 solution, which is flat, the LIP solution is spiky in character and for this reason is ideally suited to the deconvolution of a sparse spike train. Following the work of Taylor et al. (1979), Levy and Fullagar (1981) applied the ℓ_1 norm to sparse spike-train reconstruction but with the important difference that the constraining equations are formulated in the frequency domain. Thus, as before, we have

$$x_t = w_t * q_t$$

and

$$X_f = W_f\, Q_f .$$

The Levy and Fullagar approach is to minimize the objective function $\sum_n |q_n|$ subject to the constraining equations

$$\mathrm{Re}[Q_f] = \sum_{n=0}^{N-1} q_n \cos(2\pi i n/N)$$

$$\mathrm{Im}[Q_f] = \sum_{n=0}^{N-1} q_n \sin(2\pi i n/N)$$

which are defined only at those frequencies at which Q_f is sufficiently large.

Oldenburg et al. (1983) have extended the Levy and Fullagar approach in an important manner by incorporating into the algorithm the possibility of constraining the impedance function that is obtained from the deconvolved reflectivity by means of Eq. (26). Results obtained by Oldenburg et al. applying the AR and LIP methods to a real data set are shown in Fig. 6. Figure 6(a) shows the band-limited input section. The frequency band that was used in the reconstructed impedance sections was 15 to 60 Hz. The constructed reflectivity using LIP which contains frequencies in the band 0 to 60 Hz is shown in Fig. 6(b). The recovered acoustic impedance is illustrated in Fig. 6(c). Figure 6(d) shows the recovered acoustic impedance using the AR approach. In both these last figures the acoustic impedance from the well-log located 200 meters from one of the shots has been inserted for comparison. A comparison of Figs. 6(c) and 6(d) with Fig. 6(e),

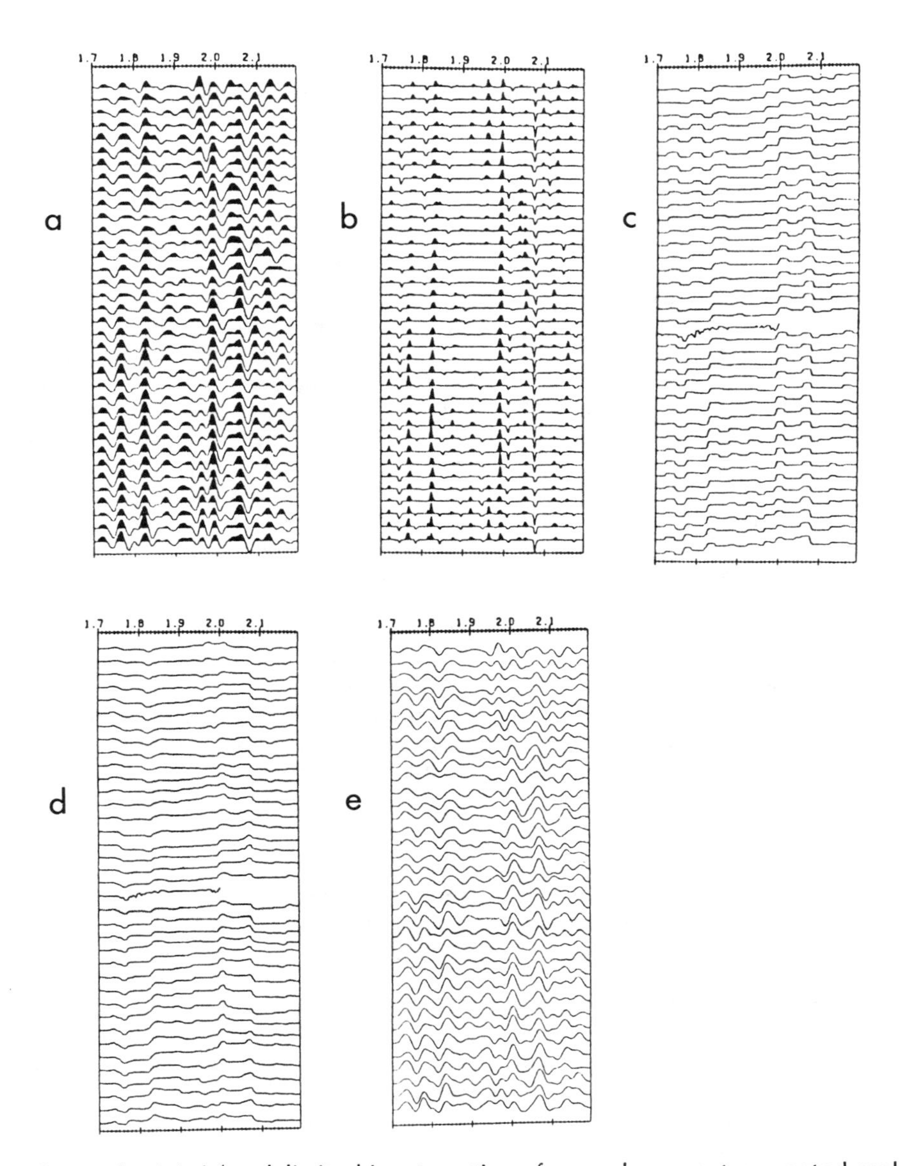

Figure 6. (a) A band-limited input section of normal moveout corrected real reflection data. (b) The deconvolved section (0-60 Hz) using the LIP method. (c) Recovered acoustic impedance using section (b). (d) Recovered acoustic impedance from a deconvolved section using the AR method. (e) Recovered acoustic impedance from a deconvolved section using the normal band-limited method.

which shows the acoustic impedance recovered using band-limited data, clearly shows the importance of the approach of Oldenburg et al. It is also encouraging that two methods, based on quite different philosophies, return such similar results.

4. Concluding Remarks

I have attempted in this paper to tie together three fields by showing the relationships of each to shared concepts. It is my experience that often such a synthesis can give fresh insights into the solution of problems that exist within each field. This synthesis is by no means exhaustive and was not meant to be so. I have left out many interesting and relevant issues simply because of time considerations. Some notable examples are the fine work of Shore and Johnson (1980) and Shore (1981) on cross-entropy approaches to spectral analysis, the phase-only techniques recently explored by Oppenheim et al. (1979) and Hayes et al. (1980) and reviewed by Oppenheim and Lim (1981), and the paper by Morf et al. (1978) which has extended the Burg algorithm to multichannel time series. I would also like to mention the work of Rietsch (1977), who has applied the principle of maximum entropy to the problem of inverting for the density structure of the earth. In this regard, it is my opinion that the application of inverse theory to signal processing, as exemplified by the work of Oldenburg (1981), is an important step in quantifying certain relationships that have hitherto been somewhat qualitative.

Finally, if in spite of the nondetailed nature of this presentation, this paper is still somewhat long, I can only paraphrase Pascal by saying that unfortunately I did not have the time to make it shorter.

References

Akaike, H. (1969) Ann. Inst. Statist. Math. **21**, 407-419.

Akaike, H. (1974) IEEE Trans. Autom. Control **AC-19**, 716-723.

Backus, G.E., and F. Gilbert (1970) Phil. Trans. R. Soc. London **A266**, 123-192.

Berkhout, A. J., and P. R. Zaanen (1976) Geophys. Prospect. **24**, 141-197.

Berryman, J. G. (1978) Geophysics **43B**, 1384-1391.

Box, G. E. P., and G. M. Jenkins (1976) Time Series Analysis, Forecasting and Control (San Francisco: Holden-Day).

Burg, J. P. (1975) Maximum Entropy Spectral Analysis, Ph.D. thesis, Stanford University.

Claerbout, J. F. (1976) Fundamentals of Geophysical Data Processing with Applications to Petoleum Prospecting (New York: McGraw-Hill).

Claerbout, J. F. (1978) Stanford Exploration Report, Report No. 15.

Claerbout, J. F., and F. Muir (1973) Geophysics **38**, 826-844.

Clayton, R. W., and T. J. Ulrych (1977) IEEE Trans. Inf. Theory **IT-23**, 262-264.

Currie, R. G. (1981) J. Geophys. Res. **86**, 11055-11064.

Donoho, D. (1981) in D. Findley, ed., Second Applied Time Series Analysis II (New York: Academic Press), pp. 565-608.

Godfrey, R. J. (1979) A Stochastic Model for Seismogram Analysis, Ph.D. thesis, Stanford University.

Gray, H. L., and F. W. Morgan (1978) in D. Findley, ed., First Applied Time Series Symposium (New York: Academic Press), pp. 39-138.

Gray, W. C. (1979) Variable Norm Deconvolution, Ph.D. thesis, Stanford University.

Griffiths, L. J., F. R. Smolka, and L. E. Trembly (1977) Geophysics **42**, 742-759.

Guarino, C. R. (1979) Proc. IEEE **67**, 957-958.

Hayes, M. A., J. S. Lim, and A. V. Oppenheim (1980) IEEE Trans. Acoust. Speech Signal Process. **ASSP-28**, 182-185.

Kay, S. M. (1979) IEEE Trans. Acoust. Speech. Signal Process. **ASSP-27**, 478-485.

Kay, S. M., and S. L. Marple, Jr. (1981) Proc. IEEE **69**, 1380-1419.

Kullback, S., and R. A. Leibler (1951) Ann. Math. Stat. **22**, 79-86.

Lacoss, R. T. (1971) Geophysics **36**, 661-675.

Landers, T. E., and R. T. Lacoss (1977) IEEE Trans. Geosci. Electron. **GE-15**, 26-33.

Levinson, N. (1947) J. Math. Phys. **25**, 261-278.

Levy, S., and P. Fullagar (1981) Geophysics **46**, 1235-1244.

Lines, L. R., and R. W. Clayton (1977) Geophys. Prospect. **24**, 417-433.

Lines, L. R., R. W. Clayton, and T. J. Ulrych (1980) Geophys. Prospect. **28**, 49-59.

Lines, L. R., and S. Treitel (1981) Paper presented at the 51st SEG Symposium, Los Angeles, California.

Lines, L. R., and T. J. Ulrych (1977) Geophys. Prospect. **25**, 512-540.

Makhoul, J. (1975) Proc. IEEE **63**, 561-580.

Makhoul, J. (1981) IEEE Trans. Acoust. Speech Signal Process. **ASSP-29**, 868-871.

Morf, M., A. Vieira, D. T. L. Lee, and T. Kailath (1978) IEEE Trans. Geosci. Electron. **GE-16**, 85-94.

Oldenburg, D. W. (1981) Geophys. J. R. Astron. Soc. **65**, 331-357.

Oldenburg, D. W., S. Levy, and K. P. Whittal (1981) Geophysics **46**, 1528-1542.

Oldenburg, D. W., T. Scheuer, and S. Levy (1983) Geophysics **48**, 1318-1337.

Ooe, M. (1978) Geophys. J. R. Astron. Soc. **53**, 445-457.

Ooe, M., and T. J. Ulrych (1979) Geophys. Prosp. **27**, 458-473.

Oppenheim, A. V., and J. S. Lim (1981) Proc. IEEE **69**, 529-541.

Oppenheim, A. V., J. S. Lim, G. Kopec, and S. C. Pohlig (1979) Proc. IEEE Int. Conf. ASSP, Washington, D. C.

Oppenheim, A. V., R. W. Schafer, and T. G. Stockham (1968) Proc. IEEE **65**, 1264-1291.

Pagano, M. (1974) Ann. Statistics **2**, 99-108.

Parker, R. L. (1977) Ann. Rev. Earth Planet. Sci., 35-64.
Parzen, E. (1974) IEEE Trans. Autom. Control **AC-19**, 183-200.
Parzen, E. (1980), in D. R. Brillinger and G. C. Tao, eds., Directions in Time Series Analysis (Institute of Mathematical Statistics), pp. 80-111.
Pisarenko, V. P. (1973) Geophys. J. R. Astron. Soc. **33**, 347-366.
Postic, A., J. Fourmann, and J. Claerbout (1980) Presented at the 50th SEG Symposium, Houston, Texas.
Rietsch, E. (1977) J. Geophys. **42**, 489-506.
Robinson, E. A. (1964) in H. Wold, ed., Econometric Model Building (Amsterdam: North-Holland), pp. 37-110.
Robinson, E. A. (1967) Statistical Communication and Detection (London: Charles Griffin), pp. 172-174.
Robinson, E. A., and S. Treitel (1981) Geophysical Signal Analysis (New Jersey: Prentice Hall).
Scargle, J. D. (1977) IEEE Trans. Inf. Theory **IT-23**, 140-142.
Scargle, J. D. (1981) Astrophys. J. Suppl. Series **45**, 1-71.
Schafer, R. W. (1969) Dept. E. E., MIT, Tech. Rep., 466.
Shore, J. E. (1981) IEEE Trans. Acoust. Speech Signal Process. **ASSP-29**, 230-237.
Shore, J. E., and R. W. Johnson (1980) IEEE Trans. Inf. Theory **IT-26**, 26-37.
Silvia, M. T., and E. A. Robinson (1979) Deconvolution of Geophysical Time Series in the Exploration for Oil and Natural Gas (Amsterdam: Elsevier).
Taylor, H. L. (1981) in A. A. Fitch, ed., Developments in Geophysical Exploration Methods - 2 (London: Applied Science Publications).
Taylor, H. L., S. C. Banks, and J. F. McCoy (1979) Geophysics **44**, 39-52.
Treitel, S., P. R. Gutowski, and E. A. Robinson (1977) in J. H. Miller, ed., Topics in Numerical Analysis (New York: Academic Press), pp. 429-446.
Treitel, S., and E. A. Robinson (1981) Geophysics **46**, 1108-1116.
Treitel, S., and T. J. Ulrych (1979) IEEE Trans. Acoust. Speech Signal Process. **ASSP-27**, 99-100.
Tribolet, J. M. (1979) Seismic Applications of Homomorphic Signal Processing (New Jersey: Prentice Hall).
Ulrych, T. J. (1971) Geophysics **36**, 650-660.
Ulrych, T. J., and T. N. Bishop (1975) Rev. Geophys. Space Phys. 183-200.
Ulrych, T. J., and R. W. Clayton (1976) Phys. Earth Planet. Inter. 188-200.
Ulrych, T. J., and M. Ooe (1979) in S. Haykin, ed., Topics in Applied Physics, 34 (Berlin: Springer Verlag), pp. 73-125.
Widrow, E., and M. E. Hoff (1960) in IRE WESCON Convention Rec., Pt. 4, pp. 96-104.
Wiggins, R. A. (1978) Geoexploration **16**, 21-35.
Willsky, A. S. (1979) Digital Signal Processing and Control and Estimation Theory (Cambridge, Mass.: MIT Press).
Wold, H. (1938) A Study in the Analysis of Stationary Time Series (Uppsala: Almqvist and Wiksell).
Woodward, W. A., and H. L. Gray (1979) ONR Tech. Rep. 134, SMU Department of Statistics.

ABSOLUTE POWER DENSITY SPECTRA

John Parker Burg

Entropic Signal Processing, Cupertino, California 95014

C. Ray Smith and W. T. Grandy, Jr. (eds.),
Maximum-Entropy and Bayesian Methods in Inverse Problems, 273–286.

1. Introduction

For various reasons, few people are concerned with absolute power density spectra. This is evidenced by the way people display plots of spectral estimates. Often, the spectra are plotted so that their maximum or minimum values are the same or the spectra all start with the same value at zero frequency. For really meaningful comparisons, spectra should be plotted in absolute terms. To do this, we need to go back to fundamentals and look at what a power density spectrum really is.

If we have the one-sided power density spectrum P(f) of an ergodic stationary time function, g(t), then the average square value or power of the time function is given by

$$\overline{g^2(t)} = \lim_{T\to\infty} \frac{1}{2T} \int_{-T}^{+T} g^2(t)dt = \int_0^\infty P(f)df .$$

If we band limit the function g(t) to frequencies between f_1 and f_2, then the average square value of this band-limited time function is

$$\int_{f_1}^{f_2} P(f)df .$$

One can also say this is the power in the f_1 to f_2 band.

For band-limited and digitized time series, where we are dealing with pure numbers, our power density spectrum is expressed in units squared per cycle per second. To relate this to the physical phenomenon that we are sampling, we need to know the power response of the measuring instrumentation that is converting the physical time function into numbers. We shall not go further into this physical consideration but shall concentrate here on correctly normalizing the power density spectrum of our numerical time series.

Assuming our sampling period is t and the time series is band limited to W = 1/(2t), we know that the zero lag value of the autocorrelation function, R(0), which is the average power or average square value of the time series, is given by

$$R(0) = \int_0^W P(f)df .$$

This equation can be used to correctly scale the power spectrum when we have only the relative shape of the spectrum. In applying this to maximum entropy spectral analysis, we find the correct absolute spectral density for the Nth-order estimate to be

$$P(f) = \frac{P_N/W}{|1 + a_1 z + a_2 z^2 + \ldots + a_N z^N|^2} ,$$

where the denominator is the power response of our Nth-order prediction error filter and P_N is the mean square error of the Nth-order residual time function. One can verify that the integral of this function is indeed R(0) by analytic integration. However, a simple way of seeing that the expression is correct is to note that the Nth-order residual is assumed to have a white spectrum of total power P_N. Thus, the absolute spectral density function of the residual is simply P_N/W. The above equation then expresses the fact that the power spectrum, P(f), of the input time function times the power response of the filter is equal to the power spectrum of the output time series.

Power spectra often have a wide dynamic range and are difficult to display on a linear scale. Also, one often wishes to multiply them by the power response of some filter. Both of these problems are handled neatly by plotting spectra in terms of decibels. A wide dynamic range can be displayed, and the power response of the filter in decibels can be simply added to the input dB spectrum to get the output dB spectrum. Actually, the naturalness of plotting spectra in terms of decibels, which are relative units, probably is the main reason that people have gotten away from thinking in terms of absolute spectra. This, in turn, leads to not making a relative comparison of two spectra correctly.

To be able to express a decibel spectrum in absolute terms, we must in effect say what the zero dB level means. This decision is quite arbitrary, but must be made. The choice we shall make here is that zero dB corresponds to unity power density in units squared per cycle per second. In this case, our correct decibel expression is simply

$$10 \log P(f) = 10 \log (P_N/W) - 20 \log |1 + a_1 z + \ldots + a_N z^N| .$$

Thus, if we had a flat spectrum at zero dB, this would correspond to white noise of power W. A person might wish to assign zero dB to unit variance white noise instead of to W variance white noise. To do this, zero dB would correspond to a power density of 1/W. This is actually not desirable since the meaning of the zero dB level then depends on the sampling rate. The spectrum of the physical process that we are sampling should not depend on our sampling rate, and thus we should not define zero dB in terms of W.

2. An Example of Observing Quantization Noise

In sampling and digitizing a physical time function, the rounding off to the nearest integer of the sampled value adds noise to the digitized data. It is usually an excellent assumption that this noise is uncorrelated sample to sample and that it is uniformly distributed from -1/2 to +1/2. The average square value of such a random variable is 1/12. Thus the quantization noise has a white spectrum with power density of 1/(12W) units squared per cycle per second. Note that the power density depends on the sampling rate.

Actually, this quantization noise is generated by the rounding, whether it is done by the analog-to-digital converter or by calculations in a computer. Often, time series data are stored as 16-bit two's complement integers before and after digital filtering. Thus, we can expect this quantization noise to be generated whenever we return our calculations to 16-bit integer form.

In Fig. 1a, we have a 255th-order maximum entropy spectrum of about 3.5 seconds of fairly monotone speech saying the numbers from one to eight.

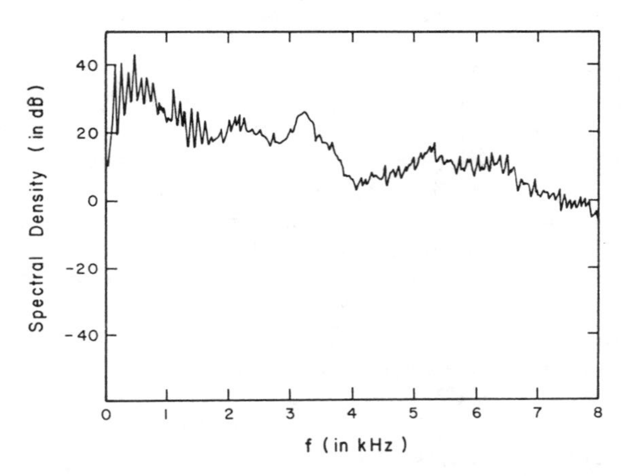

Figure 1a. Maximum entropy (order 255) spectrum of 3.5-second speech pattern.

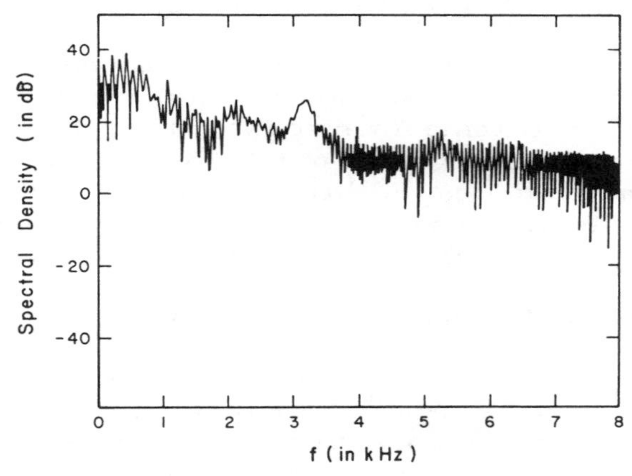

Figure 1b. Same speech pattern as 1a, but analyzed by conventional spectral analysis, without windowing.

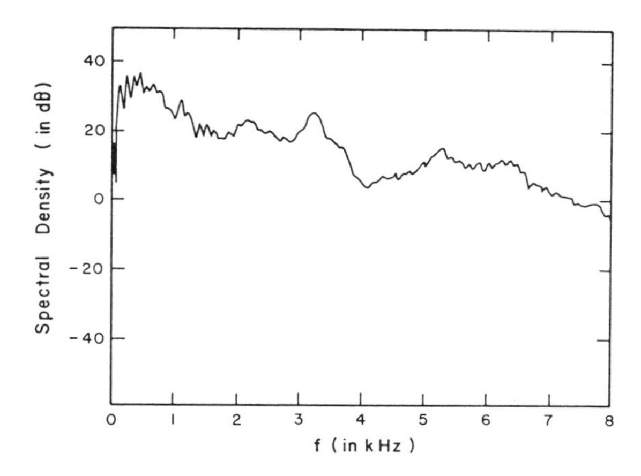

Figure 1c. Same as 1b, but using cosine weighted autocorrelations.

The autocorrelation was calculated using lag products. The sampling rate was 16,000 per second, so the foldover frequency is 8000 Hz. One can see clearly the pitch frequency and many harmonics. For comparison, Fig. 1b is conventional spectral analysis doing the Fourier transform of the autocorrelation out to lag 255 without windowing, and Fig. 1c is with a cosine window.

We wished to band-limit these data to 4000 Hz. We did this one time by convolving the time function with the boxcar truncated, sin(x)/x digital filter, whose z-transform is

$$\sum_{n=-500}^{500} \frac{\sin(n\pi/2)}{n\pi/2} z^n .$$

Figure 2 shows the Fourier transform power response of the boxcar filter. In this figure, the response in the reject band varies too rapidly to be plotted accurately. Only an envelope of the variation in power is shown, and even the ripple in the envelope is an artifact due to a beating phenomenon between the power fluctuation and the spacing between calculated points in frequency. Figure 3 shows an interesting plot obtained by taking the boxcar impulse response (extended by zeros) as a time function and doing a 255th-order maximum entropy analysis of it. This gives us a much more manageable plot than presented in Fig. 2. As is appropriate for power responses, we have set unity power response to zero dB. We see that the maximum entropy obtained plot is a smoothed version of the Fourier transform plot. Note that the reject region gets down to 66 dB from the passband.

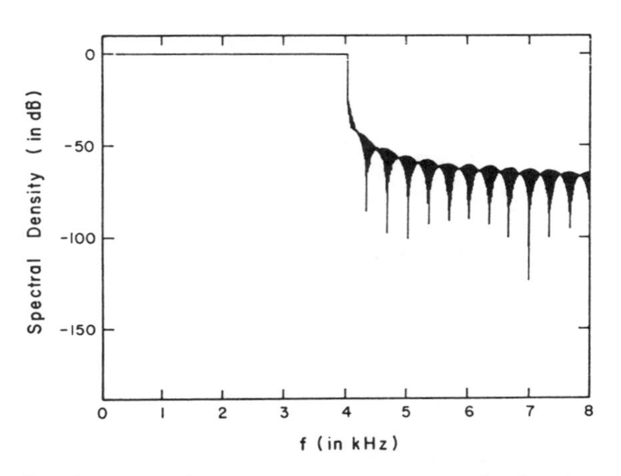

Figure 2. Fourier transform power response of the boxcar weighted sin(x)/x.

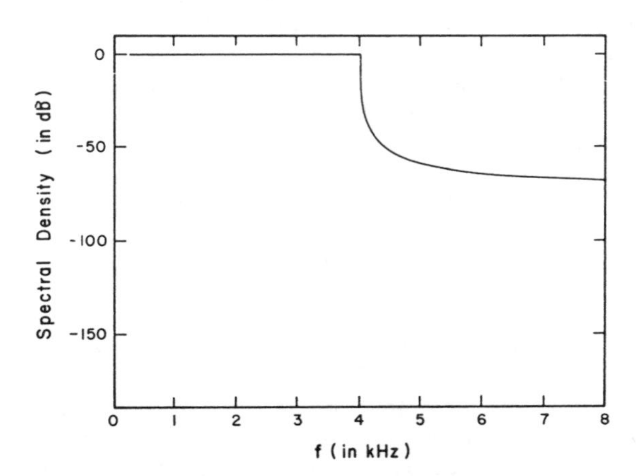

Figure 3. Maximum entropy (order 255) spectrum of the boxcar weighted sin(x)/x.

Figure 4 shows the Fourier transform power response of the cosine weighted filter with z-transform

$$\sum_{n=-500}^{500} \frac{1 + \cos(n\pi/500)}{2} \frac{\sin(n\pi/2)}{n\pi/2} z^n ,$$

and Fig. 5 shows its maximum entropy plot. We now get a huge 164-dB drop in the reject band.

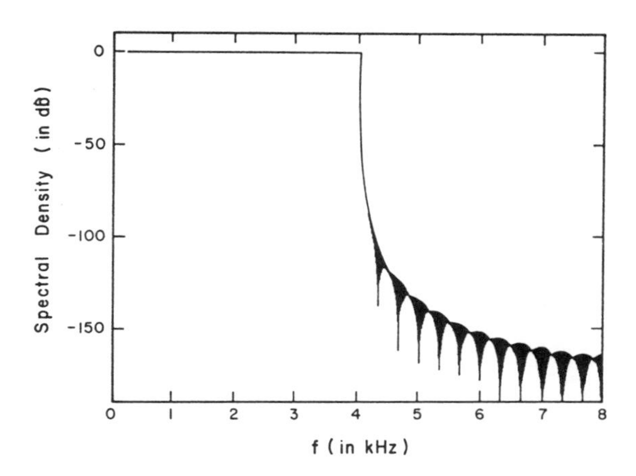

Figure 4. Fourier transform power response of cosine weighted sin(x)/x.

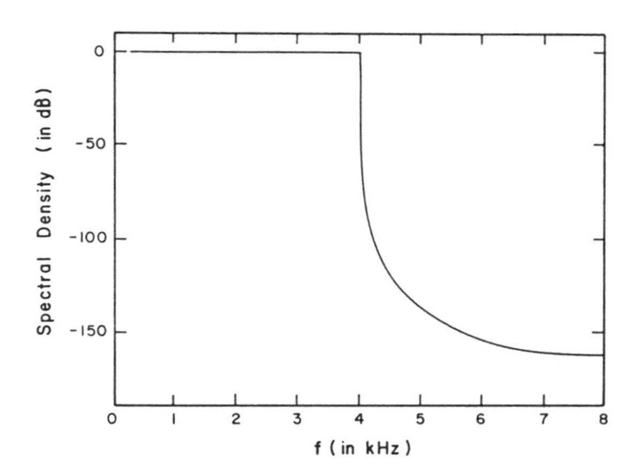

Figure 5. Maximum entropy (order 255) spectrum of cosine weighted sin(x)/x.

In doing the convolution of the speech time function, we used double precision floating point arithmetic, but rounded the result to 16-bit integers. Figure 6 shows the 255th-order maximum entropy spectrum of the boxcar sin(x)/x filtered speech, and Fig. 7a shows the spectrum of the cosine weighted filtered speech. We note in both cases that the spectrum from 0 to 4000 Hz is unchanged by the filter but that the reject band is pushed down to -10 log (12W) = -10 log (12 x 8000) = -10 log (96000) = -49.82 dB. This is of course the white quantization noise level. In Fig. 1, we see around 5200 Hz a spectral pattern that we recognize in Fig. 6, but at about 60 dB lower, agreeing with the 60-dB rejection of the boxcar filter at that

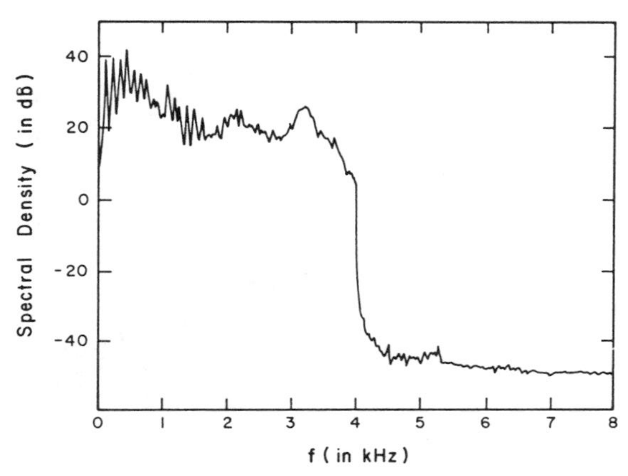

Figure 6. Maximum entropy (order 255) spectrum (length 1001) of boxcar weighted sin(x)/x filtered speech.

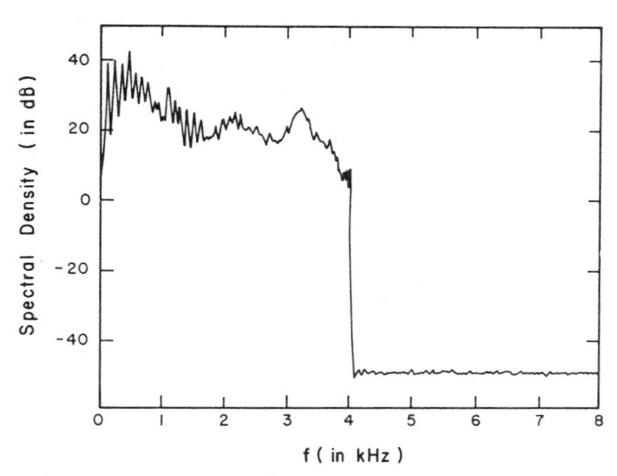

Figure 7a. Maximum entropy (order 255) spectrum (length 1001) of cosine weighted sin(x)/x filtered speech.

frequency. In Fig. 7a, the huge rejection of the cosine weighted filter has pushed the speech spectrum far below the quantization noise level, in fact, about 100 dB below the quantization noise level at 8000 Hz. By converting our double precision filtered numbers to integers, we have added 100 dB of high frequency noise to our speech! However, no one is surprised that we do not hear this noise.

Figures 7b and 7c are Fourier transform spectral estimates from the same autocorrelations. The side lobes of the windows prevent us from seeing the true spectrum in the reject band.

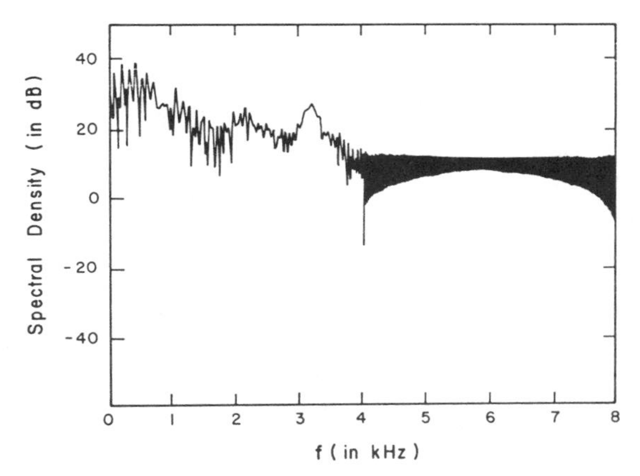

Figure 7b. FFT of 256 autocorrelations (length 1001) of cosine weighted sin(x)/x filtered speech.

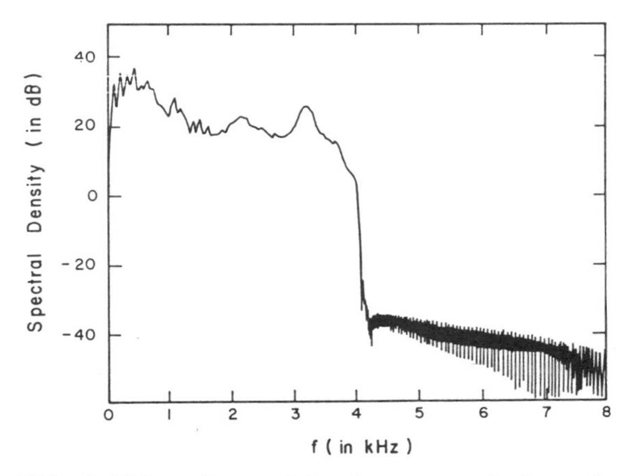

Figure 7c. FFT of 256 cosine weighted autocorrelations (length 1001) of cosine weighted sin(x)/x filtered speech.

3. Power Spectrum Distribution Functions

If we have the power spectrum, P(f), one very interesting thing to do is to calculate the power spectrum distribution function,

$$\int_0^f P(f)df \, .$$

This function is clearly montonically increasing and goes from zero to R(0). For actual plotting purposes, it is handier to use the normalized distribution function

$$D(f) = \frac{1}{R(0)} \int_0^f P(f)df \, .$$

This function now goes from zero to one. It is a valuable plot because we can easily find the power between two frequencies, f_1 and f_2, as $D(f_2) - D(f_1)$. The total power between these frequencies is just R(0) times $[D(f_1) - D(f_2)]$.

To show the value of the distribution function, we show in Fig. 8 a power density spectrum having some sinusoidal components. Looking at this figure, what would one estimate the relative power in the various components to be? Figure 9 is the normalized integral of this spectrum. One now sees that the sharpest peak, the prominent one at 0.78 W, contains only

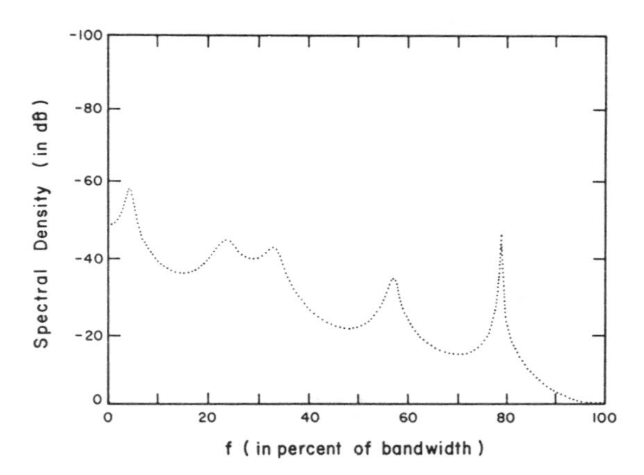

Figure 8. A power density spectrum with sinusoidal components.

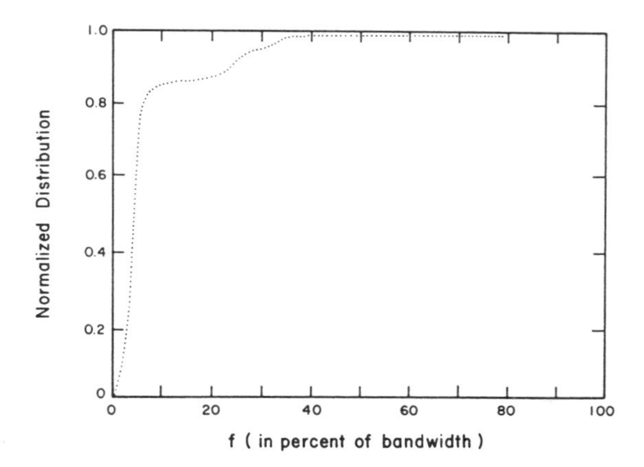

Figure 9. Normalized integral of the spectrum in Fig. 8.

about 1% of the total power. The low frequency peak accounts for about 80% of the power. Decibel plots can indeed be misleading.

In Fig. 10, we show three overlaid spectra, each containing a sine wave at W/2 of power 0.5 plus white noise. The total power of the white noise is 1.0, 0.5, and 0.25 in the three spectra. Thus, the white noise spectral levels differ by 3 dB, which is indicated in the spectral plots. In Fig. 11, we have expanded the pictures horizontally by a factor of 5 to show the shapes of the peaks more clearly. We see that the spectrum with the least white

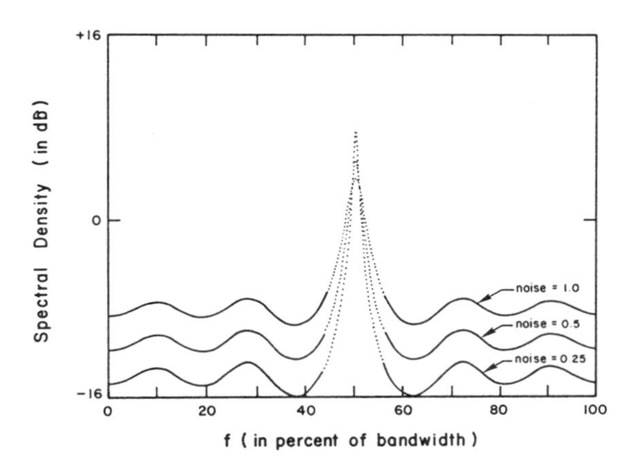

Figure 10. Three overlaid spectra, each containing a sine wave at W/2 of power 0.5. White noise powers are 1.0, 0.5, and 0.25, respectively.

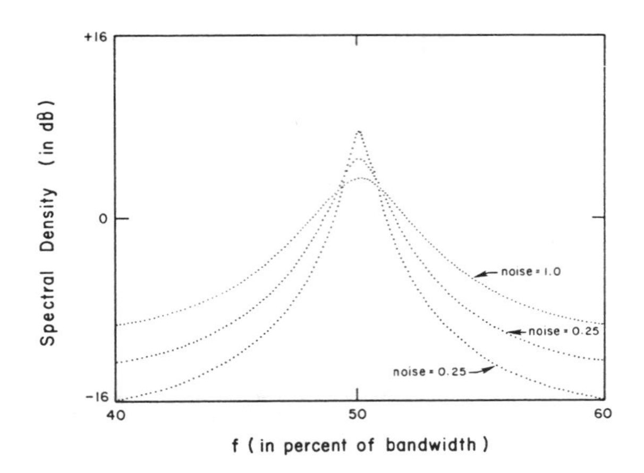

Figure 11. Spectra of Fig. 10, expanded horizontally by a factor of 5.

noise has the highest and sharpest peak while the lowest signal-to-noise spectrum has the lowest resolution peak. But remember that the sine wave has the same power in each of these cases.

Figures 12, 13, and 14 show the normalized spectral distribution functions of these three spectra. By fitting straight lines to these graphs, one can accurately see that the sine wave at W/2 contains 1/3, 1/2, and 2/3 of the total power, as it should.

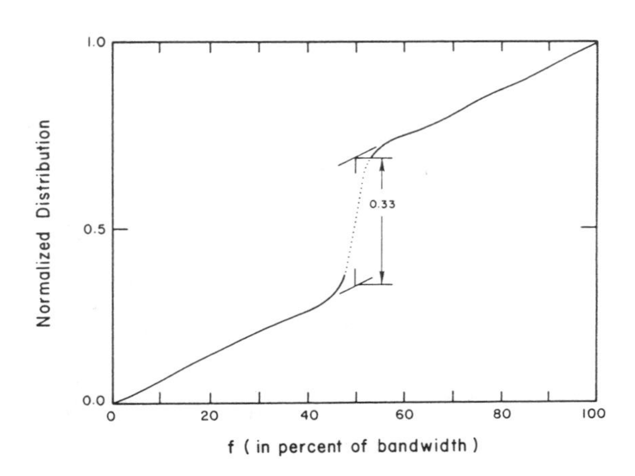

Figure 12. Normalized spectral distribution function of a single sine wave in noise of power 1.0.

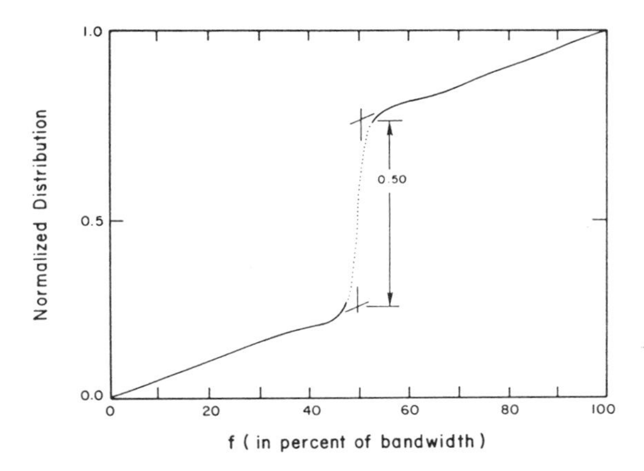

Figure 13. Normalized spectral distribution function of a single sine wave in noise of power 0.5.

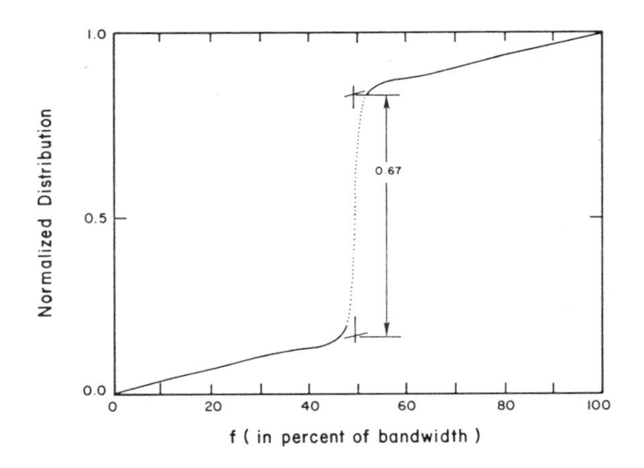

Figure 14. Normalized spectral distribution function of a single sine wave in noise of power 0.25.

In Fig. 15, we have a sine wave at W/4 whose power is 10 times that of another sine wave at 3W/4. Added to this is white noise of total power equal to the major sine wave. Figure 16 is the normalized distribution function, which shows quite accurately the relative power in each component.

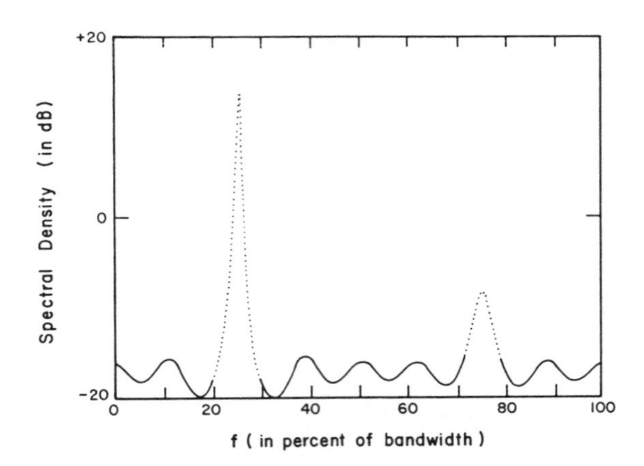

Figure 15. Sine wave at W/4, with power 10 times that of sine wave at 3W/4 in white noise of power equal to total power of major sine wave.

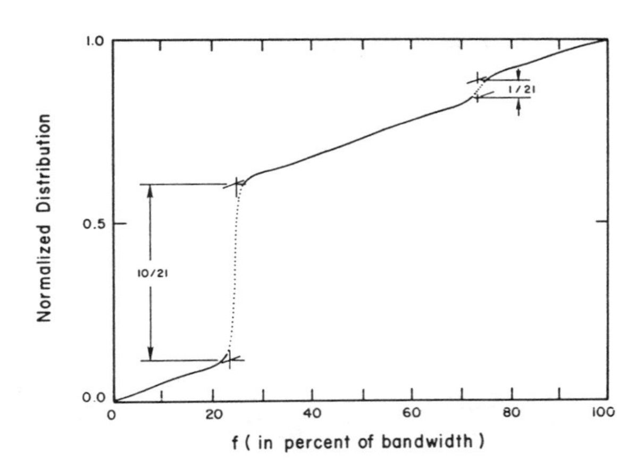

Figure 16. Normalized distribution function of spectrum in Fig. 15.

Many people do not believe that they can use spectral analysis to measure the power in sinusoidal components. If one uses maximum entropy spectral analysis, or other methods that are true spectral estimation techniques, and pays attention to getting absolute spectral estimates and integrates, accurate power measures can be made of sinusoidal and other spectral components.

THE ENTROPY OF AN IMAGE

S. F. Gull[1] and J. Skilling[2]

[1]Mullard Radio Astronomy Observatory, Cavendish Laboratory, Madingley Road, Cambridge, England
[2]Department of Applied Mathematics and Theoretical Physics, Silver Street, Cambridge, England

We investigate the entropy expressions that have been used for image reconstruction, including the spectral analysis of time-series data. We find that one should always use the Shannon formula $S = -\Sigma\ p_i \log p_i$ when attempting to reconstruct the shape of an image. This produces an image that is maximally noncommittal about the set of relative proportions p_i which one wishes to construct. Alternative forms of entropy, notably the $\Sigma \log f_i$ form recommended by Burg, refer to the physical process producing the image, and produce results that are maximally noncommittal, instead, about the probability distribution function governing individual realizations of this process. It is appropriate to use a Burg-type form when attempting to predict future samples from the physical process, but not when reconstructing the proportions p_i themselves.

C. Ray Smith and W. T. Grandy, Jr. (eds.),
Maximum-Entropy and Bayesian Methods in Inverse Problems, 287–301.

1. Introduction

Maximum entropy is being increasingly widely used as a technique of image reconstruction. The form of entropy to be used, however, has been the cause of dissention, some authors [1-7] preferring forms based on the Shannon form, and others [8-10] preferring the form recommended by Burg. Yet others [11,12] follow Kikuchi and Soffer [13] in attempting to reconcile the two. Because the Shannon form is being used with increasing success for many types of generalized inverse problems [14-17], it is important to determine whether either form of entropy is to be intrinsically preferred, or whether such questions are meaningless and should be answered pragmatically [18].

In Section 2, we give what we believe to be the correct derivation of the maximum entropy technique for determining probability distributions, following Jaynes. Positive, linear images are directly identified with probability distributions in Section 3, and this leads naturally and immediately to the Shannon/Jaynes form of entropy, $-\Sigma\, p_i \log p_i$, which should accordingly be used in image reconstruction. It answers directly the fundamental question of image reconstruction, namely, "Where would the next photon come from?"

In Section 4, we turn to the derivation of alternative entropies. These are physical entropies expressing the statistical freedom allowed to individual realizations of the physical process producing the image, and are usually of the Burg type. Using Jaynes' technique of maximizing the Shannon entropy, we derive mathematical forms for the physical entropy of basic quantum and classical processes. Turning in Section 5 to time-series analysis, we find that both forms of entropy are appropriate in different circumstances. However, the problem of power spectrum determination is an image reconstruction problem for which we recommend using the Shannon/Jaynes formula. The practical difference between the two forms is highlighted in Section 6 with a specific example.

Returning in Section 7 to the problem of image reconstruction, we find that the Kikuchi-Soffer form measures the freedom allowed to individual realizations of the set of photon occupation numbers describing a photon image. This depends nontrivially on the total number of photons observed, and is not a fully objective measure of the spatial power pattern of the photon field.

Our conclusion is firm. There is only one viable contender for the configurational entropy of an image. It is the Shannon/Jaynes form $-\Sigma\, p_i \log p_i$.

2. Basis of Maximum Entropy Method

It was pointed out by Shannon [19] that any probability distribution has a well-defined information content, which should be identified as (minus) its entropy. The algebraic form of this entropy is

$$S = -\Sigma\, p_i \log p_i \tag{1}$$

for a discrete set of probabilities $\{p_i\}$, $\Sigma\, p_i = 1$. It has been shown that this is, apart from the arbitrary base of the logarithm, a unique measure satisfying the axioms one assigns to additive information [20].

An alternative derivation of entropy starts with an ensemble embracing a large number N of equivalent random experiments, each of which can have any of m possible outcomes $i = 1,2,\ldots,m$. Let n_i be the number of occurrences of outcome "i" in the sequence ($\Sigma\, n_i = N$). In experiments of this type, the probability p_i of outcome "i" is taken to be the expectation of n_i/N, and one needs to consider the relative likelihood of the set of numbers $\{n_i\}$. If the experiments are truly random (that is, performed by the traditional team of monkeys), then the number of occurrences of $\{n_i\}$ will become proportional to the degeneracy $N!/\Pi n_i!$. In the limit $N\to\infty$, the logarithm of the degeneracy becomes NS, where S is the entropy as defined above.

Jaynes [21] was the first to use entropy as a tool for assigning probability distributions in the light of certain types of constraint, such as an ensemble average $\langle r\rangle = \Sigma\, r_i p_i$. He suggests that it is wise to choose the probability distribution in such a way as to keep as open a mind as possible concerning other quantities. To this end, he chooses that particular probability distribution consistent with the constraint(s) which has maximum entropy S. Specifically, for the above example, maximizing $S = -\Sigma\, p_i \log p_i$ under $\langle r\rangle = \Sigma\, r_i p_i$ yields $p_i = \exp(-\lambda r_i)/Z(\lambda)$, where λ is the appropriate Lagrange multiplier or "potential" (chosen to fit the constraint) and Z is its partition function.

If one is to use entropy in image reconstruction, one must identify some quality of the image with the probability distribution on which the entropy is defined. Most of the controversy in the field has stemmed from differing views about this identification.

3. Entropy of a Positive Additive Image

When reconstructing an image, one is concerned with its configurational structure in the form of a sequence of positive numbers f_i ($i = 1,2,\ldots,m$) representing the fluxes or equivalent quantities in the m pixels. The most direct way of identifying f with a probability distribution is simply to remove its dimensionality by considering only the pattern of proportions $p_i = f_i/\Sigma f_i$ that describes the "shape" (but not the overall intensity scale) of the image.

Many different types of image have similar mathematical properties in that they satisfy axioms of positivity and additivity. Such images can be directly visual, such as two-dimensional light distributions, or can be three-dimensional densities of (say) electrons in free space, or one-dimensional sequences of numbers in a computer memory or on magnetic tape, or even quantities, such as absorptivity. It is a common feature of these types of image that they can be displayed by electrons in a cathode-ray tube, by grains of silver in a photographic plate, by holes in computer cards, or abstractly as sequences of numbers. The shape $\{p_i\}$ of each of these images satisfies what are, in fact, the Kolmogorov axioms of probability theory, namely

(a) $p_i \geq 0$ (positivity)

(b) $\Sigma p_i = 1$

(c) $p_{i \cup j \cup \ldots} = p_i + p_j + \ldots$ (additivity) $(i \neq j \neq \ldots)$.

By this simple derivation, we have already arrived at the main conclusion of our paper, namely that positivity and additivity are alone sufficient to identify the shape of an image with a probability distribution. Having done this, we immediately define its (negative) information content or entropy as $S = -\Sigma p_i \log p_i$.

For an ordinary optical image, there is another straightforward interpretation of p_i as the probability that the next photon (in a separate experiment) will come from pixel i. The entropy corresponds to the uncertainty in this photon's location. Equivalently, it is the number of bits of information needed to encode its position, given the overall shape $\{p_i\}$. Maximizing this entropy keeps one's options as open as possible about the next photon, and this automatically produces the most uniform, featureless image that is consistent with the constraints. It gives a maximally noncommittal answer to the question "Where would the next photon come from?" We contend that this question lies at the very heart of image reconstruction. It is the precise formal definition of the problem of image reconstruction.

We must stress, though, that this interpretation does not depend on the physical characteristics of photons. It is equally applicable to questions such as "Where will the next electron arrive on a TV screen?", "Where will the next radioactive decay occur in a solid?" and "Where will the next quasar be discovered on the sky?" Distributions of photons, nuclei, and quasars are each valid images. Whether the particles obey classical or quantum statistics is quite irrelevant. It does not even matter that the fluxes or densities are quantized at all. We have merely used photons, with time as an extra dimension, to enumerate the realizations that comprise the classical ensemble inherent in the "team of monkeys" argument in Section 2.

4. Physical Entropies

In Section 3, we have given a simple and direct identification of the shape of an image with a probability distribution, leading immediately to the Shannon formula. It is perhaps unfortunate that other entropies can also be defined on an image. These arise because the physical generation of an image is often a statistical process, involving a probability distribution that in turn has an entropy. All the alternative forms of entropy proposed for images are of this (more complicated) physical type rather than the simple, direct form we commend.

A straightforward illustration concerns $p(f \mid \langle f \rangle)$, the probability of obtaining an individual realization of an image f, given its ensemble average

$\langle f \rangle$. The corresponding entropy [21] is

$$S = -\int p \log(p/m) \, df , \qquad (2)$$

where m is the appropriate measure of f-space. This entropy represents the flexibility allowed to realizations of f within the ensemble-average constraint. It is clear that the physics of the image generation, involving factors such as Bose clumping of photons, is included in these probabilities, and in their corresponding entropies. Although such an entropy will usually also contain terms that depend on the configurational shape, this is by no means necessary, as can be seen by considering a one-pixel image, for which there is clearly no configurational information at all. To illustrate this we re-derive the ensemble-average constrained entropy for various physical cases.

4.1 The classical image. The distribution of quasars on the sky.

Let f_i represent the number of quasars in cell i, now an integer variable, and consider the probability $p(f_i)$. Jaynes' technique allows us to determine the maximally noncommittal $p(f_i)$, given the ensemble-average constraint $\langle f_i \rangle$, by using Eq. (2) for the entropy. The appropriate measure m is the limiting degeneracy

$$m(f_i) = z^{f_i}/f_i!$$

of f_i objects distributed among z subdivisions of cell i, where z is large. This yields the familiar Poisson formula

$$p(f_i) = \frac{\exp(-\langle f_i \rangle)}{f_i!} \langle f_i \rangle^{f_i} .$$

If one were to ignore the measure m, one would assign an entropy

$$S = - \sum_{f_i=0}^{\infty} p(f_i) \log p(f_i) ,$$

which does not evaluate to any simple algebraic form, though its large $\langle f_i \rangle$ limit is

$$S \simeq \frac{1}{2} \log \langle f_i \rangle ,$$

corresponding to the $f_i^{1/2}$ standard deviation in a Poisson process of mean much greater than 1.

With the correct measure included, the entropy is

$$S = - \sum_{f_i=0}^{\infty} p(f_i) \log [p(f_i)/m(f_i)] \ .$$

Combining pixels, the entropy simply adds, leading to

$$S = F \log z - F \log F - F - F \sum_i p_i \log p_i \ ,$$

where $F = \Sigma <f_i>$ and $p_i = <f_i>/F$. The only shape-dependent term is $-\Sigma\, p_i \log p_i$.

4.2 The Bose image. Photons.

Similar analysis holds, save that there is now only one degenerate state for each $f_i = 0,1,2...$, so that $m(f_i) = 1$. Hence maximum entropy yields

$$p(f_i) = (<f_i> + 1)^{-f_i-1} <f_i>^{f_i} \ ,$$

for which the entropy is

$$S = (<f_i>+1) \log(<f_i>+1) - <f_i> \log<f_i> \ .$$

This formula has two interesting limits. For large $<f_i>$ it becomes $S \simeq \log<f_i>$. For small $<f_i>$ it becomes $S \simeq -<f_i> \log<f_i>$.

These formulas are very similar to those derived by Kikuchi and Soffer for photon degeneracy numbers, save that they needed to generalize to the case of z (greater than 1) degrees of freedom. If we do so, we merely alter the degeneracy $m(f_i)$ to

$$m(f_i) = (z+f_i-1)!/f_i!(z-1)! \ .$$

For finite z, our probability distribution takes a form intermediate between the simple Bose and the pure classical Poisson process.

4.3 The Fermi image. Electrons.

Again similar analysis holds, save that there are now only two allowed states $f_i = 0,1$, each having unit measure $m(f_i) = 1$. Maximum entropy yields

$$p(0) = 1-<f_i> \ , \quad p(1) = <f_i> \ ,$$

for which the entropy is

$$S = -<f_i>\log<f_i> - (1-<f_i>)\log(1-<f_i>) \ .$$

Here $\langle f_i \rangle$ is already a probability distribution in its own right, because $f_i = 0$ and 1 are mutually exclusive. The generalization to finite z uses the Bernoulli form

$$m(f_i) = z!/f_i!(z-f_i)! .$$

4.4 The continuous image.

An interesting case occurs when f_i is a continuous variable with uniform measure m, which could be taken to be unity, in $[0,\infty)$. It is of particular importance when estimating the power spectrum of a time series. For this case, the probability distribution is exponential,

$$p(f_i) = \exp(-f_i/\langle f_i \rangle)/\langle f_i \rangle ,$$

with the entropy of the entire power spectrum being

$$S = \sum_i (\log\langle f_i \rangle + \log m_i + 1) .$$

This is precisely the Shannon-Weaver result [22] as used in the Burg algorithm.

We emphasize that all these formulas have been simply derived from the fundamental Shannon formula by Jaynes' method of maximum entropy. In information theory terms, they measure the expected number of bits needed to encode a particular pattern $\{f_i\}$, given the ensemble-averages $\langle f_i \rangle$. Equivalently, they measure the time-bandwidth product of the communication channel needed to transmit the information.

Physical entropies have been recommended for image reconstruction by Kikuchi and Soffer [13] and by others [11,12], and for power spectrum analysis by Burg [23,24] and his followers [25-29]. They suggest that the physical entropy be used as a measure of the intrinsic merit of the long-term average image $\langle f \rangle$. In this way, they distinguish among different images $\langle f \rangle$ each of which satisfies the observational constraints: normally such constraints are not simply the values of $\langle f_i \rangle$ (for which reconstruction would be trivial) but are instead an incomplete set of functions $\phi(\langle f \rangle)$. By doing this, one does indeed produce estimates of $\langle f \rangle$ that are maximally noncommittal about individual physical realizations of f. We contend, though, that this is not the problem of image reconstruction.

5. Time Series Analysis

Which entropy should one use to analyze a time series? The answer depends on what one is interested in. The physical entropy of the Burg form $\Sigma \log f$ is maximally noncommittal about individual samples from the periodogram or, equivalently, from the time series itself. Thus, despite the fact that it may exhibit peculiarities such as line-splitting if displayed as a representation of an ensemble-average spectrum, there are nevertheless many occasions when it is entirely appropriate. An example is the prediction problem, either for filling in gaps in a time series or for extrapolation. When used for such purposes, the Burg form is entirely correct.

However, the Burg form has also been supposed to give the best spectral reconstruction from data such as a subset of the long-term autocorrelation coefficients. This is simply not true.

When reconstructing a spectrum, one is usually interested in questions such as the numbers and positions of spectral lines. These features of the spectrum are exactly what we mean by its configurational content. Indeed, the spectrum, considered as a pattern of proportions $\{p_i\}$, is an image in our sense: it should be analyzed as such. Maximizing the Shannon/Jaynes entropy will yield spectral reconstructions that are as uniform as possible, and will not show structure for which there is no evidence in the data. Maximizing the Burg entropy involves making the spectral response as large as possible, and no such guarantee then exists. As we show next, it is inherently prone to produce peaks of unreliable position and height.

6. Operational Comparison: Moment Data

Because it has often been suggested that it makes little difference what entropy or prior information is used to reconstruct an image [18], we give a specific example of the power spectrum of a time series where the operational difference between the two forms of entropy is particularly dramatic.

Consider a simple one-dimensional continuous power spectrum f(t) (for simplicity we omit ensemble-average brackets) for which f is known to have uniform prior measure on $[0,\infty)$. Instead of considering autocorrelation data, which others [23-29] have thoroughly investigated, suppose that we are given moment data

$$\int_{-\infty}^{\infty} t^n f(t)\, dt = \begin{cases} 1/(n+1), & n \text{ even} \\ 0, & n \text{ odd} \end{cases}$$

up to some maximum value N.

Suppose that we maximize the Shannon/Jaynes entropy under these constraints. Invoking a potential λ_n for each given moment yields

$$\delta \left(-\int p(t) \log p(t)\, dt + \sum_n \lambda_n \int t^n f(t)\, dt\right) = 0$$

so that

$$f(t) = \exp[-Q_N(t)] , \tag{3}$$

where $Q_N(t)$ is a polynomial of degree N. Direct numerical computation gives the results shown in Fig. 1 for N = 4,12,48. As $N \to \infty$, the results converge uniformly (piecewise) to a square wave.

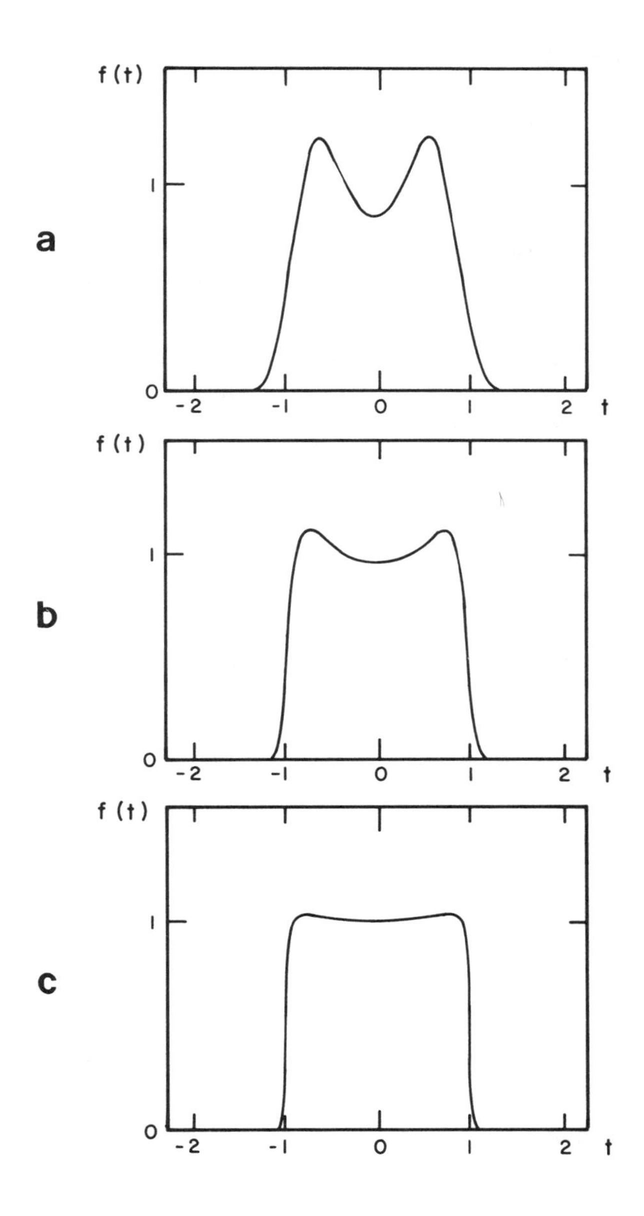

Figure 1. Shannon/Jaynes reconstructions from moment data up to order N.
a. N = 4; b. N = 12; c. N = 48.

Suppose now that we maximize the physical entropy in the form $\Sigma \log f$, as suggested by Burg [23] and (for the radio sky) by Ables [28] and by Kikuchi and Soffer [13]. Invoking potentials as before, we obtain

$$f(t) = 1/R_N(t)\,, \tag{4}$$

where $R_N(t)$ is a polynomial of degree N. There is a small difficulty in that the highest (Nth) moment is divergent at infinity for any fixed $R_N(t)$, but this can be circumvented by restricting the range of integration to some large interval $[-T,T]$. The coefficient of t^N in R_N then decreases as T^{-1}.

The character of the solution (4) is quite different from the smooth, uniformly convergent solution (3). Being the reciprocal of a polynomial, it exhibits N poles in conjugate pairs in the complex plane. (If N is odd, the single real pole lies outside the interval $[-T,T]$.) As we proceed to the limit $T \to \infty$, these poles squeeze the real axis, so the solution approaches a set of $[N/2]$ delta functions. The positions are the zeros t_k of the Legendre polynomial $P_{[N/2]}(t)$, and their areas are the weights

$$w_k = 2/(1-t_k^2)\, P'_{[N/2]}(t_k)^2$$

conventionally used for Gaussian numerical integration [30]. This result follows from the fact that

$$\Sigma\, w_k q(t_k) = \int_{-1}^{1} q(t)dt$$

for any polynomial q of degree up to and including N. In particular, by taking

$$q(t) = \begin{cases} t^n/2, & |t| < 1 \\ 0, & |t| \geq 1 \end{cases}$$

we reach, for $n \leq N$,

$$\Sigma\, w_k t_k^n = \begin{cases} 1/(n+1), & n \text{ even} \\ 0, & n \text{ odd} \end{cases}$$

so that all given moments of the set of delta functions are correct, and we have the unique solution. Figure 2 illustrates solutions for N = 4,12,48.

This spiky behavior is inherent in the $\Sigma \log f$ form, because the solution is always of the type

$$f(t) = 1/(\text{sum of constraints})$$

and the relevant sum will have zeros, normally near the real t-axis. Such

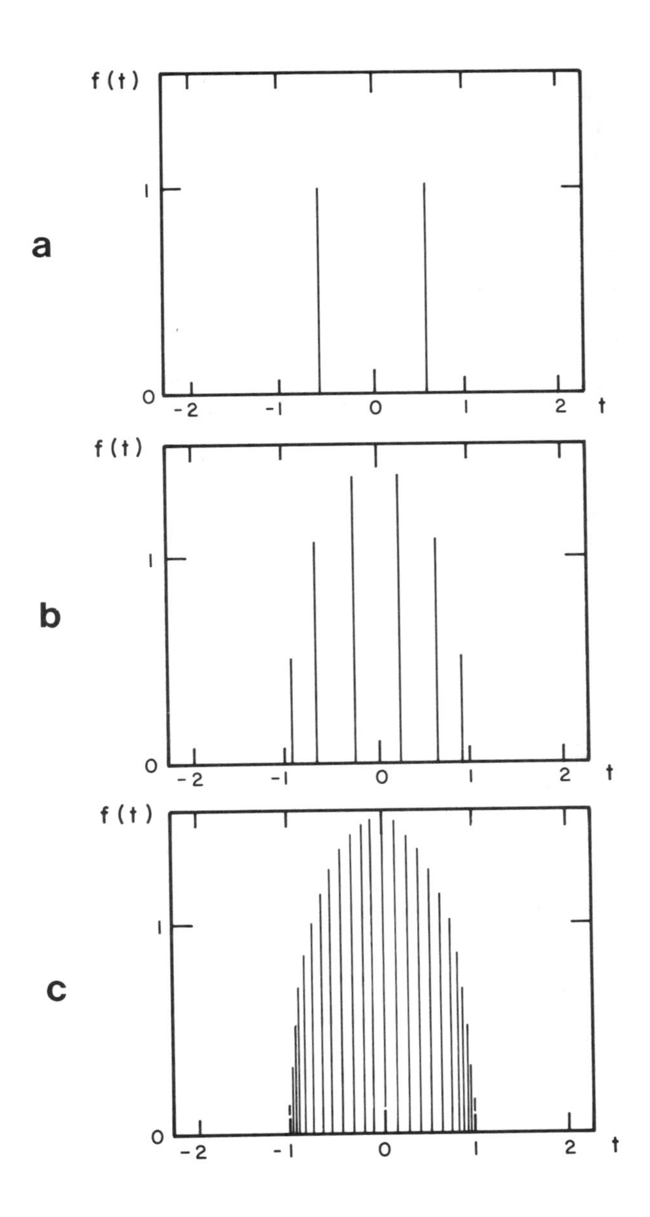

Figure 2. Burg reconstructions from moment data up to order N. a. N = 4; b. N = 12; c. N = 48.

behavior has been noted many times before in the analysis of line spectra from autocorrelation data [31-34]: it leads to line splitting. Futhermore, in our example the positions of the spikes are determined by the highest measured order. As more moments are measured, the number of spikes increases and their positions move.

Had the moments taken values different from $1/(n+1)$, the solution still would have exhibited the same qualitative behavior. Suppose we had taken moments appropriate to a positive function $\rho(t)$. Then the Shannon/Jaynes solution would have coverged uniformly to $\rho(t)$ as $N \to \infty$. The Burg solution, on the other hand, would have been a set of delta functions at the zeros of the polynomial $S_{[N/2]}(t)$, where the $S_j(t)$ are the polynomials orthogonal over $\rho(t)$.

We gave this example to illustrate dramatically the difference that can be made by changing the type of the entropy being maximized. The Shannon/Jaynes solution is noncommittal about the position of t from the distribution f(t), and it gives smooth, uniformly convergent results. The Burg solution is maximally noncommittal about samples f(t) from an ensemble-average constrained distribution, which involves maximizing the hypervolume beneath the solution $\langle f(t) \rangle$. It makes the solution as large as possible. If, for particular types of data such as autocorrelation coefficients, such results are also reasonably smooth and convergent, that is merely a fortunate accident.

7. Image Reconstruction

Our major aim in this paper is to decide which entropy should be used when reconstructing 'ordinary' images. The Shannon/Jaynes solution is particularly compelling here, since it answers the straightforward question 'Where would the next photon come from?' This entropy is the configurational information, and nothing but. It stands above such questions as the quantization of the radiation field, with the consequent physics.

However, in a significantly-referenced paper, Kikuchi and Soffer [13] recommended using the physical entropy. As they pointed out, and as we saw in Section 4.2, this reduces to $-\Sigma\, p_i \log p_i$ for nondegenerate photons (which includes most of optical astronomy) and to $\Sigma \log f_i$ for degenerate photons (radio astronomy). Using the physical entropy here is entirely equivalent to using the Burg form for time series, and is liable to the same operational difficulties.

The physical entropy of the photon arrival pattern measures the number of bits of information needed to encode the arrival pattern as a sequence of integers, given the ensemble-average constraint on the image. Although it does contain terms that depend on the shape of the image, it is really a design feature of the experiment. It gives the maximum brightness the radiating object could have before it saturated the communication channel. As such, it has only a tenuous connection with image reconstruction.

We do, of course, entirely agree with Kikuchi and Soffer in their aim of 'attempting to find the radiant spatial power pattern of the object.' Our main criticism of their logic is that, by bringing in the duration of observation via a total number N of photons, they convert the power pattern (which is not physically quantized) into a photon pattern (which is quantized). Thus, their definition of what they mean by an image involves some number N, determined by the duration of the experiment, as well as the external power pattern itself. Since this number is included in their entropy, they fail to provide full objectivity: their entropy depends partly upon an accident of observation. We can restore objectivity by focusing on the location of the next photon to arrive. Since this does not involve how long one has to wait for it, the corresponding probability distribution faithfully reflects the spatial power pattern. In fact, it does not matter that the radiation is quantized at all. This is exactly the type of argument that underlies our identification of the shape of an image with a probability distribution. Again, time is used merely as an extra dimension to provide the physical (quantized photon) realizations underlying our classical ensemble.

There is another telling criticism, this time of the recommendation to use a form of entropy having different algebraic limits for optical and radio astronomy. Consider two telescopes, one optical and the other radio, with identical response functions, observing the same astronomical object. The object is colorless in the sense that its power pattern $\{f_i\}$ is independent of the observing wavelength. The long-term data sets produced by the two telescopes would be identical, in the sense of ensemble averages. It would be wrong to use different reconstruction algorithms for the two data sets, because this would amount to prejudging the shape and structure of an astronomical object according to the observing wavelength or color. An intrinsically colorless object would be reconstructed as more spiky in the radio band than in the optical band, without any supporting evidence in the data. Of course, one may have prior opinions about such colors, but these should derive from one's experience of <u>astro</u>physics. In the absence of such prior opinions, conclusions about color should come <u>from the observational constraints</u> themselves, and not from using different reconstruction algorithms.

We conclude that the physical entropy should <u>not</u> be used in image reconstruction.

8. Conclusions

The shape of any positive, additive image can be directly identified with a probability distribution. Accordingly, whether it be for spectral analysis of time series, radio astronomy, or optical or X-ray astronomy, or for reconstruction of any other positive, additive image, there remains only one contender for the configurational entropy. It is

$$S = - \Sigma p_i \log p_i .$$

Maximization of this entropy, subject to observational constraints, yields images that are maximally noncommittal in the sense that they will not show spurious features for which there is no clear evidence in the data.

Acknowledgment

We thank Dr. G. J. Daniell for his stimulating discussions and for his continued close involvement with this work.

References

1. Frieden, B. R. (1972) Restoring with maximum likelihood and maximum entropy, J. Opt. Soc. Am. **62**, 511-518.
2. Frieden, B. R., and Burke, J. J. (1972) Restoring with maximum entropy. II: Superresolution of photographs of diffraction-blurred impulses, J. Opt. Soc. Am. **62**, 1202-1210.
3. Gull, S. F., and Daniell, G. J. (1978) Image reconstruction from incomplete and noisy data, Nature **272**, 686-690.
4. Skilling, J., Strong, A. W., and Bennett, K. (1979) Maximum entropy image processing in gamma-ray astronomy, Mon. Not. R. Astron. Soc. **187**, 145-152.
5. Bryan, R. K., and Skilling, J. (1980) Deconvolution by maximum entropy as illustrated by application to the jet of M87, Mon. Not. R. Astron. Soc. **191**, 69-79.
6. Fabian, A. C., Willingale, R., Pye, J. P., Murray, S. S., and Fabbiano, G. (1980) The x-ray structure and mass of the Cassiopeia A supernova remnant, Mon. Not. R. Astron. Soc. **193**, 175-188.
7. Willingale, R. (1981) Use of the maximum entropy method in x-ray astronomy, Mon. Not. R. Astron. Soc. **194**, 359-364.
8. Ponsonby, J. E. B. (1973) An entropy measure for partially polarised radiation and its application to estimating radio sky polarisation distributions from incomplete aperture synthesis by the maximum entropy method, Mon. Not. R. Astron. Soc. **163**, 369-380.
9. Wernecke, S. J. (1977) Two-dimensional maximum entropy reconstruction of radio brightness, Radio Sci. **12**, 831-844.
10. Wernecke, S. J., and d'Addario, L. R. (1977) Maximum entropy image reconstruction, IEEE Trans. **C-26**, 351-364.
11. Frieden, B. R., and Wells, D. C. (1978) Restoring with maximum entropy. III: Poisson sources and background, J. Opt. Soc. Am. **68**, 93-103.
12. Frieden, B. R. (1980) Statistical models for the image restoration problem, Comp. Graphics Image Process. **12**, 40-59.
13. Kikuchi, R., and Soffer, B. H. (1977) Maximum entropy image restoration. I: The entropy expression, J. Opt. Soc. Am. **67**, 1656-1665.
14. Minerbo, G. (1979) MENT: A maximum entropy algorithm for reconstructing a source from projection data, Comp. Graphics Image Process. **10**, 48-68.
15. Daniell, G. J., and Gull, S. F. (1980) Maximum entropy algorithm applied to image enhancement, Proc. IEE **127E**, 170-172.
16. Kemp, M. C. (1980) Maximum entropy reconstructions in emission tomography, in Medical Radionuclide Imaging 1980, Vol. 1 (Vienna: IAEA), pp. 313-323.

17. Burch, S. F. (1980) Comparison of image generation methods, UKAEA Harwell report **AERE-R** 9671.
18. Högbom, J. A. (1979) The introduction of a priori knowledge in certain processing algorithms, in Image Formation from Coherence Functions in Astronomy, C. Van Schooneveld, ed. (Dordrecht: Reidel), pp. 237-239.
19. Shannon, C. E. (1948) A mathematical theory of communication, Bell System Tech. J. **27**, 379-423 and 623-656.
20. Ash, R. B. (1965) Information Theory (New York: Interscience), pp. 5-12.
21. Jaynes, E. T. (1968) Prior probabilities, IEEE Trans. **SSC-4**, 227-241.
22. Shannon, C. E., and Weaver, W. (1949) The Mathematical Theory of Communication (Urbana: University of Illinois).
23. Burg, J. P. (1967) Maximum entropy spectral analysis. Presented at the 37th meeting of the Society of Exploration Geophysicists, Oklahoma City.
24. Burg, J. P. (1972) The relationship between maximum entropy spectra and maximum likelihood spectra, Geophysics **37**, 375-376.
25. Edward, J. A., and Fitelson, M. M. (1973) Notes on maximum entropy processing, IEEE Trans. Inf. Theory **IT-19**, 232-234.
26. Ulrych, T. J., and Bishop, T. M. (1975) Maximum entropy spectral analysis and autoregressive decomposition, Rev. Geophys. Space Phys. **13**, 183-200.
27. Ulrych, T. J., and Clayton, R. W. (1976) Time series modelling and maximum entropy, Phys. Earth Planet. Inter. **12**, 188-200.
28. Ables, J. G. (1974) Maximum entropy spectral analysis, Astron. Astrophys. Suppl. **15**, 383-393.
29. Newman, W. I. (1977) A new method of multidimensional power spectral analysis, Astron. Astrophys. **54**, 369-380.
30. Abramowitz, M., and Stegun, I. A. (1970) Handbook of Mathematical Functions (New York: Dover), p. 916.
31. Fougere, P. F., Zawalick, E. J., and Radoski, H. R. (1976) Spontaneous line splitting in maximum entropy power spectrum analysis, Phys. Earth Planet. Inter. **12**, 201-207.
32. Fougere, P. F. (1977) A solution to the problem of spontaneous line splitting in maximum entropy power spectral analysis, J. Geophys. Res. **82**, 1051-1054.
33. Herring, R. W. (1980) The cause of line splitting in Burg maximum entropy spectral analysis, IEEE Trans. Acoust. Speech Signal Process. **ASSP-28**, 692-701.
34. Marple, L. (1980) A new autoregressive spectrum analysis algorithm, IEEE Trans. Acoust. Speech Signal Process. **ASSP-28**, 441-454.

A REVIEW OF THE PROBLEM OF SPONTANEOUS LINE SPLITTING IN MAXIMUM ENTROPY POWER SPECTRAL ANALYSIS

Paul F. Fougere

Air Force Geophysics Laboratory, Hanscom AFB, Massachusetts 01731

C. Ray Smith and W. T. Grandy, Jr. (eds.),
Maximum-Entropy and Bayesian Methods in Inverse Problems, 303–315.

1. Introduction

The maximum entropy method of power spectral estimation was first discovered in papers presented in 1967 and 1968 by John Burg. These two seminal papers have revolutionized the subject of power spectrum estimation. Hundreds of papers have been written on the subject of high resolution spectral estimation since that time. Burg's two papers are available in the book Modern Spectrum Analysis edited by Don Childers and published by IEEE Press (1978). Another excellent source of information on the method is Burg's thesis of 1975.

Very briefly the Maximum Entropy Method (MEM) consists in maximizing the entropy rate of a process, subject to constraints that express whatever is known about the process. If the first few lags of the autocorrelation function are known, then an expression for entropy is written in terms of the power spectral density and the expression is then maximized with respect to the unknown autocorrelation lags. The constraint is that the autocorrelation function is the Fourier transform of the power spectral density.

In the so-called Burg technique (the 1968 paper) the power spectrum is determined directly from the data without any need to estimate autocorrelations and without any artificially introduced windows to corrupt the given information. The autocorrelation (or autocovariance) function is never needed, but it can be produced as a simple by-product if desired.

The topic of this paper is a review of an annoying problem that arises as a result of a very subtle constraint involving the Burg technique: the problem of spontaneous spectral line splitting and shifting, and of some methods that have been designed to ameliorate this problem.

The fascination with the Burg technique is due at least in part to its success in handling very short signals. Figure 1a shows a rather extreme case: one-half cycle of a 3-1/3-Hz sinusoid sampled four times at a Nyquist frequency of 10 Hz. The plotter program has simply connected the points with straight lines.

Figure 1b shows the Burg power spectrum obtained using four filter weights—a beautiful sharp line at 3-1/3 Hz. (The number of filter weights is analogous to the number of lags in a Blackman-Tukey spectrum.) This kind of result was first presented by Ulrych (1972).

2. The Splitting Problem

Now, however, if one more point is added, to give the picture shown in Fig. 2a, we have two-thirds of a cycle of the same signal, and its power spectrum, shown in Fig. 2b, splits into two components that roughly straddle the correct frequency. This difficulty was first pointed out by Chen and Stegen (1974), but some of their systematics were wrong and they incorrectly ascribed the unsatisfactory spectrum as due to amplification of noise peaks, caused by the use of too many prediction error filter coefficients. They also claimed incorrectly that this problem was similar to that observed by Jackson (1967) for FFT spectra. Shortly after the work of Chen and

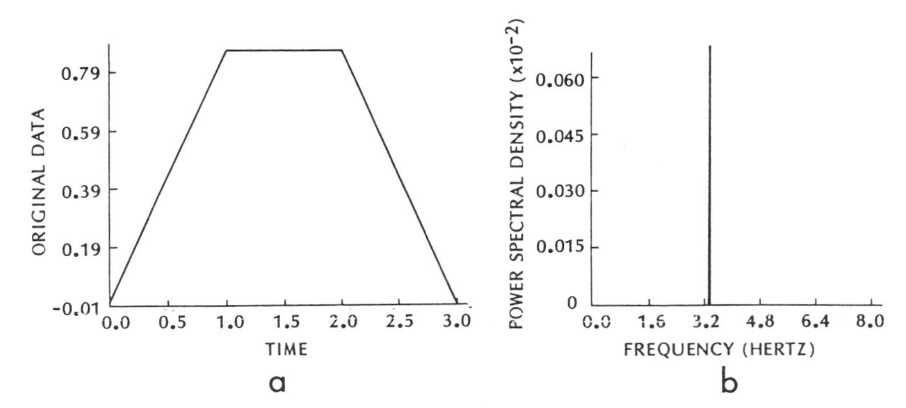

Fig. 1. a. One-half cycle of a 3-1/3-Hz sine wave sampled four times. The four points are simply connected by straight lines. b. The Burg spectrum of the signal.

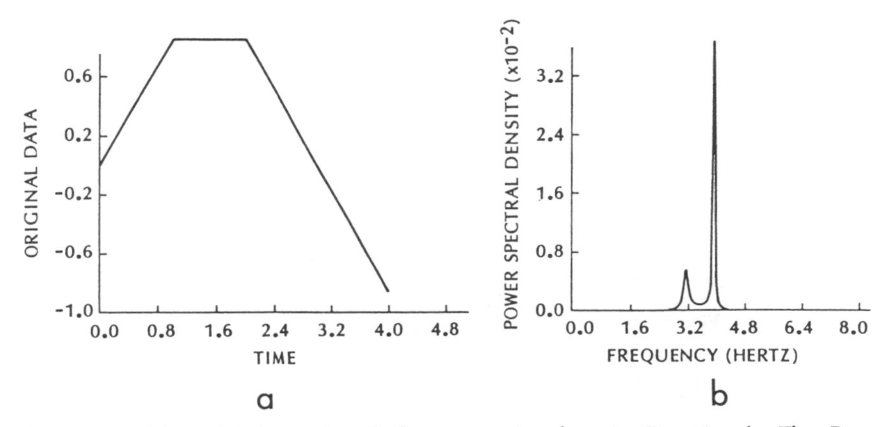

Fig. 2. a. Two-thirds cycle of the same signal as in Fig. 1. b. The Burg spectrum of the signal.

Stegen, Fougere, Zawalick, and Radoski (1976) systematically explored the problem as follows.

Figure 3 shows spectra of the following signals: unit-amplitude, 1-Hz sine waves sampled 21 times at a Nyquist frequency of 10 Hz; 10^{-4} Gaussian white noise added and 20 filter weights. This is one full cycle: the variable is initial phase. There is no splitting—every spectrum is a very sharp accurate needle—and we have used 20 of a possible 21 filter weights. Note that the classical FFT-based or Blackman-Tukey spectra are never as sharp as this!

Next, in Fig. 4, we use 1.25 cycles with varying initial phase. Every spectrum splits into a sharp doublet unless the initial phase is 0°, 90°, or 180°, that is, a multiple of $\pi/2$. But this signal is terribly short: only 1.25 cycles.

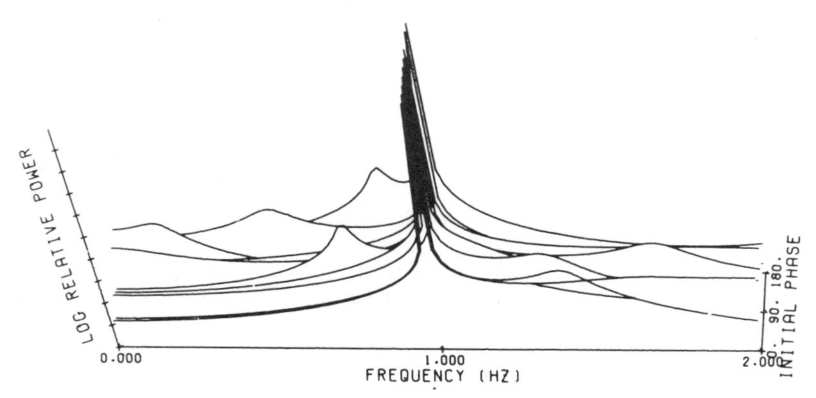

Fig. 3. Nineteen Burg spectra of one cycle of a unit-amplitude, 1-Hz sinusoidal signal, to which Gaussian noise of amplitude 10^{-4} has been added, sampled 21 times. The initial phase varies from 0° to 180° in steps of 20°. Twenty filter weights were used.

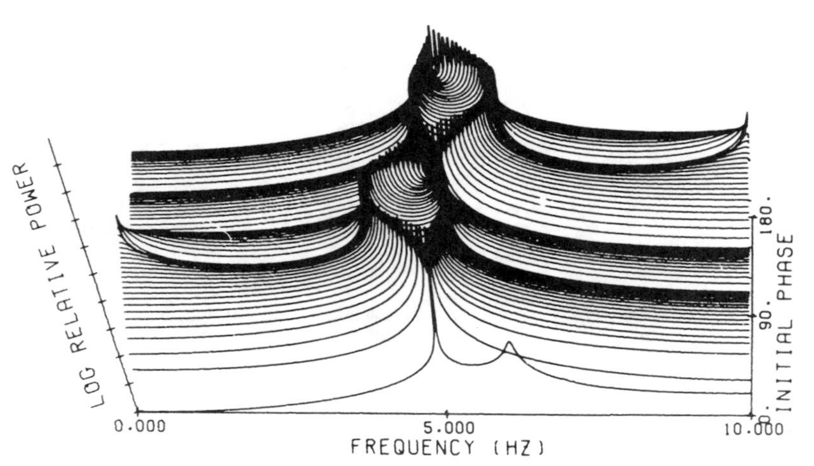

Fig. 4. Ninety-one Burg spectra of 1.25 cycles of a unit-amplitude, 5-Hz sine wave (with 10^{-4} amplitude of additive Gaussian noise), sampled six times using six filter weights. Initial phase varies from 0° to 180° in steps of 2°. Note that there is no splitting at 0°, 90°, and 180° and maximum splitting at 45° and 135°.

The next question arises: What happens when the signal is lengthened? Figure 5a shows the result. The signals are 1.25 Hz, 3.25 Hz, and so on up to 49.25 Hz with a Nyquist frequency of 50 Hz. Every one splits, even with 49.25 cycles of signal, and we have used only 25 of a possible 101 weights—not too high, by anyone's standards. Figure 5b is an enlarged view of the central region.

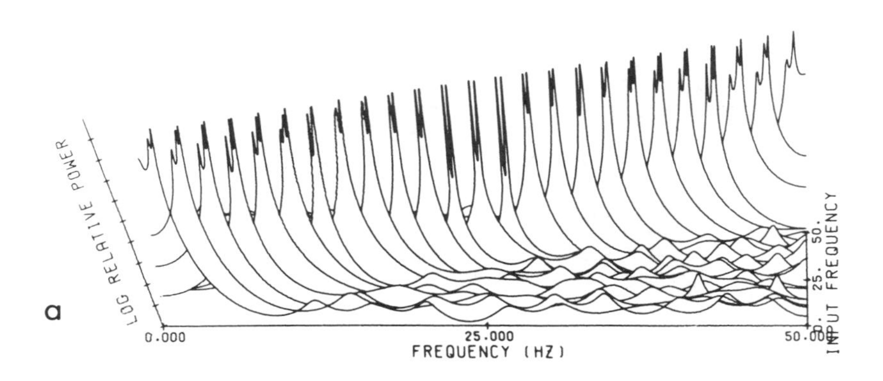

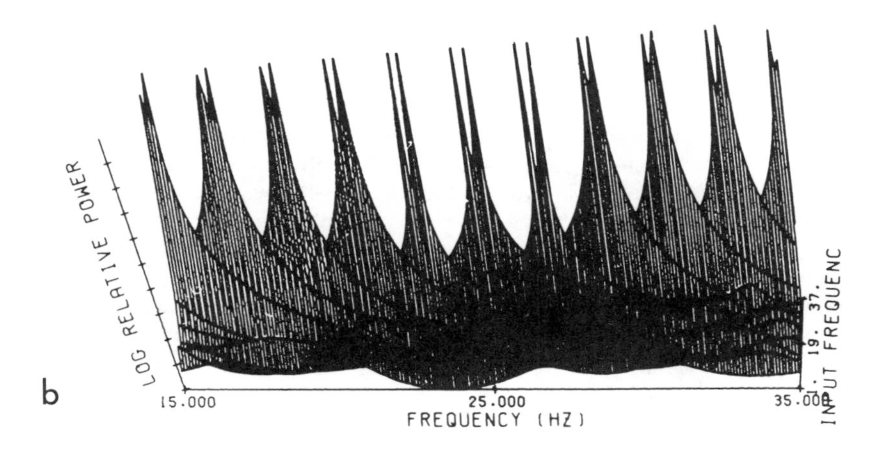

Fig. 5. a. Burg spectra of unit amplitude sinusoidal signals at 45° initial phase, sampled 101 times in 1 second. The frequencies of the signals are 1.25, 3.25, 5.25, ..., 49.25 Hz. Gaussian white noise of amplitude 10^{-4} has been added to each signal. b. Expanded view of the central region.

3. The Solution to the Splitting Problem

Now I will present some elementary mathematics to set the stage for a discussion of various attempts to solve this problem. Linear prediction is at the very heart of the Burg technique. Let the data set be

$$X_1, X_2, X_3, \ldots, X_n ,$$

and define a prediction filter G_1, G_2, ... such that a prediction at X_3, $\hat{X}_3$, is given by the simple convolution

$$\hat{X}_3 = G_1 X_2 + G_2 X_1 . \tag{1}$$

Define the prediction error at X_3 as ε_3:

$$\varepsilon_3 = X_3 - \hat{X}_3 = X_3 - (G_1 X_2 + G_2 X_1) . \tag{2}$$

Define prediction error coefficients:

$$g_0 = 1; \quad g_1 = -G_1; \quad g_2 = -G_2 ; \tag{3}$$

then

$$\varepsilon_3 = g_0 X_3 + g_1 X_2 + g_2 X_1 \tag{4}$$

and ε_3 is called a PEF for Prediction Error in the Forward direction. If the same filter is reversed and run backward, we define δ_1, the prediction error in the backward direction, or PER (Prediction Error Reverse)

$$\delta_1 = g_0 X_1 + g_1 X_2 + g_2 X_3 . \tag{5}$$

We can picture the filters moving over the data to generate prediction errors at all of our observed times in both time directions. But for a three-point filter we begin the forward prediction at X_3; we never run the filter off the data—we use all of the given data and only the given data. There is never any need to make assumptions about what the data are like outside of our given data set. There is never a periodic extension of data as in the Cooley-Tukey FFT method or a zero extension as in the Blackman-Tukey method. Also there is never any tapering—which really ought to be called tampering—of the data with resulting loss of information! The idea of running the filter in both time directions is extremely important. It explicitly takes care of the fundamental symmetry principle that the power spectrum of a stationary process is invariant under a time reversal. Note that we are limiting ourselves to stationary signals. Some authors criticize the Burg technique because it does not work well with exponentially growing or decaying signals. This is nonsense! The Burg technique was not designed for such signals. In fact, all techniques designed to estimate power spectral density require at least approximate stationarity.

Now the best linear prediction error filter minimizes some function of the forward and backward errors. Wiener (1949) chose to minimize the mean square forward error. On the other hand, the Burg norm is given as the arithmetic mean of forward and backward squared error

$$P_m = \frac{1}{2(n-m)} \sum (\overline{PEF^2} + \overline{PER^2}) , \tag{6}$$

where n is the number of observations = sample size and m is the number of g's = filter length. Let the sequence $\phi_0, \phi_1, \phi_2, \ldots$ define the autocorrelation function. Then the Wiener (1949) equation [also called the Yule-Walker equation (Yule, 1927; Walker, 1931)] for real data can be written

$$\begin{bmatrix} \phi_0 & \phi_1 & \phi_2 & \cdots & \phi_m \\ \phi_1 & \phi_0 & \phi_1 & \cdots & \phi_{m-1} \\ \phi_2 & \phi_1 & \phi_0 & \cdots & \phi_{m-2} \\ \vdots & \vdots & \vdots & & \vdots \\ \phi_m & \phi_{m-1} & \phi_{m-2} & \cdots & \phi_0 \end{bmatrix} \begin{bmatrix} 1 \\ g_{m1} \\ g_{m2} \\ \vdots \\ g_{mm} \end{bmatrix} = \begin{bmatrix} P_0 \\ 0 \\ 0 \\ \vdots \\ 0 \end{bmatrix} . \tag{7}$$

The matrix of autocorrelations is a highly symmetric form, called a Toeplitz matrix, with all of the elements on any diagonal equal to each other. This symmetric form greatly simplifies the solution of Eq. (7).

The matrix of prediction error filters, stripped of their leading unity elements, is

$$G = \begin{bmatrix} g_{11} & & & & \\ g_{21} & g_{22} & & & \\ g_{31} & g_{32} & g_{33} & & \\ \vdots & \vdots & \vdots & & \\ g_{m1} & g_{m2} & g_{m3} & \cdots & g_{mm} \end{bmatrix} . \tag{8}$$

The diagonal elements,

$$g_{21} = g_{11} + g_{22}g_{11}$$
$$g_{31} = g_{21} + g_{33}g_{22}$$
$$g_{32} = g_{22} + g_{33}g_{21}$$

called reflection coefficients, are the critically important ones because of the Levinson (1947) recursion given in Eq. (9). In general,

$$g_{nk} = g_{n-1,k} + g_{nn}g_{n-1,n-k} \cdot \tag{9}$$

Once we know g_{11} and g_{22} we can find g_{21}. Once we know g_{11}, g_{22}, and g_{33}, we can find g_{31} and g_{32} to complete the third-order prediction error filter. In general, given the set of reflection coefficients, the set of all prediction error filters is easily determined by using this simple and elegant algorithm repeatedly. Given only the first m diagonal elements of Eq. (8), the entire m×m matrix is determined using the Levinson recursion. Conversely, given the prediction error filter, the backward Levinson recursion yields all of the reflection coefficients. That is, given the mth row of Eq. (8), the entire m×m matrix can be found.

Now in the Burg technique the Levinson recursion is used to write the expression for the mean forward and backward error power in terms of the reflection coefficients. If previously determined reflection coefficients are kept fixed, the current one is found by minimizing the error power with respect to the one current reflection coefficient. That procedure guarantees that all of the reflection coefficients are bounded in magnitude by unity and that the resulting filter is a prediction error filter. The corresponding Toeplitz matrix of estimated autocorrelations is nonnegative definite and the basic autocorrelation function theorem is satisfied (Burg, 1975).

Unfortunately, the constraint of fixing previously determined reflection coefficients is not necessary to achieve these ends. It is necessary only that all reflection coefficients at any stage be bounded by unity in magnitude. The independent variables, g_{ii}, $i = 1,\ldots,m$, must be constrained such that $|g_{ii}| < 1$. Such a constraint can be satisfied identically, that is, for any choice of the variables θ_i, $i = 1,\ldots,m$, by setting $g_{ii} = U \sin\theta_i$ with U slightly less than 1; typically $U = 1 - 10^{-7}$. With the independent variables chosen to be the set of real angles θ_i, the minimization problem is now nonlinear. This is the solution to the problem that I discovered for real signals (Fougere, 1977) and for complex signals (Fougere, 1978). Minimizing the mean error power with respect to the set of θ's completely solves the line-splitting and shifting problem. It is true that the objective function P_n is a nonlinear function of the angles θ_i and that the nonlinear minimization is nontrivial, but the 1977 and 1978 papers show how to find the gradient of the objective function, and so standard nonlinear minimization techniques, which require the value of the function and its gradient, can be used to solve the problem quite effectively.

Figure 6 shows Burg spectra of the worst case: a signal consisting of a sine wave of frequency 26.25 Hz with 45° initial phase sampled 101 times at a Nyquist frequency of 50 Hz. The noise levels are shown on the right. Note that the splitting is strongest when the noise level is lowest. As the noise level is raised, the peaks broaden and finally coalesce into a single broad peak located at a shifted frequency. When the nonlinear method just described is applied to the same signals, no splitting occurs, as shown in

Fig. 7. The method is called nonlinear because the independent variables are not the reflection coefficients but arc-sines of the reflection coefficients. The independent variables enter the expression for the mean error power nonlinearly.

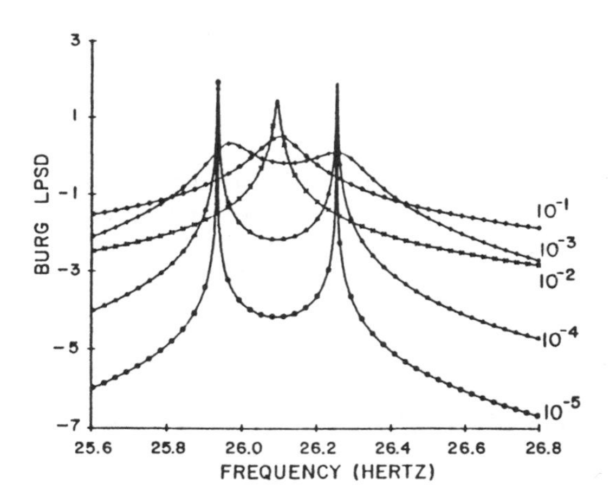

Fig. 6. Burg log power spectral density (LPSD) versus frequency for signals consisting of 101 samples, 0.01 second apart, of a unit-amplitude sine wave of 26.25-Hz frequency and 45° initial phase in additive Gaussian white noise with amplitudes 10^{-5}, 10^{-4}, 10^{-3}, 10^{-2}, and 10^{-1}. Five filter weights were used.

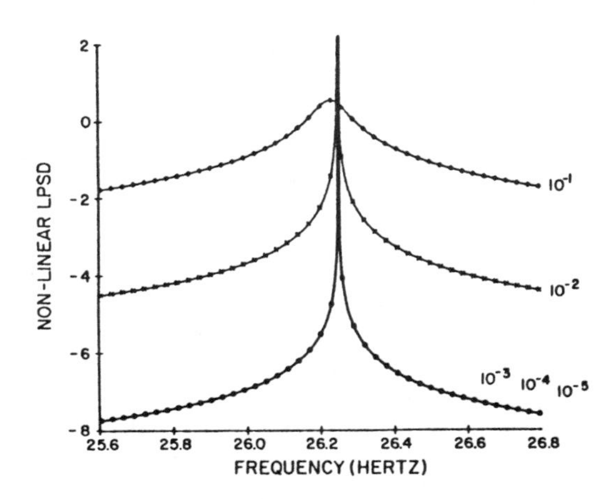

Fig. 7. Nonlinear log spectra of the same signals as in Fig. 6.

One further simple bit of arithmetic shows that if we use $m = 2/3n$ we should be able to achieve a perfect fit to any set of data. If we have m filter weights and n observations, there are $2(n-m)$ independent prediction errors (in the forward and backward time directions). If $m = 2(n-m)$, then a perfect fit can be achieved because there are as many independent variables as there are independent prediction errors. Solving the equation yields $m = 2/3n$ for a perfect fit. This is not a desirable thing to do—it is merely a test of the method. In practice, a perfect fit can be achieved under either of two conditions. If the constraints are not tight, that is, if no reflection coefficient (g_{ii}) is near unity in magnitude, then it is easy to achieve a perfect fit at $m = 2/3n$. I have observed a perfect fit when the input data set consists of white noise samples. This condition can be observed on the computer when the mean square prediction error becomes so small as to be computationally indistinguishable from zero. If the computer word length contains k significant decimal figures, then the ratio of mean square prediction error to the mean square observation is less than 10^{-k}.

If the constraints are tight, as they tend to be with sine waves in noise, it is sometimes impossible to achieve a perfect fit. In these cases, when the constraints are dropped so that some reflection coefficient is allowed to become greater than 1 in magnitude, a perfect fit can again be achieved. Of course the resulting filter is nonphysical. These remarks are not intended to suggest that achieving a perfect fit is desirable; it never is, except as a very powerful check on the correct operation of the program.

In summary, the nonlinear method employs the lightest possible constraint to the reflection coefficients—that they all are bounded by unity in magnitude; and it guarantees that the prediction error filters will always be stable minimum-phase filters and the resulting spectrum will always be positive. All of the roots of the Z-transform of the prediction error filter lie outside the unit circle.

4. The Unconstrained Least Squares Method

It is simpler, of course, to remove all constraints entirely and to solve a minimization problem in which the normal equations of least squares are linear in the prediction error coefficients.

The following authors all suggest and recommend such a method: Nuttall (1976), Ulrych and Clayton (1976), Kay and Marple (1979), Swingler (1979), Barodale and Erickson (1980), Marple (1980), Lang and McClellan (1980). The list of names should be headed by that of John Burg, who first described it as erroneous and rejected it in his 1968 paper. When the unconstrained method actually produces a prediction error filter and the backward Levinson recursion shows that in fact the reflection coefficients are all bounded by unity in magnitude, then the results must be algebraically identical to those of the nonlinear method. But there is absolutely no guarantee that this situation will obtain on any given data set. It is especially unlikely to work correctly in the low noise situations.

Some of these very authors also claim that the cause of line splitting/shifting is the constraint of using the Levinson recursion. This is totally absurd. The Levinson recursion is in no sense a constraint—but it is quite magical! Given any set of numbers whatever, determined in any way imaginable, which are alleged to be prediction error coefficients, we can apply the backward Levinson recursion. If the resulting reflection coefficients are all of magnitude less than unity, then the original numbers are indeed valid prediction error coefficients; otherwise they are not! Conversely, given any set of numbers all of which are less than unity in magnitude, we can use the direct Levinson algorithm to find a valid prediction error filter.

5. Summary

Following is a complete summary of the nonlinear technique, putting together four equations we have just seen:

$$P_m = \frac{1}{2(n-m)} \sum_{k=1}^{n-m} (\varepsilon_k^2 + \delta_k^2) , \tag{10}$$

$$\varepsilon_k = \sum_{i=0}^{m} x_{k+m-i}\, g_{mi}$$

$$\delta_k = \sum_{i=0}^{m} x_{k+i}\, g_{mi} \tag{11}$$

$$g_{nk} = g_{n-1,k} + g_{nn} g_{n-1,n-k} \tag{12}$$

$$g_{ii} = U \sin\theta_i \; ; \; i = 1,2,\ldots,m, \; U < 1 \, . \tag{13}$$

The mean error power P_m is given in Eq. (10). This is what I previously called the Burg norm. The prediction errors PEF = ε_k and PER = δ_k are given in Eqs. (11). The Levinson recursion is written out in Eq. (12), and the simple necessary and sufficient constraints to produce a stable prediction error filter are given in Eq. (13). We substitute Eq. (13) into Eq. (12); Eq. (12) into Eqs. (11); Eqs. (11) into Eq. (10) and then minimize the mean error power P_n with respect to the true independent variables, the angles θ_j. This produces the optimal correct solution to the problem first posed by Burg, and it completely solves the line-splitting and shifting problem. It is true that the objective function P_n is a nonlinear function of the angles θ_j and that the nonlinear minimization is nontrivial; but the original papers (Fougere 1977, 1978) show how to find the gradient of the objective function, and so standard nonlinear minimization techniques that require the value of the function and its gradient are used to solve the problem quite effectively.

References

Burg, J. P. (1967) Maximum entropy spectral analysis. Paper presented at 37th meeting of the Society of Exploration Geophysicists, Oklahoma City, Oklahoma.

Burg, J. P. (1968) A new analysis technique for time series data, NATO Advanced Study Institute on Signal Processing with Emphasis on Underwater Acoustics.

Burg, J. P. (1975) Maximum entropy spectral analysis, Ph.D. thesis, Stanford University, Palo Alto, Calif., 123 pp.

Chen, W. Y., and G. R. Stegen (1974) Experiments with maximum entropy power spectra of sinusoids, J. Geophys. Res. **79**, 3019.

Childers, D. G., ed. (1978) Modern Spectrum Analysis (New York: IEEE Press).

Fougere, P. F. (1977) A solution to the problem of spontaneous line splitting in maximum entropy power spectrum analysis, J. Geophys. Res. **82**, 1051.

Fougere, P. F. (1978) A solution to the problem of spontaneous line splitting in maximum entropy power spectrum analysis of complex signals, Proceedings of the RADC Spectrum Estimation Workshop, May 1978.

Fougere, P. F., E. J. Zawalick, and H. R. Radoski (1976) Spontaneous line splitting in maximum entropy power spectrum analysis, Phys. Earth Planet. Inter. **12**, 201.

Jackson, P. L. (1967) Truncations and phase relationships of sinusoids, J. Geophys. Res. **72**, 1400.

Kay, S. M., and S. L. Marple, Jr. (1979) Sources of and remedies for spectral line splitting in autoregressive spectrum analysis, p. 151 in Records, 1979 International Conference on Acoustics, Speech, and Signal Processing.

Lang, S. W., and J. H. McClellan (1980) Frequency estimation with maximum entropy spectral estimators, IEEE Trans. Acoust. Speech Signal Process. **ASSP-28**, 716.

Levinson, N. (1947) The Wiener RMS (root mean square) error criterion in filter design and prediction, J. Math. Phys. (Cambridge, Mass.) **25**, 261.

Marple, S. L., Jr. (1980) A new autoregressive spectrum analysis algorithm, IEEE Trans. Acoust. Speech Signal Process. **ASSP-28**, 441.

Nuttall, A. H. (1976) Spectral analysis of a univariate process with bad data points, via maximum entropy, and linear predictive techniques, Naval Underwater Systems Center, Tech. Rept. 5303, New London, Conn.

Swingler, D. N. (1979) A comparison between Burg's maximum entropy method and a nonrecursive technique for the spectral analysis of deterministic signals, J. Geophys. Res. **84**, 679.

Ulrych, T. J. (1972) Maximum entropy power spectrum of truncated sinusoids, J. Geophys. Res. **77**, 1396.

Ulrych, T. J., and R. W. Clayton (1976) Time series modelling and maximum entropy, Phys. Earth Planet. Inter. **12**, 188.

Walker, G. (1931) On periodicity in series of related terms, Proc. R. Soc. London, Ser. A, **131**, 518.

Wiener, N. (1949) Extrapolation, Interpolation and Smoothing of Stationary Time Series (New York: Wiley and Sons).

Yule, G. U. (1927) On a method of investigating periodicities in distributed series, with special reference to Wolfer's sunspot numbers, Phil. Trans. R. Soc. London, Ser. A, **226**, 267.

APPLICATION OF TWO-CHANNEL PREDICTION FILTERING TO THE RECURSIVE FILTER DESIGN PROBLEM

L. C. Pusey

Chevron Oil Field Research Company, La Habra, California 90631

In the recursive filter design problem, we are given a finite segment of a causal filter,

$$H(Z) = h_0 + h_1Z + \ldots + h_{N-1}Z^{N-1},$$

and it is desired to approximate the filter by a recursive filter of the form

$$G(Z) = B(Z)/A(Z)$$

where

$$B(Z) = b_0 + b_1Z + \ldots + b_mZ^m$$

$$A(Z) = 1 + a_1Z + \ldots + a_nZ^n .$$

One obvious requirement of this problem is that we want a stable solution; that is, we must require that the roots of $A(Z)$ lie outside the unit circle.

A common approach is to consider the modified least-squares problem, where we seeek to minimize

$$f_1(a,b) = \sum_{i=0}^{N-1} (h_i*a_i - b_i)^2, \qquad a_0 = 1 , \tag{1}$$

where $*$ denotes convolution.

It can be shown that the minimization problem (1) decomposes into two parts. That is, we can first choose $\{a_i\}$ to minimize

$$f(a) = \sum_{i=m+1}^{N-1} (h_i*a_i)^2 , \qquad a_0 = 1 , \tag{2}$$

C. Ray Smith and W. T. Grandy, Jr. (eds.),
Maximum-Entropy and Bayesian Methods in Inverse Problems, 317–318.

and then compute $\{b_i\}$ as the convolution

$$b_i = h_i * a_i , \quad i = 0,1,\ldots,m . \tag{3}$$

For the case where $n+m = N-1$, Eqs. (2) and (3) become the well-known Padé-approximant method for obtaining the recursive filter. When a reduced order ($n+m < N-1$) is desired, Eq. (2) becomes a standard linear least-squares problem, originally suggested by Shanks as a way of estimating $\{a_i\}$. Unfortunately, neither method guarantees a stable solution, and it is possible to obtain filters $A(Z)$ that have roots inside the unit circle.

To guarantee a stable solution, a number of authors have suggested the use of an infinite sum in Eq. (1). That is, we define $h_i = 0$, $i = N,N+1,\ldots$, and consider the minimization of

$$f_2(a,b) = \sum_{i=0}^{\infty} (h_i * a_i - b_i)^2, \quad a_0 = 1 . \tag{4}$$

For the case where $n = m$, it has been shown that Eq. (4) can be imbedded in the context of a two-channel prediction filtering problem. In fact, we can obtain the solution to Eq. (4) by formally applying the correlation method to the two-channel data

$$X_i = \begin{bmatrix} h_i \\ \delta_i \end{bmatrix}, \quad i = 0,1,\ldots,N-1 ,$$

where δ_i is the Kronecker delta. Stability then follows from the well-known results of Whittle for two-channel prediction filters. Another benefit of regarding the problem in this context is that we obtain a fast algorithm for the solution of Eq. (4). The algorithm is just the two-channel equivalent of the Levinson recursion.

Although the solution to Eq. (4) is stable, it may be unreasonable in many cases to take $h_i = 0$ for $i = N,N+1,\ldots$, and it would appear that we have sacrificed accuracy for stability. However, the main point of this paper is that there is no reason to do this. That is, by applying a two-channel equivalent of the Burg scheme, we need make no assumptions about the tail of $\{h_i\}$ and can still guarantee a stable solution.

The performance of the new method is evaluated by comparing it with the two-channel correlation method for the case where $\{h_i\}$ is the truncated impulse response of a known recursive filter. It is shown that the new method can provide a significant improvement in performance over the correlation method when $\{h_i\}$ does not decay smoothly to zero at $i = N-1$.

This work is published in IEEE Trans. Acoust. Speech Signal Process. **ASSP-31**, 1169-1177 (1983).

APPLICATION OF MAXIMUM ENTROPY AND BAYESIAN OPTIMIZATION METHODS TO IMAGE RECONSTRUCTION FROM PROJECTIONS

Gabor T. Herman

Medical Image Processing Group, Department of Radiology, University of Pennsylvania, Philadelphia, Pennsylvania 19104

The problem of image reconstruction from projections is translated, via approximation of the image by linear combination of basis pictures, into a large system of approximate equalities in the unknown coefficients of the desired linear combination. Optimization criteria for selecting these coefficients include unconstrained regularized least squares estimation based on Bayesian optimization, and constrained norm minimization and entropy maximization. Algebraic reconstruction techniques (ART) are given that converge to these different optimizers. Methods for evaluating the efficacy of reconstruction procedures are discussed and illustrated. Reasons are given for the lack of widespread use of maximum entropy and Bayesian optimization techniques in practical applications of image reconstruction from projections. The ideas are illustrated by a small sample of techniques.

C. Ray Smith and W. T. Grandy, Jr. (eds.),
Maximum-Entropy and Bayesian Methods in Inverse Problems, 319–338.

1. Image Reconstruction from Projections

The problem of image reconstruction from projections has arisen independently in a large number of scientific fields. In medicine, an important version of the problem is that of obtaining the density distribution within the human body from multiple x-ray projections. This process, referred to as computerized tomography (CT), has revolutionized diagnostic radiology over the past decade. The 1979 Nobel prize in medicine was awarded for work on computerized tomography. In 1982 the Nobel prize in chemistry was also awarded for work on image reconstruction, this time for the reconstruction of viruses and macromolecules from electron micrographs.

There is a large literature on image reconstruction from projections; for a general discussion, we refer to Ref. [1]. In this paper, we concentrate on a rather specific group of approaches to solving the image reconstruction problem.

Following the terminology and notation of Ref. [1], we define a picture function, f, as a function of two polar variables (r,ϕ) that is zero-valued outside a square-shaped picture region, the center of which is at the origin of the coordinate system. In what follows, we assume that f has additional properties, so that all integrals used below are defined.

The mathematical idealization of the image reconstruction problem is: Recover the picture function f from its line integrals. More precisely, we use the pair (ℓ,θ) to parameterize a line (see Fig. 1), and we define the Radon transform $\mathcal{R}f$ of f by

$$
\begin{aligned}
[\mathcal{R}f](\ell,\theta) &= \int_{-\infty}^{\infty} f[\sqrt{\ell^2+z^2},\ \theta+\tan^{-1}(z/\ell)]\,dz\,, \qquad \text{if } \ell \neq 0,\\
[\mathcal{R}f](0,\theta) &= \int_{-\infty}^{\infty} f(z,\ \theta+\pi/2)\,dz\,.
\end{aligned}
\tag{1}
$$

The input data to a reconstruction algorithm are estimates (based on physical measurements) of the values of $[\mathcal{R}f](\ell,\theta)$ for a finite number of pairs (ℓ_i,θ_i), $1 \le i \le I$. The output of the algorithm is an estimate, in some sense, of f. For $1 \le i \le I$, define

$$\mathcal{R}_i f = [\mathcal{R}f](\ell_i,\theta_i) \tag{2}$$

and let y denote the I-dimensional column vector whose ith component y_i is the available estimate of $\mathcal{R}_i f$. We refer to y as the measurement vector.

When designing a reconstruction algorithm, we assume that the method of data collection, and hence the set $\{(\ell_1,\theta_1),\ldots,(\ell_I,\theta_I)\}$, is fixed and known. Roughly stated, the problem is: GIVEN measurement vector y, ESTIMATE picture f.

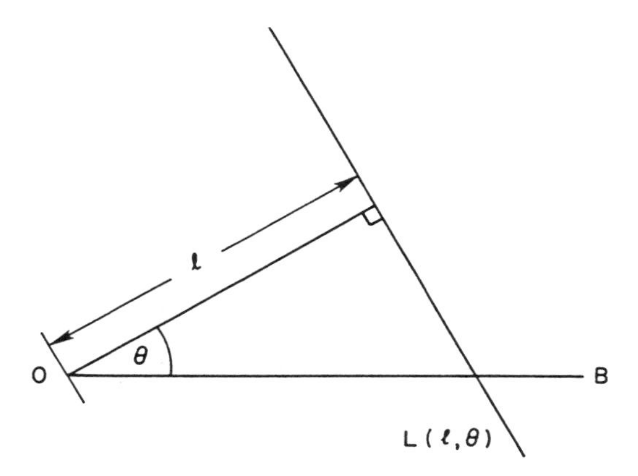

Figure 1. The line $L(\ell,\theta)$ is the one whose distance from the origin O is ℓ and the perpendicular to which makes an angle θ with the baseline OB.

One way of making this statement precise is by the so-called series expansion approach, in which we make use of a fixed set of J basis pictures $\{b_1,\ldots,b_J\}$. It is assumed that any picture f that we may wish to reconstruct can be adequately approximated by a linear combination of the b_j's.

An example of such an approach is n × n digitization, in which the picture region is subdivided into an n × n array of small squares, called pixels (for picture elements; see Fig. 2). We number the pixels from 1 to $J = n^2$, and define

$$b_j(r,\phi) = \begin{cases} 1, & \text{if } (r,\phi) \text{ is inside the jth pixel} \\ 0, & \text{otherwise.} \end{cases} \tag{3}$$

Then the n × n digitization of the picture f is the picture $\hat{f}$ defined by

$$\hat{f} = \sum_{j=1}^{J} x_j \, b_j \, , \tag{4}$$

where x_j is the average value of f inside the jth pixel.

There are other ways of choosing the basis pictures, but once they are chosen, any picture $\hat{f}$ that can be represented as a linear combination of the basis pictures $\{b_j\}$ is uniquely determined by the choice of the coefficient x_j as shown in Eq. (4). The column vector x whose jth component is x_j is referred to as the image vector. In this environment the problem "estimate the picture f" becomes the problem "find an image vector x" such that the $\hat{f}$ defined by Eq. (4) is a good approximation to f.

The functionals $\mathcal{R}_i$ defined in Eq. (2) are clearly linear. Assuming they are continuous (an assumption that may well be violated when we make precise the underlying function spaces and hence the notion of continuity), we

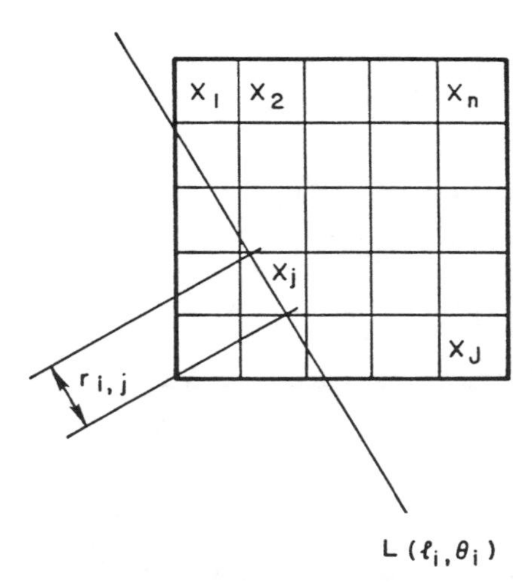

Figure 2. In an n × n digitization the picture region is subdivided into $J = n^2$ small equal-sized square-shaped pixels. The average value of the picture function in the jth pixel is x_j. The length of intersection of the ith line $L(\ell_i, \theta_i)$ with the jth pixel is $r_{i,j}$.

find, if $\hat{f}$ approximates f, that

$$\mathcal{R}_i f \simeq \sum_{j=1}^{J} x_j \mathcal{R}_i b_j . \tag{5}$$

Since the b_j are user-defined functions, $\mathcal{R}_i b_j$ is usually easily calculated. (In our digitization example it denotes the length of intersection of the ith line $L(\ell_i, \theta_i)$ with the jth pixel; see Fig. 2.) Denoting the calculated value of $\mathcal{R}_i b_j$ by $r_{i,j}$ and recalling that y_i is an estimate of $\mathcal{R}_i f$, we obtain

$$y_i \simeq \sum_{j=1}^{J} r_{i,j} x_j , \qquad 1 \leq i \leq I . \tag{6}$$

We refer to the matrix R whose (i,j)th element is $r_{i,j}$ as the <u>projection matrix</u>. Let e denote the I-dimensional <u>error vector</u> defined by

$$y = Rx + e . \tag{7}$$

The series expansion approach led us to the following <u>discrete reconstruction problem</u>: Based on Eq. (7), GIVEN measurement vector y, ESTIMATE image vector x.

The estimation is usually performed using an optimization criterion that defines the image vector of choice for a given measurement vector. In the next section we discuss Bayesian and maximum-entropy optimization criteria.

We mention here that the reconstruction methods that are commonly adopted in practical image reconstruction from projections are not series expansion methods. Standard image reconstruction procedures in various applications are usually variants of the so-called convolution method (also called filtered back-projection; for details see Ref. [1], chapters 8 and 10). The major reason is that convolution methods are computationally simple and inexpensive, but nevertheless produce reconstructions that are often as good as or better than, and hardly ever more than slightly worse than, reconstructions produced by more cumbersome methods. When we refer to standard reconstruction methods in the following, we shall always mean a convolution method. Much of the following discussion is motivated by the question: Are there clearly defined situations in which series expansion methods can be shown to be definitely more efficacious than standard methods?

2. Optimization Criteria

In the Bayesian approach to image reconstruction (first introduced by Hurwitz [2]) we consider both the image vector x and the error vector e to be samples of random variables with probability density functions p_X and p_E, respectively. Then the Bayesian estimate is the image vector that maximizes

$$P_E(y - Rx)\, P_X(x)\,. \tag{8}$$

A serious difficulty with using Bayesian estimation is that it presupposes knowledge of P_E and P_X, an unlikely situation in image reconstruction. For reasons that have to do more with the mathematical and computational ease of optimization than with any physical justification, multivariate Gaussian distributions have been commonly assumed. Denoting by V_E and V_X the covariance matrices of the two distributions and by μ_X the expected image vector, and assuming that the expected error is the zero vector, we find that the x that maximizes expression (8) is the same as the x that minimizes

$$(y - Rx)^T V_E^{-1}(y - Rx) + (x - \mu_X)^T V_X^{-1}(x - \mu_X)\,. \tag{9}$$

A special case of this criterion is when both V_E and V_X are assumed to be multiples of identity matrices of appropriate size. In this case the problem reduces to the minimization of

$$r^2 \| y - Rx \|^2 + \| x - \mu_X \|^2\,, \tag{10}$$

where

$$\|y - Rx\| = \left[\sum_{i=1}^{I}\left[y_i - \sum_{j=1}^{J} r_{i,j}\, x_j\right]^2\right]^{1/2} \tag{11}$$

is referred to as the residual norm,

$$\|x - \mu_X\| = \left[\sum_{j=1}^{J} [x_j - (\mu_X)_j]^2\right]^{1/2}, \tag{12}$$

and r, the signal-to-noise ratio, is determined by the diagonal entries of V_E and V_X. Note that a larger value of r indicates greater confidence that the measurements should override the a priori expectation. The solution that minimizes expression (10) is often referred to as a regularized least-squares estimate.

So far we have discussed the global, or unconstrained, Bayesian optimization criterion (8) and its two progeny, (9) and (10). Typically in image reconstruction we do not seek an unconstrained optimum, but restrict the region of "acceptable" solutions x to Eq. (6), the so-called feasible region, by requiring x to satisfy one or more constraints. Typical constraints that have been used in image reconstruction are the following three:

C_1. The x must satisfy

$$-\varepsilon_i \leq y_i - \sum_{j=1}^{J} r_{i,j}\, x_j \leq \varepsilon_i, \qquad 1 \leq i \leq I, \tag{13}$$

where ε_i is a known tolerance to the accuracy of the ith measurement. A theoretically important special case of this constraint is when $\varepsilon_i = 0$, for $1 \leq i \leq I$ (that is, all measurements are to be exactly satisfied simultaneously). We shall refer to this special case as condition $C_{\bar{1}}$.

C_2. The x must satisfy

$$\sum_{j=1}^{J} x_j = S. \tag{14}$$

The assumption that such an S is known to us is justified as follows. For any image vector x, not necessarily satisfying Eq. (14), we define

$$\bar{x} = \frac{1}{J} \sum_{j=1}^{J} x_j \, . \tag{15}$$

If Eq. (14) is satisfied, then $\bar{x} = S/J$. In the pixel model of Eq. (3), $\bar{x}$ is the average value of f in the picture region. It can be estimated from the measurements as follows. For $1 \leq i \leq I$, the value of $\mathcal{R}_i f$ divided by the length of intersection of the ith line with the picture region gives an estimate of the average value of f along that line. If we have many such lines, providing a fairly uniform and dense covering of the picture region, then the sum of the $\mathcal{R}_i f$'s divided by the sum of the lengths is a reasonable estimate of the average value of f, and hence, in the pixel model, of $\bar{x}$. It has been shown (see, for example, Ref. [1], p. 106) that in typical situations in computed tomography all of the measurements combined together lead to a very accurate estimate of $\bar{x}$.

C_3. The components of x must be positive, that is,

$$x_j \geq 0 \, , \qquad 1 \leq j \leq J \, . \tag{16}$$

This constraint is usually justified by the physical interpretation of the x_j. For example, in x-ray computerized tomography and the pixel model, x_j denotes the necessarily nonnegative average x-ray linear attenuation in the jth pixel.

In what follows we shall use C with multiple subscripts. For example, "x satisfies $C_{1,3}$" will mean that x satisfies C_1 and C_3 simultaneously.

In addition to the unconstrained optimization criteria discussed above, there have been a number of criteria proposed in image reconstruction, specifically to be used for optimization constrained to a feasible region. Two of these, both first proposed in image reconstruction by Gordon, Bender, and Herman [3], are as follows:

Minimum norm. It can be reasonably argued that, among the x that satisfy C_1, one should choose the solution for which the variance,

$$\sum_{j=1}^{J} (x_j - \bar{x})^2 \, , \tag{17}$$

is minimal. If the feasible region is defined by $C_{1,2}$, then it is easily shown (see, for example, Ref. [4]) that the minimum variance solution is also the minimum norm solution, that is, that it minimizes

$$\| x \| = \left[\sum_{j=1}^{J} x_j^2 \right]^{1/2} . \tag{18}$$

Maximum entropy. In $C_{1,2,3}$ the maximizer of

$$-\sum_{j=1}^{J} x_j \ln x_j \tag{19}$$

(0 ln 0 is defined to be 0) also maximizes

$$-\sum_{j=1}^{J} (x_j/J\bar{x}) \ln(x_j/J\bar{x}) . \tag{20}$$

Presumably, no reader of these proceedings needs justification at this point as to why the use of such a maximum-entropy criterion is reasonable. It is curious, however, that many recent publications cite papers from the late 1970s as sources for the maximum-entropy method for image reconstruction, although the criterion had been clearly stated, philosophical justifications had been discussed, and a valid optimization algorithm had been published by Gordon, Bender, and Herman in 1970 (see, for example, Ref. [3] and its references).

3. An Unconstrained Optimization Algorithm

In this section we discuss an algorithm for finding the regularized least-squares estimate (10). The algorithm of choice is influenced by the fact that the projection matrix R is very large (typically 10^4 to 10^6 elements) and very sparse (typically only 0.1% to 1% of elements are nonzero), with no obvious structure but with the location and size of the nonzero entries for each row of R easily calculable.

Under such circumstances, a class of methods that the image reconstruction literature calls ARTs (algebraic reconstruction techniques) has been found quite efficacious [1,3,4,5,6]. ARTs are iterative methods, producing a sequence $x^{(0)}, x^{(1)}, x^{(2)}, \ldots$ of vectors. The kth iterative step (producing $x^{(k+1)}$ from $x^{(k)}$) would typically have the following properties:

(a) An i_k is selected, $1 \leq i_k \leq I$. Typically, and in particular in this paper, $i_k = (k \bmod I) + 1$.

(b) $x_j^{(k+1)} = x_j^{(k)}$ if $r_{i_k,j} = 0$; that is, a pixel value is not changed if the i_kth line misses the pixel.

(c) A large bulk of the work in computing $x^{(k+1)}$ from $x^{(k)}$ is taken up by calculating $\langle r_{i_k}, x^{(k)} \rangle$. Here r_{i_k} is the transpose of the i_kth row of R, and hence

$$\langle r_{i_k}, x^{(k)} \rangle = \sum_{j=1}^{J} r_{i_k,j}\, x_j^{(k)} . \tag{21}$$

For computational considerations of what is involved, see Ref. [1], p. 182.

There is a misconception, repeatedly stated in the literature, that ART is applicable only to finding exact solutions of consistent systems of equations (constraint $C_{\overline{1}}$ in the terminology of this paper). This is not so. ART can handle unconstrained Bayesian optimization as well as entropy maximization over a feasible region determined by the estimated inaccuracy of the measurement, as will be illustrated by the algorithms discussed below.

The ART algorithm that converges to the regularized least-squares estimate makes use of a sequence of I-dimensional auxiliary vectors $u^{(0)}, u^{(1)}, u^{(2)}, \ldots$. Only one component of $u^{(k)}$ is changed to get $u^{(k+1)}$. In what follows, t_i is the I-dimensional vector whose ith component is 1 and all other components are 0, and θ_I denotes the I-dimensional zero vector.

Algorithm I

$$x^{(0)} = \mu_X ,$$

$$u^{(0)} = \theta_I ,$$

$$x^{(k+1)} = x^{(k)} + r\, c^{(k)}\, r_{i_k} ,$$

$$u^{(k+1)} = u^{(k)} + c^{(k)}\, t_{i_k} ,$$

where

$$c^{(k)} = \lambda^{(k)} \frac{r[y_{i_k} - \langle r_{i_k}, x^{(k)} \rangle] - u_{i_k}^{(k)}}{1 + r^2 \| r_{i_k} \|^2} ,$$

with $\lambda^{(k)}$ a relaxation parameter such that

$$0 < \gamma_1 \leq \lambda^{(k)} \leq \gamma_2 < 2$$

for some fixed γ_1 and γ_2.

Theorem I (see Ref. [1], section 11.3)

The sequence $x^{(0)}, x^{(1)}, x^{(2)}, \ldots$ generated by Algorithm I converges to a vector that minimizes Eq. (10), that is, to the regularized least-squares estimate.

The important question is: Does such an approach, based on Bayesian optimization theory, improve the clinical efficacy of reconstructed images (as compared to what is in standard use today)? In the literature, such a question typically is "answered" by the showing of a few images, which usually illustrate the superiority of the technique that the article is trying to sell. Although I have engaged (and will continue to engage) in such behavior myself, I firmly believe that such illustrations do not and cannot prove the superiority of one reconstruction procedure over another. Since the relative efficacy of two reconstruction procedures is very much dependent on application, data collection, and picture [5], the only way to illustrate the superiority of one is by a study of observer performance in a large number of reconstructions typical of the application area of interest.

Such a study has been carried out for testing whether Bayesian optimization provides an improvement over standard reconstruction methods for the detection of low-contrast tumors in a low x-ray dose computed tomography environment.

Low x-ray dose increases the unreliability of measurements (that is, increases the size of e in Eq. (7)). This is demonstrated in Fig. 3, in which reconstructions (a) and (b) are identical in every way except that the dose used in obtaining (b) is half of that used in obtaining (a). (For details see Ref. [6].) Figure 3(c) is a reconstruction from the low-dose data based on Algorithm I.

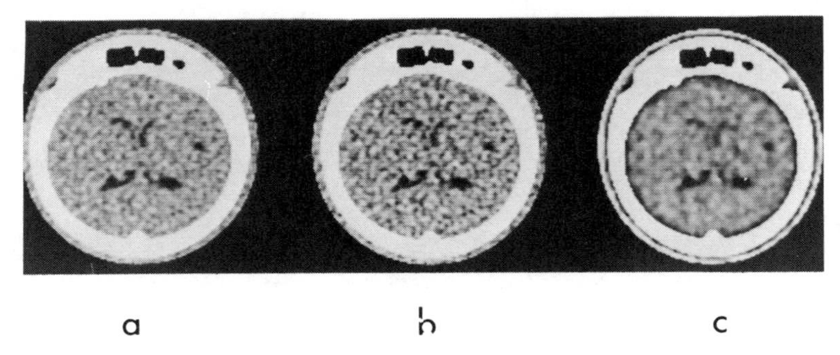

Figure 3. Examples of images used for an ROC study, reconstructed from two different sets of projection data by two different algorithms. Note the simulated tumor in the brain at right center of each image, between the ventricles and skull bone. a. Reconstruction from normal-dose data by a standard algorithm. b. Reconstruction from low-dose data by the same algorithm. c. Reconstruction from the same low-dose data by a variant of Algorithm I. (Reproduced from Ref. [6], with the publisher's permission.)

The exact variations to Algorithm I that were used to produce Fig. 3(c) are not of interest here (see Ref. [6] for details). The important point is that the detectability of the "tumor" and the rejectability of tumorlike nontumors due to noise in the data are not easily quantifiable on the basis of a single example. We in fact produced more than 100 reconstructions with each method, using randomly generated noise in the data and randomly placed tumors (in half of the reconstructions). Then we asked human observers to indicate their level of confidence that a tumor was or was not present. Such an approach gives, for each observer and each reconstruction method, a number of points indicating, at different levels of bias toward calling a tumor, how the ratio of false positives (tumor called when there is not one) to the total number of negatives is related to the ratio of true positives (tumor called when there is one) to the total number of positives [7].

The line that connects these points is called the ROC (receiver operating characteristic) curve. Such curves are shown in Fig. 4 for our three reconstruction methods and three observers. Random guesses would result in an

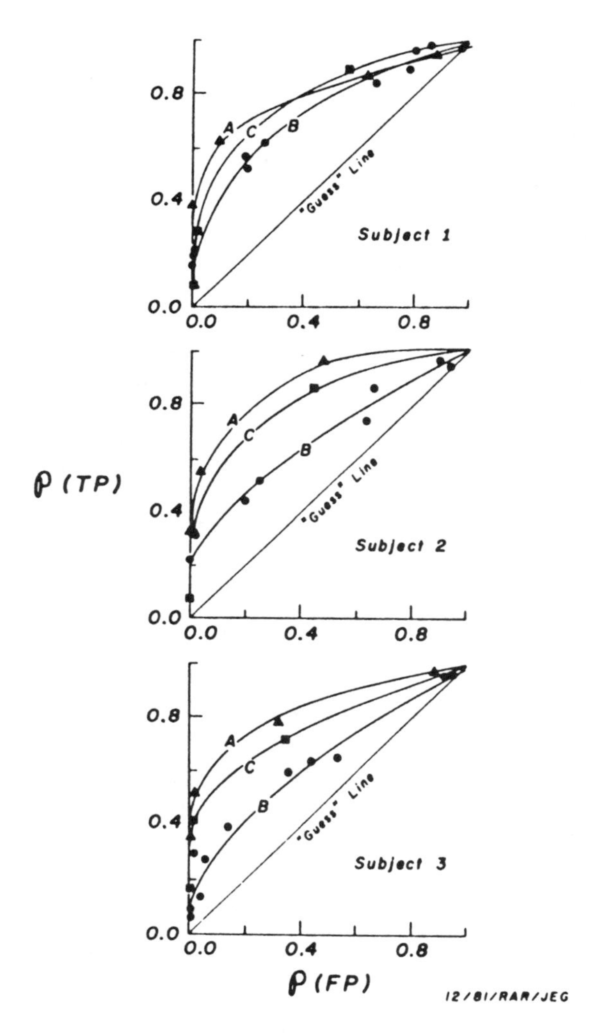

Figure 4. ROC curves for three observers. Curves A represent the response to images reconstructed from normal-dose data by a standard convolution algorithm. Curves B represent the response to images reconstructed from low-dose data by the same algorithm. Curves C show the response to images reconstructed from the same low-dose data by a variant of Algorithm I. (Reproduced from Ref. [6], with the publisher's permission.)

ROC curve along the 45° diagonal (the "guess line"). Improved performance is indicated by a higher curve (more true positives without increase in false positives). Figure 4 indicates that the Bayesian approach provided better tumor detectability on low-dose data than the standard approach, but the improvement is not quite as good as can be obtained by doubling the dose (a potentially harmful procedure).

The production of a large number of measurement vectors and reconstructions, and the observer performance tests on all of them, require a great deal of effort. Even then, the results may not be reproducible: on the same set of reconstructions, a different group of observers at a different site have not indicated a significant difference between the Bayesian and the standard approach on low-dose data [6]. This is why I, and most others, will retain the illustration of the virtues of new reconstruction methods by a few examples, but, when we really want to prove our point, we need to carry out a complex study of the kind discussed above, with, we hope, much more definite results.

In the study just described, μ_X was taken to be the output of a reconstruction algorithm of the kind commonly used in CT scanners. We may view Algorithm I as an attempt to iteratively improve this estimate. In another application [8], where a time-varying object (the beating heart) was imaged over a period from a large number of directions, μ_X was taken to be a reconstruction from all the data collected over time (a blurred heart inside a well defined thorax), but the projection matrix R and the measurement vector y were based on a small set of instantaneously obtained measurements. The regularized least-squares estimate provided a good picture: the details of the thorax were as good as could be expected from a large number of measurements, and the instantaneous location of the heart was also well indicated. Figure 5 is an example from this study.

One final comment regarding Algorithm I: While the algorithm converges in the limit to the regularized least-squares solution irrespective of the choice of $\lambda^{(k)}$, this choice has an important influence on the early behavior of the algorithm. This is illustrated in Fig. 6, in which the values of Eq. (10) are plotted for iterative sequences that differ only in the choice of the (constant) $\lambda^{(k)}$ (for details, see Ref. [9]). Such significant differences in the early behavior of the algorithm depending on the choice of this parameter makes its evaluation in a particular field even more difficult, especially since there is as yet no theory on how to choose the $\lambda^{(k)}$ to ensure rapid initial convergence.

In this section we have discussed unconstrained optimization based on Eq. (10). The question arises whether we can do better with the more general criterion (9). A detailed experimental study based on a single picture is reported in Ref. [10]. There it was found that, although a nondiagonal matrix in the second term of Eq. (9) improved things, there was no improvement in using other than a multiple of the identity matrix in the first term. In particular, choosing entries of a diagonal matrix V_E based on the noise due to statistical fluctuation of x-ray photon counts did not improve matters. The reason is that in x-ray computerized tomography, most of the error e in Eq. (7) is due to the discretization process. For example, during the study reported in Ref. [6], the residual norm (11) with y the noiseless projection data (of the phantom prior to digitization) was 9.60, while the residual norm with y the noisy projection data (including variation due to photon statistics) was 10.06. The first of these numbers is the size of the error due to discretization, and the second includes both discretization error and photon noise. Thus, basing the entries of V_E on photon noise alone is not justified in the x-ray CT application.

There is also evidence that the effect of the nondiagonal matrix in the second term of Eq. (9) can be achieved by using Algorithm I in combination with selective smoothing [1,6,10].

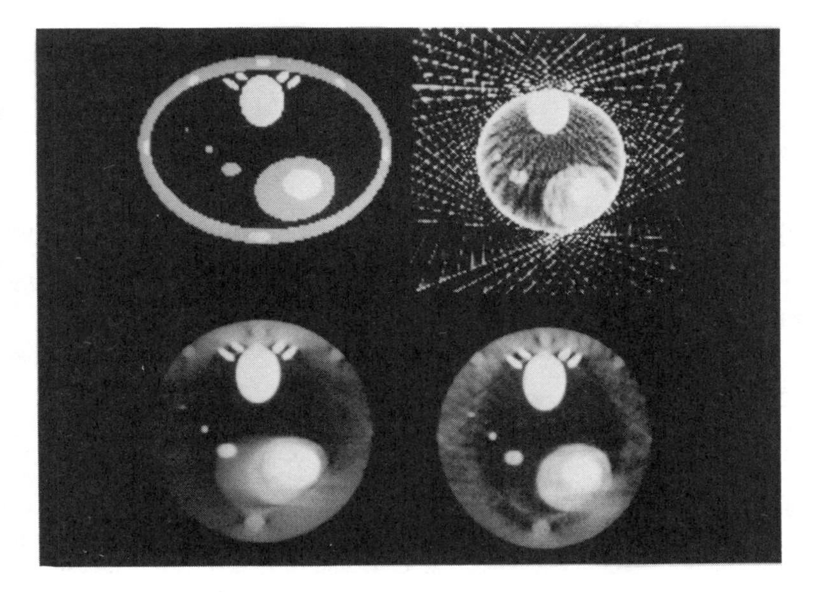

Figure 5. Top left shows a cross-section of a simulated time-varying thorax phantom at a particular instant. Top right shows a reconstruction obtained by a fast standard algorithm from instantaneously collected data. The reconstruction shows the effect of a limited field of view: the bright ring is at the edge of the region over which x-ray projection data are collected. The bottom left figure shows a reconstruction of the same cross-section using <u>all</u> the data collected during the simulation period, with each projection extended (so that it is no longer limited in field of view) by a mathematical extrapolation method. Since the total data set is collected over two heartbeats, the reconstruction of the dynamic heart-region is blurred. This blurred image was used as μ_X in Algorithm I for reconstruction from the instantaneously collected data. The result is shown on the lower right. (Reproduced from Ref. [8], with the publisher's permission.)

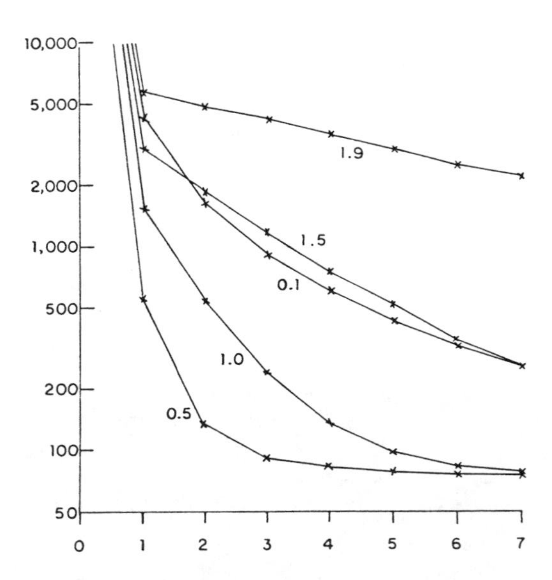

Figure 6. Logarithmic plots of the values of the functional in Eq. (10) as Algorithm I proceeds with different constant values of $\lambda^{(k)}$ (between 0.1 and 1.9). The integers on the abscissa indicate the number of cycles through all the equations; that is, t represents the (tI)th iteration.

4. Constrained Optimization Algorithms

We first discuss the special constraint $C_{\bar{1}}$ in which the tolerances ε_i in Eq. (13) are zero. This means that we need to satisfy a system of equations $Rx = y$. Since errors in the measurement vector y (due to both physical inaccuracies and the discretization process) are inevitable, $C_{\bar{1}}$ does not make good physical sense. However, one can generate error-free measurement vectors on a computer and, using such data, investigate the relative behavior of algorithms aimed at different optimization criteria. The absence of the error vector e may help us in observing the behavior of the algorithms.

To simplify the description of the algorithms, we combine $C_{\bar{1}}$ and C_2 and simply assume that Eq. (14) is the Ith equation of (13). For purely technical reasons, we shall assume that all equations have been normalized and that superfluous equations and unknowns have been removed (see Ref. [11] for details), and for the remaining system we have, for $1 \le i \le I$,

$$y_i > 0 ,$$

$$0 \le r_{i,j} \le 1, \qquad \text{for all } j, \quad 1 \le j \le J ,$$

$$r_{i,j} = 1, \qquad \text{for at least one } j, \quad 1 \le j \le J .$$

We now describe two algorithms, which in the literature have been called ART2 and MART [12,13].

Algorithm II

$$x_j^{(0)} = \hat{x}_j^{(0)} = S/J, \qquad \text{for } 1 \leq j \leq J\,,$$

$$\hat{x}^{(k+1)} = \hat{x}^{(k)} + \frac{y_{i_k} - \langle r_{i_k}, x^{(k)} \rangle}{\| r_{i_k} \|^2}\, r_{i_k}\,,$$

$$x_j^{(k+1)} = \begin{cases} \hat{x}_j^{(k+1)}, & \text{if } \hat{x}_j^{(k+1)} \geq 0 \\ 0, & \text{otherwise}\,. \end{cases}$$

Theorem II (see Ref. [11], p. 103)

If there exists an x satisfying $C_{\bar{1},2,3}$, then $x^{(0)},x^{(1)},x^{(2)},\ldots$ generated by Algorithm II converge to a vector that minimizes the norm in Eq. (18) over the feasible region determined by $C_{\bar{1},2,3}$, that is, to the constrained minimum norm solution.

Algorithm III

$$x_j^{(0)} = S/J\,, \qquad \text{for } 1 \leq i \leq J\,,$$

$$x_j^{(k+1)} = x_j^{(k)} \left[\frac{y_{i_k}}{\langle r_{i_k}, x^{(k)} \rangle} \right]^{r_{i_k,j}}, \qquad \text{for } 1 \leq j \leq J\,.$$

Theorem III (see Ref. [11], p. 103)

If there exists an x satisfying $C_{\bar{1},2,3}$, then $x^{(0)},x^{(1)},x^{(2)},\ldots$ generated by Algorithm III converge to a vector that maximizes the entropy in Eq. (19) over the feasible region determined by $C_{\bar{1},2,3}$, that is, to the constrained maximum entropy solution.

The comparative behavior of Algorithms II and III is quite different from what one might assume at first glance. The entropy maximizer and norm minimizer subject to $C_{2,3}$ are the same; both algorithms start with that image. At this point $C_{\bar{1}}$ is badly violated: The residual norm (11) is large. Both algorithms proceed so that, at the end of each cycle (I steps), $C_{2,3}$ is satisfied and the residual norm is considerably reduced (in the limit, it is of course zero). In an example given in Ref. [11], pp. 108-109, the residual norm is initially 13.6185, and it is reduced to 0.1540 in seven cycles of Algorithm II and to 0.1536 in ten cycles of Algorithm III. However, there is also precious little difference between the norms and the entropies of the two image vectors at these stages of the two algorithms; the norms are 168.5117 and 168.9938 and the entropies are 846.0574 and 847.1995, for Algorithms II and III respectively. If we consider reduction of the residual norm to about 1% of its initial value as an acceptable stopping criterion, we see that the image vectors produced by Algorithms II and III differ little from each other either by the norm or by the entropy measure. It is currently unknown whether this is due to the algorithms or, as is more likely, to the fact that the minimum norm picture in image reconstruction is likely to have nearly maximal entropy and vice versa.

The situation is even worse if the measurement vector y is contaminated by even a small amount of noise. In such a case convergence is not guaranteed, but early rapid reduction of the residual norm is nevertheless observed. However, it has been demonstrated [11] that in such a case, at points where the residual norms match, the entropy of the image produced by Algorithm II can be actually higher than that of the image produced by Algorithm III.

Such results have caused us to abandon a tumor detectability ROC study comparing the minimum norm and maximum entropy criteria when applied to highly underdetermined situations. Figure 7 shows pictures produced by Algorithms II and III that were intended to be part of this ROC study.

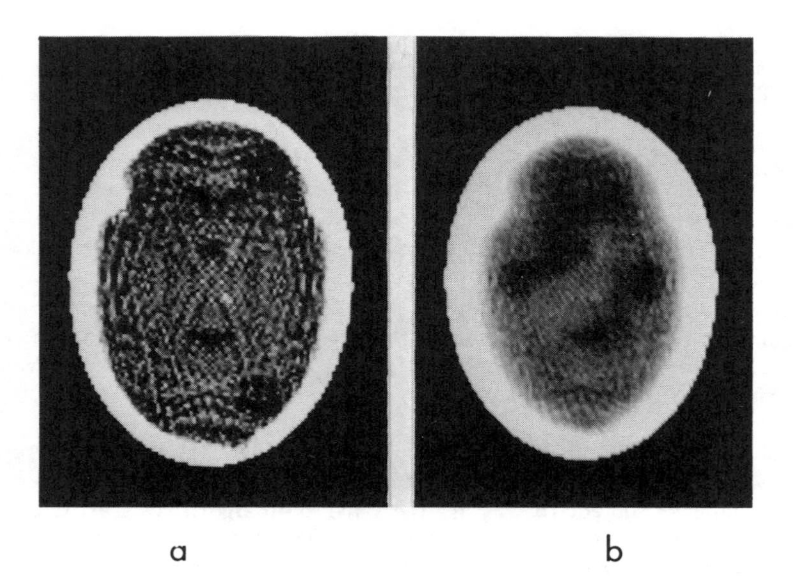

Figure 7. Reconstructions produced by (a) a variant of Algorithm II and (b) a variant of Algorithm III. The inferior quality of both reconstructions as compared to what we are used to in CT is due largely to these reconstructions having been produced from a much smaller number of measurements than is customary in CT.

We now consider the general constraints $C_{1,3}$. Again we ignore C_2 since if needed it can be made one of the constraints in C_1. ART methods exist both for norm minimization and for entropy maximization.

Algorithm IV

$$\hat{x}^{(0)} = \theta_J ,$$

$$u^{(0)} = \theta_I ,$$

$$\hat{x}^{(k+1)} = x^{(k)} + c^{(k)} r_{i_k} ,$$

$$u^{(k+1)} = u^{(k)} - c^{(k)} t_{i_k} ,$$

$$x_j^{(k+1)} = \begin{cases} \hat{x}_j^{(k+1)} , & \text{if } \hat{x}_j^{(k+1)} \geq 0 \\ 0 , & \text{otherwise} , \end{cases}$$

where

$$c^{(k)} = \text{mid}\left[u_{i_k}^{(k)}, \frac{y_{i_k} + \varepsilon_{i_k} - \langle r_{i_k}, x^{(k)}\rangle}{\| r_{i_k} \|^2}, \frac{y_{i_k} - \varepsilon_{i_k} - \langle r_{i_k}, x^{(k)}\rangle}{\| r_{i_k} \|^2}\right],$$

and mid [a,b,c] denotes the median of the three numbers a, b, and c.

Theorem IV (see Ref. [14], p. 25)

If there exists an x satisfying $C_{1,3}$, then $x^{(0)},x^{(1)},x^{(2)},\ldots$ generated by Algorithm IV converges to a vector that minimizes the norm in Eq. (18) over the feasible region determined by $C_{1,3}$, that is, to the constrained minimum norm solution.

Algorithm V

$$x_j^{(0)} = \exp(-1) , \quad \text{for } 1 \leq j \leq J ,$$

$$u^{(0)} = \theta_I ,$$

$$x_j^{(k+1)} = x_j^{(k)} \exp[c^{(k)} r_{i_k,j}] , \quad \text{for } 1 \leq j \leq J ,$$

$$u^{(k+1)} = u^{(k)} - c^{(k)} t_{i_k} ,$$

where

$$c^{(k)} = \text{mid}\,[u_{i_k}^{(k)} , \delta_-^{(k)} , \delta_+^{(k)}] ,$$

with $\delta_-^{(k)}$ and $\delta_+^{(k)}$ defined as the unique real numbers satisfying

$$\sum_{j=1}^{J} r_{i_k,j}\, x_j^{(k)} \exp[\delta_+^{(k)}\, r_{i_k,j}] = y_{i_k} - \varepsilon_{i_k}$$

and

$$\sum_{j=1}^{J} r_{i_k,j}\, x_j^{(k)} \exp[\delta_-^{(k)}\, r_{i_k,j}] = y_{i_k} + \varepsilon_{i_k},$$

respectively.

Theorem V (see Ref. [15], p. 453)

If there exists an x satisfying $C_{1,3}$, then $x^{(0)}, x^{(1)}, x^{(2)}, \ldots$ generated by Algorithm V converge to a vector that maximizes the entropy in Eq. (19) over the feasible region determined by $C_{1,3}$, that is, to the constrained maximum entropy solution.

Algorithms IV and V have been implemented, and their performance is illustrated (under matching conditions) in Refs. [14] and [16], respectively. The conclusion drawn in Ref. [16] is that the two algorithms produce comparable results after an equal number of iterations although Algorithm V (the entropy optimizer) is slower because of the extra work involved in the inner loop in which δ_- and δ_+ are calculated.

5. Discussion

In this paper we restricted our attention to certain types of constraints, certain types of optimization criteria, and ART methods for achieving the optimization. Our aim was to give a sample of methods that form a logical whole and with respect to which our ideas can be naturally discussed and illustrated. Those seeking a more general survey of the field should look at Ref. [1] for an overall discussion of image reconstruction methods and at Ref. [17] as a more nearly, but by no means totally, complete up-to-date tutorial on series expansion methods (including a discussion of the potential advantages of series expansion methods in certain situations).

The general ideas we wished to convey are the following. Bayesian optimization and maximum entropy methods appear very attractive, for basic scientific and philosophical reasons. However, there have been few authoritative studies comparing their performance in practical image reconstruction from projections with those methods that are commonly in current use. In those cases where careful studies have been attempted, the performance of the optimization methods has been patchy. Also, it has not been demonstrated that these "sophisticated" optimizers are likely to outperform in practice "simple" optimizers such as the constrained minimum norm; if anything, evidence indicates otherwise. For these reasons (as well as others, including just general inertia), Bayesian optimization and maximum entropy methods have not been widely adopted, in spite of the demonstration of their spectacular success on isolated examples.

Acknowledgments

The author's research is supported by grants from the National Science Foundation (ECS-8117908) and the National Institutes of Health (HL 28438 and HL 4664). He is grateful to Drs. Y. Censor and T. Elfving for their comments on the first draft of the paper and to Mrs. M. Blue for preparing the manuscript.

References

1. Herman, G. T. (1980) Image Reconstruction from Projections: The Fundamentals of Computerized Tomography (New York: Academic Press).
2. Hurwitz, H. (1975) Entropy reduction in Bayesian analysis of measurements, Phys. Rev. A **12**, 698-706.
3. Gordon, R., R. Bender, and G. T. Herman (1970) Algebraic reconstruction techniques (ART) for three-dimensional electron microscopy and x-ray photography, J. Theor. Biol. **29**, 471-481.
4. Herman, G. T., A. Lent, and P. H. Lutz (1978) Relaxation methods for image reconstruction, Commun. ACM **21**, 152-158.
5. Herman, G. T., and S. W. Rowland (1973) Three methods for reconstructing objects from x-rays: a comparative study, Comp. Graph. Image Process. **2**, 151-178.
6. Herman, G. T., R. A. Robb, J. E. Gray, R. M. Lewitt, R. A. Reynolds, B. Smith, H. Tuy, D. P. Hanson, and C. M. Kratz (1982) Reconstruction algorithms for dose reduction in x-ray computed tomography, in Proceedings MEDCOMP 82 (Silver Spring, Md.: IEEE Computer Society Press), pp. 448-455.
7. Swets, J. A., and R. M. Pickett (1982) Evaluation of Diagnostic Systems: Methods from Signal Detection Theory (New York: Academic Press).
8. Altschuler, M. D., Y. Censor, P. P. B. Eggermont, G. T. Herman, Y. H. Kuo, R. M. Lewitt, M. McKay, H. K. Tuy, J. K. Udupa, and M. M. Yau (1980) Demonstration of a software package for the dynamically changing structure of the human heart from cone beam x-ray projections, J. Med. Syst. **4**, 289-304.
9. Herman, G. T., A. Lent, and H. Hurwitz (1980) A storage-efficient algorithm for finding the regularized solution of a large, inconsistent system of equations, J. Inst. Math. Its Appl. **25**, 361-366.
10. Artzy, E., T. Elfving, and G. T. Herman (1979) Quadratic optimization for image reconstruction, II, Comp. Graph. Image Process. **11**, 242-261.

11. Herman, G. T. (1982) Mathematical optimization versus practical performance: a case study based on the maximum entropy criterion in image reconstruction, Math. Prog. Study **20**, 96-112.
12. Herman, G. T., A. Lent, and S. W. Rowland (1973) ART: mathematics and applications (a report on the mathematical foundations and on the applicability to real data of the algebraic reconstruction techniques), J. Theor. Biol. **42**, 1-32.
13. Lent, A. (1976) A convergent algorithm for maximum entropy image restoration, with a medical x-ray application, in Shaw, R., ed., Image Analysis and Evaluation (Washington, D.C.: Society of Photographic Scientists and Engineers), pp. 249-257.
14. Herman, G. T., and A. Lent (1978) A family of iterative quadratic optimization algorithms for pairs of inequalities, with application in diagnostic radiology, Math. Prog. Study **9**, 15-29.
15. Censor, Y. (1982) Entropy optimization via entropy projections, in Drenick, W. R. F., and F. Kozin, eds., System Modeling and Optimization (Berlin: Springer-Verlag), pp. 450-454.
16. Censor, Y., A. V. Lakshminarayanan, and A. Lent (1979) Relaxational methods for large scale entropy optimization problems, with application in image reconstruction, in Wang, P. C. C., et al., eds., Information Linkage Between Applied Mathematics and Industry (New York: Academic Press).
17. Censor, Y. (1983) Finite series expansion reconstruction methods, Proc. IEEE **71**, 409-419.

MULTIVARIATE EXTENSIONS OF MAXIMUM ENTROPY METHODS

James H. Justice

Chair in Exploration Geophysics, University of Calgary, Alberta, Canada

C. Ray Smith and W. T. Grandy, Jr. (eds.),
Maximum-Entropy and Bayesian Methods in Inverse Problems, 339–349.

1. Introduction

Since almost all of the papers presented at these workshops deal with the single-channel, or one-dimensional, maximum entropy spectral estimation problem, it seemed appropriate that someone should review the interesting work being reported on the analogous multivariate problem. This survey is intended to acquaint the reader with this area of research.

Spectral estimation is an important procedure in many diverse areas of application. That is, we often need to estimate the spectral characteristics of a data set, given that we have observed only a finite segment of that set. We are forced, then, to form our estimate in the face of uncertainty. The question "What would have happened next if the process could have been observed longer?" can generally not be answered with certainty, and yet a perfect knowledge of the process spectrum would imply that this question could be answered and so our estimate can seldom be known to be precise.

There are many ways to look at the problem, but most classic spectral estimation procedures make assumptions about the data outside the observation window (zero, periodic, etc.) or try to minimize the effects of truncation implied by our finite observation window by minimizing the "smearing" or "blurring" of the spectrum which is necessarily implied by a finite observation window.

These procedures could be applied either to the data themselves or, in the methods developed by Blackman and Tukey [1], to the autocorrelations of the data sequence, which could be estimated in various ways. In the Blackman and Tukey procedures, the autocorrelation lags that could not be estimated from the data were assumed to be zero, and procedures were used to minimize the effect of this truncation.

Clearly, in the absence of information, any assumption about the data or their autocorrelation beyond our ability to observe is unjustified, and will influence our spectral estimate.

In 1967, John Burg suggested a procedure that effectively honored the "observed" autocorrelations but allowed a nonzero extension of these values. The principle employed was that the spectrum should have the maximum entropy of all spectra that are consistent with the known autocorrelations. In a sense, this method makes the least assumption concerning the extension of the autocorrelation lags or the data outside our observation window. The spectral estimates so obtained were shown in many cases to be superior to estimates obtained by other methods.

Specifically, Burg's suggestion [2] was this: Find the power spectrum P(f) that maximizes the value of

$$\int_{-W}^{W} \ln P(f)\, df$$

subject to the constraints

$$R_n = \int_{-W}^{W} P(f) \exp(2\pi i f n \Delta t)\, df, \qquad -N \leq n \leq N ,$$

where Δt is the sampling period, $W = 1/2\Delta t$ is the Nyquist frequency, and the sequence $\{R_n\}$ are the known lags of the autocorrelation function.

The resulting spectral estimate is called the maximum entropy spectral estimate, since maximizing the integral above effectively maximizes the entropy.

Burg went on to suggest an algorithm for this analysis that could be applied directly to the data without estimating the autocorrelation function. This algorithm is called the "Burg algorithm."

2. Multichannel Maximum Entropy Spectral Analysis

In his thesis, Burg [2] considered the extension of the maximum entropy method to the multichannel case. In this case, he defined the multichannel entropy to be

$$\int_{-W}^{W} \ln[\det P(f)]\, df ,$$

where $P(f)$ is the multichannel power spectrum matrix. This definition was naturally suggested for the case of independent channels, and Burg showed that, in fact, this form is invariant under "change of coordinates" and so is a reasonable definition for a nondiagonal power spectrum matrix as well.

Burg solves the problem of maximizing the multichannel entropy subject to the constraints

$$\int_{-W}^{W} P(z) \exp(2\pi i f n \Delta t)\, df = R(n) ,$$

where $R(n)$ is a square cross-correlation matrix at lag n, by using Lagrange multipliers.

Needless to say, interest soon developed in extending the Burg algorithm (wherein the correlation matrices do not have to be estimated) to the multichannel case. Several methods were suggested for doing this, but the extension was not exactly straightforward since certain matrices that occur in the recursion equations for the forward and backward residuals are different [3,4] although they must yield equivalent spectral estimates. Various solutions to this problem were suggested, but not all resulted in stable predictors. Finally, several successful extensions have been found [4,5] that reduce to the single-channel Burg algorithm in the single-channel case.

3. Multidimensional Maximum Entropy Spectral Analysis

Whereas the multichannel extension of the maximum entropy spectral estimation method for the known correlation case as well as the Burg algorithm was carried out without too much difficulty, the multidimensional extension of these methods has been quite a different story. As has been the case with attempts to extend other procedures to the multidimensional case, numerous and almost insurmountable difficulties have been encountered. It is well known that the Fundamental Theorem of Algebra breaks down for more than one variable, so that the ability to factor polynomials is no longer guaranteed. Indeed, Hayes and McClellan [6] have shown that the set of reducible multivariate polynomials is a set of measure zero. The implication is, then, that very few traditional tools are available and so it is quite difficult to proceed.

Nonetheless, these difficulties have not kept researchers from attempting to break new ground in the multidimensional case, and numerous attempts have been made to extend Burg's methods to this setting.

Woods [7] reports that the form of the 2-D maximum entropy spectrum was given by Burg in an unpublished report, using the Lagrange multiplier approach. However, the existence of a solution was left in doubt. The analog to the one-dimensional approach [8] could not be used since it invoked the Fundamental Theorem of Algebra.

Woods then considered the problem of the existence of the two-dimensional maximum entropy spectral estimate in terms of discrete Markov random fields. Following Woods, let ξ be a random field on $I \times I$. Then ξ is said to be Markov-p if $\{\xi|_{G^+}$ given $\xi|_{\partial G}\}$ is independent of $\{\xi|_{G^-}\}$, where G^- is the "past" and G^+ is the "future," separated by a band, ∂G, of width p (see Fig. 1). If ξ is homogeneous, zero-mean, and Gaussian, then it

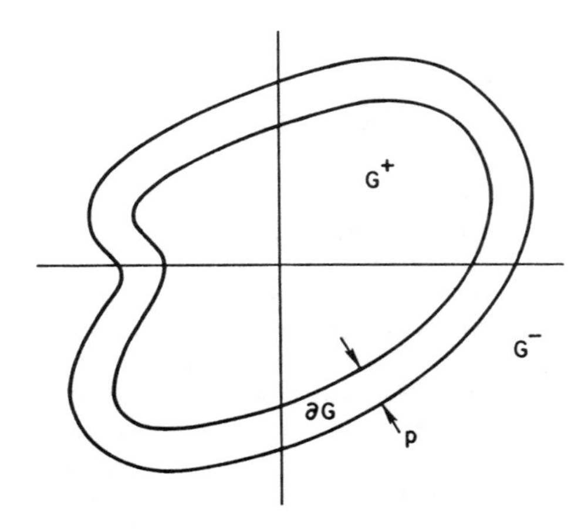

Figure 1. Markov-p random field.

is possible to show [9] that

$$\xi_{ij} = \sum_{D^0} h_{k\ell}\, \xi_{i-k,j-\ell} + e_{ij} ,$$

where

$E[\xi_{ij}\, e_{k\ell}] = C\, \delta_{ik}\, \delta_{j\ell}$,

$D^0 = \{(k,\ell)/(k^2 + \ell^2) \leq p^2 \text{ and } (k,\ell) \neq (0,0)\}$,

e_{ij} is a Gaussian zero-mean homogeneous random field with correlation function of bounded support.

In this case, it follows that the spectral estimate is given by

$$S_\xi(u,v) = C / [1 - \sum_{D^0} h_{k\ell} \exp(2\pi i(ku + \ell v)] ,$$

which is the form derived by Burg for the 2-D ME spectrum.

To complete the proof of the existence of the ME spectral estimate for a homogeneous random field, x, we need only find a Markov random field, y, such that the observed autocorrelation coefficients, R_x, of x agree with those of y, R_y, on the set χ where they are known. Woods takes χ to be a nearest-neighbor array (contains all points within its perimeter) that is symmetric through the origin. He shows that the Markov random field y exists, provided R_x is positive definite (on all of I × I).

Woods goes on to consider the question of uniqueness and gives a procedure for approximating the coefficients $C, \{h_{k\ell}\}$ so that $R_x = R_y$ on χ.

Dickinson [10] pointed out that the Markov spectrum need not necessarily exist since the observed autocorrelation values R_x may not be extendable to all of I x I.

Krein [11] had considered the problem of extendability of functions positive definite on open intervals in R and had shown that they do indeed extend to all of R, and that the extension is generally not unique. Rudin [12], however, had shown that Krein's result on extendability can fail in more than one dimension, and Calderon and Pepinsky [13] had proved the same result for I × I.

As a result, it is critical that a criterion be established for the extendability of an observed set of autocorrelations before we can conclude the existence of the ME spectrum.

In addition to the problems already mentioned, it has been well known for some time [14] that the multidimensional analog of the Levinson recursion, which Burg used as a constraint to guarantee the stability of the spectral estimator, does not guarantee stability in the two-dimensional case.

Marzetta [15] suggests a way around this difficulty, but this requires an infinitely long computation to implement and so has not led to a practicable solution to the problem. Finally, we could point out that spectral factorization is possible in higher dimensions [16], but it does not share the properties of its one-dimensional counterpart. The multidimensional factors can be infinite in extent.

Several direct approaches to the ME estimation problem have been considered. Newman [17] defines the two-dimensional entropy to be

$$H = \frac{1}{8U_NV_N}\int_{-U_N}^{U_N}\int_{-V_N}^{V_N} \ln[S(u,v)]\,du\,dv + \ln(2\pi e)^{1/2}$$

and also justifies the relationship between the maximum entropy spectral estimate and the autoregressive form

$$S(u,v) = \left[\sum_m \sum_n A_{mn} \exp[-2\pi i(m\Delta xu + n\Delta yv)]\right]^{-1} .$$

Wernecke and D'Addario [18] also consider a Lagrange multiplier formulation for the solution to the 2-D ME problem with known autocorrelations, and point out that no known explicit solution exists for this formulation. They then suggest an iterative numerical scheme to approximate the solution. They acknowledge that the constrained optimization problem may, in fact, not have a solution.

Returning to the concept of obtaining the maximum entropy spectral estimate by performing a maximum entropy extension of the autocorrelation function, Roucos and Childers [19] consider the extension of this technique to the multidimensional case and suggest an iterative algorithm based on this approach.

Recently, Lang [20] considered the multidimensional ME spectral estimation problem. He has given conditions that guarantee the extendability of the autocorrelation sequence and the existence of the ME spectral estimate. In addition, he suggests algorithms for solving a dual optimization problem to obtain this estimate.

Lang's method makes no assumption on the set of lags for which the autocorrelation estimates are known.

Lang formulates the 2-D ME problem as follows. Let the autocorrelation function, r, be known on the co-array

$$\Delta \subset R^D$$

where

$$\Delta = \{\delta_0 \equiv 0, \pm\delta_1, \ldots, \pm\delta_m\} .$$

The entropy, H, is defined by

$$H(S) = \int_K \ln[S(k)]\, d\nu ,$$

where S is continuous, $S > 0$, and K is the support set for the power spectra S to be considered. K is assumed to be compact (closed and bounded) in R^D. Lang further assumes that the set $\{\exp(ik\cdot\delta) \mid \delta \in \Delta\}$ is linearly independent over K.

The problem is now to maximize the entropy, H, subject to the constraint

$$r(\delta) = \int_K S(k) \exp(ik\cdot\delta)\, d\nu$$

for $\delta \in \Delta$.

The first requirement for the existence of the ME spectrum is that the observed autocorrelation, r, should be extendable. Lang gives the first general solution to this problem by showing that r is extendable if and only if $(r,p) \geq 0$ for all positive δ_- polynomials, p, that is, polynomials of the form

$$p(k) = \sum_{\delta \in \Delta} p(\delta) \exp(-ik\cdot\delta)$$

that satisfy $p(k) \geq 0$ for $k \in K$.

Lang shows that this criterion reduces to well-known criteria in special cases. The constrained maximization problem to obtain the ME spectrum may now be viewed as an optimization problem to maximize a concave functional (the entropy, H) over a convex set $C \cap D$, where C is the set of continuous functions, S, on K that satisfy

$$r(\delta) = \int_K S(k) \exp(ik\cdot\delta)\, d\nu$$

and D is the set of positive continuous functions on K. Since r is assumed to be extendable, $C \cap D$ is not empty.

The Fenchel duality theorem [21] is then used to reformulate the problem in the dual form

$$\underset{C^* D^*}{\text{maximize}} \left[\int_K d\nu + \int_K \ln p(k)\, d\nu - \sum_{\delta \in \Delta} C_j\, r(\delta_j) \right] ,$$

where p is the δ-polynomial,

$$p(k) = \sum_{\delta_j \varepsilon \Delta} C_j \exp(-ik \cdot \delta_j) \,.$$

The set C^*D^* is a subset of the δ-polynomials.

Lang then shows that if the solution to the (finite-dimensional) dual problem (which exists) is strictly positive, then $1/p(k)$ is the ME spectral estimate and the maximum value above is

$$\int_K \ln p(k) \, d\nu \,.$$

The condition that guarantees a positive solution (and the existence of the ME spectral estimate) is

$$\int_K [p(k)]^{-1} \, d\nu = \infty$$

for all $p \varepsilon \partial P$, where

$$\partial P = \{p \geq 0 \text{ on } K \,|\, p(k) = 0 \text{ somewhere on } K\} \,.$$

This condition is, in fact, necessary and sufficient for the existence of the ME spectral estimate.

Several important questions relating to multidimensional ME spectral estimation have now been resolved. First, the question concerning extendability of the observed autocorrelations has been dealt with, and a condition has been given for its validity. It may be expected that many other tests will result from the application of this criterion. Second, a necessary and sufficient condition for the existence of the multidimensional ME spectrum, given the extendability of the autocorrelations, has been established.

The dual problem is again an optimization problem for a convex function over a convex set, but has the advantage that only the set $\{C_j\}$ has to be determined, so the problem is finite dimensional.

This is essentially the point of departure for many of the suggested iterative algorithms designed to approximate the ME spectral estimate, including the ones already mentioned. Lang points out that these approaches can be unified if they are written in the form

$$p_{n+1} = p_n - \alpha_n d_n \,,$$

where d_n is the update direction and α_n is the step size. Burg had proposed an algorithm of this type, and Lim and Malik [22] have recently proposed another iterative procedure to obtain the ME estimate.

Some of the difficulties mentioned by Lim and Malik for finding a closed-form solution for the 2-D ME spectral estimate are accounted for by Lang. However, it remains true that a closed-form solution has not been found.

Lim and Malik assume that we have autocorrelations R_x given on a set A and that R_x is extendable to a positive definite function, R_y. Defining the ME power spectrum by

$$P_y(\omega_1,\omega_2) = \left[\sum_{-\infty}^{\infty}\sum_{-\infty}^{\infty} \lambda(n_1,n_2)\exp(-i\omega_1 n_1)\exp(-i\omega_2 n_2)\right]^{-1}$$

$$\equiv \sum_{-\infty}^{\infty}\sum_{-\infty}^{\infty} R_y(n_1,n_2)\exp(-i\omega_1 n_1)\exp(-i\omega_2 n_2) ,$$

we choose an initial estimate, $\lambda^{(0)}(n_1,n_2)$. This generates an estimate

$$R_y^{(0)}(n_1,n_2) .$$

This estimate is corrected (replaced) by the values $R_x(n_1,n_2)$ on A. The resulting function is used to generate $\lambda^{(1)}(n_1,n_2)$. We then truncate $\lambda^{(1)}(n_1,n_2)$ to the desired size and repeat the process. The algorithm, as stated, needs some modification owing to the possibility of spectral zeros. A detailed description of the suggested modifications and convergence are given in the Lim and Malik paper.

4. Conclusion

Several outstanding questions relating to the extension of the ME spectral estimation concept to the multivariate case have been answered. Several different approaches have thrown light on the issues, and Lang has finally cleared up the extendability issue for positive definite functions and provided a necessary and sufficient criterion for the existence of the ME spectrum once the extendability issue is decided.

Whereas the multichannel extensions of the autocorrelation matching maximum entropy procedure as well as the Burg algorithm have been carried out, the multidimensional extension of the Burg algorithm still eludes us, and even the correlation matching problem can be solved today only with a generally infinite iterative process.

Lim and Malik consider the problem of finding a closed-form solution to the 2-D ME problem, and indeed the restricted degrees of freedom in the multidimensional problem versus the corresponding multichannel problem [23] may be a part of the difficulty.

In any event, some important advances have been made, and perhaps we have come a few steps closer to a satisfactory solution to the problem of using maximum entropy methods for multidimensional spectral estimation.

Postscript

In view of the "new" Burg technique to appear in this volume, which apparently takes care of multichannel and multidimensional cases, perhaps we should refer to the above discussion as the "classic" maximum entropy problem.

References

1. Blackman, R. B., and J. W. Tukey (1959) The Measurement of Power Spectra (New York: Dover).
2. Burg, J. P. (1975) Maximum Entropy Spectral Analysis, Ph.D. thesis, Stanford University.
3. Jones, R. H. (1978) Multivariate autoregression using residuals, in D. Findley, ed., Applied Time Series Analysis (New York: Academic Press).
4. Morf, M., A. Vieira, D. T. L. Lee, and T. Kailath (1978) Recursive multichannel maximum entropy spectral estimation, IEEE Trans. Geosci. Electron. **GE-16**, 85-94.
5. Strand, O. N. (1977) Multichannel complex maximum entropy (autoregressive) spectral analysis, IEEE Trans. Autom. Control **AC-22**, 634-640.
6. Hayes, M. H., and J. H. McClellan (1982) Reducible polynomials in more than one variable, Proc. IEEE **70**, 197-198.
7. Woods, J. W. (1976) Two-dimensional Markov spectral estimation, IEEE Trans. Inf. Theory **IT-22**, 552-559.
8. Edward, J. A., and M. M. Fitelson (1973) Notes on maximum entropy processing, IEEE Trans. Inf. Theory **IT-19**, 232-234.
9. Woods, J. W. (1972) Two-dimensional discrete Markovian fields, IEEE Trans. Inf. Theory **IT-18**, 232-240.
10. Dickinson, B. (1980) Two dimensional Markov spectrum estimates need not exist, IEEE Trans. Inf. Theory **IT-26**, 120-121.
11. Krein, M. G. (1940) Sur le problème du prolongement des fonctions hermitiennes positives et continues, C. R. Dokl. Acad. Sci. URSS (N.S.) **26**, 17-22.
12. Rudin, W. (1963) The extension problem for positive-definite functions, Ill. J. Math **7**, 532-539.
13. Calderon, A., and R. Pepinsky (1952) On the phases of Fourier coefficients for positive real periodic functions, Computing Methods and the Phase Problem in X-Ray Crystal Analysis (Department of Physics, Pennsylvania State College).
14. Genin, Y., and Y. Kamp (1975) Counter example in the least square inverse stabilization of 2-D recursive filters, Electron. Lett. **11**, 330-331.
15. Marzetta, T. (1978) A linear prediction approach to two-dimensional spectral factorization and spectral estimation, Ph.D. dissertation, MIT.
16. Ekstrom, M., and J. Woods (1976) Two-dimensional spectral factorization with applications in recursive digital filtering, IEEE Trans. Acoust. Speech Signal Process. **ASSP-24**, 115-128.

17. Newman, W. I. (1977) A new method of multidimensional power spectral analysis, Astron. Astrophys. **54**, 369-380.
18. Wernecke, S. J., and L. R. D'Addario (1977) Maximum entropy image reconstruction, IEEE Trans. Comput. **C-26**, 351-364.
19. Roucos, S., and D. Childers (1979) A two-dimensional maximum entropy spectral estimator, Proc. IEEE ICASSP.
20. Lang, S. W. (1981) Spectral estimation for sensor arrays, Ph.D. Thesis, MIT.
21. Luenberger, D. G. (1969) Optimization by Vector Space Methods (New York: Wiley), p. 21.
22. Lim, J. S., and N. A. Malik (1981) A new algorithm for two-dimensional maximum entropy power spectrum estimation, IEEE Trans. Acoust. Speech Signal Process. **ASSP-29**, 401-413.
23. Justice, J. H. (1982) Array processing in exploration seismology, manuscript.

INDUCTIVE INFERENCE AND THE MAXIMUM ENTROPY PRINCIPLE

N. C. Dalkey

Cognitive Systems Laboratory, University of California at Los Angeles, Los Angeles, California 90024

C. Ray Smith and W. T. Grandy, Jr. (eds.),
Maximum-Entropy and Bayesian Methods in Inverse Problems, 351–364.

1. Figures of Merit for Induction

A conventional schema for an inference consists of a set of premises and a conclusion, as in the diagram:

		Deduction	Induction
Premises	{ ——— · ———	True	?
	↓	↓	↓
Conclusion	———	True	?

If the inference is a valid deduction, then, if the premises are true, the conclusion must be true. If the inference is an induction, the question arises, what is the figure of merit that furnishes a basis for the justification of the inference?

A frequent suggestion is to use probability (or some variant) as the figure of merit for induction, giving the schema

Premises	True
↓	↓
Conclusion	Probable

where the degree of probability presumably is a function of the form of the premises and of the conclusion [1].

The suggestion has not received general acceptance by the relevant communities of scholars (logicians, probability theorists, statisticians, decision theorists) for a variety of reasons. One reason that appears to me to be the most crucial has not received much attention. That reason is the fact that the degree of probability is not a proper scoring rule [2].

The notion of a proper scoring rule (admissible score, reproducing score, honesty-promoting score, etc.) can be most easily explicated via the honesty-promoting feature. Suppose you are contemplating hiring a consultant to furnish you with the probabilities of some events you are interested in. One thing you would want to avoid is formulating the contract so it would be to the advantage of the consultant to lie—to report a probability distribution that he did not believe.

Let's say you are interested in knowing the probability distribution on a partition (exclusive and exhaustive set) of events $E = \{e_1,\dots,e_n\}$. The distribution you want is thus the set $P = \{P(e_1),\dots,P(e_n)\}$, where

$$\sum_E P(e) = 1\,, \qquad 0 \leq P(e) \leq 1\,.$$

Suppose you intend to reward the consultant with a score $S(P,e)$, which is a function of the consultant's report P and of the event e that happens. If the consultant believes P, but reports a distribution R, his (subjective) expectation will be

$$\sum_{E} P(e)S(R,e) \ .$$

To motivate the consultant to be honest, the function S would have to fulfill the condition

$$\sum_{E} P(e)S(R,e) \leq \sum_{E} P(e)S(P,e) \ ; \tag{1}$$

that is, his expected score would have to be a maximum when he reported the probability distribution he believed.

Equation (1) is the definition of a proper scoring rule. It has a great deal more content than just the honesty-promoting feature. For example, if P is the objectively correct distribution on E, then Eq. (1) mandates that the objective expectation of the score be a maximum when the correct distribution is reported.

From Eq. (1) we can conclude that the probability estimate is not itself a proper score. Set $S(R,e) = R(e)$:

$$\sum_{E} P(e)R(e)$$

is a maximum when $R(e) = 1$ for the e such that

$$P(e) = \max_{E} P(f) \ .$$

In short, if a probability is used as the figure of merit for an induction, the analyst would maximize his expected score by setting $R(e) = 1$ for the conclusion having the highest probability, and $R(f) = 0$ for all the others.

Note that the traditional true-false evaluation is a proper score of a particularly simple form. Consider the matrix

	State of the World	
Statement	p	$\overline{p}$
"p"	True	False
"$\overline{p}$"	False	True

where the overline indicates negation. If an individual says "p" and the state of the world is p, then he receives the score <u>true</u>. If the state of the world is $\overline{p}$, he receives the score <u>false</u>. True is <u>of course</u> a higher score than false. Thus, if an individual <u>believes</u> p, his expectation is maximized if he says "p", and vice versa if he believes "$\overline{p}$".

A more appropriate schema for induction, then, is that in which the conclusion is formulated as a probability statement, and the figure of merit is a proper scoring rule:

Premise	—	Figure of merit
Conclusion	$\{P(e)\}$	$\sum_E P(e)\, S(P,e)$

The preceding remarks indicate that the operant figure of merit is not the score, but the expected score, in the case both of deduction and of induction. In the case of deduction, however, the expectation is generally truncated because of the all-or-none quality of true-false. I have not indicated a figure of merit for the premises in the case of induction; depending on the way the premises are formulated, the figure of merit could be the traditional true, or an expected, score.

A significant feature of proper scores is that they give a clear procedure for the verification of assertions of the probability of a single case. In traditional logic, a statement of the form "the probability of rain tomorrow is 0.6" is compatible with either rain or not-rain tomorrow. Thus the statement is not verifiable by the relevant event. This fact has led most investigators interested in the foundations of probability to claim that assertions concerning the probability of one-of-a-kind events are meaningless. However, with a proper score rule, a statement concerning the probability of a single event can be verified in an unambiguous fashion. The analyst waits until the relevant event occurs, and then assigns a figure of merit that depends both on the asserted probability and on the event that happens. The dependence on the occurrence of the relevant event gives the necessary tie to reality, and the dependence on the asserted probability furnishes the required relation to the "content" of the statement.

There is a large infinity of functions that fulfill Eq. (1). Among them are (1) the discriminant score,

$$S(R,e) = \begin{cases} 1, & \text{if } P(e) = \max_E P(f) \\ 0, & \text{otherwise} \end{cases}$$

(2) the quadratic score,

$$S(R,e) = 2R(e) - \sum_E R(e)^2 ;$$

(3) the spherical score,

$$S(R,e) = \frac{R(e)}{[\sum_E R(e)^2]^{1/2}} ;$$

(4) the logarithmic score,

$$S(R,e) = \log R(e) ;$$

and (5) the family of decisional scores.

For the last-named, let U_{ae} designate a payoff matrix expressing the outcome if action a is taken and the event e occurs, and $U^*(R,e)$ designate the payoff if the optimal action for the distribution R is selected and the event e occurs; then $S(R,e) = U^*(R,e)$. The discriminant score is a special case of a decisional score with U_{ae} a square matrix, $U_{ae} = 1$ on the principal diagonal and 0 elsewhere.

The large array of potential scores is an advantage from the standpoint of decision theory; it allows one to streamline many problems by formulating an appropriate score rule. However, the nonuniqueness is awkward for scientific purposes. The solutions to problems in statistics, for example, can be highly sensitive to the score rule adopted. It would clearly be advantageous to have a single score rule that in some sense is "most appropriate" for scientific investigations. The score most often nominated for this status is the logarithmic score. It has a number of unique features: (a) It is the only score that is a function solely of the probability asserted for the event that occurs. (b) The expectation of the log score is the negative of Shannon's entropy, which makes available a wide collection of results from information and communication theory. (c) It is symmetrical, and the minimum of the expectation occurs at the traditional uniform distribution. (d) It is closely related to the notion of independence; for problems where independence can be formally defined, the minimum expectation of the log score coincides with independence.

Despite these and other notable features, the logarithmic score has not received the full endorsement of the relevant communities, mainly because of the obvious significance of decisional scores for applied decision theory. However, there is one property of the logarithmic score that appears definitive for scientific purposes. If we want a score rule to play the role of the analog of truth values for inductive inference, then the score must have a certain flexibility. In Eq. (1), the defining property is expressed relative to a fixed (but unspecified) <u>partition</u>. However, structures that are sufficiently general to form the basis of a logic are much richer than partitions. One more or less canonical form for such a structure is an algebra of events, that is, a set of events closed under disjunction (or combinations) and negation. An algebra is a convenient event basis for representing elementary probability theory, as well as a convenient representation of an elementary two-valued logic. It seems unlikely that any analog for truth-values could be generally applicable unless it could be extended at least to an algebra of events.

Let $E = \{e,f,\ldots\}$ now represent an algebra of sets, and let $P = \{P(e),P(f),\ldots\}$ represent a probability <u>measure</u> on the algebra; that is, P assigns a probability to each event in E such that for any exclusive pair of events e,f, $P(e+f) = P(e) + P(f)$ and $0 \leq P(e) \leq 1$. Consider a score rule

$S(P,e)$ which is a function of probability measure P and event e; that is, S is defined for every event e in E and for every probability measure P on E.

We will say that a score rule $S(P,e)$ is coherent if, for every partition F definable in E, S is a proper score on F. Let

$$G_F(P,R) = \sum_F P(e)S(R,e) \quad \text{and} \quad H_F(P) = \sum_F P(e)S(P,e) .$$

S is coherent on E if

$$G_F(P,R) \leq H_F(P) \quad \text{for all partitions F in E .} \tag{2}$$

Two measures P and P' are said to agree on a partition F if $P(e) = P'(e)$ for all e in F.

Lemma 1. If S is coherent, and if P and P' agree on F, then $H_F(P) = H_F(P')$.

Proof:
$$G_F(P,P') \leq H_F(P) , \quad H_F(P') \geq G_F(P',P) ,$$
$$G_F(P,P') = H_F(P') , \quad H_F(P') = G_F(P',P) .$$

The equalities follow from the fact that P and P' agree on F, and the inequalities from Eq. (2).

The lemma states, in effect, that if S is coherent, then for any partition F there is a function H_F that depends only on the probabilities for events in that partition.

Lemma 2. $G_F(P,R)$ is linear in P.

Proof: Let $P = aQ + (1-a)Q'$, $0 \leq a \leq 1$. Then

$$G_F(P,R) = \sum_F [aQ(e) + (1-a)Q'(e)] S(R,e)$$

$$= a \sum_F Q(e)S(R,e) + (1-a) \sum_F Q'(e)S(R,e)$$

$$= aG_F(Q,R) + (1-a)G_F(Q',R) .$$

Lemma 3. $H_F(P)$ is convex in P; that is, if $P = aQ + (1-a)Q'$, $0 \leq a \leq 1$, then $H_F(P) \leq aH_F(Q) + (1-a)H_F(Q')$.

Proof:

$$H_F(P) = \sum_F [aQ(e) + (1-a)Q'(e)] S(P,e)$$

$$= a \sum_F Q(e)S(P,e) + (1-a) \sum_F Q'(e)S(P,e)$$

$$= aG_F(Q,P) + (1-a)G_F(Q',P).$$

From Eq. (2), $G_F(Q,P) \leq H_F(Q)$ and $G_F(Q',P) \leq H(Q')$; hence, $H_F(P) \leq aH_F(Q) + (1-a)H_F(Q')$.

The lemma asserts that if P is the average of two measures, $H_F(P)$ is less than the average H of the two measures. The result clearly extends to the case of averages of several measures.

A fundamental property of convex functions is that at any given point R there is a supporting hyperplane, that is, a linear function T tangent at R, such that H lies uniformly above T. For the function H, the corresponding linear function, from Lemma 2, is G. This property is illustrated in Fig. 1, for the case of a binary partition.

If H(P) does not have a corner at R (continuous first derivatives at R), then G(P,R) is unique. We will call a score rule S exact if $H_F(P)$ is strictly convex with continuous first derivatives on every binary partition F.

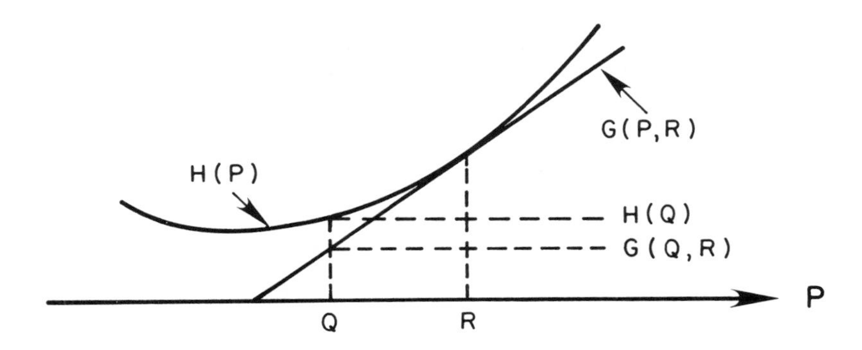

Figure 1. Graphical representation of the functions H(P) and G(P,R).

Lemma 4. If S is coherent and exact, then $S(P,e)$ is a function solely of $P(e)$.

Proof: Let $F = \{e,\bar{e}\}$. Since S is exact, $G_F(P,R)$ is unique at each point $R(e)$. Let P and P' agree on F; then $G_F(Q,P) = G_F(Q,P')$ for every Q. Since $G_F(Q,P)$ is linear in Q, the coefficients $S(P,e)$ and $S(P,\bar{e})$ must equal the coefficients $S(P',e)$ and $S(P',\bar{e})$ respectively.

Theorem 1. If S is exact, the only coherent form for S is $S(P,e) = a \log P(e) + b_e$, where a is a constant independent of e and P, and b_e is a constant dependent only on e.

Proof: From Lemma 4, $S(P,e)$ depends only on $P(e)$. The theorem then follows from a result demonstrated by Shuford, Albert, and Massengill [3].

The theorem mandates that for score rules that are coherent on an algebra of events, and that are exact, the only possible form is a linear transformation of the logarithm of the probability. Exactness appears to be a reasonable requirement for a scientific figure of merit. Thus, there is a relatively strong motivation for adopting the logarithmic score for scientific inference.

The family of scores allowed by Theorem 1 includes many that are not symmetric. Call a score rule S normal if, for any partition F, $H_F(P)$ is a minimum at the uniform distribution $\tilde{P}$, $\tilde{P}(e) = \tilde{P}(f)$ for all e and f in F.

Theorem 2. If S is coherent, exact, and normal, then S is symmetric.

Proof: If H has a minimum at $\tilde{P}$, then $G(Q,\tilde{P})$ is a horizontal hyperplane, and thus $S(\tilde{P},e) = S(\tilde{P},f)$ for all e and f. But $S(\tilde{P},f) = a \log(1/n) = b_f$; hence $b_e = b_f$.

The assumption of normality is generally perceived as appropriate for scientific purposes. It is closely related to notions such as the neutrality of nature. Combined with coherence and exactness, it essentially specifies a unique figure of merit since it is straightforward to show that two score rules that are linear transforms of each other are strategically equivalent; that is, any comparison based on one gives the same result as the corresponding comparison using the other.*

---

*However, it probably should not be overlooked that asymmetrical score rules allow the possibility of including a common "fact of life" in scientific investigations, namely, that some events are much more "interesting" than others. For example, most geologists would consider a "major" earthquake of greater significance than a "minor" one. Put in other terms, it would be considered more important to be correct in the prediction of a major earthquake than to be correct with a minor one. This has sometimes been

2. Inductive Inference

Given a figure of merit and the assurance that probabilistic conclusions are verifiable, we can turn to the problem of devising inference rules for induction. Consider the case where what is wanted as a conclusion is a probability distribution P on some set of events E, where E is a partition. One situation of relatively wide interest is that in which the distribution P is unknown but can be assigned to some class K of distributions. For example, the mean of some random variable on E may be known; or some bounds on the distribution P may be known. In the extreme case (complete ignorance), K could be the entire simplex of all possible distributions on n events.

In this situation, we can take the available knowledge—or its equivalent, the class K—as the premises, and some distribution R selected out of K as the conclusion.

Although P is unknown, we can entertain the possibility that a given P in K is the correct distribution, in which case the expectation, given that a distribution R is selected as an estimate, is G(P,R). The analyst can readily compute max(R) min(P) G(P,R).† If he selects an R^* that generates max min, then he is guaranteed to obtain at least max min whatever P may be in fact; that is, max(R) min(P) G(P,R) $\leq$ G(Q,R) for any Q in K. To this extent, R^* carries a "hard" guarantee of at least max min.

The crucial question is whether the analyst has to "settle for" max min, or whether there is a more optimistic conclusion he could adopt which at the same time is justifiable in some clear way.

One argument that has been used to suggest that the analyst should settle for max min is to invoke the so-called "game against nature" [4]. If we assume that the process which determines P is rational and has interests directly opposed to those of the analyst (constant-sum payoff), then this rational "opponent" can guarantee that the analyst cannot achieve more than min(P) max(R) G(P,R). In many cases relevant to inference problems, max min coincides with min max, and thus max min is both a floor and a ceiling to what the analyst can expect.

---

treated as a "practical" concern, rather than a theoretical one—a matter of human values rather than logic. However, once we treat logical indices as figures of merit, it is no longer clear that treating all events on the same footing is a correct representation of the "scientific spirit." Coherence is pretty much a fundamental requirement: an index would be crippled if it did not fulfill that condition. But normality is perhaps more a matter of taste. At all events, the constants b_e allow some freedom for putting in a plug for your favorite event.

†In the general case, the maxima or minima may not exist, and (for bounded G) we might have to examine sup(R) inf(P).

Critics of this argument have contended that there is no justification for the assumption that the process determining P—often personified as "Nature"—is either rational or hostile [5]. In the vernacular, there is no reason for assuming that Mother Nature is a crafty, hostile bitch. The most that can be charged is that she is neutral, she doesn't care. In this case, there is no reason for assuming min max is an upper limit to what can be hoped for.

I'm not sure I agree with the critics; the issue strikes me as being more profound. Luckily, however, there is a way of resolving the problem without recourse to anthropomorphisms like "rational" and "hostile."

One of the fundamental classical results in decision theory is the so-called positive value of information (PVI) principle. The result states that information is a good thing, or in a variant statement, that—disregarding costs—additional information is always helpful, or at worst, nonharmful. The result can be demonstrated for the case of complete information, the case in which all the relevant probabilities are known.* The principle follows directly from the convexity of H(P). The receipt of information transforms the prior distribution on a set of events of interest to a conditional distribution given the information. Prior to the receipt of information, the expected score is the average of the score for the conditional distributions over the possible items of information, which by Lemma 3 is greater than the expected score for the prior distribution [7].

The result cannot be demonstrated for the case of partial knowledge in which some or all of the relevant probabilities are unknown. However, it seems highly plausible that the principle should apply a fortiori to the case of partial knowledge. If additional information is valuable when the distributions are known, then it would appear to be even more valuable if one starts with less knowledge.

Since it is clearly important that the principle appears to be true for the case of partial knowledge, but cannot be demonstrated, the obvious tactic is to impose it as a postulate. To do so requires finding an explication for the term "additional knowledge" in the context of partial knowledge. That turns out to be difficult if complete generality is demanded. There is one special case, however, in which the notion seems to be quite clear. Suppose we have two knowledge states, one a class K of distributions, and another, K', where K' is wholely contained in K. It seems reasonable to claim that you know more if you know K' than if you know only K.

Let V(K,S) designate the expected value of a decision if you know K and if S is the score rule. V is, so to speak, the maximum reasonable expectation given you know K. Our PVI principle can then be formulated as: If $K' \subset K$, then $V(K',S) \geq V(K,S)$. It follows directly from PVI that:

---

*There are some caveats attached to the PVI principle that are often overlooked in conventional presentations, but they are "second-order" with respect to the present discussion [6].

Theorem 3. $V(K,S) \leq V(P,S)$ for every P in K, with the immediate corollary: Corollary 3.1. $V(K,S) \leq \min(K)\ V(P,S)$.

We can translate V(P,S) as H(P). We have, from Eq. (2), $\min(P)\ \max(R)\ G(P,R) = \min(P)\ H(P)$. Thus, $V(K,S) \leq \min(P)\ \max(R)\ G(P,R)$. Finally, then, we have

Theorem 4. $\max(R)\ \min(P)\ G(P,R) \leq V(K,S) \leq \min(P)\ \max(R)\ G(P,R) = \min(K)\ H(P)$.

Thus, the PVI principle imposes the same bounds on V(K,S) as the game against nature. The upper bound, min(P) H(P), is imposed by the requirement that additional information be used constructively, that is, that additional information will not be ignored on the grounds that it might be harmful.

For a given score rule S, the preceding leads to what could be called the min-score induction rule: Given a set K of potential distributions, select the distribution R^* that generates min(P) H(P). For the logarithmic score, the rule is: Select the R^* with maximum entropy.

The PVI principle does not determine a unique conclusion in case max min < min max. In general, if K is convex and closed (contains all weighted averages of members of K and contains its boundary), then max min = min max, and the rule specifies a unique conclusion. In other cases, it merely determines lower and upper bounds.

3. Min-Score Rules and Conservatism

A principal objection to min-score rules (usually directed against the game-against-nature analysis of induction) has been that the rule is ultraconservative. For example, if the score is based on the decision matrix

	e	$\bar{e}$
a	2	3
b	1	x

then the rule mandates selecting act a, whatever x is. If x were for example 10^6, it is not intuitively persuasive that it should be ignored. However, this result holds only for the case of complete ignorance concerning the probabilities on E. For the example, if it is known that $P < (10^6-3)/(10^6-2)$, then b would not be ignored.

For scientific inference, the conservatism is much less obtrusive. In the case of complete ignorance, the maximum-entropy rule calls for R^* = the uniform distribution.

A much more impressive example of the not-so-conservative nature of min-score inference is given by the aggregation of information systems. An information system (signal system, experiment, information source, inquiry system, . . .) is a process that generates information (data, messages, signals, observations, . . .) on the basis of which the conditional probabilities of events of interest (target events, hypothesis, etc.) can be established. Thus, an information system is a set $I = \{i,j,k,...\}$ of information items, a set $E = \{e,f,g,...\}$ of interesting events, and a joint probability distribution $P(e \cdot i)$ on information and events. The expected score $V(I)$ of an information system is

$$V(I) = \sum_I P(i) \sum_E P(e|i) S(i,e) ,$$

where $S(i,e)$ is shorthand for the score given that the conditional probability distribution $P(e|i)$ is asserted, and e occurs.

Suppose a set of information systems is available, $I = \{I_1,...,I_n\}$. I could, for example, be the array of information sources (radars, radio communications, visual sightings, etc.) available to the controllers at an airport, and E the pattern of air traffic in the vicinity of the airport. Or I might be the various types of input (electronic, agent reports, open sources) available to an intelligence analyst, and E some hypothesis concerning enemy behavior. If the separate sources are well known, that is, the joint distributions $P(e \cdot i_k)$ are known for each k, and the interdependencies (correlations) of the systems are known, then the calculus of probabilities determines the appropriate joint distribution $P(e \cdot i)$ of the composite system. For example, if it is known that the separate systems are completely independent, then the aggregate joint distribution can be computed. However, if the interdependencies are not known, there is no immediate way to specify the composition.

The set $\{P(e \cdot i_k)\}$ of individual joint distributions determines a set K of composite distributions, $P(e \cdot i)$, each of which is compatible with the individual distributions. If we designate by I^* the min-score composite, we can derive the result:

<u>Theorem 5.</u> $V(I^*) \leq V(I_k)$ <u>for every k.</u>

<u>Proof</u>: Any composition $I = \{I_1,...,I_n\}$ has the property $V(I) \geq V(I_k)$, every k, for the case of complete knowledge, by an obvious extension of PVI for complete knowledge. I^* is a composition, and thus, also has this property. The inductive justification adds that $V(I^*)$ is guaranteed and the best that can be obtained.

Except for trivial cases, the inequality in Theorem 5 is strict. We can diagram the inference as:

		Figure of Merit
	I_1	$V(I_1)$
Premises	⋮	⋮
	I_n	$V(I_n)$
Conclusion	I^*	$V(I^*)$

The inference is constructive; the figure of merit for the conclusion is greater than the figure of merit for any of the premises. It is, of course, not possible for a deduction to be constructive in this sense; colloquially, nothing can be truer than true. The constructive nature of min-score inference holds for any proper score rule. For the logarithmic score rule, the average entropy of the maximum-entropy composite information system is lower than the average entropy of any of the separate systems. If negentropy is taken as a measure of the amount of information in a system, then the information in the composite is greater than the information in any single component system. This, of course, is a property that one would hope to see in an inductive inference rule.

4. Comments

The min-score inference rule appears to put both induction and the maximum-entropy principle on a sound footing. The min-score rule has a built-in guarantee of the sort that appears necessary for the justification of an inference rule, and in addition is the only rule that supports the positive value of information for the case of partial knowledge.

A number of issues remain for further study. The min-score rule does not furnish a basis for asserting that the min-score probability distribution is the (for example, physically operative) probability distribution in a given information context. Rather, the most it assures is that the min-score distribution is the most reasonable estimate (or "best guess"), given the information K. Thus, the min-score rule is not sufficient to establish knowledge, in the usual sense of that phrase. This suggests that there are additional inference rules operative in the solidification of scientific theories.

Although the log score furnishes a coherent score rule for a probability measure on an algebra, it does not furnish a global figure of merit for the measure; instead it offers a collection of figures of merit (the expected scores on each partition in the algebra). For example, given an algebra of events E and a set K of probability measures on E, the min-score rule does not specify a best-estimate measure P^*. To conduct an inference, given K, it is necessary to select some partition F, find the min-score distribution P_F^* on F, and then select any P^* out of K that agrees with P_F^*. Thus, the min-score estimate may not lead to a unique measure from K, and more troublesome, P^* may depend strongly on the partition F.

For a finite algebra, one partition that can be singled out is the product partition Π, which consists of all logical products of members of E or their negations. Π consists of the "smallest" or "most specific" or "most detailed" descriptions that can be formulated in E. As is well known, if the probability distribution $P(\Pi)$ is known, then the probabilities for any event in E are determined by summation. One might be inclined to claim—again for scientific purposes—that the most appropriate partition for deriving a measure P^* is Π. Although this suggestion has a certain appeal, I must admit I have not found a persuasive argument for its adoption.

Notes and References

1. This appears to be the approach followed by the "logical" school of probability, exemplified by H. Jeffries (1939, Theory of Probability, Oxford Univ. Press), J. M. Keynes (1921, A Treatise on Probability, London: MacMillan), and R. Carnap (1950, Logical Foundations of Probability, University of Chicago Press). The distinction between probability as a physical property of systems and as a logical construct was recognized by Carnap, who labeled the logical notion "degree of confirmation."
2. Proper scoring rules are an increasingly important topic in a variety of information sciences. Associated with important early work on the subject are the names B. de Finetti, I. J. Good, L. J. Savage, E. H. Shuford, E. H. Massengill, G. W. Brier, M. Toda, and R. W. Winkler.
3. Shuford, E. H., A. Albert, and E. H. Massengill (1966) Admissible probability measurement procedures, Psychometrika, **31**, 2.
4. The game against nature as a statistical inference tool was initiated by A. Wald (1950, Statistical Decision Functions, New York: John Wiley and Sons) and pursued by D. Blackwell and M. A. Girshick (1954, Theory of Games and Statistical Decisions, New York: John Wiley and Sons).
5. The game against nature has been criticized from all sides, by objectivists, e.g., H. Reichenbach (1949, The Theory of Probability, Berkeley: Univ. of Calif. Press), subjectivists, e.g., B. de Finetti (1975, Theory of Probability, vols. I and II, New York: John Wiley and Sons), and logical probabilists, e.g., Carnap (op. cit.).
6. The caveats relate to a kind of decoupling that must be assumed between actions and events. One aspect of this decoupling is examined by I. H. LaValle (1980, On value and strategic role of information in semi-normalized decisions, Operations Research, **28** (1), 129-138). A somewhat more general treatment is pursued in N. Dalkey (to be published, Group Decision Theory, Addison-Wesley).
7. Compare N. Dalkey (1980, The aggregation of probability estimates, UCLA-ENG-CSL-8025).

TOWARD A GENERAL THEORY OF INDUCTIVE INFERENCE

John F. Cyranski

Physics Department, Clark College, Atlanta, Georgia 30314

This paper outlines an information processing scheme for using real (possibly "fuzzy") observations to predict the results of other experiments within the context of a "theory." The approach is based on the dichotomy between observer and system. It includes the explicit (geometrical) construction of "theories" from a group defining observational "degrees of freedom." Empirical evidence is also analyzed fundamentally. Both observations and predictions involve the use of a "communication channel" linking system with observer. The quantum or classical nature of the theory is apparent in terms of channel distortion. The MEP provides the inference for the appropriate channel probability. The scheme allows the inference of stochastic processes and accommodates particle physics.

Notation: The notation $\mathcal{B}(A)$ refers to the Borel algebra (generated from the topology) of the topological space A. Absolute continuity of measure μ with respect to measure λ is denoted by $\mu \ll \lambda$, and $d\mu/d\lambda$ is the Radon-Nikodym derivative. "Almost everywhere with respect to λ" is denoted by a.e.$[\lambda]$; operators are distinguished by carets (^); and $i = \sqrt{-1}$.

C. Ray Smith and W. T. Grandy, Jr. (eds.),
Maximum-Entropy and Bayesian Methods in Inverse Problems, 365–375.

1. Introduction: The Desideratum

What is a "general theory of inductive inference"? In my mind, it is a universal scheme for using real observations to make predictions about future observations. Clearly, such a scheme requires the context of a "theory" or model and should involve in a fundamental way information, since inference involves a processing of information.

The dichotomy between system and observer is the clue suggesting that the scheme involves a communication channel between "theory" and observer. We elaborate on the geometric construction of the "theory" and note that the relationship between system and observer is what distinguishes classical from quantum descriptions. Empirical evidence is associated with the basic independent observables, albeit often indirectly via (classical) Borel functions of these; also, it includes a fidelity criterion on the channel. Minimizing "cross-entropy" subject to these constraints yields the "channel" (joint probability on system-observer space). Predictions involve subsequent channel use. We conclude with an apologia.

2. The Clue

The clue to inference is the inescapable conceptual division of the universe into two disjoint parts: the system under scrutiny and the observer. This suggests that the process of measurement be viewed as a transmission of information over a communication channel. Formally, a channel is characterized by two measure spaces, $(\Omega,\mathcal{B}(\Omega))$ and $(\Lambda,\mathcal{B}(\Lambda))$, and a joint probability measure p on $\mathcal{B}(\Omega)\times\mathcal{B}(\Lambda)$ [Ref. 1]. Thus, the first step of our program must be an identification of the system and the observer as appropriate measure spaces.

The channel is further characterized by the distortion of the input space by the output. A distortion measure is a function $d:\Omega\times\Lambda\to[0,\infty]$ [Ref. 2]. Its expectation on a given channel (p) defines the fidelity of the output realization of the input. If it is impossible to code Ω into Λ in a 1-1 way, then no channel should allow perfect fidelity (zero expected distortion). This suggests that the difference between a classical and a quantum description of a system lies in the nature of the system-observer channel.

3. The System

The system is defined by the set of its "attributes." A scientific theory characterizes the system by establishing the "attributes" that are empirical and indicating how these are to be observed. Guided by physics [3,4,5], we postulate that underlying any theory is a Lie group G that specifies the "motions" that the observer can apply to distinguish the system's attributes. A Lie subgroup G_0 serves to define the (coset) manifold $X = G/G_0$ of distinct observational viewpoints. Indeed, if G is the Poincaré group and G_0 the Lorentz transformations, then X is the Minkowski space-time manifold.

At each point $x \in X$ we construct a "skeleton" upon which the description of the system will be incorporated. We may use the tangent space at x, but more generally we assume a separable Hilbert space B_x. Consistency requires that as we flit from point to point in X, we find at each x essentially the same "skeleton." That is, if $x \to x' = gx$, then there is a unitary isomorphism (preserving the inner products $\langle \, , \, \rangle_x$) from B_x to B_{gx}. This map D_g is actually defined on a big "skeleton" B that is a suitable collage of the B_x's. If $\Pi : B \to X$ is such that $\Pi^{-1}(x) = B_x$, then

$$\Pi(D_g b) = g\,\Pi(b) , \qquad b \in B,\ g \in G , \tag{1}$$

defines how the B_x's are related. This construct, a G-Hilbert space bundle, generalizes the Cartesian product

$$\underset{x \in X}{\times} B_x$$

to accommodate "twisting."

We contend that the system can be completely described by a finite set of <u>independent infinitesimal motions</u> at each $x \in X$. These motions tell us, in effect, how the system changes (at x) as we change our basic perspectives slightly. For a Lie group, its Lie algebra $A(G)$ is a linear vector space so that at any $x_0 \in X$ we can choose a <u>basis</u> $\{\tau_n^0 \in A(G),\ n = 1,2,\ldots,N\}$ to represent the basic observations at x_0.

Let $\Gamma(x)$ denote the set of all $g \in G$ for which $x = gx_0$. That is, $\Gamma(x)$ is the set of "routes" from x_0 to x. Now $\Gamma(x) = \beta^{-1}(x)$, where $\beta : G \to X$ is the canonical projection $\beta(g) \equiv gG_0$, because $x_0 = \beta(e) \equiv eG_0$ (e is the identity). Let V be the space of Borel functions $c : X \to G$ such that $\beta \cdot c(x) = x$. Clearly each $c \in V$ chooses a particular route from x_0 to x, as $c(x) \in \beta^{-1}(x)$. We may now consider how the basis $\{\tau_n^0\}$ at x_0 transforms to a new basis at x using the adjoint representation of G on the linear space $A(G)$:

$$\tau_n^0 \to \tau_n[x,c] \equiv \sum_{n'=1}^{N} \mathrm{Ad}[c(x)]_{nn'} \tau_{n'}^0 , \qquad n = 1,2,\ldots,N . \tag{2}$$

Thus, the basic measurements vary over X in a natural way depending on the section $c \in V$.

In order to derive real-valued effects of such motions, we introduce another cross-section: $\psi : X \to B$ such that $\Pi \cdot \psi(x) = x$ (that is, $\psi(x) \in B_x$). Assuming Borel sections, we note that $\langle \psi(x), \psi'(x) \rangle_x$ is a Borel function. It is a fact that there exists on $B(x)$ a quasi-invariant (σ-finite) measure. That is, if $\alpha^g(E) = \alpha[g^{-1}(E)]$, then $\alpha(E) = 0$ iff $\alpha^g(E) = 0$. It also is a fact that <u>all</u> such measures define the same null sets (a priori "impossible"). Thus, we may arbitrarily select one such α and define the space H of all sections (modulo α-equivalence) such that

$$\int_X d\alpha \langle \psi(x), \psi(x) \rangle_x < \infty .$$

This is a separable Hilbert space. The effect of our infinitesimal transformations is to take one section into another. This can be described formally in terms of the unitary representation of G over H:

$$[\hat{U}_g \psi](x) \equiv \sqrt{\frac{d\alpha}{d\alpha^{g^{-1}}}} (g^{-1}x) \; [D_g \psi] \; (g^{-1}x) , \qquad g \varepsilon G . \tag{3}$$

Thus, if $\psi \to \hat{U}_g \psi$, then $\langle [\hat{U}_g - \hat{1}] \psi(x), \psi(x) \rangle_x \div \langle \psi(x), \psi(x) \rangle_x$ defines the effect of this motion at x. Now given $\tau_n(x,c) \varepsilon A(G)$, the exponential map $\exp[a\tau_n(x,c)]$, "a" real, defines a curve in G, and $\hat{U}_{\exp[a\tau_n(x,c)]}$ is a one-parameter unitary group that represents (for small a) the "motion" $\tau_n(x,c)$. Thus, we are naturally led to define

$$T_n(x,\psi,c) \equiv \lim_{a \downarrow 0} \frac{\langle \left[\hat{U}_{\exp[a\tau_n(x,c)]} - \hat{1} \right] \psi(x), \psi(x) \rangle_x}{ia \langle \psi(x), \psi(x) \rangle_x} . \tag{4}$$

By Stone's theorem, this becomes

$$T_n(x,\psi,c) = \frac{\langle \hat{T}_n(x,c) \psi(x), \psi(x) \rangle_x}{\langle \psi(x), \psi(x) \rangle_x} , \tag{5}$$

$\psi \varepsilon H$ such that the limit in Eq. (4) exists for $\psi(x) \varepsilon B_x$, where $\hat{T}_n(x,c)$ is self-adjoint. But this can be extended naturally to a Borel function on all H [6] so that the infinitesimal motions determine N "random fields" $(X \to \{f: H \to R^1\})$ dependent on $c \varepsilon V$. Note that, although steeped in Hilbert space lore, the "theory" is essentially classical [7].

In a departure from custom, we propose that whether a "system" is deemed classical or quantum depends on the nature of the relationship between the system and the observer, as noted above. If the distortion measure $d(c,\psi,x,\mathbf{z})$ is measurable with respect to $B(V) \times B(H) \times B(X) \times B(\mathbf{R}^N)$, then the channel admits a classical relationship. For example, consider

$$d(c,\psi,x,\mathbf{z}) \equiv N^{-1} \sum_{n=1}^{N} [T_n(x,c,\psi) - z_n]^2 . \tag{6}$$

Note that the expected distortion vanishes for any channel p concentrated on $\{(c,\psi,x,\mathbf{z}) \mid T_n(x,c,\psi) = z_n, \; n = 1,2,\ldots,N\}$.

However, a "quantum" relationship is defined by a distortion function that is measurable with respect to $B(V) \times P(H) \times B(X) \times B(\mathbf{R}^N)$, where $P(H) \subset$

$B(H)$ is the family of all closed subspaces of H. In case $\hat{T}_n(x,c)$ are defined and bounded on all $\psi \varepsilon H$, the analog of Eq. (6) is

$$d(c,\psi,x,\mathbf{z}) \equiv N^{-1} \sum_{n=1}^{N} \frac{\langle [\hat{T}_n(x,c) - z_n\hat{I}]^2 \psi(x), \psi(x) \rangle_X}{\langle \psi(x), \psi(x) \rangle_X} . \qquad (7)$$

Since the terms in Eq. (7) are essentially quantum dispersions, the expected channel distortion generally cannot be made to vanish.

Once we have specified the type of system-observer relationship, we choose $D > 0$ and assert the fidelity criterion

$$E(d) = \int_{V\times H\times X\times \mathbf{R}^N} dp\; d(c,\psi,x,\mathbf{z}) < D . \qquad (8)$$

Typically we suppose that this D will be as small as possible.

4. Evidence

The observer, of course, is associated with $(\Lambda, B(\Lambda)) = (X\times\mathbf{R}^N, B(X)\times B(\mathbf{R}^N))$. In what follows it may be helpful to imagine a real scattering experiment in which a finite set of detectors, with characteristic sizes and response times, is deployed over a finite region of space and operates for a finite duration. Direct data for quantum systems are analyzed classically [8].

An experiment consists, first, of a finite partition of a subspace

$$X_0 \subseteq X:\ \Pi_X \equiv \{E_j \varepsilon B(X) \mid \bigcup_{j=1}^{J} E_j = X_0,\ E_j \cap E_{j'} = \emptyset\} .$$

This defines the precision and range of the observations. Second, one (generally) uses indirect data in the form of M Borel functions $g_m: \mathbf{R}^N \to \mathbf{R}$, with associated range and precision given (for each E_j observation) by

$$\Pi_j \equiv \{\Delta_{k(j)} \varepsilon B(\mathbf{R}^M) \mid \bigcup_{k(j)=1}^{K(j)} \Delta_{k(j)} = M_j \varepsilon B(\mathbf{R}^M);\ \Delta_{k(j)} \cap \Delta_{k(j')} = \emptyset\} .$$

This induces a partition on the range $\mathbf{R}^N$ (of the "basic" variables) via $\mathbf{g}^{-1}(\Delta_{k(j)}) \equiv \{\mathbf{z}\varepsilon\mathbf{R}^N \mid \mathbf{g}(\mathbf{z}) \varepsilon \Delta_{k(j)}\}$. Thus, the experiment finally involves the partition

$$\Pi \equiv \{E_j \times \mathbf{g}^{-1}[\Delta_{k(j)}] \mid E_j \varepsilon \Pi_X,\ \Delta_{k(j)} \varepsilon \Pi_j\} .$$

Fundamentally, an observation associated with the experiment provides a count of the number of "events" in each $E_j \times \mathbf{g}^{-1}[\Delta_{k(j)}] \varepsilon \Pi$. Associated with each $E_j \times \mathbf{g}^{-1}[\Delta_{k(j)}]$ is a membership function $\chi_{E_j \times \mathbf{g}^{-1}[\Delta_{k(j)}]}: X \times \mathbf{R}^N \to [0,1]$ that defines whether $(x, \mathbf{g}(\mathbf{z}))$ is in $E_j \times \mathbf{g}^{-1}[\Delta_{k(j)}]$ or not. The counter may not be reliable in general, so this function could be "fuzzy" [9] (that is, "subjective"). If we let p^{0B} denote the marginal of p on $\mathcal{B}(X) \times \mathcal{B}(\mathbf{R}^N)$, then the number of "events" is given by

$$N_{jk(j)} \propto \int_{X \times \mathbf{R}^N} dp^{0B} \, \chi_{E_j \times \mathbf{g}^{-1}[\Delta_{k(j)}]}(x, \mathbf{z}) \tag{9a}$$

or equivalently by

$$\rho_{jk(j)} = \frac{N_{jk(j)}}{\sum_j \sum_{k(j)} N_{jk(j)}} . \tag{9b}$$

If the counters are reliable, Eq. (9b) becomes

$$\rho_{jk(j)} = \frac{p^{0B}(E_j \times \mathbf{g}^{-1}[\Delta_{k(j)}])}{\sum_{j'} p^{0B}(E_{j'} \times \mathbf{g}^{-1}[M_{j'}])} . \tag{9'}$$

The "histogram" data defined by Eq. (9) are often summarized by averaging the functions over all (or some) of the elements of the partition. Thus we could have as (reliable) evidence

$$\langle g_m \rangle_j^{0B} \equiv \sum_{k(j)} \left[\rho_{jk(j)} \, g_m(\mathbf{z}^*_{k(j)})\right] \simeq \int_{E_j \times \mathbf{R}^N} dp^{0B} \, g_m(\mathbf{z}) , \tag{10}$$

where $g_m(\mathbf{z}^*_{k(j)})$ is a typical value of $g_m(\mathbf{z})$ in $\Delta_{jk(j)}$. Aside from this common situation, one frequently assumes data that are not actually observed. For example, one may assert certain symmetries or stationarity of the process. Verification of such hypotheses involves (in principle) infinitely precise observations on all of X (for example, stationarity requires "constancy" for all time) and is thus more a hope than an observed fact. Such hopes can be expressed as constraints on p^V, the marginal of the channel on the space of "processes," V.

5. Information

Let the "prior" measure on the channel be

$$p_0 = p_0^V \times p_0^H \times p_0^X \times p_0^{\mathbf{R}^N} . \tag{11}$$

Here (a) $p_0^{\mathbf{R}^N}$ is the Lebesgue (translation invariant) σ-finite measure on $\mathcal{B}(\mathbf{R}^N)$; (b) $p_0^X = \alpha$, a σ-finite measure on $\mathcal{B}(X)$ that is quasi-invariant under G; (c) p_0^H is a quasi-invariant σ-finite measure on $\mathcal{B}(H)$ such that its restriction to any finite dimensional subspace is equivalent to Lebesgue measure [10]; (d) p_0^V is a σ-finite measure on $\mathcal{B}(V)$. (Details about $\mathcal{B}(V)$ and measures thereon are unavailable at this time.) Note that Eq. (11) reflects well our knowledge before the channel is used (no linkage) and defines the a priori "impossible" sets. If $p \ll p_0$ is any probability (channel), then

$$H(p,p_0) \equiv c \int_{V\times H\times X\times \mathbf{R}^N} dp \log \frac{dp}{dp_0} \qquad (c > 0 \text{ and constant}) \tag{12}$$

is characterized as the unique "dispersion measure" $p \to H(p,p_0) \in \overline{R}$ that is invariant under density preserving maps (T such that $dp/dp_0 = (dp/dp_0)\cdot T$ a.e.$[p_0]$), quasi-subadditive, additive, and restrictedly continuous [11].

Inference is accomplished via the MEP: minimize Eq. (12) subject to Eqs. (8) and (9) [or (10), etc.]. If one assumes all observations are reliable, the formal solution, based on Eqs. (8) and (9), becomes (a.e.$[p_0]$)

$$\frac{dp}{dp_0}(c,\psi,x,\mathbf{z}) = e^{-sd(c,\psi,x,\mathbf{z})} \times \frac{\displaystyle\sum_j \sum_{k(j)} \rho_{jk(j)} \frac{\chi_{E_j\times \mathbf{g}^{-1}[\Delta_{k(j)}]}(x,\mathbf{z})}{\sigma_s(E_j\times \mathbf{g}^{-1}[\Delta_{k(j)}])} + \Gamma_{s,A}(c,\psi,x,\mathbf{z})}{1+\Gamma^0_{s,A}}, \tag{13}$$

where

$$\Gamma_{s,A}(c,\psi,x,\mathbf{z}) \equiv A^{-1}\left[\sum_j \chi_{E_j\times \mathbf{g}^{-1}(M_j^c)}(x,\mathbf{z}) + \chi_{X_0^c\times \mathbf{R}^N}(x,\mathbf{z})\right] e^{-sd(c,\psi,x,\mathbf{z})}, \tag{14}$$

$$\sigma_s(\Delta) \equiv \int_\Delta d(p_0^X\times p_0^{\mathbf{R}^N}) \int_{V\times H} d(p_0^V\times p_0^H)\, e^{-sd(c,\psi,x,\mathbf{z})}, \tag{15}$$

$$\Gamma^0_{s,A} \equiv \int_{V\times H\times X\times \mathbf{R}^N} dp_0 \Gamma_{s,A}(c,\psi,x,\mathbf{z}) . \tag{16}$$

Notice that if $\Gamma_{s,A} \ll 1$ a.e.$[p_0]$, Eq. (13) reduces to

$$\frac{dp}{dp_0}(c,\psi,x,\mathbf{z}) \simeq \sum_j \sum_{k(j)} \frac{\rho_{jk(j)}\, e^{-sd(c,\psi,x,\mathbf{z})}\, \chi_{E_j\times\mathbf{g}^{-1}[\Delta_{k(j)}]}(x,\mathbf{z})}{\sigma_s(E_j\times\mathbf{g}^{-1}[\Delta_{k(j)}])} . \tag{13'}$$

Here A is an arbitrary real parameter—arbitrary because the conditions in Eqs. (9) are not independent:

$$\sum_j \sum_{k(j)} \rho_{jk(j)} = 1.$$

To illustrate the scheme, let us suppose that the theory is classical with distortion (6). Then Eq. (8) can be written as

$$D = -[1+\Gamma^0_{s,A}]^{-1}\left[\sum_j \sum_{k(j)} \rho_{jk(j)} \frac{\partial}{\partial s}\{\log \sigma_s(E_j\times\Delta_{k(j)})\} - \frac{\partial}{\partial s}\Gamma^0_{s,A}\right] . \tag{8'}$$

To evaluate this, we will make a plausible (but unjustified) assertion to simplify matters:

$$\int_X dp_0^{th} \exp\left[-s\sum \frac{[T_n(x,c,\psi)-z_n]^2}{N}\right]$$

$$= k(x,\mathbf{z}) \int_{\mathbf{R}^N} d\boldsymbol{\xi} \exp\left[-s\sum \frac{(\xi_n - z_n)^2}{N}\right] , \tag{17}$$

where $k(x,\mathbf{z})$ is known and

$$\Gamma(\Delta) \equiv \int_\Delta dp_0^{OB}\, k(x,\mathbf{z}) \tag{18a}$$

$$\Gamma(X\times\mathbf{R}^N) < \infty . \tag{18b}$$

Let us finally choose A so that

$$A \equiv \sum_j \Gamma\left[E_j \times \mathbf{g}^{-1}(\,{}^c_j)\right] + \Gamma[X_0^c \times \mathbf{R}^N] . \tag{19}$$

Then, using Eq. (17), we find

$$\sigma_s(E_j \times \Delta_{k(j)}) = \left(\frac{\pi N}{2s}\right)^{N/2} \Gamma\left[E_j \times \mathbf{g}^{-1}(\Delta_{k(j)})\right] \tag{20}$$

$$D = \frac{N}{2s} \tag{21}$$

and therefore

$$\frac{dp}{dp_0}(c,\psi,x,\mathbf{z}) = \left[\prod_{n=1}^{N} \frac{e^{-[T_n(x,\psi,c)-z_n]^2/2D}}{\sqrt{\pi D}}\right]\left[1 + (\pi D)^{N/2}\right]^{-1}$$

$$\times \left[\sum_j \sum_{k(j)} \rho_{jk(j)} \frac{\chi_{E_j \times \mathbf{g}^{-1}[\Delta_{k(j)}]}(x,\mathbf{z})}{\Gamma\left[E_j \times \mathbf{g}^{-1}[\Delta_{k(j)}]\right]} + \frac{\sum_j \chi_{E_j \times \mathbf{g}^{-1}(M_j^c)}(x,\mathbf{z}) + \chi_{X_0^c \times \mathbf{R}^N}(x,\mathbf{z})}{\sum_j \Gamma\left[E_j \times \mathbf{g}^{-1}(M_j^c)\right] + \Gamma\left[X_0^c \times \mathbf{R}^N\right]}\right]. \tag{22}$$

6. Prediction

From Eq. (13) we obtain p^{th} on $B(V) \times B(H)$ by integration over p_0^{0B}. Predictions may be made for a <u>different</u> channel link d^* and/or different fidelity bounds $D^* > 0$. One applies the MEP with p^{th} fixed and $E(d^*) \leq D^*$. This problem is well known in rate distortion theory [2,12]. If $d^* = d$ and $D^* \geq D$, the original channel can be used.

Once the (new) channel is determined, one applies p^{0B*}, the marginal of p^* on $B(X) \times B(\mathbf{R}^N)$, to calculate the predicted frequencies $\rho^*_{jk(j)}$ on the future experiment Π^*.

7. Apologia

In outline we have presented a fairly general theory of inductive inference. The theorist a priori supplies the groups G and G_0, fixes x_0, B_{x_0}, and a basis set of infinitesimal motions at x_0, and chooses the relevant distortion measure and fidelity criterion. He may also restrict the space of admissible "processes" (constrain V). The experimentalist provides the partition and data, and the statistician solves the MEP to make predictions.

In defense of the hair-raising mathematics, let me note that typically when we seek answers to fundamental questions (What is matter? Why do apples fall downward? What causes weather? etc.) we require difficult and esoteric methods. The construction of "theories" is based on a naive geometric idea, but in fact incorporates (and accommodates) the deepest theories of physics [3-5,13]. The general problem of "inverse scattering" is in fact a rather special case (linked to inference of $c \in V$ for the Hamiltonian field, the time-generator, with other space-time fields fixed). Yet, this case includes the description of unstable particles (via scattering) [14]. In effect we infer the "random potentials" consistent with exponential decay, and so forth [15].

Note also that the MEP evidence input has been extended to account directly for precision and domain limitations of the experiment. We also see how a classical analysis of quantum mechanical scattering is a natural result of this scheme. The fact that "fuzzy" data can be employed suggests application to social science (psychology, etc.) where observations involve subjectivity.

The philosophy underlying this program can be applied to the many problems of interpretation suffered by quantum mechanics [16]. Indeed, the "molecular structure" controversy [17] can be resolved once one admits that neither "state" nor "Hamiltonian" are properties intrinsic to the molecule. We have already noted that our construction does not associate such variables as position, angle, time, etc., with the system. (These are observer parameters!) Thus we avoid the questionable practice of defining position [18], angle, time [19], etc., operators in quantum mechanics.

In short, the scope of application justifies the investment in high-level mathematics. Nevertheless, the program is still not completely general. Basic results from "measurement theory" indicate that the appropriate structure at each $x \in X$ should be a complete orthocomplemented lattice (which defines the system directly via its basic "attributes") [20]. The resultant "bundle" structure must be described by abstract "informations" that generalize probability [20]. Preliminary results suggest a possible application to modeling language translation [21].

In conclusion, I believe my scheme is a viable step toward a general theory of inductive inference. Yet, despite the 34 years since information measures and their applications began to be discovered [22], much remains to be done.

Acknowledgments

Major influences on my thinking have been E. T. Jaynes, A. Shimony, N. S. Tzannes, and C. A. Nicolaides. I wish also to acknowledge helpful discussions with W. Hammel and financial assistance from Hobart and William Smith Colleges.

References

1. R. G. Gallager (1968) Information Theory and Reliable Communication (New York: Wiley).
2. T. Berger (1971) Rate Distortion Theory (Englewood Cliffs, N.J.: Prentice-Hall).
3. R. Hermann (1966) Lie Groups for Physicists (Reading, Mass.: Benjamin/Cummings).
4. V. S. Varadarajan (1970) Geometry of Quantum Theory (II) (New York: Van Nostrand Reinhold).
5. G. W. Mackey (1976) The Theory of Unitary Group Representations (Chicago: University of Chicago Press).
6. J. F. Cyranski (1982) J. Math. Phys. **23** (6), 1074-1077.
7. The classical family of all "random fields" $X \rightarrow \{f:H \rightarrow R^1\}$ certainly includes those parametrized by V. Whether or not the classical notion is vaster is immaterial, as our procedure defines the natural fields that can be geometrically described. See C. DeWitt-Morette and K. D. Elworthy (1981) Phys. Rep. **77** (3), 125-167.
8. See, for example, G. E. Chamerlain, S. R. Mielczarek, and C. E. Kuyatt (1970) Phys. Rev. A **2** (5), 1905-1922.
9. L. A. Zadeh (1965) Inf. Control **8**, 338-353.
10. A. V. Skorohod (1974) Integration in Hilbert Space (New York: Springer-Verlag).
11. W. Ochs (1976) Rep. Math. Phys. **9** (3), 331-354.
12. J. F. Cyranski (1981) Inf. Sci. **24**, 217-227.
13. S. K. Bose (1981) Phys. Rev. D **24** (8), 2153-2159.
14. R. Newton (1966) Scattering Theory of Waves and Particles (New York: McGraw-Hill).
15. L. Fonda (1976) A critical discussion on the decay of quantum unstable systems, Trieste Preprint IC/76/20.
16. J. Rayski (1973) Found. Phys. **3**, 89; (1977) **7**, 151; (1979) **9**, 217.
17. R. G. Wooley (1980) Israel J. Chem. **19**, 30-46.
18. H. Laue (1977) Found. Phys. **8** (1/2), 1-30.
19. Angle and time operators have no satisfactory self-adjoint realization in quantum theory. For angle variables in quantum mechanics, see P. Carruthers and M. M. Nieto (1968) Rev. Mod. Phys. **40** (2), 411-440. For time operators, see M. Jammer (1974) The Philosophy of Quantum Mechanics (New York: Wiley).
20. J. F. Cyranski (1981) J. Math. Phys. **22** (7), 1467-1478.
21. J. F. Cyranski (1980) pp. 3-10 in D. G. Lainiotis and N. S. Tzannes, eds., Advances in Communications (Dordrecht: Reidel).
22. C. Shannon (1948) Bell Syst. Tech. J. **27**, 379-423, 623-656.

GENERALIZED SCATTERING

E. T. Jaynes

Visiting/Adjunct Professor, 1982-1983, Department of Physics, University of Wyoming, Laramie; Permanent Address: Department of Physics, Washington University, St. Louis, Missouri 63130

C. Ray Smith and W. T. Grandy, Jr. (eds.),
Maximum-Entropy and Bayesian Methods in Inverse Problems, 377–398.

1. Historical Background

In much of maximum-entropy inference (MAXENT) and its ancestor, Gibbsian thermodynamics, we are concerned with a single solution, that is, imposing one set of constraints and examining the resulting distribution and predictions. However, Gibbs (1875) had already shown, in thermodynamics, that new and important facts appear when we consider the relationship between two different solutions. In this paper we show that the same is true for the MAXENT generalization. Indeed, much of conventional physics, and for that matter, conventional wisdom, is contained in a general theorem relating two different MAXENT predictions, before and after adding a new constraint.

Gibbs (1875) gave two relations connecting neighboring thermal equilibrium states. The linear one is the familiar

$$T\delta S - \delta U - P\delta V + \Sigma\, \mu_i \delta n_i = 0\,, \tag{1}$$

where we use the conventional symbols for temperature, entropy, energy, pressure, volume, chemical potentials, and mole numbers, respectively. This "Gibbs relation" is in constant use in chemical thermodynamics.

The less familiar but more instructive quadratic relation (Gibbs, 1875, Eq. 171)

$$\delta T\delta S - \delta P\delta V + \Sigma\, \delta\mu_i \delta n_i \geq 0 \tag{2}$$

expresses a basic convexity property, from which Gibbs derived all his conditions for stability. But this convexity may fail at certain critical points, and then we have some kind of "catastrophe"—a phase transition, bifurcation, or other instability.

(We remark parenthetically that René Thom's modern catastrophe theory may be given an alternative mathematical form in terms of convexity of an entropylike function. This was anticipated by Gibbs, whose first published work (1873) determined conditions for thermodynamic stability as a geometrical convexity property of the entropy, the condition for coexistence of two phases in equilibrium being a local nonconvexity that makes it possible for a supporting tangent plane to make contact with the entropy surface at two points. Gibbs' choice of variables has the advantage of avoiding multiple-valued functions; a catastrophe is explained in terms of dimples in a single-valued function instead of folds in a multiple-valued one.)

These relations were retained implicitly in Gibbs' final work (1902), but the work was left unfinished and they were not emphasized. Few readers since then have been aware that the Gibbs "canonical ensemble" formalism contains such convexity relations as Eq. (2).

These properties are still present in the modern MAXENT formalism, but now they apply to problems of inference in general. Jaynes (1980) has hinted rather cryptically at some of these relationships; in the present talk we have time to develop only one example, but shall try to do it explicitly.

2. Inferential Scattering

The usual scattering theory of physics is also concerned with a relationship between two solutions, rather than with a single solution. In physical scattering, a wave field is modified by an obstacle that imposes new constraints (for example, new boundary conditions) on the field. By the "scattered" wave we mean the difference between the two solutions.

By analogy we may define "inferential scattering," in which an inference is modified by new information that imposes new constraints on the entropy maximization. The resulting change in our predictions is a kind of "scattering" off the new information. Highly relevant information is information that scatters strongly.

But inferential scattering is a more general phenomenon than physical scattering; it need not involve an influence traveling in physical space and time but may, for example, take place in a more abstract thermodynamic state space (or, just to illustrate the range of possible applications, in a study of political stability—in a space whose coordinates include the popularity of the leader, the resources of the opposition, and the amount of foreign investment).

Generally, new information about any quantity, A, will change our predictions of any other quantity, B, that was correlated with A in the original MAXENT distribution. But there is an old adage in statistics: "Correlation does not imply causation!"

In particular, when inferential scattering does take place in physical space and time, it need not be "causal" in the physicist's sense of that word. That is, while physical scattering proceeds only forward in time (a perturbation at time t affecting the later state but not the earlier one), inferential scattering runs equally well forward or backward. New information about the present can change our estimates of the past as well as the future; such "backward traveling" inferences are essential for geology.

(As another parenthetic remark, our present quantum field theory (QFT) expresses a kind of mixture of principles of physics and principles of inference; but the latest mathematical formulations of QFT look remarkably like the most general (functional integral) MAXENT formalism. The Feynman propagators, with parts that seem to run backward in time, have been puzzling conceptually—but perhaps they may be understood eventually in this way: what is traveling backward is not a physical influence, but only an inference.)

Indeed, in purely classical physics the "causality" by which new interactions influence the future, but not the past, appears in our equations only because of the unsymmetrical information we have put into them. In specifying definite initial conditions without making any allowance for uncertainty about them, we are in effect claiming exact information about the past—so firmly established that new information about the present cannot change it.

But if we were more honest and admitted some uncertainty about the initial state (representing it by an ensemble of possible past states), then new information about what is happening now would, obviously, change our estimates of what had happened in the past, as well as what will happen in the future. Viewed in this way, we see that backward-traveling terms in the equations of physics are not necessarily paradoxical; indeed, in statistical mechanics they seem natural and necessary.

We believe, as firmly as anyone else, that "you can't change the past." But you can improve your knowledge of the past; that is the goal of virtually everything that is done under the label of "education."

On the other hand, if our inference is about a causal wave process taking place in space and time with the antecedent state specified, and if the new information tells us of a scattering obstacle, we might expect some relationship between the change in the MAXENT prediction of the wave function and conventional physical scattering theory.

Our original aim here was only to investigate this for the particular case of Rayleigh acoustical scattering. However, it developed that we are running into other deep conceptual problems that have plagued statistical mechanics for two generations and that, indeed, lie behind many of the attacks on the principle of maximum entropy itself. These problems must be cleared up first. Otherwise what we are about to do will seem incomprehensible or worse to those with conventional statistical training.

Operationally, what we are going to do when given new information is simply to remaximize the entropy, subject to the new constraints. But this evokes howls of protest from some, who say, "This procedure cannot succeed because you are ignoring the dynamics. Merely remaximizing the entropy is a completely arbitrary procedure, and one can have no reliance at all in the results. I would as soon trust the predictions of a crystal-ball gazer."

This is the attitude we have to answer first. We have to explain in clear, physical terms why remaximizing the entropy does, after all, lead us to reliable predictions of reproducible phenomena (which are, we hope, the only ones experimentalists are recording for publication in our scientific journals).

Therefore, we must take a rather long detour to deal with these conceptual problems, which will occupy the next six sections. Then we develop the general inferential scattering formulas in Section 9, and return finally, in Section 10, to the MAXENT version of Rayleigh scattering.

3. Generalized First Law

First, let us note some MAXENT relations reminiscent of the linear Gibbs relations. Some physical quantity A is capable of taking on the values $\{A_1, A_2, \ldots, A_n\}$, where the indices may refer to quantum states, but we need not commit ourselves to any particular meaning. It is enough that we can assign corresponding probabilities $\{p_1, \ldots, p_n\}$. Thus the expectation of A is

$$\langle A \rangle = \sum_{i=1}^{n} p_i A_i \ . \qquad (3)$$

A small change in the problem might involve independent changes in the possible values $\{A_i\}$ and in the assigned probabilities $\{p_i\}$. The change in expectation is then

$$\delta\langle A \rangle = \Sigma\, p_i \delta A_i + \Sigma\, \delta p_i A_i \ . \qquad (4)$$

But we recognize the first sum as the expected change in A: $\langle \delta A \rangle = \Sigma\, p_i \delta A_i$. Therefore we can write Eq. (4) in the form

$$\delta\langle A \rangle - \langle \delta A \rangle = \delta Q_A \ , \qquad (5)$$

where $\delta Q_A \equiv \Sigma\, \delta p_i A_i$.

We call Eq. (5) a generalized first law, for the following reason. Suppose $A = E$ is the energy of a system, E_i its value in the ith quantum state. Then $\langle E \rangle$ is the predicted thermodynamic energy function U. On a small change of state (caused, for example, by a change in volume, magnetic field, etc.) the work done on the system will be δE_i if it is in the ith state; thus $\delta W = - \langle \delta E \rangle$ is the predicted work done by the system. Equation (5) then has the form

$$\delta U + \delta W = \delta Q_E \qquad (6)$$

and since δU and δW are unambiguously identified, δQ_E is identified as representing heat.

The first law of thermodynamics (or at least a relation that in textbooks is often called the "first law") is seen here as a special case of a general rule: A small change in the predicted value of any quantity—whatever its physical meaning—may be resolved into parts arising from a change in the distribution (the "generalized heat") and from a change in the physical quantity (the "generalized work"). And of course, this holds for any small change in the ensemble, however it was specified.

Evidently, the change in information entropy, $-\Sigma\, p_i \log p_i$, can arise only from the components δQ_A and not from the δW_A. Thus, facts that were first unearthed by lifetimes of effort now correspond to a mathematical relation so trivial that it would pass unnoticed if not pointed out.

But it is just because of its mathematical triviality that we need to stress the physical importance of Eq. (5), which has nontrivial implications. In phenomenological thermodynamics the first law relation $dQ = dU + dW$ is, of course, used with complete freedom: given any two of these quantities, the third is determined. It has not always been recognized that we have the same freedom in the statistical theory for any quantity, whether conserved or not, and for any situation, equilibrium or nonequilibrium. In any of these circumstances, there are three basically different means by which our knowledge about a system might change:

(I) Measurements on the system may show a macroscopic change in pressure, magnetization, etc. Thus a term of the form $\delta\langle A\rangle$ is known.

(II) We may know from theory that a physical quantity A has changed—for example, according to the equations of motion, or because we have varied an external parameter such as magnetic field—or we may know that A has not changed. Thus a quantity of the form $\langle\delta A\rangle$ is known.

(III) We may know from measurements on another system coupled to the one of interest that a flux of heat, charge, particles, angular momentum, etc. has taken place. As we shall see presently, this means that a "source term" of the form δQ_A is known.

The content of Eq. (5) is that, given any two of these pieces of information, the third is also known. But the notion of "macroscopic sources" δQ_A has also given rise to conceptual difficulties that have retarded development of statistical mechanics for many years, and we need to clean up some of this unfinished old business before turning to the new.

4. The Basic Dilemma

Suppose we put a pot of water on an electric stove and turn on the burner. How shall we account for the heating of the water in terms of statistical mechanics?

Whether we use quantum theory or classical theory will not matter for the point to be made here. We use the quantum-mechanical notation because it is more concise, but wherever our quantum-mechanical density matrix ρ appears, we may equally well think of it as a classical probability distribution over coordinates and momenta, and our Schrödinger equation of motion, Eq. (8), below, as the classical Newtonian equations of motion.

It is usually taken as axiomatic, in either quantum or classical statistical mechanics, that our probability distributions must evolve in time according to the equations of motion. As a result, as noted in the review article of Zwanzig (1965), "thermal driving" was long an awkward topic, workers trying constantly to replace a heat source δQ with some kind of dynamical driving that would have similar effects. In principle, of course, this is quite correct because the real process is indeed dynamical. Let us see what this would entail.

The process takes place by converting the energy carried by two macroscopic coordinates (voltage and current supplied by the electric company) into excitation of an increased atomic/molecular motion in an enormous number of microscopic coordinates, and transferring that increased motion down a chain of interacting atoms of the burner and pot, to the water molecules.

To describe this process according to statistical mechanics as usually taught, we should introduce all the microscopic coordinates of water, pot, and burner and their interactions, giving a grand total Hamiltonian:

$$H_{tot} = H_{water} + H_{pot} + H_{burner} + H_{interactions} . \qquad (7)$$

Then we should put the applied voltage and current into a given "externally applied" Hamiltonian $H_{ext}(t)$, which is added to H_{tot} whenever the switch is turned on. Then we should solve the Schrödinger equation of motion

$$i\hbar\dot{\rho} = [H_{tot} + H_{ext}(t), \rho] \qquad (8)$$

with an initial density matrix $\rho(0)$ given, if the water is initially in thermal equilibrium, by a canonical distribution at the initial water temperature T_i:

$$\rho(0) = Z^{-1} \exp(-H_{tot}/kT_i) , \qquad (9)$$

where $Z \equiv Tr[\exp(-H_{tot}/kT_i)]$ is inserted for normalization, $Tr(\rho) = 1$.

One would expect, naively, that the solution of Eq. (8) after we turn off the switch should tend to a final density matrix that is again canonical,

$$\rho(t) \rightarrow \rho_c \equiv Z_f^{-1} \exp(-H_{tot}/kT_f) , \qquad (10)$$

representing the final higher temperture T_f of the water.

The fact that we cannot actually do this calculation (the density matrix has more rows and columns than the number of microseconds in the age of the universe) is the least of our worries. Even to think about doing it gets us into paradoxes that are not resolved in even the latest textbooks on statistical mechanics. (See, for example, the work of Akhiezer and Peletminskii (1981) and my review of it (Jaynes, 1982).)

It is a theorem that, under the equations of motion (8), each individual eigenvalue of $\rho(t)$ is constant. This has two disturbing consequences: (a) the quantity $-Tr(\rho \log \rho)$, usually interpreted as the thermodynamic entropy, cannot change, and (b) since the eigenvalues of ρ_c in Eq. (10) are different from those of $\rho(t)$, the density matrix at a later time $t > 0$ can never become canonical at a higher temperature!

Yet we know, as about the most familiar experimental fact in thermodynamics, that the water temperature rises by an amount that we can predict correctly without knowing a thing about those microscopic coordinates and interaction forces. Instead of all those billions of microscopic details, we need in practice only two macroscopic numbers: the total energy supplied, and the total heat capacity of the system.

So here is the basic dilemma of conventional statistical mechanics: If we deny the validity of the $\rho(t)$, evolved from the dynamics according to Eq. (8), we are denying that the system obeys the Schrödinger equation. If we deny the validity of the Gibbsian canonical ρ_c in Eq. (10), we are denying experimental facts. Yet it is a theorem that $\rho(t)$ and ρ_C are incompatible, in the sense that $\rho(t)$ can never become equal to ρ_C.

Each writer must find his own way around this circumstance, and we are not surprised to find that no two writers have done it in the same way. Surely it ought to be considered a major scandal that statistical mechanics, as usually taught, is helpless to account for the most familiar of all thermodynamic experiments.

Still, we must not be too harsh. Conventional statistical mechanics is very far from being a failure. Throughout this century it has been the vehicle for some of the most important advances in physics. But to acknowledge this only deepens the mystery; why does the theory succeed brilliantly on many recondite problems, only to stumble on such a simple one?

To understand this, we need to examine more closely: on which type of problems have the methods typified by Eqs. (8) and (10) been successful?

For predicting the behavior of a system initially in thermal equilibrium, when an external perturbation takes it into a nonequilibrium state, we have full confidence in the dynamically evolved $\rho(t)$ generated by Eqs. (8) and (9). For example, it gives all the intricate details of multiple spin echoes (Slichter, 1978). Indeed, our confidence in $\rho(t)$ is so great that discovery of a single case where it can be proved to fail would shake the foundations of physics and merit a dozen Nobel Prizes.

But we have an almost equal confidence in the Gibbsian ρ_C of Eq. (10). In every case where one has succeeded in doing both the calculations and the experiments, it has led us to quantitatively correct predictions of equilibrium properties. The intricate details of ortho- and para-hydrogen provide an impressive example of this success.

In short, ρ_C has never failed us for the case of thermal equilibrium; and $\rho(t)$ has never failed us for small departures from thermal equilibrium. We do not expect either to fail us here.

The dilemma appeared only because workers had expected $\rho(t)$ to predict final equilibrium in the same way that ρ_C does. That is, we were making an unconscious assumption, rather like that of absolute simultaneity in pre-relativity physics. There is a paradox only if we suppose that a density matrix (a probability distribution) is something "physically real" and "absolute."

But now the dilemma disappears when we recognize the "relativity principle" for probabilities. A density matrix (or, in classical physics, a probability distribution over coordinates and momenta) represents, not a physical situation, but only a certain state of knowledge about a range of possible physical situations.

The results $\rho(t)$ and ρ_C are both "correct" for the two different problems that they solve. They represent different states of knowledge about the final condition of the water; but that does not mean they make different predictions of the observable properties of the hot water.

Thus in Eq. (8) the constancy of

$$S_I = -k\,\mathrm{Tr}(\rho \log \rho) \qquad (11)$$

ceases to be paradoxical as soon as we recognize that S_I is not in general the same as the phenomenological thermodynamic entropy. It is rather the information entropy, essentially (by Boltzmann's $S = k \log W$) the logarithm of the number W of "reasonably probable" quantum states in whatever ensemble we may have before us, however defined. In the MAXENT principle, S_I is the thing we maximize to define our initial density matrix.

On the other hand, the phenomenological entropy S_E of the experimenter is by construction a function $S_E(P,T,M,...)$ of the experimentally observed macroscopic quantities (P,T,M,...). The relationship between these entropies has been demonstrated before (Jaynes, 1963, 1965). Only after it has been maximized subject to the constraints of the experimenter's data does S_I become equal to S_E.

Therefore, as we have stressed before, the constancy of Eq. (11) under the equations of motion, far from presenting a paradox for the second law, is precisely the dynamical property we need to demonstrate that law, in the Clausius adiabatic form $S_{final} \geq S_{initial}$.

Given any ensemble ρ, to ask "What information is contained in this ensemble?" is the same as asking "With respect to which constraints does this ensemble have maximum S_I?" We can answer this at once for both $\rho(t)$ and ρ_C.

For $\rho(t)$ the initial density matrix (9) has maximum S_I for prescribed initial energy E_i of the cold water. We call this maximum S_i. The multiplicity $W_i = \exp(S_i/k)$ is therefore essentially the number of quantum states that have energy near E_i ("near" meaning within the range of thermal fluctuations).

For all practical purposes, we could think of the density matrix $\rho(t)$ as assigning uniform probabilities to these W_i states, zero probability to all others; this corresponds to the "asymptotic equipartition theorem" of information theory.

The dynamical evolution (8) induces a unitary transformation of $\rho(t)$ that does not lose any information, and therefore always defines a "high-probability set" containing the same number W_i of states, each being the time development of one of the initial states.

But the dynamically evolved $\rho(t)$ at later times $t > 0$ would indicate, by an increase in the predicted energy $\langle H_{water} \rangle$, that the water is being heated. Although we cannot actually carry out the calculation (8), we believe with absolute confidence that it would yield the correct final energy E_f of the hot water. (At least, anyone who can disprove this will be a cinch for one of those Nobel prizes.)

In contrast, to determine the canonically assigned final density matrix ρ_C in Eq. (10) we need no microscopic details, only the amount of heat δQ delivered to the water. The result has maximum S_I for the prescribed final energy $E_f = E_i + \delta Q$, and its high-probability region of phase space would contain about $W_f = \exp(S_f/k)$ states, the number that are "near" E_f.

The two ensembles $\rho(t)$ and ρ_C agree on the value of E_f. In what way are they different?

The difference is that the calculation (8) would tell us much more than the final energy E_f of the water. It would also indicate, out of all the W_f quantum states that have energy near E_f, a small subspace of only W_i states. These are the particular states that could have arisen from the exact history (initial temperature and details of heating) by which that final state was reached.

Now all our experience tells us that the reproducible properties of hot water depend only on its present temperature, and not on the details of the particular history by which it got to that temperature. Therefore, although the calculation (8) of $\rho(t)$ is not in any way "wrong" for this problem, it is inefficient. It requires us to calculate some microscopic details that are irrelevant to our purpose.

Let us get some idea of how much more detail is contained in the dynamically evolved $\rho(t)$ than in the canonically assigned ρ_c.

5. Those Numbers

Every morning I heat about a quart, or 2 x 453/18 = 50 moles, of water to the boiling point to make coffee. The molar heat capacity of water is about 9R, where $R = 6 \times 10^{23}k$ is the gas constant and k is Boltzmann's constant. So the water absorbs about $\delta Q = 50 \times (373-293) \times 9R = 72$ kilocalories of heat, and its entropy increases by about

$$S_f - S_i = 50 \times 9R \log(373/293) = 6.5 \times 10^{25}k \ . \qquad (12)$$

Therefore, the ratio of the number of states in the two density matrices is about

$$W_f/W_i = \exp[(S_f - S_i)/k] = \exp(6.5 \times 10^{25}) \ . \qquad (13)$$

By contrast, the number of microseconds in the age of the universe is only about $10^{24} = \exp(55)$. Had I heated only a cubic millimeter of water through one degree C, the ratio would still be about $\exp(10^{18})$.

The appearance of such numbers in statistical mechanics was noted long ago by both Boltzmann and Planck, and it was stressed in the textbook of Mayer and Mayer (1940). For reasons we cannot explain, these numbers seldom appear in modern works; yet it is essential to know about them in doing practical calculations.

Thanks to these numbers, an experienced practitioner of the art can get away with approximations that would appear horrendously bad to one not in on the secret. For example, if $W = \exp(10^{25})$, then if we make an error by a factor of 10^{100} in the calculation of W, this leads to an error of only 1 part in 10^{23} in the value of log W.

Planck called this phenomenon "the insensitivity of the thermodynamic functions." In our present problem it means that in setting up ρ_c we need not bother with specifying the exact width $\delta E = (\langle E^2\rangle - \langle E\rangle^2)^{1/2}$ of that range of thermal energy fluctuations within which we are counting the number of states W_f.

In fact, for a one-mole system, δE is of the order of $kT\sqrt{n}$, where n is the number of "effective degrees of freedom" of the system, about 10^{24}. But if we took δE as a million times too large or too small, it would have an absolutely negligible effect on our calculation of the experimental entropy $S_E = k \log W_f$, and therefore the heat capacity and equation of state, of that hot water. For remaximizing the information entropy S_I, a single constraint on $\langle E \rangle$ (which already implies about the right δE) will suffice to accomplish all that we could get by using two constraints, specifying also $\langle E^2 \rangle$ and therefore δE.

6. So Why Does MAXENT Work?

We have just seen, in this water heating episode, that although $\rho(t)$ and ρ_C predict the same energy for the hot water, the ratio W_f/W_i, or

$$\frac{\text{number of high-probability states in } \rho_C}{\text{number in } \rho(t)} ,$$

is fantastically large; in other words, $\rho(t)$ contains enormously more information about the state of the hot water than does ρ_C. This makes the entropy remaximization that leads to ρ_C appear, if anything, even more precarious than crystal-ball gazing.

Yet the experimental fact is that ρ_C works, yielding the correct predictions of observable properties of that hot water by a calculation that, while not exactly trivial, is simpler by many orders of magnitude than that for $\rho(t)$. This is the fact that is not understood in the conventional statistical mechanics of our textbooks. But if we can learn how to understand it, we shall see why MAXENT works in much more general situations.

Under closer examination we see that the useful predictions we can make from ρ_C are not greatly different from those we could make if we had the greater information contained in $\rho(t)$. Indeed, if we are interested in predicting only reproducible effects, we expect no difference at all in their predictions. For when we repeat the experiment, we do not repeat all the microscopic details that were assumed known in $\rho(t)$: The pot is never put on the stove twice in exactly the same position to atomic accuracy, and it is never filled twice with exactly the same number of water molecules. Therefore H_{tot} is never the same twice. The water is never in exactly the same state of turbulent motion at the instant the switch is turned on; therefore $\rho(0)$ is never the same twice. The switch is never turned on at exactly the same point in the AC cycle, and the electric company never supplies exactly the same voltage and current; therefore the unitary transformation of the equations of motion generated by Eq. (8) is never the same twice.

In short, there would be an entirely different $\rho(t)$ for every repetition of the experiment. Indeed, in view of the smallness of the high-probability sets W_i compared to W_f, we could repeat this water heating every day for millions of times the age of the universe, with almost no chance that any specific quantum state would appear in the W_i set on two different days. On repetitions of the experiment, the tiny sets W_i would be scattered about at random, like stars in the sky, within the MAXENT set W_f.

How then could the effect of the heating δQ be reproducible? Evidently, it must be true that all those intricate details contained in $\rho(t)$, which determine a particular set W_i, are irrelevant for predicting reproducible effects of the heating δQ. We could as well have used the big set W_f, which is the union of all the little sets W_i.

At this point, we finally see why remaximizing entropy is superior to crystal-ball gazing. While it does not take into account all those billions of microscopic details that do not matter and that we never possess anyway, it does take into account all the information that is actually relevant for predicting reproducible phenomena.

In the laboratory, a reproducible result can depend only on properties of the microstate that are the same on successive repetitions of the experiment; in the cases we are considering, the only such constant thing is the source strength δQ_A itself. We expect, then, that in the theory, information about that source strength should suffice to predict any reproducible effects that are caused by it. That is, any such effect should be predictable from the ρ_c that incorporates the information about that source strength.

This concludes our rather lengthy sermon. Now let's get back to the constructive development of the mathematics that realizes this program in real situations, and see whether it actually works as just supposed.

7. Macroscopic Predictions

What macroscopic effects may result from operation of a source δQ_A? In general this will cause internal readjustments, in the course of which some other quantity, B, may be changed. Supposing δQ_A to be so small that the ensemble is only slightly modified, the amount $\delta\langle B\rangle$ of that change is given by the general variational property of neighboring canonical ensembles, given by Gibbs. However, as Mitchell (1967) showed, the answer can be reasoned out heuristically but more generally, without invoking canonical distributions.

A really careful exposition would have to discuss a number of technical qualifications on the following, but lacking the time and space for it, we ask the reader's indulgence for our aim of expounding only the essential ideas. We believe that anyone who perceives the need for some qualifications here and there will also see how to supply them for himself.

If in the original ensemble ρ_0 the quantities A and B are positively correlated, that is, have a positive covariance

$$K_{AB} = [\langle AB\rangle - \langle A\rangle\langle B\rangle] > 0 , \tag{14}$$

then in the high-probability set (HPS) of W_0 states picked out by ρ_0, microstates of higher than average A tend to be also states of higher than average B; and vice versa. Evidently, if we now learn that $\delta Q_A > 0$, the new ensemble ρ_1 with remaximized information entropy will assign higher probability to states of high A. Clearly, this will lead us to expect that B has also increased although it may not be obvious by how much.

But now, consider another quantity C. If it is uncorrelated with A in the initial ensemble, $K_{CA} = 0$, then in the HPS there is no tendency for states of high A to have either higher or lower than average C. Then knowing δQ_A gives us no reason to expect that C has increased rather than decreased, and our prediction of C should be unchanged: $\delta\langle C\rangle = 0$.

This holds for any quantity C that is uncorrelated with A. Therefore, make the choice $C = B - xA$, where x is any fixed number. Then $K_{CA} = K_{BA} - x\, K_{AA}$, which vanishes if $x = K_{BA}/K_{AA}$, and then $\delta\langle C\rangle = \delta\langle B\rangle - x\delta\langle A\rangle = 0$.

So we have the rule: when a small microscopic source δQ_A operates and thereby affects any other quantity B internally, the predicted change in B is

$$\delta\langle B\rangle = (K_{BA}/K_{AA})\, \delta Q_A \, . \tag{15}$$

This agrees with a more rigorous perturbation treatment of canonical ensembles, but also holds more generally. Indeed, much of the theory of regression in statistics textbooks is based on a result formally identical with Eq. (15) although differently interpreted.

Now let us indicate the MAXENT distribution more explicitly. We have a number of physical quantities $(A_1,\ldots,A_m)$ and associated Lagrange multipliers, or "potentials" $(\lambda_1,\ldots,\lambda_m)$. For brevity, write their inner product as

$$\boldsymbol{\lambda}\cdot\mathbf{A} \equiv \sum_{i=1}^{m} \lambda_i A_i \, . \tag{16}$$

As written, this form includes all those treated by Gibbs. But now these quantities might depend on time and/or position, and the quantity A_k may have a "source region" consisting of some space-time domain R_k.

The covariance of two quantities A,B is a function of whatever parameters are in A,B, so we may have a space-time covariance function $K_{AB}(x,t;x',t')$ that now begins to resemble a Green function of physics. With such space-time dependences, the partition function and entropy functions of Gibbs become promoted to functionals (Jaynes, 1980), and the MAXENT formalism strongly resembles that of quantum field theory. But for present simpler purposes, we may accomplish nearly the same thing while retaining a Gibbs-like form (16) of our equations, by defining our physical quantities to be localized to small space-time regions.

Our partition function is then

$$Z(\lambda_1,\ldots,\lambda_m) = \mathrm{Tr}\exp(-\boldsymbol{\lambda}\cdot\mathbf{A}) \, , \tag{17}$$

and the MAXENT density matrix is

$$\rho_0 = Z^{-1} \exp(-\boldsymbol{\lambda} \cdot \mathbf{A}) \,, \tag{18}$$

the entropy of which is

$$S = (S_I)_{max} = \log Z + \boldsymbol{\lambda} \cdot \mathbf{A} \,, \tag{19}$$

and the potentials λ_k are determined from the experimenter's data $(A_1', \ldots, A_m')$ by the m simultaneous equations

$$A_k' = \langle A_k \rangle = -(\partial/\partial \lambda_k) \log Z, \quad 1 \leq k \leq m \,. \tag{20}$$

These relations merely summarize the standard MAXENT formalism, still another time, but in our present notation.

The two Gibbs relations (1),(2) are now generalized to two identities connecting neighboring MAXENT distributions:

$$\delta S = \boldsymbol{\lambda} \cdot (\delta \langle \mathbf{A} \rangle - \langle \delta \mathbf{A} \rangle) = \boldsymbol{\lambda} \cdot \delta \mathbf{Q} \,, \tag{21}$$

$$\delta \boldsymbol{\lambda} \cdot \delta \langle \mathbf{A} \rangle \leq 0 \,. \tag{22}$$

8. Meaning of the Gibbs Convexity

To see the meaning of Eq. (22), suppose our original MAXENT ensemble is based on knowledge of only two physical quantities, and write $A_1 = A$, $A_2 = B$. Now a "heatlike" source δQ_A operates. That is, the generalized work $\langle \delta A \rangle$ is zero, and B is unconstrained; it is allowed to adjust itself in response to this source. As expounded above, we shall remaximize the entropy to take account of this new information. But we know only the change in A, $\delta \langle A \rangle = \delta Q_A$, and have to infer that of B, from the MAXENT principle. In going to this slightly different MAXENT distribution we might expect both potentials λ_a and λ_b to change, so we would have

$$\delta \langle A \rangle = \frac{\partial \langle A \rangle}{\partial \lambda_a} \delta \lambda_a + \frac{\partial \langle A \rangle}{\partial \lambda_b} \delta \lambda_b \,, \tag{23a}$$

$$\delta \langle B \rangle = \frac{\partial \langle B \rangle}{\partial \lambda_a} \delta \lambda_a + \frac{\partial \langle B \rangle}{\partial \lambda_b} \delta \lambda_b \,. \tag{23b}$$

But by the general MAXENT reciprocity theorem these coefficients are just the covariances

$$K_{AB} = K_{BA} = -\frac{\partial \langle A \rangle}{\partial \lambda_b} = -\frac{\partial \langle B \rangle}{\partial \lambda_a} \,, \tag{24}$$

so Eq. (23) is, in matrix form,

$$- \begin{bmatrix} \delta\langle A\rangle \\ \delta\langle B\rangle \end{bmatrix} = \begin{bmatrix} K_{AA} & K_{AB} \\ K_{BA} & K_{BB} \end{bmatrix} \begin{bmatrix} \delta\lambda_a \\ \delta\lambda_b \end{bmatrix} . \qquad (25)$$

Now if we substitute this into the Gibbs convexity relation (22), it reduces to the statement that when the inequality holds for all small but nonzero changes, the covariance matrix in Eq. (25) is positive definite. Thus Eq. (25) can be inverted, and the potentials are uniquely determined by $\langle A\rangle$ and $\langle B\rangle$. This is just the statement that our MAXENT conditions (20), determining the potentials, have a unique solution.

We could imagine more general constraints on MAXENT than specifying expectations $\langle A_k\rangle$. The constraints might themselves take the form of inequalities rather than equalities. But if the constraints confine us to any convex set in the $\langle A_k\rangle$, the solution is still unique.

More generally, the Gibbs convexity relation tells us that, when the inequality holds, the eigenvalues of the covariance matrix of any number of quantities are all positive. This makes it clear why Gibbs found—in the phenomenological theory of heterogeneous equilibrium, 25 years before his Statistical Mechanics—that relation (2) was the fundamental key to understanding thermodynamic stability, however many components and phases a thermodynamic system may have.

9. Mitchell's Relations

But there is a still more interesting result contained in the above relations. Notice that if $\delta\lambda_b = 0$, Eq. (25) reduces to

$$\delta\langle A\rangle = - K_{AA}\delta\lambda_a = \delta Q_A \qquad (26)$$

$$\delta\langle B\rangle = - K_{BA}\delta\lambda_a$$

or

$$\delta\langle B\rangle = (K_{BA}/K_{AA})\, \delta Q_A , \qquad (27)$$

which is identical to the prediction rule (15) that we reasoned out in a different way!

To make a long story short, the situation uncovered by Mitchell (1967) shows that when a source δQ_A operates and an unconstrained quantity B readjusts itself as a result, to predict the amount of that readjustment there are three principles:

(I) Quantities C uncorrelated with A are unchanged.
(II) Potentials λ_b of unconstrained quantities B are unchanged.
(III) The information entropy is remaximized.

Mitchell discovered the remarkable fact that these three conditions are mathematically equivalent. The fact is remarkable because they seem so different to our untutored intuition. Almost everybody finds (I) so intuitive that he will accept it at once, without demanding any formal proof. But to many, (II) and (III) are so far from being intuitive that they will scarcely believe them even after seeing the proof. This shows how much our intuition can be educated by studying the MAXENT formalism and thinking hard about why and how it works.

Mitchell's next relation introduces us to inferential scattering. Introduce a third variable C, so that on a small change in the MAXENT distribution

$$-\begin{bmatrix} \delta\langle A\rangle \\ \delta\langle B\rangle \\ \delta\langle C\rangle \end{bmatrix} = \begin{bmatrix} K_{AA} & K_{AB} & K_{AC} \\ K_{BA} & K_{BB} & K_{BC} \\ K_{CA} & K_{CB} & K_{CC} \end{bmatrix} \begin{bmatrix} \delta\lambda_a \\ \delta\lambda_b \\ \delta\lambda_c \end{bmatrix} . \tag{28}$$

If the source δQ_A operates and both B and C are left free to readjust to this, we have by the above principles

$$\begin{aligned} \delta\langle B\rangle &= (K_{BA}/K_{AA})\, \delta Q_A \\ \delta\langle C\rangle &= (K_{CA}/K_{AA})\, \delta Q_A \end{aligned} \tag{29}$$

amounting to two independent applications of our rule. But now let us impose a new constraint, that $\delta\langle C\rangle = 0$. Writing out the bottom line of Eq. (28), we can solve for $\delta\lambda_c$:

$$K_{CC}\delta\lambda_c = - K_{CA}\delta\lambda_a - K_{CB}\delta\lambda_b \tag{30}$$

and substituting this into Eq. (28), we find that the changes in A and B are still related by an equation like (25), but with a new "renormalized" covariance matrix:

$$-\begin{bmatrix} \delta\langle A\rangle \\ \delta\langle B\rangle \end{bmatrix} = \begin{bmatrix} K'_{AA} & K'_{AB} \\ K'_{BA} & K'_{BB} \end{bmatrix} \begin{bmatrix} \delta\lambda_a \\ \delta\lambda_b \end{bmatrix} \tag{31}$$

with the new matrix elements

$$\begin{aligned} K'_{AA} &= K_{AA} - K_{AC}K_{CC}^{-1}K_{CA} \\ K'_{AB} &= K_{AB} - K_{AC}K_{CC}^{-1}K_{CB} \end{aligned} \tag{32}$$

and so on. By our principles, $\delta\lambda_b$ will still be zero if B is left free to readjust, so the predicted change in B due to the source δQ_A is now

$$\delta\langle B\rangle = (K'_{BA}/K'_{AA})\, \delta Q_A . \tag{33}$$

The difference between Eqs. (33) and (25) represents inferential scattering; the logical connection between A and B is altered by the new constraint $\delta\langle C\rangle = 0$.

We shall examine the meaning of every term in this difference. Define the correlation coefficient of A and C:

$$R_{AC} \equiv K_{AC}/(K_{AA}K_{CC})^{1/2} . \tag{34}$$

Then we can write Eq. (33) as

$$\delta\langle B\rangle = [(K_{BA}/K_{AA}) - (K_{BC}/K_{CC})(K_{CA}/K_{AA})]\ \delta Q'_A , \tag{35}$$

where

$$\delta Q'_A \equiv \delta Q_A/(1 - R_{AC}^2) \tag{36}$$

is a "renormalized source strength," whose significance will appear presently. In Eq. (35) we have two terms. The first term represents the effect we would have without the constraint on C, if the renormalized source strength had operated. The second term can be interpreted by rewriting it as

$$(K_{CB}/K_{CC})\ \text{"}\delta Q_C\text{"} . \tag{37}$$

This is the response to a fictitious source strength

$$\text{"}\delta Q_C\text{"} \equiv -(K_{CA}/K_{AA})\ \delta Q_A \tag{38}$$

which we recognize as minus the change $\delta\langle C\rangle$ in Eq. (29) that would be produced by the renormalized source δQ_A if C were unconstrained.

The constraint $\delta\langle C\rangle = 0$ has therefore modified our predicted relationship between A and B in two ways:

(I) The source δQ_A is renormalized. Intuitively, if A and C are correlated (positively or negatively), then holding $\langle C\rangle$ fixed makes the system "stiffer" against an attempt to change A, and a given actual change δQ_A has a greater effect on B because of this. As an analogy, if the input impedance to a network is increased owing to the blocking of an internal current path, then to inject a given current into it will in general result in increased voltages at other points.

(II) A new scattering term appears which, as seen from B, appears to come from a fictitious source at C. Our network analogy still holds; the point C, where the current was blocked, becomes a new voltage source whose effects appear at other points of the network.

But the relationships just found are of far more general meaning than those of the network. A,B,C may stand for any physical quantities, not necessarily localized in space or time.

10. Acoustics—Direct Propagation

Finally, we are ready for the promised specific case. In an acoustical problem, take our initial ensemble as the conventional canonical ρ_C representing the air in thermal equilibrium at some temperature T. In the following, expectations of the form $\langle X \rangle$ are over this ensemble, and we examine the effect of modifying it by new information. Let n(x,t) be the particle density (number of molecules per unit volume) and choose A to be the number N of particles in a small volume V_A about the point x', at time t', while B is the air pressure at a different space-time point (x,t):

$$A \equiv n(x',t')\, V_A = N\,, \tag{39}$$

$$B \equiv P(x,t)\,. \tag{40}$$

We make the nonessential but simplifying assumption that the region V_A is small compared to a mean free path, so that for all practical purposes the fluctuations in N are those of the ideal gas law, as used by Einstein long ago:

$$K_{AA} = \langle A^2 \rangle - \langle A \rangle^2 = \overline{\delta N^2} = N_0 = n_0 V_A\,, \tag{41}$$

where n_0 is the equilibrium particle density. The region V_A is to act as an acoustical source during a short time interval about t', in which

$$\delta Q_A = \delta n\, V_A = \text{number of particles injected}\,. \tag{42}$$

In conventional acoustics a source δs is usually defined instead in terms of volume of fluid injected; so they are related by

$$\delta s = \delta Q_A / n_0\,. \tag{43}$$

With these preliminaries, our general prediction rule $\delta\langle B \rangle = (K_{BA}/K_{AA})\delta Q_A$ becomes: The predicted sound pressure is

$$\delta\langle P(x,t) \rangle = \frac{[\langle P(x,t)n(x',t') \rangle - \langle P \rangle \langle n \rangle] V_A}{\langle n \rangle V_A}\,[\langle n \rangle \delta s]$$

$$= [\langle P(x,t)n(x',t') \rangle - P_0 n_0]\delta s\,, \tag{44}$$

where $P_0 = \langle P \rangle$, $n_0 = \langle n \rangle$ are the equilibrium pressure and particle density, supposed independent of x and t.

Comparing this with the conventional acoustic Green function solution for a prescribed source distribution $s(x',t')$,

$$P(x,t) = \int_{-\infty}^{t} dt' \int d^3x' \, G(x,t;x',t') \, s(x',t') \, , \tag{45}$$

we see that the MAXENT prediction of the acoustic Green function is

$$G(x,t;x',t') = \langle \delta P(x,t) n(x',t') \rangle = (1/kT) \langle \delta P(x,t) \delta P(x',t') \rangle \, , \tag{46}$$

where now we are writing (in a notation perhaps slightly inconsistent with our previous usage), $\delta P \equiv P - P_0$, the departure from equilibrium pressure.

The evident symmetry in Eq. (46) is recognized as just the Helmholtz-Rayleigh reciprocity theorem. All the known reciprocity principles seem to appear automatically in MAXENT, without our ever having to make any special effort to get them.

In principle, we could calculate the pressure-pressure covariance function in Eq. (46), but this is a complicated problem in many-body theory that would itself require a separate long article. Our point is made more quickly if we just note that we already know the Green function G from ordinary acoustical theory. A velocity potential $\phi(x,t)$ generates the velocity and density fields through $\mathbf{v} = \nabla \phi$, $\delta n = -n_0 \dot{\phi}/c^2$, where c is the velocity of sound and $\mathbf{v}$ the mass velocity of the fluid. The point source solution of the acoustical wave equation is spherically symmetric:

$$\phi(r,t) = -(1/4\pi r) \dot{s}(t - r/c) \, ; \tag{47}$$

and if the source operates as a short pulse, our δQ_A is

$$\delta Q_A = n_0 \delta s = n_0 \int \dot{s}(t') dt' = \int \dot{Q}_A dt' . \tag{48}$$

At this point it is easier mathematically—and also more general—to go into the frequency domain by taking time Fourier transforms of Eq. (47). Using Eq. (48) this gives

$$\phi(r,\omega) = (i\omega/4\pi r) \exp(i\omega r/c) \, Q_A(\omega) \, . \tag{49}$$

Therefore, using the above relations, we predict density and pressure variations at B given by (now we write r_{AB} for the distance $|x - x'|$):

$$\delta n(x,\omega) = \frac{n_0 \omega^2}{4\pi r_{AB} c^2} \exp(i\omega r_{AB}/c) \, Q_A(\omega) \, , \tag{50}$$

$$\delta P(x,\omega) = \frac{i\omega}{4\pi r_{AB}} \exp(i\omega r_{AB}/c) \, Q_A(\omega) \, . \tag{51}$$

This completes our derivation of the direct propagation term, corresponding to Eq. (27) and (neglecting for the moment the renormalization of the source strength) the first term $(K_{BA}/K_{AA})\delta Q_A$ in Eq. (35). We now try to relate the inferential scattering indicated by the last term of Eq. (35) to Rayleigh scattering.

11. The Rayleigh Scattering Term

Let us introduce that third quantity C as representing, like A, the number of particles in a small volume V_C, again supposed small compared to a mean free path (and therefore small compared to the wavelength $\lambda = 2\pi c/\omega$):

$$C \equiv n(x'',t'')\, V_C \, . \tag{52}$$

To impose the constraint $\delta\langle C\rangle = 0$ is, in effect, to replace the boundary of V_C by a rigid wall that allows no particles to cross it. This is just the problem that Rayleigh (1877) solved as a boundary-value problem of mathematical physics, and we now try to relate it to our statistical result (35).

Comparing the two terms in Eq. (35) enables us to define the scattering cross-section. The energy radiated from the source A is $U(\delta Q_A)^2$, where U is a factor we could easily calculate, but need not because the same factor appears in the energy scattered from C:

$$\sigma \left[\frac{U\delta Q_A{}^2}{4\pi r_{AC}{}^2}\right] \tag{53}$$

which defines the cross-section σ. We relate this to the fictitious source strength "δQ_C" by noting that the energy density arriving at B from A is $U(\delta Q_A)^2/4\pi r_{AB}{}^2$, so the scattered flux at B from C is

$$\frac{U(\text{"}\delta Q_C\text{"})^2}{4\pi r_{BC}{}^2} = \sigma \left[\frac{U\delta Q_A{}^2}{4\pi r_{AC}{}^2}\right] \frac{1}{4\pi r_{BC}{}^2} \, . \tag{54}$$

Therefore the predicted scattering cross-section is given in terms of our covariance functions by

$$\sigma = 4\pi r_{AC}{}^2 \, |K_{CA}/K_{AA}|^2 = 4\pi r_{AC}{}^2 \, |\text{"}\delta Q_C\text{"}/\delta Q_A|^2 \, . \tag{55}$$

But the required ratio is, from our previous equations,

$$K_{CA}/K_{AA} = (\text{"}\delta Q_C\text{"}/\delta Q_A) = \left[\frac{n_0\omega^2 V_C}{4\pi r_{AC} c^2}\right] \exp(i\omega r_{AC}/c) \, , \tag{56}$$

so our predicted cross-section is

$$\sigma = \frac{\omega^4 V_C{}^2}{4\pi c^4} \propto \frac{V_C{}^2}{\lambda^4} \, , \tag{57}$$

which is just Rayleigh's formula, with the λ^{-4} dependence that he used to explain the blue color of the sky.

This little test of the MAXENT relations is admittedly rather trivial, but even a trivial problem is educational if one does not see in advance how every detail is going to work out. With more effort we could have removed our assumption about the smallness of V_C and derived more elaborate (t-matrix) scattering formulas of more general validity.

One bit of unfinished business remains: Until now we have ignored that prime on δQ_A in Eq. (35). But that is hiding the most interesting part of our story.

12. Meaning of the Renormalized Source

We noted before that, intuitively, source renormalization is something like increased "stiffness" of the kind we are familiar with in mechanics or electrical network theory, where imposing a constraint on motion or current increases the resistance to other motions or currents. But in inferential scattering this "stiffness" may take a very unexpected form.

Let us expand the renormalization factor in Eq. (36):

$$(1 - R_{AC}^2)^{-1} = 1 + R_{AC}^2 + R_{AC}^4 + \dots \qquad (58)$$

and substitute the result into Eq. (35). We shall need a more compact notation, so define the "propagators"

$$X_{BA} = K_{BA}/K_{AA} . \qquad (59)$$

Then our full MAXENT prediction (35) expands into

$$\delta\langle B\rangle = [X_{BA} - X_{BC}X_{CA} + X_{BA}X_{AC}X_{CA} - X_{BC}X_{CA}X_{AC}X_{CA} + \dots]\delta Q_A . \qquad (60)$$

Each of these terms has a simple meaning. The first is just the standard regression result (25) that held before the constraint $\delta\langle C\rangle = 0$ was imposed. The second, as we have just seen, represents the Rayleigh scattering of the constraint. But that is only the first-order term in the full effect of the new constraint. A moment's contemplation of the third term will reveal its meaning: it is the amplitude of a double scattered wave that has propagated from A to C, scattered off C back to A, then scattered off A on to B. We might represent this by

$$(A \to C \to A \to B) .$$

Likewise the fourth term is the effect, as seen at B, of the triple scattering process

$$(A \to C \to A \to C \to B) ,$$

and so on!

So what the source renormalization has done for us, in this particular case, is to put in every possible multiple scattering effect in addition to the direct propagation and Rayleigh terms. At first glance it may seem surprising that arbitrarily high-order scatterings are given already by what is only the first order of MAXENT perturbation theory. But we can understand it as follows.

This phenomenon was noted before in Heims and Jaynes (1962), where MAXENT was applied to calculation of gyromechanical and gyromagnetic effects. All terms of the famous susceptibility formula of van Vleck, which he derived by second-order energy-level perturbation theory, appeared in the first order of our calculation. The reason was that the expansion parameter was different in the two calculations.

In a conventional physics calculation where one expands in powers of the interaction forces, nth-order multiple scattering would appear only in the 2nth order of the perturbation. But we are expanding in powers of the source strength, and the MAXENT formalism gives in first order the exact part of the response that is linear in the source strength, however high order it may be in the interaction forces.

References

Akhiezer, A. I., and S. V. Peletminskii (1981) Methods of Statistical Physics (New York: Pergamon Press).

Gibbs, J. W. (1873, 1875) in The Scientific Papers of J. Willard Gibbs (New York: Dover, 1961).

Gibbs, J. W. (1902) Elementary Principles in Statistical Mechanics (New York: Dover, 1961).

Heims, S., and E. T. Jaynes (1962) Theory of gyromagnetic effects and some related magnetic phenomena, Rev. Mod. Phys. **34**, 143-165.

Jaynes, E. T. (1963) Information theory and statistical mechanics, in K. W. Ford, ed., Statistical Physics (New York: Benjamin), pp. 181-218.

Jaynes, E. T. (1965) "Gibbs vs Boltzmann entropies," Am. J. Phys. **33**, 391-398.

Jaynes, E. T. (1980) The minimum entropy production principle, Ann. Rev. Phys. Chem. **31**, 579-601.

Jaynes, E. T. (1982) Review of Akhiezer and Peletminskii, Phys. Today, **35**, 57.

Mayer, J. E., and M. G. Mayer (1940) Statistical Mechanics (New York: van Nostrand).

Mitchell, William C. (1967) Thermal Driving, Ph.D. Thesis, Washington University, St. Louis, Missouri.

Rayleigh (1877) The Theory of Sound (New York: Dover, 1945, reprint of 2nd (1894) edition).

Slichter, C. P. (1978) Principles of Magnetic Resonance (Berlin: Springer Verlag).

Zwanzig, R. (1965) Time-correlation functions and transport coefficients in statistical mechanics, Ann. Rev. Phys. Chem. **16**, 67-102.

DETECTION FOR ACTIVE SONARS BY MEANS OF AUTOREGRESSIVE NOISE MODELING

Steven Kay

Electrical Engineering Department, University of Rhode Island,
Kingston, Rhode Island 02881

The problem of detecting a known signal in colored Gaussian noise of unknown covariance is addressed. The noise is modeled as an autoregressive process of known order but unknown coefficients. By use of the theory of generalized likelihood ratio testing, a detector structure is derived and then analyzed for performance. It is proven that for large data records the detection performance is identical to that of an optimal prewhitener and matched filter, and therefore the detector itself is optimal. Simulation results indicate that the data record length necessary for the asymptotic results to apply can be quite small. Thus, the proposed detector is well suited for practical applications.

C. Ray Smith and W. T. Grandy, Jr. (eds.),
Maximum-Entropy and Bayesian Methods in Inverse Problems, 399–411.

1. Introduction

The problem of detecting a known signal in colored Gaussian noise with known covariance has a well-known solution [1]: One need only prewhiten the noise and then implement a filter matched to the signal at the output of the prewhitener. However, if the covariance of the noise is not known, then the appropriate prewhitener cannot be determined. This lack of noise structure information is illustrated in the fields of active sonar and radar, in which reverberation and clutter, respectively, tend to obscure the target return. Both these noise processes have unknown covariances or power spectral densities (assuming wide-sense stationarity), since the character of the noise is highly dependent upon unknown and in general time-varying environmental conditions. The usual approach to this problem has been either to assume the noise is white or nearly so and use a matched filter [2] or to assume an average covariance for the noise in order to employ an average prewhitener [3,4]. In the first case, a detectability loss will be incurred whenever the noise is nonwhite and especially when the noise is narrowband. In the second case, detection losses will occur whenever the character of the noise deviates from the assumed average noise covariance.

An alternative approach to this problem is to measure on line the covariance of the noise and use this information to design the prewhitener. This approach can be especially valuable when the noise covariance is changing in time—as in a reverberation or clutter environment. To successfully do this one must be able to accurately measure the noise covariance from a limited data set. The length of data to be analyzed is such that the noise process can be considered wide-sense stationary over the analysis time interval. Furthermore, to accurately estimate the covariance from a limited data set it is necessary to parametrize the covariance, that is, to express the covariance as a function of a small number of parameters. A popular model for the noise that accomplishes this goal is the autoregressive (AR) model [5]; that is, the noise is represented as

$$n_t = -\sum_{k=1}^{p} a_k n_{t-k} + \varepsilon_t , \tag{1}$$

where ε_t is zero mean discrete white noise with variance σ_ε^2, and $\{a_1,a_2,\ldots,a_p,\sigma_\varepsilon^2\}$ are termed the AR parameters. Note that n_t is a zero mean wide-sense stationary (WSS) discrete random process. It can be shown [5] that there is a one-to-one relationship between the AR parameters and the covariance function (autocorrelation function) of n_t. Furthermore, any WSS random process can be modeled as in Eq. (1) if p is taken to be large enough [5]. Finally, the approximate maximum likelihood estimators (MLEs) of the AR parameters are easily obtained as the solution of a set of linear equations [6]. Thus the modeling given by Eq. (1) is well suited to on-line estimation of the noise covariance.

Several researchers have proposed the use of an AR prewhitener, measured on line, to approximate the optimal detector [7,8]. Implicit in their approach is the use of an AR model. These approaches are ad hoc, however, in that they have no firm statistical foundation. Also, it is not clear how the AR parameters are actually estimated if a signal is also present in the data. The latter condition will lead to a severely biased prewhitener estimate. In this paper, we take a more fundamental approach to the problem by employing AR modeling for the noise and a generalized likelihood ratio test (GLRT) to arrive at a detector. The GLRT has a number of optimal properties for large data records [9]. In fact, the GLRT has been applied by Helstrom [10] and later by Nuttall and Cable [11] to the somewhat simpler problem of detecting a signal known except for amplitude in white noise of unknown variance. The results obtained for that detection problem are analogous to the ones presented here. Also, the latter problem has been approached using invariance techniques as reported by Scharf and Lytle [12]. However, the invariance principles do not appear to be readily extendable to the colored noise case. It should be noted that for the problem considered here no uniformly most power (UMP) test exists. This is because the use of the Neyman-Pearson criterion does not lead to a detector that is independent of the unknown parameters [1].

The paper is organized as follows: In Section 2 we derive the GLRT detector, and in Section 3 we analyze its performance for large data records. The performance is shown to be optimal if the data record is large enough. In Section 4, the optimal detector performance is compared to the GLRT detection for short data records via a simulation. Only a very slight degradation in performance from the optimal detector is noted for data records as short as 20 data points. Section 5 gives a summary and conclusions.

2. Derivation of Generalized Likelihood Ratio Test Detector

We consider the following detection problem:

$$\begin{aligned} H_0&: \; r_t = n_t \\ H_1&: \; r_t = s_t + n_t , \qquad t = 1,2,\ldots,N , \end{aligned} \tag{2}$$

where s_t is a known signal and n_t is an AR(p) wide-sense stationary Gaussian random process as given by Eq. (1); p is the order of the AR process. We assume that $\{a_1,a_2,\ldots,a_p,\sigma_\varepsilon^2\}$ are unknown but fixed constants and that they are the <u>same</u> under either hypothesis. We can reformulate the problem as

$$\begin{aligned} H_0&: \; \mathbf{r} = \mu_0\mathbf{s} + \mathbf{n} \\ H_1&: \; \mathbf{r} = \mu_1\mathbf{s} + \mathbf{n} \end{aligned}$$

where $\mathbf{r} = [r_1,r_2,\ldots,r_N]^T$, $\mathbf{s} = [s_1,s_2,\ldots,s_N]^T$, $\mathbf{n} = [n_1,n_2,\ldots,n_N]^T$, $\mu_0 = 0$, and $\mu_1 = 1$.

The theory of GLRT can be applied here [9,13]. This technique forms the modified likelihood ratio as

$$\lambda_G = \frac{p_1(\mathbf{r};\mu_1,\hat{\mathbf{a}}_1,\hat{\sigma}^2_{\varepsilon_1})}{p_0(\mathbf{r};\mu_0,\hat{\mathbf{a}}_0,\hat{\sigma}^2_{\varepsilon_0})}, \tag{3}$$

where p_i is the probability density function (PDF) under hypothesis H_i, and $\hat{\mathbf{a}}_i,\hat{\sigma}^2_{\varepsilon_i}$ are the maximum likelihood estimates (MLEs) under hypothesis H_i. The GLRT detector then compares λ_G to a threshold η, and

$$\begin{aligned} &\text{if } \lambda_G > \eta, \quad \text{decide } H_1 , \\ &\text{if } \lambda_G < \eta, \quad \text{decide } H_0 . \end{aligned} \tag{4}$$

The results of this procedure produce the detector of Fig. 1. This detector is now derived.

Since n_t is assumed to be Gaussian, r_t and hence $\mathbf{r}$ are Gaussian. It can be shown [14] for AR processes that are not too narrowband and for data records of reasonable length that

$$p_1(\mathbf{r};\mu_1,\mathbf{a},\sigma^2_\varepsilon) \simeq \frac{1}{(2\pi)^{N/2}\sigma^N_\varepsilon} \exp\left\{-\frac{1}{2\sigma^2_\varepsilon} \sum_t \left[\sum_{k=0}^{p} a_k(r_{t-k} - s_{t-k})\right]^2\right\}, \tag{5a}$$

$$p_0(\mathbf{r};\mu_0,\mathbf{a},\sigma^2_\varepsilon) \simeq \frac{1}{(2\pi)^{N/2}\sigma^N_\varepsilon} \exp\left\{-\frac{1}{2\sigma^2_\varepsilon} \sum_t \left[\sum_{k=0}^{p} a_k r_{t-k}\right]^2\right\}, \tag{5b}$$

where $a_0 \equiv 1$. Note that the range of summation on t is purposely left unspecified. If it is chosen to be $t = p+1$ to $t = N$, then the MLE of the AR parameters will be given by the covariance method of linear prediction [5]. If we choose $t = -\infty$ to ∞ and assume r_t and $s_t = 0$ for $t < 1$ and $t > N$, then the MLE will be obtained using the autocorrelation method of linear prediction [5].

Now to find $\hat{\sigma}^2_{\varepsilon_1}$, we maximize $\ln p_1$:

$$\frac{\partial}{\partial\sigma^2_\varepsilon} \ln p_1 = \frac{\partial}{\partial\sigma^2_\varepsilon}\left\{-\frac{N}{2}\ln 2\pi - \frac{N}{2}\ln\sigma^2_\varepsilon - \frac{1}{2\sigma^2_\varepsilon}\sum_t\left[\sum_{k=0}^{p} a_k(r_{t-k} - s_{t-k})\right]^2\right\},$$

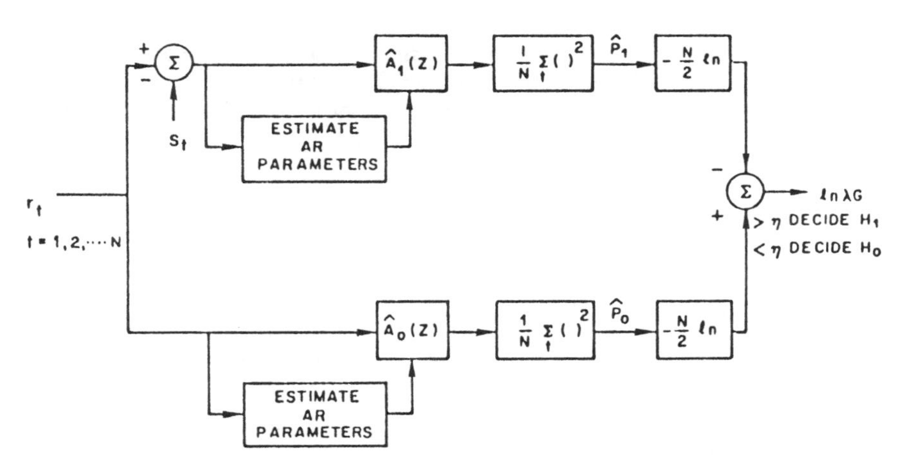

Figure 1. Generalized likelihood ratio test detector.

whence

$$\sigma_\varepsilon^2 = \frac{1}{N} \sum_t \left[\sum_{k=0}^{p} a_k(r_{t-k} - s_{t-k}) \right]^2 . \tag{6}$$

When the $\{a_1, a_2, \ldots, a_p\}$ are replaced by their MLEs, then σ_ε^2 as given by Eq. (6) becomes $\hat{\sigma}_{\varepsilon_1}^2$. To find $\hat{\mathbf{a}}_1$, substitute Eq. (6) into Eq. (5a) and maximize over $\mathbf{a}$:

$$\ln p_1 = -\frac{N}{2} \ln 2\pi - \frac{N}{2} \ln \left\{ \frac{1}{N} \sum_t \left[\sum_{k=0}^{p} a_k(r_{t-k} - s_{t-k}) \right]^2 \right\} - \frac{N}{2} .$$

To maximize $\ln p_1$ over $\{a_1, a_2, \ldots, a_p\}$, we must minimize

$$\sum_t \left[\sum_{k=0}^{p} a_k(r_{t-k} - s_{t-k}) \right]^2 . \tag{7}$$

But this is just the standard linear prediction formulation except that we must first subtract the known signal from the data. As mentioned previously, by assuming various limits for the summation on t, we can use either the covariance or autocorrelation method of linear prediction. In fact, since Eq. (5a) is only an approximation, any approximate MLE may be used. Assuming we have minimized (7) to produce $\hat{\mathbf{a}}_1$, then

$$\ln p_1(\mathbf{r};\mu_1,\hat{\mathbf{a}}_1,\hat{\sigma}^2_{\varepsilon_1}) = -\frac{N}{2}\ln 2\pi - \frac{N}{2} - \frac{N}{2}\ln \hat{\sigma}^2_{\varepsilon_1}, \tag{8}$$

where

$$\hat{\sigma}^2_{\varepsilon_1} = \frac{1}{N}\sum_t \left[\sum_{k=0}^{p} \hat{a}_{1k}(r_{t-k} - s_{t-k})\right]^2 .$$

Similarly under H_0 (let $s_t = 0$):

$$\ln p_0(\mathbf{r};\mu_0,\hat{\mathbf{a}}_0,\hat{\sigma}^2_{\varepsilon_0}) = -\frac{N}{2}\ln 2\pi - \frac{N}{2} - \frac{N}{2}\ln \hat{\sigma}^2_{\varepsilon_0}, \tag{9}$$

where

$$\hat{\sigma}^2_{\varepsilon_0} = \frac{1}{N}\sum_t \left[\sum_{k=0}^{p} \hat{a}_{0k}\, r_{t-k}\right]^2 .$$

The $\{\hat{a}_{0k}\}$ is found using a standard linear prediction approach on the data $\{r_1,r_2,\ldots,r_N\}$. For ease of notation let $P_i = \hat{\sigma}^2_{\varepsilon_i}$.

Note that P_i is just an estimate of the prediction error power under hypothesis H_i. Then

$$\ln \lambda_G = -\frac{N}{2}\ln \hat{P}_i + \frac{N}{2}\ln \hat{P}_0 \begin{cases} > \ln \eta, \text{ decide } H_1, \\ < \ln \eta, \text{ decide } H_0. \end{cases} \tag{10}$$

An estimate of a prewhitener is formed under both hypotheses. The powers of the whitened time series are then estimated, transformed by a logarithm operator, and compared. As an example, if $P_1 < P_0$ (that is, $\ln \lambda_G$ is a large positive number exceeding the threshold), H_1 would be chosen.

The detector of Fig. 1 does not resemble the ad hoc detectors of references [7] and [8]. However, a shortcoming of the GLRT detector is that its extension to the case of a signal known except for a few parameters is not straightforward. (This case will be considered in a future paper.) The ad hoc detector, on the other hand, appears to be able to handle the detection of a signal known except for amplitude and phase. To the extent that the prewhitener can be estimated accurately in the presence of a signal, the ad hoc detector can utilize a quadrature matched filter [15].

3. Asymptotic Performance of Generalized Likelihood Ratio Detector

We show in this section that, for large data records, the GLRT detector of Fig. 1 has the same detection performance as an optimal matched filter. Thus, asymptotically the GLRT detector is optimum.

The optimal detector for the case of a known signal in colored noise of known covariance is found by comparing the log likelihood ratio to a threshold, γ [15]:

$$\ln \lambda = \ln \frac{p_1(\mathbf{r})}{p_0(\mathbf{r})} \begin{cases} > \gamma, \text{ decide } H_1 , \\ < \gamma, \text{ decide } H_0 . \end{cases} \tag{11}$$

For Gaussian noise with zero mean, this can be shown [15] to be

$$\ln \lambda = \mathbf{r}^T \mathbf{R}_n^{-1} \mathbf{s} - \frac{1}{2} \mathbf{s}^T \mathbf{R}_n^{-1} \mathbf{s} , \tag{12}$$

where $\mathbf{R}_n$ is the $N \times N$ covariance matrix of the noise (assumed to be known).

It can be shown further [15] that the probabilities of false alarm, P_{FA}, and of detection, P_D, are

$$P_{FA} = Q \left[\frac{\gamma + \frac{1}{2} \mathbf{s}^T \mathbf{R}_n^{-1} \mathbf{s}}{(\mathbf{s}^T \mathbf{R}_n^{-1} \mathbf{s})^{1/2}} \right]$$

$$P_D = Q \left[\frac{\gamma - \frac{1}{2} \mathbf{s}^T \mathbf{R}_n^{-1} \mathbf{s}}{(\mathbf{s}^T \mathbf{R}_n^{-1} \mathbf{s})^{1/2}} \right] \tag{13}$$

where

$$Q(x) = \int_x^\infty \frac{1}{\sqrt{2\pi}} e^{-t^2/2} \, dt .$$

It can be shown [16] that for large data records

$$\ln \lambda_G \rightarrow \ln \lambda . \tag{14}$$

Note, however, that the use of the GLRT does not lead to a constant false alarm rate (CFAR) detector since from Eq. (13) the P_{FA} clearly depends upon the unknown parameters.

For the asymptotic results to hold, that is, for Eq. (14) to be satisfied, the data record length should satisfy [16]

$$N \gg \mathbf{s}^T \mathbf{R}_n^{-1} \mathbf{s} . \tag{15}$$

4. Performance of GLRT for Finite Data Records

In this section we verify the asymptotic results of the previous section for finite data records. Consider the case

$$H_0: \; r_t = n_t ,$$

$$H_1: \; r_t = A \cos(2\pi f_0 t) + n_t , \qquad t = 1,2,\ldots,N , \tag{16}$$

where n_t is an AR(1) process with parameters a and σ_ε^2. It can be shown [16] that

$$\mathrm{SNR} \equiv \mathbf{s}^T \mathbf{R}_n^{-1} \mathbf{s} = \frac{1}{\sigma_\varepsilon^2} \left[(1 - a^2) s_1^2 + \sum_{t=2}^{N} (s_t + a s_{t-1})^2 \right] , \tag{17}$$

where

$$s_t = A \cos(2\pi f_0 t) .$$

For purposes of the simulation, σ_ε^2 equals 1 and A is adjusted to yield the given SNR. In the case of a known noise covariance matrix, it is customary to choose H_1 if

$$l = \mathbf{r}^T \mathbf{R}_n^{-1} \mathbf{s} > \gamma' , \tag{18}$$

where

$$\gamma' = \gamma + \frac{1}{2} \mathbf{s}^T \mathbf{R}_n^{-1} \mathbf{s} .$$

For this detector

$$P_{FA} = Q\left[\frac{\gamma'}{(\mathbf{s}^T \mathbf{R}_n^{-1} \mathbf{s})^{1/2}} \right] ,$$

$$P_D = Q\left[\frac{\gamma' - \mathbf{s}^T \mathbf{R}_n^{-1} \mathbf{s}}{(\mathbf{s}^T \mathbf{R}_n^{-1} \mathbf{s})^{1/2}} \right] . \tag{19}$$

To compare the GLRT detector to the optimal detector, we choose H_1 if

$$l_G = -\frac{N}{2}\ln(\hat{P}_1/\hat{P}_0) + \frac{1}{2}\mathbf{s}^T\mathbf{R}_n^{-1}\mathbf{s} > \gamma' . \qquad (20)$$

In Fig. 2, P_{FA} and P_D are plotted versus the threshold γ'. The Burg estimate has to be used to find $\underline{a}$ [5]. The GLRT detector is shown for N = 20 and N = 100, as are the optimal detector curves as given by Eq. (19). Good agreement is noted for N = 100, which satisfies the condition N >> SNR. If we increase the SNR to 15 dB ($\mathbf{s}^T\mathbf{R}_n^{-1}\mathbf{s} = 31.6$), then as shown in Fig. 3 the agreement is poorer, as expected. The effect of changing $\underline{a}$ is to improve the agreement for larger $-\underline{a}$ as shown in Fig. 4. This result is rationalized by observing that the asymptotic variance of the MLE of $\underline{a}$ can be shown [17] to be

$$\mathrm{Var}(\hat{a}) = (1/N)(1 - a^2) , \qquad (21)$$

which approaches 0 as $|a|$ approaches 1.

5. Summary and Conclusions

A detector has been derived for the case of a known signal in colored noise of unknown covariance. The detector assumes the noise is an autoregressive process and estimates the unknown parameters by maximum likelihood estimates for use in a generalized likelihood ratio test. The performance of the new detector approaches that of the optimal prewhitener and matched filter detector for moderately sized data record lengths. One drawback is that the detector is not CFAR, that is, the threshold cannot be set to maintain a given probability of false alarm. (This restriction will be addressed in future work.) If, however, the receiver is intended for a communications application, as in coherent binary data transmission in the presence of a jammer, then this is not a problem. In this case, assuming equal a priori probabilities of either signal transmission, the optimum threshold is always zero [15].

The results reported here can easily be extended to the case of autoregressive-moving average (ARMA) noise. For this assumption one needs to find the maximum likelihood estimators of the parameters of an ARMA process, for which many approaches are available [5]. Finally, future research will concentrate on the gains obtained using the GLRT over a conventional matched filter and also on extending the results to the case of a signal known except for a few parameters.

Acknowledgments

The author would like to express his appreciation to Louis Scharf of the University of Rhode Island for the many helpful discussions of the work reported herein. This work was supported by the Office of Naval Research under ONR grant N00014-81-K-0144.

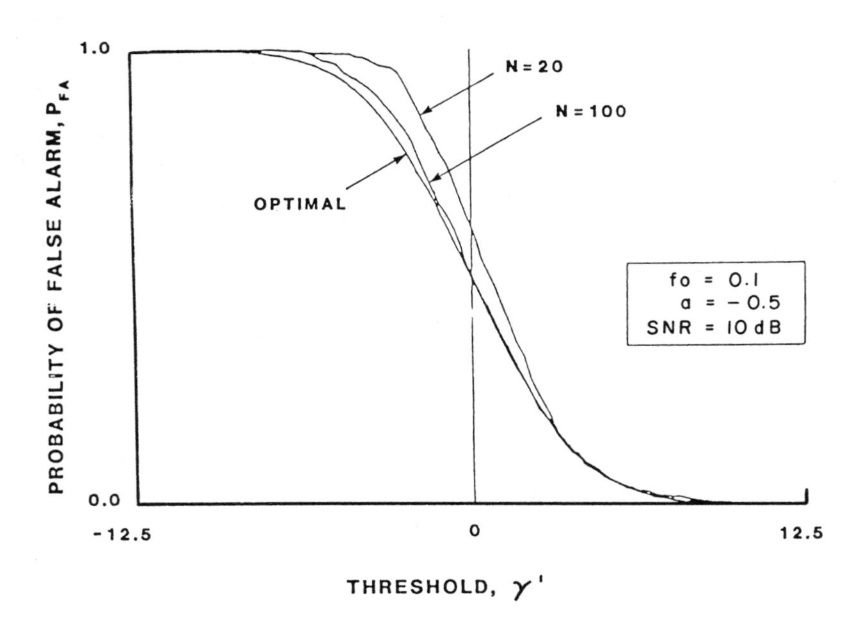

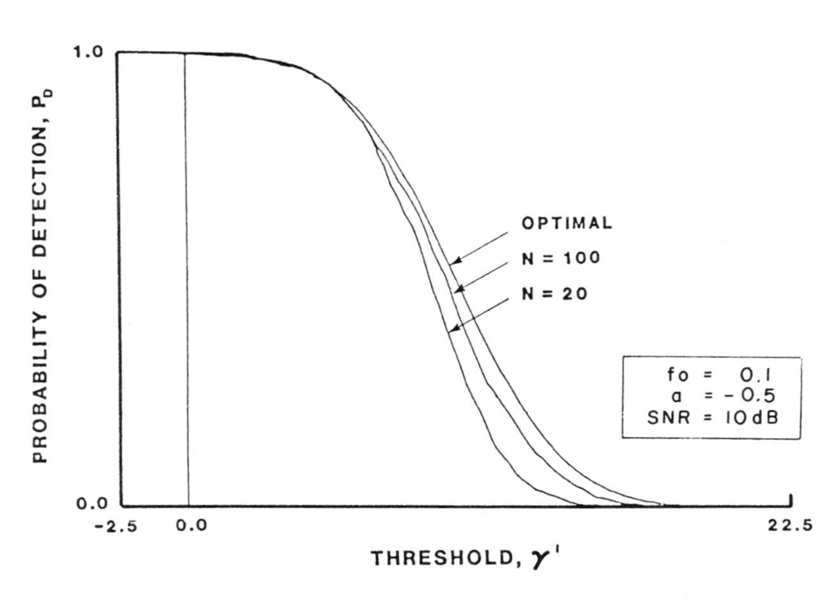

Figure 2. Performance of GLRT detector vs. optimal detector performance: $a = -0.5$, SNR = 10 dB.

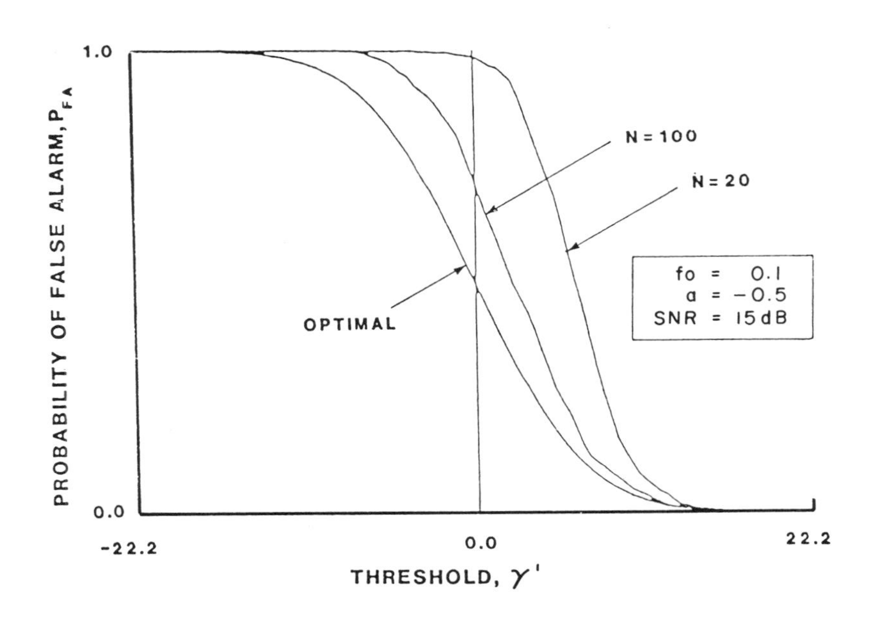

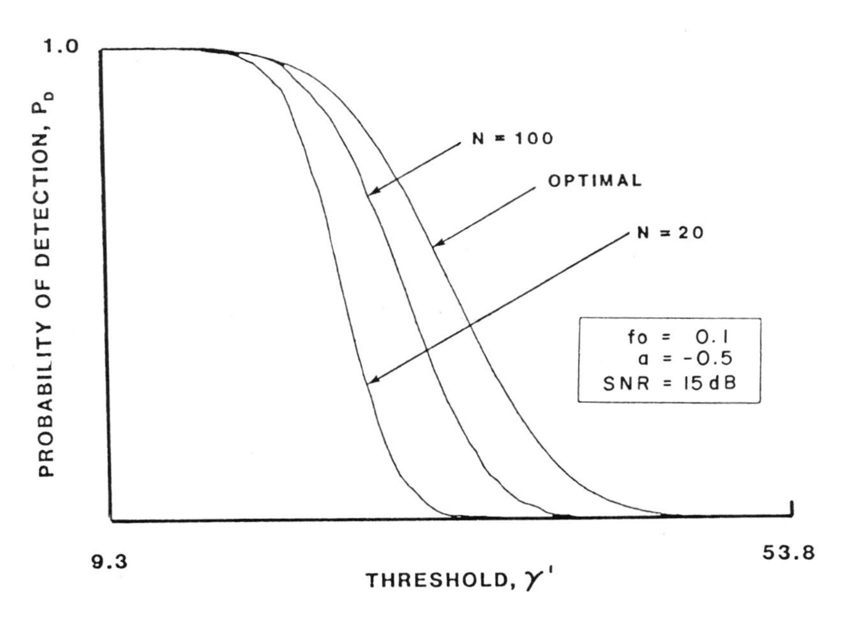

Figure 3. Performance of GLRT detector vs. optimal detector performance: a = -0.5, SNR = 15 dB.

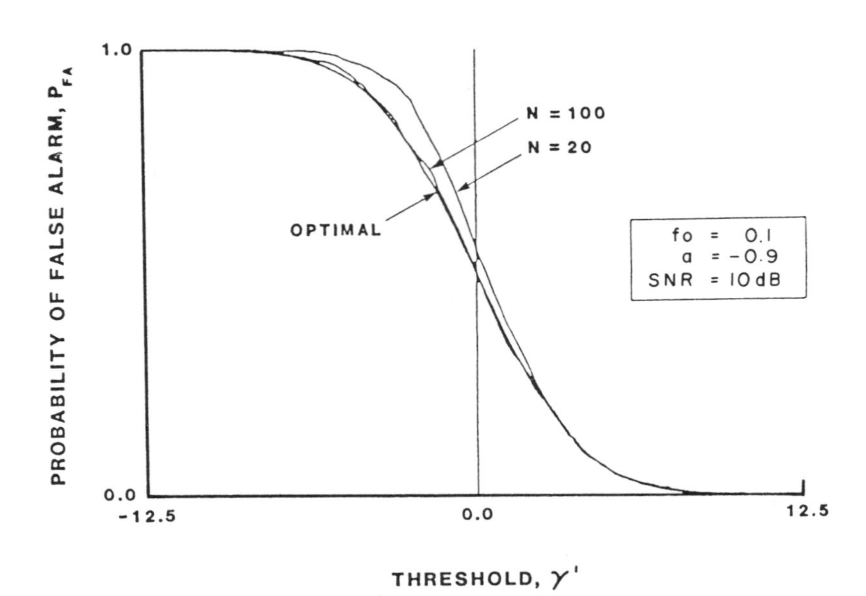

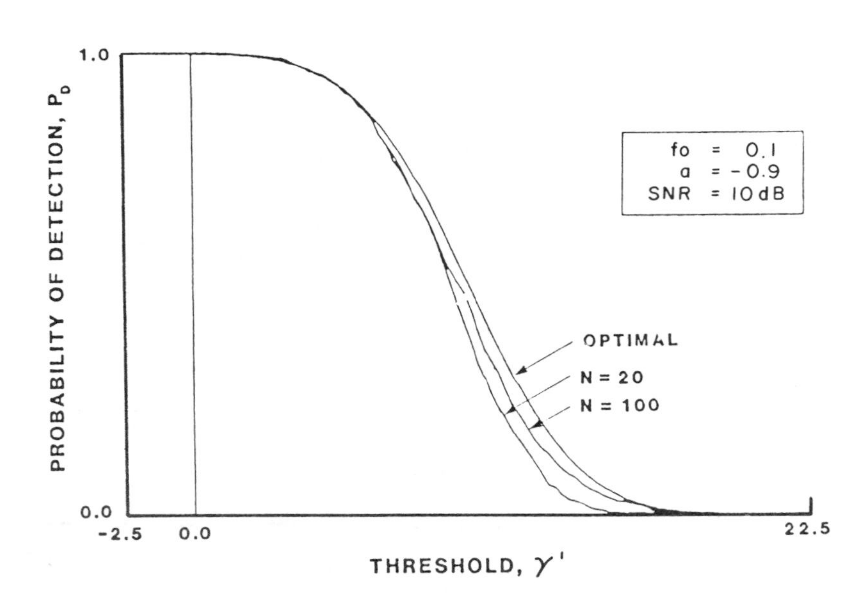

Figure 4. Performance of GLRT detector vs. optimal detector performance: a = -0.9, SNR = 10 dB.

References

1. H. L. Van Trees (1968) Detection, Estimation, and Modulation Theory, vol. I (New York: McGraw-Hill).
2. W. C. Knight, R. G. Pridham, and S. M. Kay (1981) Digital signal processing for sonar, Proc. IEEE **69**, 1451-1506.
3. A. A. Winder (1975) Sonar system technology, IEEE Trans. Sonics Ultrason. **SU-22**, 291-332.
4. M. Skolnik (1980) Introduction to Radar Systems (New York: McGraw-Hill).
5. S. M. Kay and S. L. Marple, Jr. (1981) Spectrum analysis—a modern perspective, Proc. IEEE **69**, 1380-1419.
6. G. J. Jenkins and D. G. Watts (1968) Spectral Analysis and Its Applications (San Francisco: Holden-Day).
7. D. E. Bowyer et al. (1979) Adaptive clutter filtering using autoregressive spectral estimation, IEEE Trans. Aerosp. Electron. Syst. **AES-15**, 538-546.
8. C. Gibson et al. (1979) Maximum entropy (adaptive) filtering applied to radar clutter, IEEE ICASSP, Washington, D.C., April 2-4.
9. Sir M. Kendall and A. Stuart (1977) The Advanced Theory of Statistics, vol. II (New York: Macmillan).
10. C. W. Helstrom (1968) Statistical Theory of Signal Detection, ed. 2 (New York: Pergamon Press).
11. A. Nuttall and P. Cable (1972) Operating characteristics for maximum likelihood detection of signals in Gaussian noise of unknown level, NUSC TR 4242, March 27.
12. L. Scharf and D. Lytle (1971) Signal detection in Gaussian noise of unknown level: an invariance application, IEEE Trans. Inf. Theory **IT-17**, 404-411.
13. P. Hoel, S. Port, and C. Stone (1971) Introduction to Statistical Theory (Boston: Houghton Mifflin).
14. S. Kay (1981) More accurate autoregressive parameter and spectral estimates for short data records, ASSP Workshop on Spectral Estimation, Hamilton, Ontario, Aug. 17-18.
15. A. T. Whalen (1971) Detection of Signals in Noise (New York: Academic Press).
16. S. Kay (1983) Asymptotically optimal detection in unknown colored noise via autoregressive modeling, IEEE Trans. Acoust. Speech Signal Process. **ASSP-31**, 927-940.
17. G. E. P. Box and G. J. Jenkins (1970) Time Series Analysis: Forecasting and Control (San Francisco: Holden-Day).

APPLICATION OF AUTOREGRESSIVE MODELS TO THE DOPPLER SONAR PROBLEM

W. S. Hodgkiss and D. S. Hansen

Marine Physical Laboratory, Scripps Institution of Oceanography, San Diego, California 92152

The application of pth-order autoregressive time series analysis models to the estimation of the time (range)-evolving Doppler sonar reverberation power spectrum is reported. In the Doppler sonar problem, scatterers suspended neutrally buoyant in the water column are used as tracers for the remote sensing of water mass motion at successive range intervals from the transducer. With the aid of a reverberation model, it is shown that the spatial transfer function characteristics of a transducer have a significant influence on the time (range)-evolving shape of the reverberation power spectrum. Thus, it is suggested that the use of first-order models is inadequate for the purpose of Doppler shift estimation. The results of an exploratory analysis of pings from a 70-kHz Doppler sonar appear to confirm this caution.

C. Ray Smith and W. T. Grandy, Jr. (eds.),
Maximum-Entropy and Bayesian Methods in Inverse Problems, 413–427.

1. Introduction

Doppler sonars use the perceived shift in carrier frequency between the outgoing pulse and the returning echo to make an estimate of scattering radial velocity at several ranges of interest. Inherently, the problem is one of high resolution (distinguishing small differences in velocity) spectral estimation across segments (range bins) that are too short to provide the desired resolution via conventional methods (fast Fourier transform) of spectral analysis.

In the Doppler sonar problem, it is of interest to follow the time-evolving (corresponding to range) spectral characteristics of a returning echo. A simplistic model of the corresponding range-Doppler map (a three-dimensional surface) consists of a single symmetrical hump whose track as a function of time is indicative of the radial velocity of scatterers in successive range cells. As shown in Section 2, however, a more careful consideration of both the acoustic backscattering process and the sonar transducer's spatial response characteristics suggests that a more complex model of the time-frequency surface is required.

The rest of this paper deals with the derivation and use of multiple-hump power spectrum estimation models. Section 3 provides the theoretical background, and Section 4 presents the results of applying such models to a 70-kHz Doppler sonar data set. Section 5 is a brief summary.

2. The Time-Evolving Power Spectrum

In this section, a characterization is provided of the time-evolving reverberation power spectrum expected at the output of the receiving transducers in a practical Doppler sonar system. It is shown that a multiple-hump trajectory in time-frequency space is appropriate. These results are conveyed qualitatively via several figures.

A simplified graphic representation of the volume being probed by a Doppler sonar mounted on the bottom of FLIP is provided in Fig. 1. In the case of the 70-kHz Doppler sonar designed and fabricated by Rob Pinkel (MPL), the main beam width of the transducer is approximately 1 degree and the range of the sonar is approximately 1.5 km [1,2]. A more detailed description of how a practical sonar actually views the ocean needs to consider both the transducer's sidelobes and the boundaries (surface and bottom) of the volume being probed. These considerations are taken into account in the reverberation model geometry shown in Fig. 2 [3].

The remaining figures in this section are taken from computer simulations modeling the reverberation received by a specific, autonomous, undersea vehicle [3]. Although the beam pattern of the transducer is relatively broad (approximately 15 degrees) as shown in Figs. 3a and 3b, the results are still relevant to the Doppler sonar problem in a qualitative sense as illustrations of the phenomena of interest.

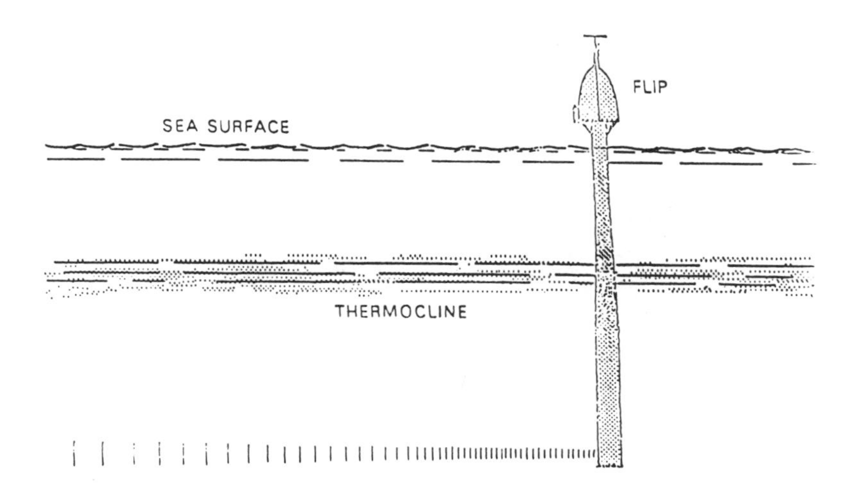

Figure 1. Doppler sonar mounted on the bottom of FLIP.

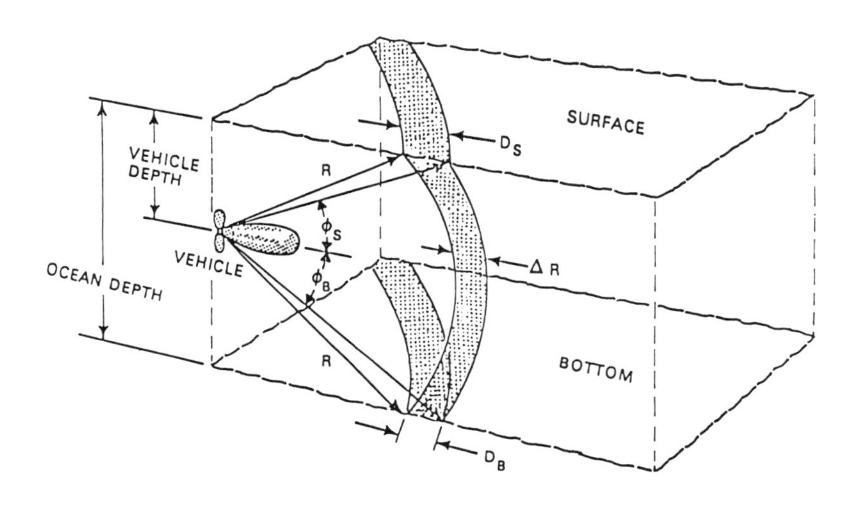

Figure 2. Reverberation model geometry.

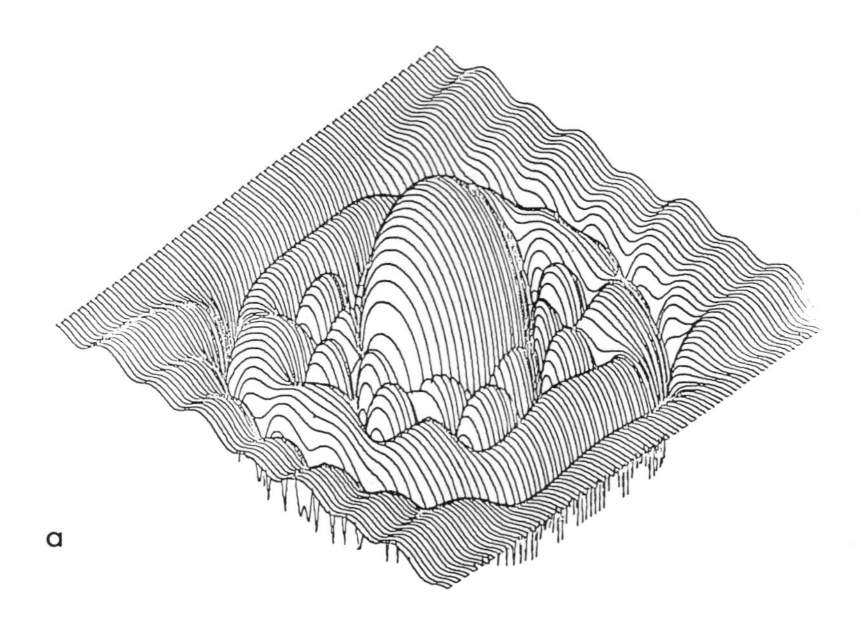

a

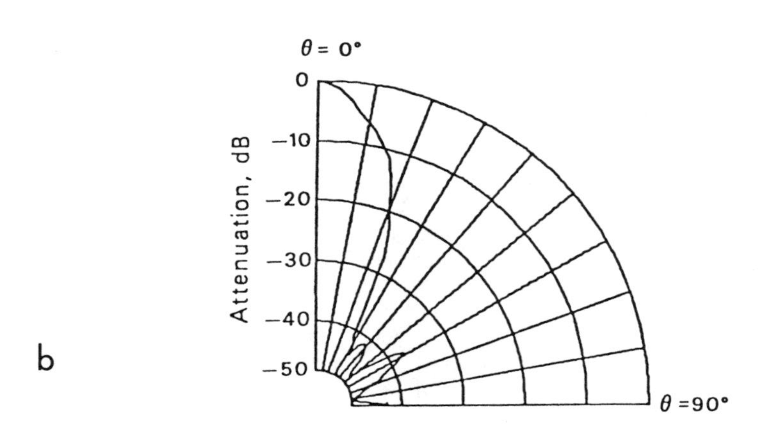

b

Figure 3. Forward beam pattern.

First, Fig. 4 shows how the total reverberation power level as a function of time (range) is partitioned between the surface, volume, and bottom returns with the vehicle at a depth of approximately 200 m. Note the sudden onset, decay, then rise again in the contributions from the surface and bottom. The explanation for this character can be seen clearly with the aid of Fig. 5 [4]. The surface projection of the beam pattern provides insight into the potential contaminating influence of surface reverberation. The manifestation of such backscatter in the frequency domain is shown in Fig. 6, where a specific range slice (R = 0.914 km) has been excised from the range-Doppler map. The spectrum has been left-shifted to correct for vehicle speed (that is, the echo from a target on transducer boresight moving with zero absolute velocity will appear at zero Doppler). The distinct second hump at negative Doppler is due to surface reverberation.

The point to be conveyed by this sequence of figures is that the spatial transfer function characteristics of a transducer have a significant influence on the time (range)-evolving shape of the reverberation power spectrum. The distinctly different sources of backscatter (for example, surface, bottom, plankton in the volume, and discrete scatterers such as fish) viewed through that spatial transfer function yield a spectrum that is complex and highly variable. Thus, in general, the use of a single, symmetrical hump model of the reverberation power spectrum is simply inadequate for Doppler shift estimation purposes.

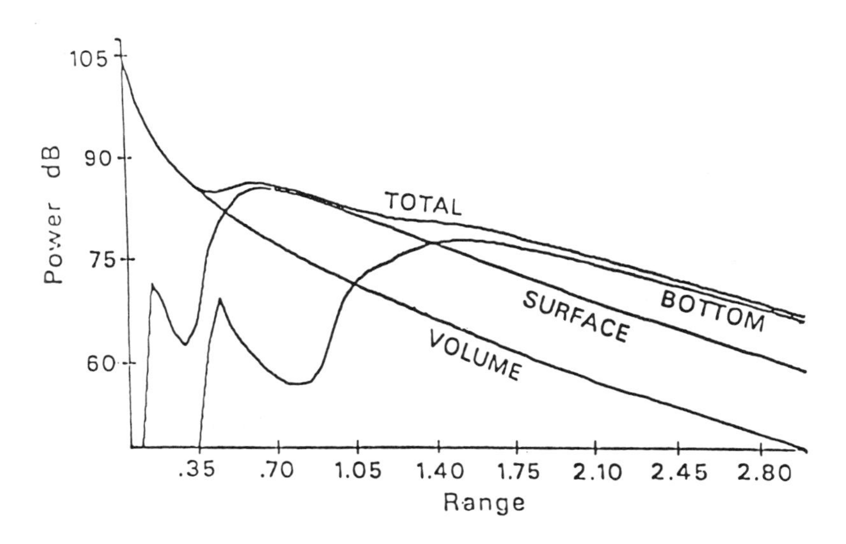

Figure 4. Reverberation power level (dB) vs. range (km) (receiver input).

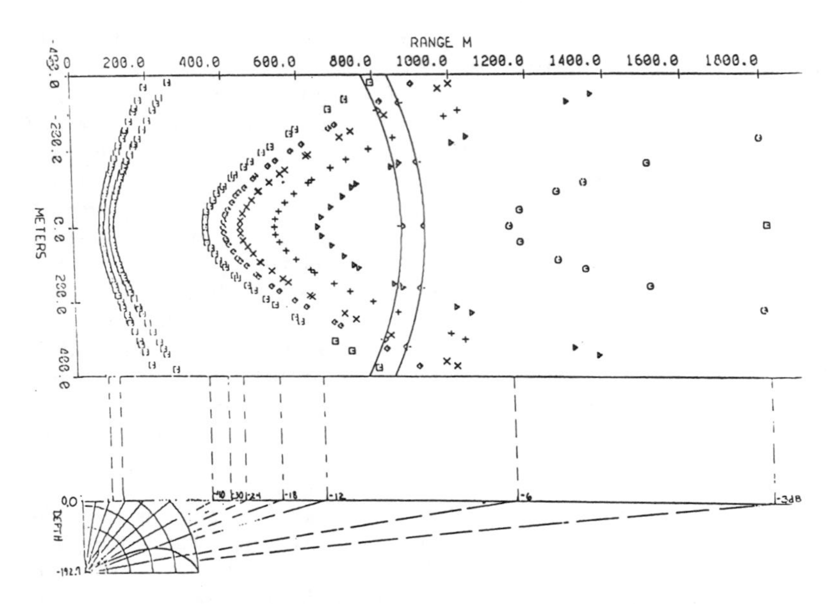

Figure 5. Surface projection of a forward-looking beam with a pulse intersection contour for a slant range of 914 m and a pulse length of 0.08 sec.

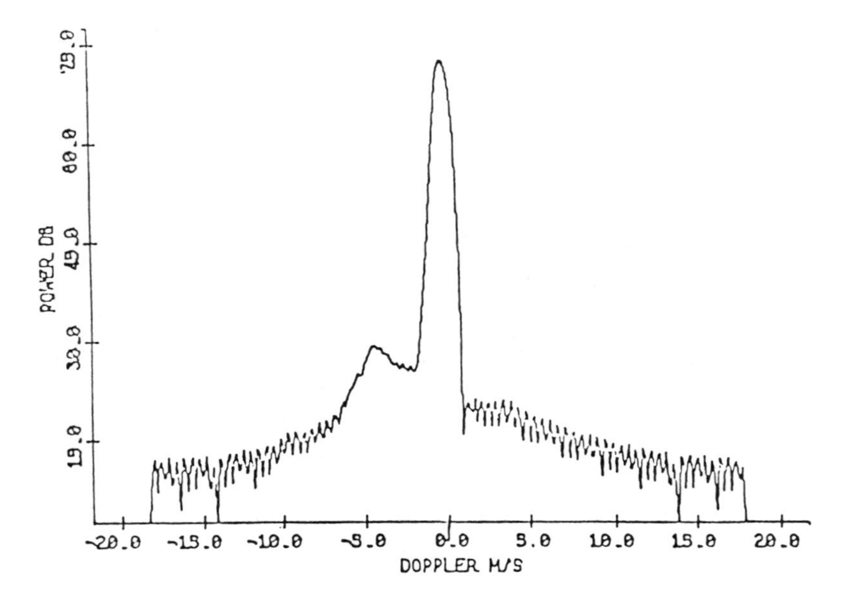

Figure 6. Surface reverberation; R = 0.914 km (receiver input).

3. High Resolution Spectral Analysis

An argument was presented in Section 2 based on both the acoustic backscattering process and the sonar transducer's spatial response characteristics suggesting that multiple-hump power spectral estimation models are needed for practical Doppler sonar systems. Now, an overview of one approach to the generation of such models will be provided.

Autoregressive Models

In recent years, much interest has been shown in high resolution spectral estimation techniques (see [5,6] for numerous references). This interest typically has been motivated by the desire to resolve narrowband frequency components in data records too short for adequate frequency separation via standard fast Fourier transform (FFT) techniques. By incorporating into the estimation problem assumptions about how the observed data were generated, rather remarkable results have been obtained.

For a number of physical reasons, modeling the observed data as an autoregressive (AR) process has been accepted widely in several applications of time series analysis. As portrayed in Fig. 7, an AR process of order p is obtained by passing the white noise sequence $\omega(n)$ having zero mean and variance σ_ω^2 through a p pole filter. The corresponding input-output relationships are

$$x(n) = -a_1 x(n-1) - a_2 x(n-2) - \dots - a_p x(n-p) + \omega(n)$$

$$= -\sum_{k}^{P} a_k x(n-k) + \omega(n) , \qquad (1)$$

where $\omega(n)$ is called the innovation of the process. By evaluating the all-pole filter's z-transform on the unit circle, we obtain the power spectrum of the AR process x(n) as

$$S_x(\omega) = \left. \frac{\sigma_\omega^2}{|A(z)|^2} \right|_{z=e^{j\omega}}$$

$$= \frac{\sigma_\omega^2}{\left|1 + \sum_{k=1}^{P} a_k e^{-j\omega k}\right|^2} . \qquad (2)$$

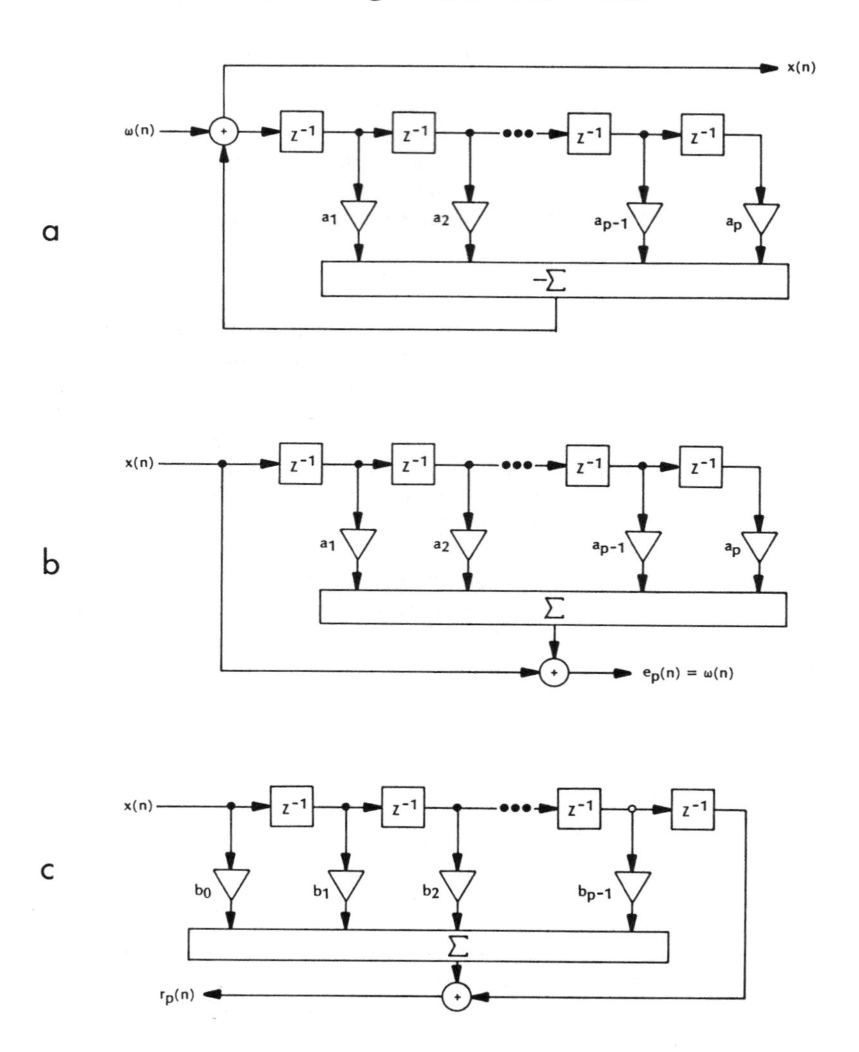

Figure 7. a. AR process generation model. b. Corresponding whitening or inverse filter (one-step forward prediction error filter). c. One-step backward prediction error filter.

Notice that the all-zero filter of order p that will recover $\omega(n)$ from $x(n)$ is $A(z)$. Appropriately, this filter has been called the whitening or inverse filter for the AR process. It is illustrated in Fig. 7b. A closer examination of Fig. 7b reveals that the structure is in the form of a one-step forward linear predictor. The most recent p samples of the AR process

$\{x(n-1), x(n-2), \ldots, x(n-p)\}$ are linearly combined to form an estimate of $-x(n)$. Removing the predictable components from $x(n)$ [or, correspondingly, the coloring from $S_x(\omega)$] yields the white forward prediction error sequence $e_p(n) = \omega(n)$. A companion to Fig. 7b is the one-step backward linear predictor (coefficients $b_0, b_1, \ldots, b_{p-1}$) shown in Fig. 7c along with the backward prediction error sequence $r_p(n)$.

The forward and backward linear predictors can be realized equivalently in the form of a lattice structure. Shown in Fig. 8, the lattice parameters $K_i^{e*} = K_i^r = K_i$ are known either as reflection coefficients (based on their correspondence to physical parameters of the AR process generation model) or partial correlation coefficients (based on their statistical interpretation as correlation coefficients). The $e_i(n)$ and $r_i(n)$ are the ith order forward and backward prediction error sequence, respectively. Thus, the pth order predictor is created on a stage-by-stage basis, a single order at a time.

Now, given that the time series of interest is pth-order AR, the spectral estimation task becomes equivalently a problem of determining the pth-order linear predictor, $A(z)$, along with the power, σ_ω^2, of its prediction error output sequence. This problem has been formulated both statistically (as a Wiener filtering problem) and deterministically (as a least-squares problem) using a minimum squared error optimality criterion.

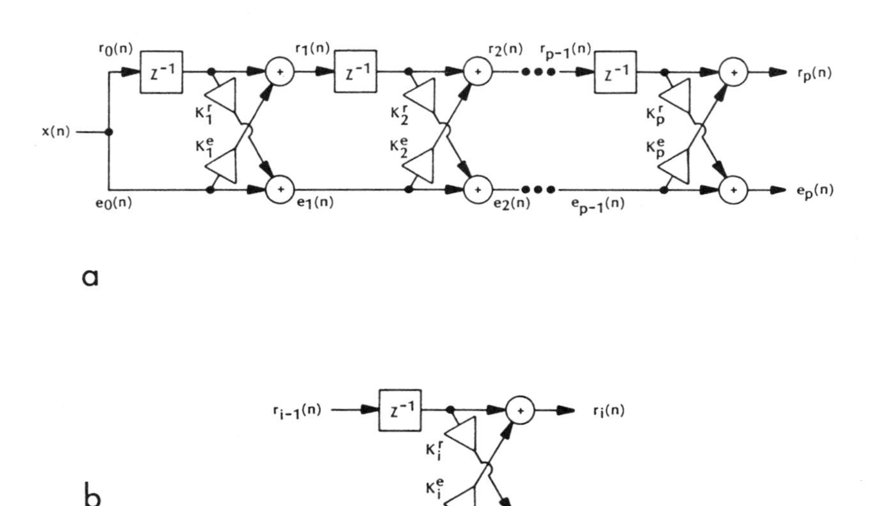

Figure 8. a. Inverse filter $A(z)$ realized in a lattice structure. b. The ith stage of the lattice.

Multiple Ping Analysis

Generally, the more data that are available, the better the results will be for the estimation of Doppler shift (velocity). The deterministic formulations of the least-squares problem have become known as the autocorrelation and covariance methods of linear prediction [7-9]. The solution algorithms require an estimate of the data autocorrelation matrix. In the case of a multiple ping analysis, the products involved in that estimate are averaged both across time (range) and from ping to ping. Clearly, tradeoffs exist regarding the time (range) extent and the number of pings over which averaging is more beneficial than detrimental.

Consider the following estimate of the autocorrelation matrix of x(n):

$$c_{ik} = \sum_j \sum_n x_j(n-i)\, x_j(n-k)^* \,, \tag{3}$$

where j is a ping index. Assuming local stationarity (the autocorrelation method) and thus forcing the resulting autocorrelation matrix to be both Hermitian and Toeplitz, $c_{ik} = c(k-i) = c(i-k)^*$ and the p+1 unknowns $\{a_1, a_2, \ldots, a_p, E_p^e\}$ can be determined in a computationally efficient, order recursive fashion:

Initialization

$$E_0^e = c(0) \,. \tag{4a}$$

Order update (i = 1,2,...,p)

$$K_i = \frac{\left[-\sum_{k=0}^{i-1} a_k^{i-1}\, c(i-k) \right]}{E_{i-1}^e} \,, \qquad a_0 = 1, \tag{4b}$$

$$a_i^{(i)} = K_i \,, \tag{4c}$$

$$a_k^{(i)} = a_k^{(i-1)} + K_i a_{i-k}^{(i-1)*} \,, \qquad 1 \le k \le i-1, \tag{4d}$$

$$E_i^e = (1-|K_i^2|)\, E_{i-1}^e \,. \tag{4e}$$

In Eqs. (4b) through (4d), the superscript i indicates the ith-order linear predictor.

The autocorrelation and covariance methods of linear prediction permit pth-order (multiple-hump) modeling of the reverberation power spectrum.

Setting p = 1 is exactly equivalent to the first moment (mean value) approach [10, 11]. It is appropriate when the spectrum to be estimated can be modeled as a single, symmetrical hump as a function of frequency (a one-pole AR model). However, the discussion in Section 2 suggests that such a model is too simplistic for a practical Doppler sonar.

4. Data Analysis

The exploratory data analysis reported in this section concerns data collected off of FLIP by Rob Pinkel (MPL) in November 1980. During this experiment, sonars at two center frequencies (67 kHz and 70.1 kHz) were viewing simultaneously the same ocean volume. The sonar transducers were pointed 45 degrees downward. Pulse lengths of two durations (40 ms and 10 ms) were launched at different times during the experiment. Here, only an illustrative sample of the results from processing the 67 kHz/40 ms data set will be discussed.

Figure 9 plots the radial velocity estimate as a function of slant range derived from a p = 1 AR model. The range interval displayed is only a small fraction of the total range of the sonar (approximately 1.5 km). For this pulse length of 40 ms, range resolution cells are approximately 30 m apart in slant range.

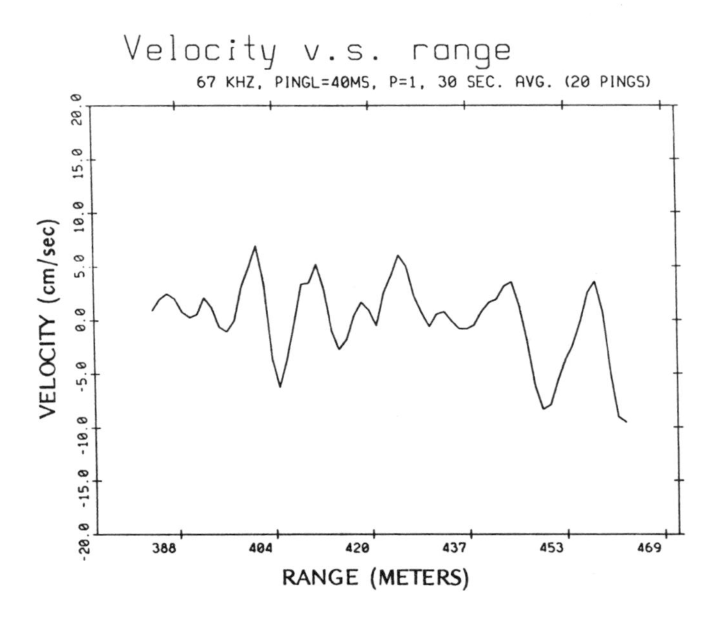

Figure 9. Radial velocity versus slant range, derived from a p = 1 AR model.

For the same segment of data, Fig. 10 displays successive (normalized) spectral estimates derived from a p = 2 AR model. Although predominantly unimodal (one-hump), at times a more complex character is noted.

One such complex range slice is observed at 444 m. For that range, a sequence of spectral estimates derived from p = 1,...,4 AR models is provided in Figs. 11 through 14. In all cases, the dashed vertical line indicates the p = 1 velocity estimate. Note the distinct influence on the first moment (p = 1) velocity estimate from what appears to be a consistent second hump in the reverberation power spectrum.

5. Summary

In the Doppler sonar problem, it is of interest to follow the time-evolving (corresponding to range) spectral characteristics of a returning echo. In Section 2, a careful consideration of both the acoustic backscattering process and the sonar transducer's spatial response characteristics suggested that a multiple-hump model of the time-frequency surface is required for a practical Doppler sonar system. An overview of one approach to the generation of such models then was provided in Section 3. Pth-order autoregressive models were discussed, and a solution algorithm corresponding to the autocorrelation method of linear prediction was presented. Lastly, Section 4 reported on the exploratory data analysis of a Doppler sonar data set. A comparison between the first moment (p = 1) velocity estimate and a higher-order characterization of the reverberation power spectrum appears to substantiate the usefulness of more complex models.

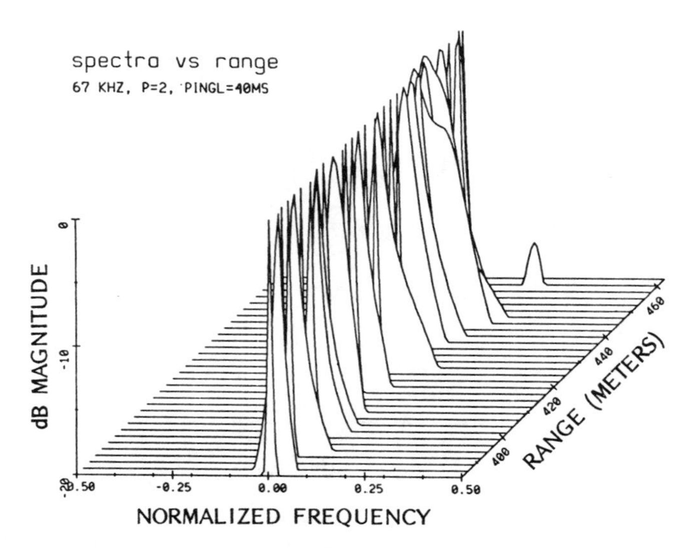

Figure 10. Successive spectral estimates, derived from a p = 2 AR model.

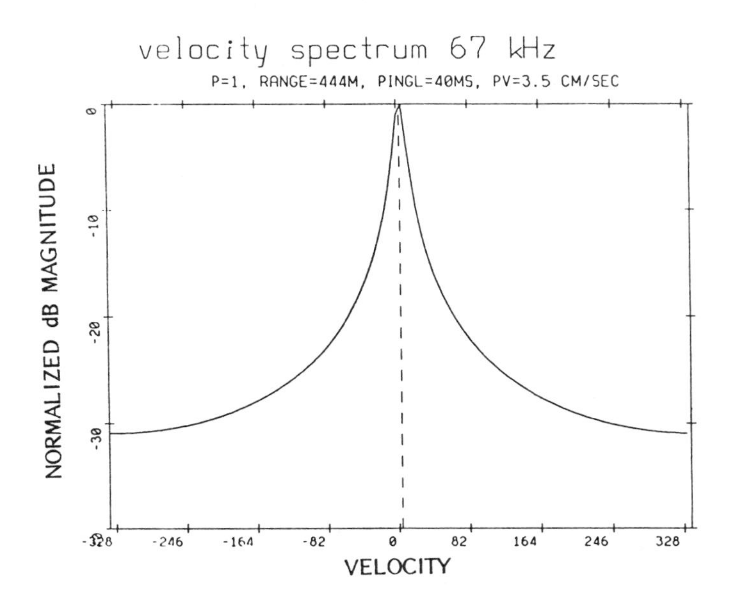

Figure 11. Spectral estimate at 444 m, derived from a p = 1 AR model.

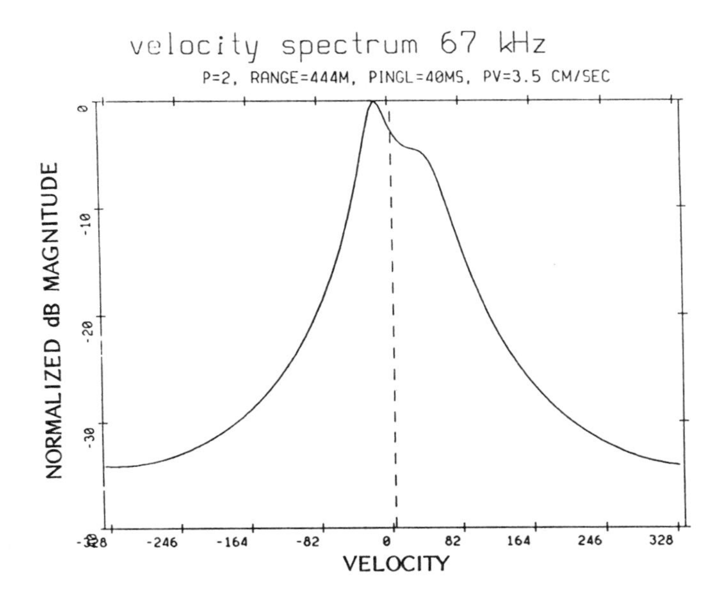

Figure 12. Spectral estimate at 444 m, derived from a p = 2 AR model.

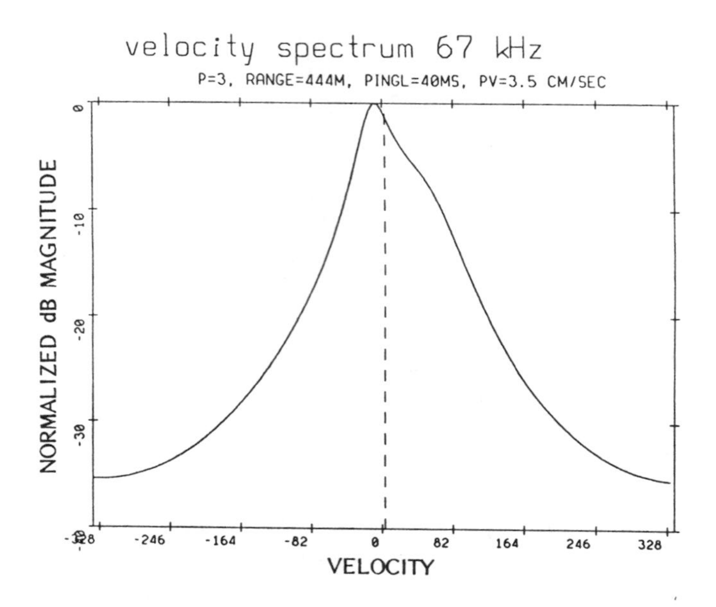

Figure 13. Spectral estimate at 444 m, derived from a p = 3 AR model.

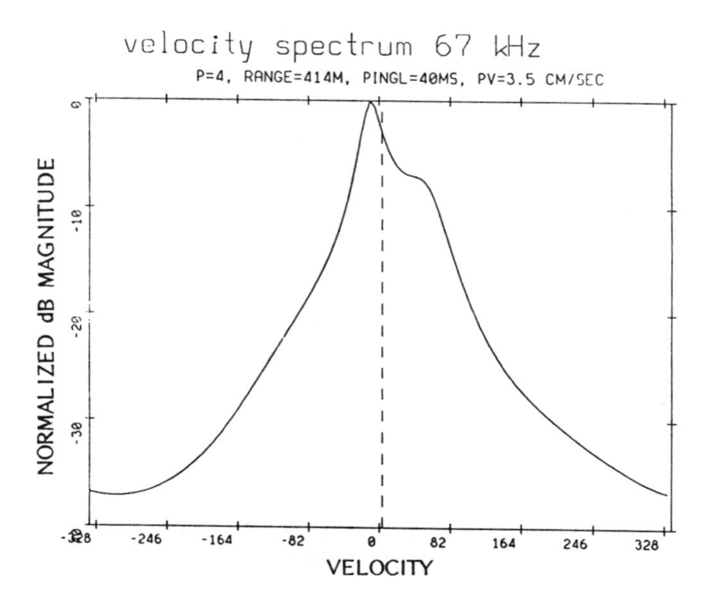

Figure 14. Spectral estimate at 444 m, derived from a p = 4 AR model.

Acknowledgments

This work was supported in part by the following sponsors: NOAA grant NA81AA-D-00082; APL/JH contract N00024-81-C-5301 (subcontract 601572-0); and ONR contract N00014-81-K-0191.

References

1. E. Slater and R. Pinkel (1979) A 32-kW Doppler sonar, Proceedings OCEANS '79, 17-19 Sept. 1979, San Diego, Calif., pp. 137-141.
2. L. Occhiello and R. Pinkel (1979) A Doppler sonar controller, Proceedings OCEANS '79, 17-19 Sept. 1979, San Diego, Calif., pp. 148-152.
3. W. Hodgkiss (1980) Reverberation model: I. Technical description and user's guide, MPL TM-319, 31 July 1980, Marine Physical Laboratory, Scripps Institution of Oceanography, San Diego, Calif.
4. J. Oltman (1981) Surface projection package, MPL TM-327, 27 February 1981, Marine Physical Laboratory, Scripps Institution of Oceanography, San Diego, Calif.
5. W. Hodgkiss and J. Presley (1981) Adaptive tracking of multiple sinusoids whose power levels are widely separated, IEEE Trans. Acoust. Speech Signal Process. **ASSP-29** (3), 710-721.
6. W. Hodgkiss and J. Presley (1982) The complex adaptive least-squares lattice, IEEE Trans. Acoust. Speech Signal Process. **ASSP-30** (2), 330-333.
7. D. G. Childers, ed. (1978) <u>Modern Spectrum Analysis</u> (New York: IEEE Press).
8. S. Haykin, ed. (1979) <u>Nonlinear Methods of Spectral Analysis</u> (New York: Springer-Verlag).
9. S. M. Kay and S. L. Marple (1981) Spectrum analysis—a modern perspective, Proc. IEEE **69**, 1380-1419.
10. K. Miller and M. Rochwaiger, (1972) A covariance approach to spectral moment estimation, IEEE Trans. Inf. Theory **IT-18**, 588-596.
11. R. Doviak, D. Zrnic, and D. Sirmans (1979) Doppler weather radar, Proc. IEEE **67**, 1522-1553.

MAXIMUM-ENTROPY AND DATA-ADAPTIVE PROCEDURES IN THE INVESTIGATION OF OCEAN WAVES

Leon E. Borgman

Professor of Geology and Statistics, University of Wyoming, Laramie 82071

The analytic decomposition of ocean waves into a mixture of wave trains with different travel directions and frequencies is based on cross-spectral analysis of time measurements of multiple wave properties. The Norwegian government, with funding by the Chevron Oil Field Research Company, is planning an extensive comparison of various systems for producing the ocean wave decompositions from wave data collected on a North Sea offshore platform. The comparisons will involve both instrumentation and analysis systems.

As a preliminary guide for planning, a survey has been made of various procedures appropriate for directional wave analysis. Where possible the techniques have been generalized and extended to treat the special case of ocean wave directional structure. Techniques concerned with maximum likelihood, maximum entropy, variational fitting, linear programming, and classical Fourier analysis all have been used or show promise for ocean wave investigations.

C. Ray Smith and W. T. Grandy, Jr. (eds.),
Maximum-Entropy and Bayesian Methods in Inverse Problems, 429–442.

1. Introduction

The Marine Board of the National Research Council convened a workshop on ocean wave measurement technology in Washington, D.C., August 22-24, 1981. During several days of review, the 65 scientists and engineers present developed a list of the ten highest-priority research areas relative to wave investigations. The top two priority items were concerned with the development of satisfactory instrument systems for measuring directional wave spectra and of accurate, efficient methods for processing the data. The third priority item was concerned with a field measurement program for intercomparison of instrument systems (Dean, 1981).

Partly as an outgrowth of this meeting, the Chevron Oil Field Research Company is funding a project through the Norwegian government to investigate these topics in conjunction with wave measurements in the North Sea. Both hardware and software comparisons are to be included. The project is in a very preliminary planning stage now. The following review of available analysis procedures and tentative theoretical extensions is a part of that preliminary investigation.

2. An Ocean Wave Model

The random-phase model is the most widely used basis for ocean wave analysis (Borgman, 1970; Borgman and Yfantis, 1979). This may be stated as

$$\eta(x,y,t) = \text{L.i.m.} \sum_{m=1}^{J} \sum_{\ell=0}^{L} a_{m\ell} \times \cos[k_m x \cos\theta_\ell + k_m y \sin\theta_\ell - 2\pi f_m t + \Phi_{m\ell}] \quad (1)$$

with

$$(2\pi f_m)^2 = g k_m \tanh k_m d , \quad (2)$$

where

$\eta(x,y,t)$ = water level elevation above mean water level

(x,y) = horizontal coordinates in a right-handed system with z positive upward

t = time

L.i.m. denotes the probabilistic limit in quadratic mean as $J \to \infty$, $L \to \infty$, and difference mesh $\to 0$

$a_{m\ell}$ = wave amplitudes

k = wave number = 2π/wavelength

f = wave frequency (cycles per unit time)

θ = direction toward which wave is traveling

Φ = random phase uniformly distributed over $(0,2\pi)$ radians and independent for each different set of subscripts

The wave amplitudes are taken so as to keep a finite amount of variance in a given small frequency interval and direction interval. The variance associated with waves in $f \pm \Delta f/2$ and $\theta \pm \Delta\theta/2$ is characterized by a spectral density $S(f,\theta)$ through the relationship

$$a_{m\ell}^2/2 = 2S(f,\theta)\,\Delta\theta\,\Delta f \tag{3}$$

or

$$a_{m\ell} = 2\sqrt{S(f,\theta)\,\Delta\theta\,\Delta f}\ . \tag{4}$$

The cosine argument just relates a long-crested, progressive wave traveling in the θ direction to the time-space coordinates. It may be shown from a central limit theorem that $\eta(x,y,t)$ is a Gaussian field.

Equations (1) and (4) are often combined symbolically as

$$\eta(x,y,t) = 2\int_0^{\infty}\int_0^{2\pi} \sqrt{S(f,\theta)\,d\theta\,df}$$

$$\times \cos[kx\cos\theta + ky\sin\theta - 2\pi ft + \Phi]\ . \tag{5}$$

This is not, of course, an integral in deterministic calculus. It is rather a symbolic statement of probabilistic integral defined as a limit in quadratic mean. However, it is a convenient notation and will be used in what follows.

3. Properties Measured in In Situ Instrument Systems

Various wave characteristics in space arrays or at a point have been measured to study the directional spectral density. Representative examples are

(a) water level elevations, $\eta(x,y,t)$, measured at an array of locations;
(b) orthogonal components of sea surface slope, $\eta_x(x,y,t)$ and $\eta_y(x,y,t)$, from a floating buoy;
(c) slopes and elevations in a spatially constrained array of buoys; and
(d) horizontal orthogonal components of water particle velocity $V_x(x,y,t)$ and $V_y(x,y,t)$, together with either the water level elevation, $\eta(x,y,t)$, or the pressure fluctuation, $\phi(x,y,z,t)$, at vertical coordinate z.

The random-phase models for these properties are as follows:

<u>Water level elevation:</u>

Given by Eq. (1)

Sea surface slopes:

$$\begin{bmatrix} \eta_x(x,y,t) \\ \eta_y(x,y,t) \end{bmatrix} = -2 \int_0^{\infty} \int_0^{2\pi} \sqrt{S(f,\theta)\,d\theta\,df} \;\; k \begin{bmatrix} \cos\theta \\ \sin\theta \end{bmatrix}$$

$$\times \sin[kx\cos\theta + ky\sin\theta - 2\pi ft + \Phi] \qquad (6)$$

Horizontal water particle velocities:

$$\begin{bmatrix} V_x(x,y,t) \\ V_y(x,y,t) \end{bmatrix} = 2 \int_0^{\infty} \int_0^{2\pi} \sqrt{S(f,\theta)\,d\theta\,df} \;\; (2\pi f) \; \frac{\cosh[k(d+z)]}{\sinh kd}$$

$$\times \begin{bmatrix} \cos\theta \\ \sin\theta \end{bmatrix} \cos[kx\cos\theta + ky\sin\theta - 2\pi ft + \Phi] \qquad (7)$$

Pressure fluctuations about static pressure:

$$\phi(x,y,z,t) = 2 \int_0^{\infty} \int_0^{2\pi} \sqrt{S(f,\theta)\,d\theta\,df} \;\; \rho g \; \frac{\cosh[k(d+z)]}{\cosh kd}$$

$$\times \cos[kx\cos\theta + ky\sin\theta - 2\pi ft + \Phi] \qquad (8)$$

All of these properties may be written in a standard form. Let $p_j(t)$ denote the time history of the jth wave property measured at location (x_j,y_j,z_j). An array of measurements corresponds to a set of time series, $p_j(t)$, $j = 1,2,3,\ldots,J$. The wave properties, $p_j(t)$, may be measured at the same location or at different points in space. A standard form is

$$p_j(t) = 2 \int_0^{\infty} \int_0^{2\pi} \sqrt{S(f,\theta)\,d\theta\,df} \;\; \tilde{W}_j(f)\, G_j(\theta) \begin{bmatrix} \sin \\ \cos \end{bmatrix}$$

$$\times [kx\cos\theta + ky\sin\theta - 2\pi ft + \Phi] \,, \qquad (9)$$

where $\begin{bmatrix} \sin \\ \cos \end{bmatrix}$ means that one or the other is used. If the sine function is used for the jth wave property, let

$$A(f,\theta)\Delta f = \begin{cases} \sqrt{S(f,\theta)\,d\theta\,df}\; e^{-i\Phi}, & \text{if } f > 0 \\ \sqrt{S(|f|,\theta)\,d\theta\,df}\; e^{i\Phi}, & \text{if } f < 0 \end{cases} \tag{10}$$

and $W_j(f) = i\,\tilde{W}_j(f)$. If the cosine function is used, let

$$A(f,\theta)\Delta f = \begin{cases} \sqrt{S(f,\theta)\,d\theta\,df}\; e^{-i\Phi}, & \text{if } f > 0 \\ \sqrt{S(|f|,\theta)\,d\theta\,df}\; e^{+i\Phi}, & \text{if } f < 0 \end{cases} \tag{11}$$

and $W_j(f) = \tilde{W}_j(f)$. Define

$$\int_{-\infty}^{\infty} \quad \text{and} \quad \int_{0}^{2\pi}$$

as symbols denoting a very fine mesh summation over f and θ respectively. It is also understood that k is negative whenever f is negative. The jth wave property may be written

$$p_j(t) = \Delta f \int_{-\infty}^{\infty} \int_{0}^{2\pi} A(f,\theta)\; W_j(f)\; G_j(\theta)$$

$$\times \exp\{-i[kx_j\cos\theta + ky_j\sin\theta] + i2\pi ft\}\,. \tag{12}$$

The probabilistic differential mesh over (f,θ) space is implicitly understood to be contained within $A(f,\theta)$ and the summation over frequency has been extended to negative frequencies to accommodate the complex sinusoids.

At any very fine, but fixed, mesh of summation, Eq. (12) may be approximated by its finite summation form

$$p_j(t) = \Delta f \sum_{m=-M}^{M} \sum_{\ell=0}^{L} A(f_m,\theta_\ell)\; W_j(f_m)\; G_j(\theta_\ell)$$

$$\times \exp\{-i[k_m x_j\cos\theta_\ell + k_m y_j\sin\theta_\ell] + i2\pi mn/N\}\,, \tag{13}$$

where the following definitions hold:

Δt = time increment
$t_n = n\Delta t,\ 0 \leq n \leq N-1$
Δf = frequency increment
$f_m = m\Delta f,\ -M \leq m \leq M$
$\Delta\theta$ = angular increment
$\theta_\ell = \ell\Delta\theta,\ 0 \leq \ell \leq L$ covering $(0,2\pi)$ radians

and Δf, Δt, and N are chosen so that

$$M < N/2 , \quad N \, \Delta t \, \Delta f = 1.0 . \tag{14}$$

This last condition leads to

$$2\pi f_m t_n = 2\pi (m\Delta f)(n\Delta t) = 2\pi mn/N , \tag{15}$$

which was used in Eq. (13).

It is convenient to repeat the integrand in Eq. (13) periodically relative to the subscript m using period N. This moves the terms for $-M \leq m < 0$ to $N-M \leq m < N$. Inspection shows that the integrand at N-m is the complex conjugate of the integrand at m. For future reference, it is also noted that $G_j(\theta)$ is real valued.

The discrete Fourier transform may be applied to $p_j(t_n)$ to compute

$$P_{jm} = \Delta t \sum_{n=0}^{N-1} p_j(n\Delta t) \exp(-i2\pi mn/N)$$

$$= \sum_{\ell=0}^{L} A(f_m, \theta_\ell) \, W_j(f_m) \, G_j(\theta_\ell)$$

$$\times \exp\{-i[k_m x_j \cos\theta_\ell + k_m y_j \sin\theta_\ell]\} . \tag{16}$$

The discrete Fourier transform eliminates t and isolates the waveform at frequency f_m. The data version of Eq. (16) is obtained by applying the fast Fourier transform to the data time series. Equation (16) summarizes the model theoretical structure of the data transform.

4. Types of Analysis Techniques

The most commonly used procedures for analyzing the time series output of wave property measurement systems are based on equations relating the co- and quad-spectral densities to the Fourier series expansion of $S(f,\theta)$ relative to θ at each fixed reference frequency. A summary of such procedures is given by Borgman (1970) and will not be repeated here. The Fourier techniques are moderately successful provided the function $S(f,\theta)$ is unimodal and relatively slowly and smoothly varying with respect to θ. However, difficulties arise if bimodality or directional spikes are present.

It is difficult and costly to maintain elaborate multi-sensor systems in the hostile marine environments. Successful in situ systems have been both sparse and relatively simple. This suggests that data adaptive techniques may be useful in processing the P_{jm}, $j = 1,2,3,\ldots,J$, in Eq. (16). Indeed this has proven true for space arrays of wave staffs or bottom pressure cells. However, the methods have mostly not been extended to multiple measurements at the same locations. The exceptions to this are the variational method developed by Long and Hasselmann (1979), which has been applied to buoy data, and the linear programming procedure, which has been used to analyze ship motion (Webster and Dillingham, 1981).

The possible data adaptive procedures that have been used or appear to be capable of application to the directional wave problem are:

(a) the maximum likelihood method (Oakley and Lozow, 1977; Baggeroer, 1979; Clarke and Gedling, 1981; Jeffreys, Wareham, Ramsden, and Platts, 1981; Goda, 1981; Davis and Regier, 1977; Regier and Davis, 1977);
(b) the maximum entropy technique (Baggeroer, 1976a);
(c) a variational fitting procedure (Long and Hasselmann, 1979; Long, 1980);
(d) a linear programming method (Oakley, 1973; Webster and Dillingham, 1981).

Good overall summaries of the procedures and applications are given by Oakley (1973) and Baggeroer (1979).

The maximum likelihood method has primarily been introduced after the discrete Fourier transform has been applied to isolate energy at each frequency. The published applications noted are restricted to space arrays of measurements, mostly water level elevations. Standard spectral estimation procedures based on the discrete Fourier transform typically have been used to move from time to frequency domain. The data adaptive procedures are then introduced to treat the wave number data.

Various attempts have been made to extend the maximum entropy technique to space arrays of the Fourier coefficients. The main problems are well summarized by Baggeroer (1976b), who gives a wave number extension and further discusses the problems in a later paper (1979).

In the Long and Hasselmann variational technique, various measures of goodness of the directional estimate in terms of nearness to a subjectively selected model and desired spectral properties are introduced, and relationships from the calculus of variations are used to determine the estimation function.

For the linear programming methods, a tent function representation of the directional spectrum is combined with a minimization of the maximum difference between the data and the theoretical Fourier transforms.

5. A Maximum Likelihood Method

At frequency $m\Delta f$, define

$$\varepsilon_m = \sum_{j=1}^{J} \alpha_{jm} P_{jm} / W_j(f_m) , \qquad (17)$$

where α_{jm} are complex-valued coefficients that will be specified later. In terms of the structure outlined in Eq. (16), this gives

$$\varepsilon_m = \sum_{\ell=0}^{L} A(f_m,\theta_\ell) \left\{ \sum_{j=1}^{J} \alpha_{jm} G_j(\theta_\ell) \times \exp[-i(k_m x_j \cos\theta_\ell + k_m y_j \sin\theta_\ell)] \right\} . \quad (18)$$

Let z^* denote the complex conjugate of z, and $\underline{v}'$ denote the matrix transpose of the vector $\underline{v}$, where the underbar designates column vectors. The variance of ε_m is given by

$$\sigma^2_m = E[\varepsilon^*_m \varepsilon_m] = \underline{\alpha}^{*\prime} S \underline{\alpha} \quad (19)$$

where

$$\underline{\alpha} = \begin{bmatrix} \alpha_{1m} \\ \alpha_{2m} \\ \cdot \\ \cdot \\ \cdot \\ \alpha_{jm} \end{bmatrix}, \quad (20)$$

and

$$S = E[\underline{P}^* \underline{P}'] , \qquad \underline{P} = \begin{bmatrix} P_{1m}/W_1(f_m) \\ P_{2m}/W_2(f_m) \\ \cdot \\ \cdot \\ \cdot \\ P_{jm}/W_j(f_m) \end{bmatrix} . \quad (21)$$

In general, S is proportional to the cross-spectral matrix for the J time series at frequency $m\Delta f$.

If the expression within the { } brackets in Eq. (18) behaves like the Kronecker delta

$$\delta_{\ell\ell_0} = \begin{cases} 1, & \text{if } \ell = \ell_0 \\ 0, & \text{otherwise} \end{cases} \quad (22)$$

then (with $\theta_0 = \ell_0 \Delta\theta$)

$$\varepsilon_m = \sum_{\ell=0}^{L} \delta_{\ell\ell_0} A(f_m,\theta_\ell) = A(f_m,\theta_0) . \quad (23)$$

Hence σ^2_m would be proportional to the variance of the wave at (f_m,θ_0).

Suppose the expression within the { } brackets is required to equal a constant c when $\ell = \ell_0$ but the coefficients α_{jm} are adjusted otherwise to minimize σ_m^2. Let

$$\underline{G}^* = \text{vector whose jth component is}$$
$$G_j(\theta_0) \exp[+i(k_m x_j \cos\theta_0 + k_m y_j \sin\theta_0)] \ . \qquad (24)$$

All of the above can be summarized by the following. Choose α_{jm} to

$$\text{minimize } \underline{\alpha}^{*\prime} S \underline{\alpha} \qquad (25)$$

subject to the constraint

$$\underline{G}^{*\prime}\underline{\alpha} = c, \quad \text{with } |c| = 1 \ . \qquad (26)$$

This is the standard form for the maximum likelihood method of ocean wave processing (Jeffreys et al., 1981, p. 206; Kay and Marple, 1981, p. 1407) except that the G vector does not necessarily coincide with the $\underline{x}$ vector of space array positions typically used.

The solution to Eqs. (25) and (26) is given in both of the previous references as

$$\hat{S}(f_m, \theta_0) \propto \hat{\sigma}^2_m = (\underline{G}^{*\prime} S^{-1} \underline{G})^{-1} \ . \qquad (27)$$

The use of $\underline{G}$ in place of the space array exponentials does not affect the mathematics of optimization.

Equation (27), when computed for a number of different θ_0 around the circle, provides values that are proportional only to the spectral estimates. Therefore, it appears reasonable to estimate the total variance at f_m by the usual frequency-only procedures and then normalize the directional estimates to make them give the correct total variance when summed over directions. However, this assumes the same constant of proportionality for each θ_0. If the array presents an approximately similar configuration to all directions, this may be reasonable. Such questions would need to be investigated for each array.

The vector of functions $\underline{G}$ should be selected so that the Kronecker delta is moderately well approximated for most directions of interest. Presumably several different space locations could be combined with orthogonal pairs of wave properties, such as sea surface slopes or subsea water velocity components, to give reasonable approximations.

6. A Maximum Entropy Method

The structure of the matrix S in Eq. (21) can be developed from Eq. (16) and the definitions of $A(f,\theta)$ in Eqs. (10) and (11):

$$E[P_{jm}^{*}P_{hm}] = \sum_{\ell=0}^{L} S(f_m,\theta_\ell)\, W_j(f_m)\, W_h(f_m)\, G_j(\theta_\ell)\, G_h(\theta_\ell)$$

$$\times \exp\{-[k_m(x_h-x_j)\cos\theta_\ell + k_m(y_h-y_j)\sin\theta_\ell]\}\, \Delta\theta\, \Delta f\ . \quad (28)$$

The uniform distribution and independence of Φ_{mj} and Φ_{mh} have been used through the relationships

$$E[\exp\{i(\Phi_{mh} - \Phi_{mj})\}] = \begin{cases} 0, & \text{if } h \neq j \\ 1, & \text{if } h = j \end{cases} \quad (29)$$

to reduce $\sum_{\ell=0}^{L} \sum_{\ell=0}^{L}$ in the product to the single summation.

The mesh over each direction may be made infinitely small to give

$$S_{jh} = E[P_{jm}^{*}P_{hm}]/W_j(f_m)W_h(f_m)$$

$$= \Delta f \int_0^{2\pi} S(f_m,\theta)\, G_j(\theta)\, G_h(\theta)$$

$$\times \exp\{-i[k_m(x_h-x_j)\cos\theta + k_m(y_h-y_j)\sin\theta]\}\, d\theta\ . \quad (30)$$

These are the elements of S in Eq. (21). In terms of $\underline{G}$ in Eq. (24), the last equation can be written

$$S = \Delta f \int_0^{2\pi} S(f_m,\theta)\, \underline{G}^{*}\underline{G}' d\theta\ . \quad (31)$$

The matrix S is Hermitian symmetric ($S^{*\prime} = S$), so the real and imaginary parts of Eq. (31) can be separately listed in J^2 real equations of the form

$$U_n(f_m) = \int_0^{2\pi} S(f_m,\theta)\, T_n(\theta)\, d\theta\ , \quad n = 1,2,\ldots,N,\ \ N = J^2, \quad (32)$$

where the U_n are spectral densities or co- or quad-spectral densities. The functions $T_n(\theta)$ are implicitly also functions of frequency.

A type of maximum entropy framework can be established. Let $S(f_m,\theta)$ be the function of θ that minimizes

$$\xi = -\int_0^{2\pi} \ln [S(f_m,\theta)]\, d\theta \tag{33}$$

subject to the constraints

$$U_n(f_m) = \int_0^{2\pi} S(f_m,\theta)\, T_n(\theta)\, d\theta\ , \qquad n = 1,2,\ldots,N\ . \tag{34}$$

By variational procedures (Rustagi, 1976) the solution satisfies the system of equations

$$-\frac{1}{S(f_m,\theta)} + \sum_{n=1}^{N} \lambda_n T_n(\theta) = 0 \tag{35}$$

$$\int_0^{2\pi} S(f_m,\theta)\, T_n(\theta)\, d\theta = U_n(f_m)\ , \quad n = 1,2,\ldots,N\ , \tag{36}$$

where λ_n are Lagrangian coefficients. Consequently

$$S(f_m,\theta) = \left[\sum_{n=1}^{N} \lambda_n T_n(\theta)\right]^{-1} \tag{37}$$

$$\int_0^{2\pi} \left[T_n(\theta)\Big/\sum_{j=1}^{N} \lambda_j T_j(\theta)\right] d\theta = U_n(f_m)\ . \tag{38}$$

Various techniques can be used to search for the solution $(\lambda_1,\lambda_2,\ldots,\lambda_N)$ of Eq. (38). Then the directional spectrum at $f = f_m$ may be obtained from Eq. (37).

7. Concluding Comments

1. This report represents a very preliminary review of what has been done and what might be done relative to directional ocean wave analysis. The direct Fourier procedures, the linear programming method, and the Long and Hasselmann variational procedures are not covered here because the published acounts are fairly complete.

2. The modifications of the maximum likelihood and maximum entropy techniques to apply to arrays of assorted measurement devices appear to offer reasonable approaches to analysis. The maximum likelihood method will probably prove the most useful of all of these. The maximum entropy procedure still involves a few unresolved computational problems.

3. Field testing of wave measurement systems requires some standard for ground truth. Photogrammetric measurements with two cameras, although laborious to analyze, probably provide the best measurement for what is actually present in the wave systems. New developments in computerized remote-sensing analysis may help greatly in the processing of such data.

4. As another source of ground truth information, the complete documentation of meteorological conditions during the preceding measurement intervals should be collected. With wave hindcasting procedures, principal directions of the wave trains present and some measure of the directional dispersion about each direction can be deduced from the meteorological map data.

5. Intercomparison of results from different instrument systems provides some information on accuracy. While agreement between systems would suggest satisfactory performance, some disagreement is almost always present. Then the two delicate problems of establishing significant difference standards and criteria to determine which results are most nearly correct must be examined.

6. The methods surveyed were restricted to those that have been used or suggested for directional ocean analysis. It is worth noting that the whole area of spectral analysis is undergoing substantial development in the area of "high-resolution" techniques. The survey paper by Kay and Marple (1981) gives a good perspective on the present situation. This intense research activity should be monitored carefully to ensure that promising new methods are considered for the wave analysis.

7. An obvious way to study the directional resolution of instrument array systems and their modes of analysis is through computer simulation (Borgman, 1982). This represents a later stage of work in the continuation of these reported studies.

References

Baggeroer, A. B. (1976a) Confidence intervals for regression (MEM) spectral estimates, IEEE Trans. Inf. Theory **IT-22**, 534-545.

Baggeroer, A. B. (1976b) Space/Time Random Processes and Optimum Array Processing, Rept. NUC TP 506, Dept. Ocean Eng., MIT, Cambridge, Mass.

Baggeroer, A. B. (1979) Recent advances in spectral and directional wave spectra analysis and their application to offshore structures, Conference on Behaviour of Offshore Structures, London.

Borgman, L. E. (1970) Directional spectra from wave sensors, Ocean Wave Climate, **88**, 269-300, Marine Science Series (New York: Plenum Press).

Borgman, L. E. (1982) Techniques for computer simulation of ocean waves, Topics in Ocean Physics (Bologna, Italy: Italian Physical Society), pp. 387-417.

Borgman, L. E., and Yfantis, E. (1979) Three-dimensional character of waves and forces, Proc. Civil Engineering in the Oceans, Vol. 4 (New York: American Society of Civil Engineers), pp. 791-804.

Clarke, D., and Gedling, P. (1981) MLM estimation of directional wave spectra, Proc. Directional Wave Spectra Applications (New York: American Society of Civil Engineers), pp. 21-41.

Davis, R. E., and Regier, L. A. (1977) Methods for estimating directional wave spectra from multi-element arrays, J. Mar. Res. **35**, 453-477.

Dean, R. G. (1981) The NRC workshop on wave measurement technology, Proc. Directional Wave Spectra Applications (New York: American Society of Civil Engineers), pp. 220-232.

Goda, Yoshimi (1981) Simulation in examination of directional resolution, Proc. Directional Wave Spectra Applications (New York: American Society of Civil Engineers), pp. 387-406.

Jeffreys, E. R., Wareham, G. T., Ramsden, N. A., and Platts, M. J. (1981) Measuring direction spectra with the MLM, Proc. Directional Wave Spectra Applications (New York: American Society of Civil Engineers), pp. 203-218.

Kay, S. M., and Marple, S. L., Jr. (1981) Spectrum analysis—a modern perspective, Proc. IEEE **69**, 1380-1419.

Long, R. B. (1980) The statistical evaluation of directional spectrum estimates derived from pitch/roll buoy data, J. Phys. Oceanog. **10**, 944-952.

Long, R. B., and Hasselmann, K. (1979) A variational technique for extracting directional spectra from multicomponent wave data, J. Phys. Oceanog. **9**, 373-381.

Oakley, O. H. (1973) Directional wave spectra measurements and analysis systems, Proc. 20th Anniversary of the Pierson-St. Denis Paper, Webb Institute of Naval Architecture, Soc. Naval Arch. and Mar. Eng.

Oakley, O. H., and Lozow, J. B. (1977) The Resolution of Directional Wave Spectra Using the Maximum Likelihood Method, Rept. 77-1, Dept. of Ocean Eng., MIT, Cambridge, Mass.

Regier, L. A., and Davis, R. E. (1977) Observations of power and directional spectrum of ocean surface waves, J. Mar. Res. **35**, 433-451.

Rustagi, J. S. (1976) Variational Methods in Statistics (New York: Academic Press), pp. 36-38.

Webster, W. C., and Dillingham, J. T. (1981) Directional seaway determined from ship motions, Proc. Directional Wave Spectra Applications (New York: American Society of Civil Engineers), pp. 1-19.

ENTROPY AND SEARCH THEORY

E. T. Jaynes

Arthur Holly Compton Laboratory of Physics, Washington University, St. Louis, Missouri 63130

C. Ray Smith and W. T. Grandy, Jr. (eds.),
Maximum-Entropy and Bayesian Methods in Inverse Problems, 443–454.

1. Introduction

A recent article by Pierce (1978) has brought search theory to the attention of workers in related fields that also use statistical theory. In recounting history, he noted that early workers tried to relate detection probability p_D and search effort to the posterior entropy H_{ND} conditional on nondetection [Eq. (4) below] or to the "expected posterior entropy" $H_E = p_D H_D + (1-p_D) H_{ND}$, discovered quickly that no general relation exists, and concluded that information theory has no useful connection with search theory.

As Pierce stated: "These negative findings had a clearly inhibiting effect on research, and relatively little effort has been devoted to the connections between information and search for the past fifteen years. Nonetheless, the intuitive appeal of information theory remains strong." He then presented some numerical analyses showing that in some cases maximum posterior entropy did, after all, correspond closely with maximum detection probability, although exceptions were also found. After analyzing the available evidence, Pierce concluded that the relation between search theory and information theory remains complex, but that the situation is promising enough to justify further study.

To an information theorist, that intuitive appeal is so strong that one is convinced from the start: there must exist a close relation between information and optimal search policy, and not just a numerical coincidence holding in some cases. There must be an exact, analytically demonstrable, and very general relation pertaining not only to search theory, but to optimal planning for any objective, for any optimal strategy is only a procedure for exploiting our prior information in order to achieve whatever goal is set, as quickly (or as efficiently, as measured by any cost assignment) as possible.

Indeed, Shannon's original creation of information theory arose from a special case of this: optimal encoding of a message so as to transmit it most efficiently by the cost assignment of channel capacity. As we have pointed out (Jaynes, 1978), all of presently known Statistical Mechanics is included in the solution that Shannon proposed for this problem. In any such problem, the attainable efficiency must be related to—because it is determined by—the amount of prior information available. If past efforts to find this relation have failed, it can be only from a technical failure to ask the right questions.

We show here that such a relation does indeed exist, but it involves different entropies than H_{ND}. We develop it by analyzing the simple search model studied by Pierce (single stationary target, no false alarms, independent detection probability for successive looks), and then speculate on generalizations. One of our entropy connections was given by Barker (1977); the other is possibly new. However, our purpose here is "introductory tutorial" rather than reporting new research.

2. The Simple Model

There is a hidden "target" in region R. Each time we look at R we have, independently, the probability q of detecting it, so the probability that it will have been detected in k looks is $[1 - (1-q)^k]$. We generalize by replacing the discrete number of looks k by a continuous "search effort" variable z, and define a "search parameter" s by $(1-q) = \exp(-s^{-1})$. Then the probability that a search effort z will result in detection is

$$p(D|z) = 1 - \exp(-z/s) . \tag{1}$$

Now consider that, instead of a single region R, the target is known to be in one and only one of n different "cells" with various search parameters $\{s_1 \ldots s_n\}$. With prior probability P_i that it is in cell i $(1 \leq i \leq n)$, the predictive prior probability that the search allocations $\{z_1 \ldots z_n\}$ will result in detection is

$$P_D = \sum_{i=1}^{n} P_i[1 - \exp(-z_i/s_i)] . \tag{2}$$

If, after this search, the target has not been located, the posterior probability that it is in cell i will be

$$p_i = \frac{P_i \exp(-z_i/s_i)}{\sum P_j \exp(-z_j/s_j)} . \tag{3}$$

In this notation, we use $p_i = p_i(z_1 \ldots z_n)$ as the "running variable" cell probabilities that evolve continuously throughout the search, and $P_i = p_i(0)$ for their fixed initial values. The aforementioned entropy is then

$$H_{ND} = - \sum p_i \log p_i . \tag{4}$$

3. Relative Entropy

The cell parameter s_i is a measure of the search effort required to achieve a given detection probability in cell i. If the cells consist of various areas to be searched, then one expects that a cell of twice the area will require twice the search effort; thus we may think of s_i quite generally as the "size" of cell i. Of course, in different problems this size may be measured in various terms: not only area, but equally well man-days, gasoline consumption, number of microscope slides, film footage, computer time, etc. In whatever units cell size s is measured, search effort z will be measured in the same units.

Now the original definition of our cells arose presumably in some natural way out of circumstances of the problem; but in principle their definition is arbitrary. They may be combined or subdivided in various ways, and the cell sizes are additive. In particular, we can find integers N_i such that

$$\frac{s_i}{\sum s_i} = \frac{N_i}{\sum N_i}, \qquad 1 \leq i \leq n, \tag{5}$$

to any desired accuracy (by choosing N_i sufficiently large). But then a cell with size, probability, and search allocation $\{s_i, p_i, z_i\}$ may be subdivided into N_i equal cells, each of size, probability, and search allocation

$$r_k = s_i/N_i, \quad w_k = p_i/N_i, \quad y_k = z_i/N_i, \qquad 1 \leq k \leq N_i, \tag{6}$$

and the detection probability $p_i[1 - \exp(-z_i/s_i)]$ may be written equally well as

$$\sum_{k=1}^{N_i} w_k[1 - \exp(-y_k/r_k)] . \tag{7}$$

But by construction all the new cells (k) are the same size, from whatever old cell (i) they were derived. Therefore we have refined the problem to one where we have

$$N = \sum_{i=1}^{n} N_i \tag{8}$$

equal cells. At this point, we could generalize further by relaxing the requirement of equal w_k, y_k in Eqs. (6).

Clearly, in view of the symmetry, the correct entropy that measures our information about the refined cells must be

$$H = -\sum_{k=1}^{N} w_k \log w_k = -\sum_{i=1}^{n} N_i(p_i/N_i) \log(p_i/N_i) , \tag{9}$$

which has the upper bound $H_{max} = \log N$. It is customary to subtract off this irrelevant additive constant by defining the new entropy $I \equiv H - \log N$, or

$$I(z) = \sum_{i=1}^{n} p_i \log(m_i/p_i) , \tag{10}$$

where

$$m_i \equiv \frac{N_i}{N} = \frac{s_i}{S}, \qquad S \equiv \sum s_i, \tag{11}$$

are the cell sizes, normalized to $\Sigma m_i = 1$. We may, equally well, define an entropy in which the roles of the distributions $\{p_i\}$, $\{m_i\}$ are interchanged:

$$J(z) \equiv \sum_{i=1}^{n} m_i \log(m_i/p_i) \ . \qquad (12)$$

These satisfy the Gibbs inequalities $I \leq 0$, $J \geq 0$, with equality in each case if and only if $\{p_i = m_i, 1 \leq i \leq n\}$.

The quantity I may be called the entropy of the distribution $\{p_i\}$ relative to the basic "measure" m_i (so called to suggest still further generalizations not needed here), while -J is the entropy of $\{m_i\}$ relative to $\{p_i\}$. These quantities go by various other names—"cross entropy," "directed divergence," "minimum discrimination information statistic," "essergy," etc.—but we think those terms should be discouraged because they imply that I is a different kind of object than H_{ND}. In fact, I is simply the entropy over the symmetric refined cells, and is every bit as much a "true entropy" as is H_{ND}.

Two problems, or two quantities, that can be transformed into each other by a mere change of variables are not really different. To minimize "cross-entropy" using one choice of variables is the same thing as maximizing entropy using a different choice. To use different terminology in the two cases—in particular, terminology that suggests that two different principles of reasoning are being used—seems to us highly misleading.

Now we show that it is the entropies I and J, rather than H_{ND}, that have a simple and general relation to search theory. Consider a search that starts from initial values I(0),J(0), which are measures of our prior information about the target location. At any subsequent stage $\{z_1 \ldots z_n\}$ of the search effort—whether optimal or not—the present values are I(z),J(z). The change [I(z) - I(0)] is the measure of the amount of prior information utilized up to that point, while [J(0) - J(z)] is the measure of the saving in search effort thereby achieved. The optimal policy is then the one that trades off initial information for reduced search effort, as quickly as possible.

The connection of I(z) with information was indicated in the derivation of Eq. (10). To demonstrate the connection of J(z) with search effort, note that from Eq. (2), the denominator of Eq. (3) is just $(1 - p_D)$. Therefore, at any stage where we have allocated the search effort $\{z_i\}$, J(z) is, from Eq. (12),

$$J(z) = \sum_{i=1}^{n} m_i \log\left[\frac{m_i}{p_i}(1-p_D)\exp(z_i/s_i)\right] = J(0) + \log(1-p_D) + \frac{z}{S}\ , \qquad (13)$$

where $z = \Sigma z_i$ is the total search effort used. But Eq. (13) states only that at this stage the detection probability is

$$p_D = 1 - \exp\left[-\frac{z + z^*}{S}\right], \tag{14}$$

where

$$z^* \equiv S[J(0) - J(z)] . \tag{15}$$

Since $J(z) \geq 0$, if we start from prior ignorance, $J(0) = 0$, then clearly the best we can do is to conduct the search so as to keep $J(z) = 0$; then the detection probability will be

$$p_D = 1 - \exp(-z/S) , \tag{16}$$

that is, just the original detection function (1), in which we have lumped all cells together into one large cell of size $S = \Sigma s_i$. Thus, z^* is precisely the saving in search effort, for a given detection probability, that has been achieved up to that point by exploiting the prior information.

All details of an optimal policy are easily set forth if we may assume the following property.

4. Dynamical Consistency

In a real-life situation the problem of deciding on a search allocation will be almost hopelessly complicated, or even indeterminate, unless the following property holds. Consider two different problems:

(A) You are allotted a total search effort $\Sigma z_i = C$. Decide on the optimal allocation $\{\hat{z}_1 \ldots \hat{z}_n\}$ to maximize the probability of detection.

(B) The authorities have divided the search effort C into two portions, $C_1 + C_2 = C$. You are allotted first the amount C_1, and must decide on the optimal allocation $\{\hat{z}_i^{(1)}\}$ with $\Sigma \hat{z}_i^{(1)} = C_1$ on the assumption that no further search effort is available. This fails to locate the target; but you then learn that you may apply for permission to use the additional search effort C_2. If this is granted, you must then decide on the optimal allocation $\{\hat{z}_i^{(2)}\}$ with $\Sigma \hat{z}_i^{(2)} = C_2$ for the second try.

The problem has dynamical consistency if the optimal total search allocation is the same in problems (A) and (B), that is, if

$$\hat{z}_i^{(1)} + \hat{z}_i^{(2)} = \hat{z}_i , \qquad 1 \leq i \leq n , \tag{17}$$

for all C_1 in $(0 \leq C_1 \leq C)$. This is a highly desirable property for psychological, practical, and mathematical reasons.

Psychologically, it is a comfort to the decision maker, for then he can face his problems one at a time, making at each step the decision that is optimal for the search effort being then committed—secure in the knowledge that whatever the final outcome, no critic full of hindsight can later accuse him of blundering (this remains true even if he has inherited the job from a blundering predecessor). Put differently, we are supposing that "global" optimization can be found by a sequence of "local" optimizations.

Practically, it is a useful property, for even if one knows in advance exactly how much total search effort can be used, it may be necessary to search the cells one at a time. Then one must in any event decide on the optimal order of searching, which amounts to a sequence of problems of type (B). Without dynamical consistency, the optimal action for today would in general depend on imaginary contingencies that might or might not arise tomorrow, and a "global" optimum would be very hard to find.

Mathematically, dynamical consistency reduces the problem for any amount of search effort to successive allocations of infinitesimal amounts δz, for which the optimal allocation is obvious. Given any previous search allocation $\{z_1 \ldots z_n\}$, whether optimal or not, that has reduced the cell probabilities (3) to $\{p_1 \ldots p_n\}$, if the new increment δz is used in cell j, the probability that it will result in detection is

$$p_j[1 - \exp(-\delta z/s_j)] = (p_j/s_j)\delta z \, . \tag{18}$$

But if detection does not result, then according to Eq. (3) the probability of the ith cell is changed by

$$\delta p_i = (p_i - \delta_{ij})(p_j/s_j)\delta z \tag{19}$$

and from Eqs. (10) and (12) the entropies $I(z), J(z)$ will receive the increments

$$\delta I = [1 + \log(p_j/m_j)](p_j/m_j)\delta z \, , \tag{20}$$

$$\delta J = S^{-1}[1 - (p_j/m_j)]\delta z \, . \tag{21}$$

Since $\Sigma p_i = \Sigma m_i = 1$, we have always $(p_j/m_j)_{max} \geq 1$, and from Eq. (10), $[1 + \log(p_j/m_j)_{max}] \geq 0$. Thus the allocation of δz that maximizes the detection probability (18) leads to $\delta I \geq 0$, $\delta J \leq 0$; from Eqs. (20) and (21) it therefore also maximizes the posterior entropy I and minimizes J. By all three criteria, the optimal policy allocates each new increment δz to whatever cell has at that time the greatest value of (p_j/m_j); as noted, this is the optimal present policy even if the previous allocation $\{z_i\}$ was not optimal.

This optimal search policy always takes us toward the condition of "complete ignorance" $I = J = 0$; and thus (as noted by Pierce, in agreement with an earlier conjecture of Richardson) it "uses up" the prior information, as rapidly as possible. The limiting state $I = J = 0$ is actually reached, at a finite total search effort, if the search continues long enough without detection [Eq. (37) below]. Such a search therefore has a fundamental division into an "early phase" in which $I < 0$, $J > 0$ and the prior information is being used to determine policy, and a "final phase" in which it is all used up: $I = J = 0$, and the optimal policy is independent of the prior information.

We now examine in some detail the course of the optimal search policy for this model. In this we necessarily repeat a few facts well known in the literature of search theory; our object is to point out their interpretation in the light of the entropies $I(z), J(z)$. The search process now appears very much like an irreversible process in thermodynamics, in which an initially nonequilibrium state relaxes into the equilibrium state of maximum entropy. But now it is only our state of knowledge that relaxes to the "equilibrium" condition of maximum uncertainty, $I = J = 0$.

5. An Example of Optimal Search

On the assumption of dynamical consistency, the entire course of the optimal search effort is clear; since according to Eq. (19) the probability of the searched cell is always lowered, that of the others raised, the optimal strategy is the one that equalizes the numbers

$$a_i \equiv p_i/m_i \,, \tag{22}$$

starting from the top, as quickly as possible. We follow the aforementioned notation of writing $a_i = a_i(z_1 \ldots z_n)$ for the "running variables" that evolve during the search, and $A_i = a_i(0)$ for their fixed initial values. Number the cells according to those initial values, so that

$$A_1 \geq A_2 \geq \ldots \geq A_n \,. \tag{23}$$

Then the optimal search proceeds as follows:

Stage 1. All the initial effort should go into cell 1 until its probability is reduced to the point where $a_1 = a_2$. The search effort required to do this is, from Eq. (3),

$$z_1^{(1)} = s_1 \log(A_1/A_2) \,, \tag{24}$$

and the prior probability of detection at or before this point is

$$p_D^{(1)} = P_1[1 - \exp(-z_1^{(1)}/s_1)] = m_1(A_1 - A_2) \,. \tag{25}$$

Thus from Eq. (13) the entropy J has changed by

$$J^{(1)} - J(0) = \log[1 - m_1(A_1 - A_2)] + m_1 \log(A_1/A_2) \,. \tag{26}$$

That this must be negative if $A_1 > A_2$ is evident from Eq. (21); to prove it directly from Eq. (26) one must take into account also the inequalities $\{A_2 \geq A_k,\ 3 \leq k \leq n\}$.

At any stage in the search, the entropy $I(z)$ may be written in the form

$$I(z) = \log(1 - p_D) + (1 - p_D)^{-1} K(z) \,, \tag{27}$$

where $K(z)$ is an analytically simpler expression. Therefore we indicate the entropy changes by giving the values of K at each stage of the search. Initially, $K(0) = I(0)$, but after the search effort (24) we find

$$K^{(1)} = I(0) - P_1 \log(m_1/P_1) + (m_1/m_2)P_2 \log(m_2/P_2) \,. \tag{28}$$

That is, the right-hand side of Eq. (28) is the expression (10) for $I(0)$ in which the first term $[P_1 \log(m_1/P_1)] = -m_1A_1 \log A_1$ has been replaced by $-m_1A_2 \log A_2$. In effect, this lumps the first two cells together into a single cell of measure $(m_1 + m_2)$.

Stage 2. According to Eqs. (18), (20), and (21), cells 1 and 2 are now equally search-worthy, a further small search effort yielding equal detection probability and equal entropy increase in either. The next efforts should therefore be allocated to both, in the ratio that maintains the equality $a_1 = a_2$, that is, in the ratio $m_1{:}m_2$, which amounts to equal allocation to the $(N_1 + N_2)$ refined cells derived from cells 1, 2. The second stage continues until $a_1 = a_2 = a_3$, at which point we have used the additional search effort

$$(s_1 + s_2) \log(A_2/A_3) \,, \tag{29}$$

and the total amounts expended in cells 1 and 2 up to this point are

$$z_1^{(2)} = z_1^{(1)} + s_1 \log(A_2/A_3) = s_1 \log(A_1/A_3) \tag{30}$$

$$z_2^{(2)} = s_2 \log(A_2/A_3) \,. \tag{31}$$

The prior probability of detection at or before this point is

$$p_D^{(2)} = m_1(A_1 - A_3) + m_2(A_2 - A_3) \,, \tag{32}$$

and the entropy $I^{(2)}$ is given by Eq. (27) with

$$K^{(2)} = I(0) - P_1 \log(m_1/P_1) - P_2 \log(m_2/P_2) + \frac{m_1 + m_2}{m_3} P_3 \log(m_3/P_3) \,, \tag{33}$$

that is, by $I(0)$ with the first two terms replaced, in effect lumping the first three cells into a single cell of measure $(m_1 + m_2 + m_3)$.

Stage 3. At this point, cells 1, 2, and 3 are equally search-worthy, so the next effort is allocated to them in the ratios $m_1{:}m_2{:}m_3$ until $a_1 = a_2 = a_3 = a_4$, at which point $K^{(3)}$ is given by $I(0)$ with the first three terms replaced, and so on.

This initial "equalization phase" continues until for the first time $a_1 = a_2 = \cdots = a_n$, at which point we have used up the total search effort

$$z' = \sum z_i = \sum_{i=1}^{n-1} s_i \log(A_i/A_n) = -S[J(0) + \log A_n] \,, \tag{34}$$

but have not searched at all in cell n: $z_n = 0$. The prior probability of detection has reached

$$p_D^{(n-1)} = 1 - A_n , \quad (35)$$

and $K^{(n-1)}$ is $I(0)$ with the first (n-1) terms replaced, that is,

$$K^{(n-1)} = - A_n \log A_n . \quad (36)$$

From Eqs. (27) and (35), the entropy $I(z)$ is now reduced to

$$I^{(n-1)} = 0 , \quad (37)$$

and from Eqs. (13), (34), and (35) we have also $J^{(n-1)} = 0$. The posterior probabilities (3) have completed the relaxation into their "equilibrium" values $\{p_i = m_i, 1 \leq i \leq n\}$; that is, the refined cells now have equal probabilities $\{w_k = N^{-1}, 1 \leq k \leq N\}$.

Final Phase. All cells are now equally search-worthy, so if detection is not yet achieved, any further search effort z'' is allocated to all cells in the proportions $z_i'' = m_i z''$ which maintain that condition; that is, it is allocated equally to the refined cells. The posterior probabilities remain at their equilibrium values, the entropies I,J remain zero, and the detection probability with any further amount of search effort (that is, for total effort $z = z' + z'' \geq z'$) is

$$p_D(\infty) = 1 - \exp\left[- \frac{z + z^{**}}{S} \right] , \quad (38)$$

where, comparing with Eqs. (14) and (15),

$$z^{**} \equiv S\, J(0) \quad (39)$$

is the maximum possible saving in search effort that can be "bought" with the prior information.

6. Conclusion—The Moral

We have shown how entropy maximization is related to optimal strategy in one simple case. This can of course be generalized in many different ways. In fact, the situation is open-ended because there is no end to the variety of new problems that could arise. So it is impossible to give a "most general" case once and for all. But before one can extend the theory to some particular new case, it is necessary to understand the moral of what we have just learned.

Why did it require nearly 30 years after Shannon's work to find this (maximum entropy)-(optimal search) connection, in spite of the fact that many workers suspected its existence and tried to find it? The answer was given about 130 years ago by George Boole, who remarked: "I think it one of the peculiar difficulties of probability theory, that its difficulties sometimes are not seen." What has not always been seen here is that the simple, unqualified term "entropy" is meaningless; entropy is always defined with respect to some basic "measure," and the result of maximizing it depends not only on the constraints but also on the measure.

The difficulty in applying maximum entropy to problems outside thermodynamics is not in deciding what constraints should be applied, but in deciding what is the underlying measure—or, as I prefer to call it, what is the "hypothesis space" on which our entropies are defined? More informally, what is the field on which our game is to be played?

This problem does not loom large in thermodynamics—in fact, most writers seem hardly aware of it—but this is only because it was solved over 100 years ago by Liouville. Classical phase volume is invariant under canonical transformations (of which the equations of motion are a special case), and so equal weighting to equal phase volumes was the field on which Gibbs' game was played.

This leads to many correct predictions (equations of state, susceptibilities, high-temperature specific heats of solids and monatomic gases), but at low temperatures Nature persisted in giving lower specific heats—and therefore states of lower entropy—than Gibbs predicted. In Nature, therefore, there must be further constraints operative, beyond those imposed by Gibbs. This was the first clue pointing to quantum theory.

The resolution, found by Einstein, Debye, von Neumann, and Brillouin, was quite simple. It seems that not all classically allowed energies are used by Nature, and equal weighting to orthogonal quantum states of a system—which goes asymptotically into Liouville's weighting—is the new field on which we play the game of quantum statistics. According to all present knowledge, maximum entropy on this hypothesis space leads unerringly to correct predictions. Still, I keep trying to find a case where it fails, because then we would have a clue pointing to the new theory that will someday replace our present quantum theory, and history would be repeated.

In applications outside thermodynamics we are still at a phase corresponding to (if one can imagine it) statistical mechanics before the discovery of Liouville's theorem. The originally tried entropy H_{ND} of Eq. (4) was defined with respect to uniform weighting of all search cells regardless of their size. Such a weighting simply ignores the cogent information about cell sizes. Our proceeding to the refined cells of equal size restored the symmetry of our hypothesis space—and corresponded to the discovery of Liouville's theorem. As soon as we play our game on the field defined by Eq. (9), the connection of entropy with optimal search appears immediately.

Moral: In any new problem, one must face anew, what is the underlying symmetrical hypothesis space on which our entropies are defined? The strategy is:

(1) Think hard about the appropriate hypothesis space. Look for some symmetry/invariance property.

(2) Try out your best choice. If the desired kind of useful results appear, then well and good: there is no evidence pointing to a

different hypothesis space and you are done, at least for the time being.

(3) If you get unsatisfactory results, then if you are convinced that all relevant constraints have been taken into account, this is evidence that Nature is using a different hypothesis space than yours. Go to step (1).

In spectrum analysis, the Burg solution implied independent uniform weighting to all possible values of $\{y_0 \dots y_N\}$. Its success thus far indicates that we are now at step (2). However, the future may bring some surprise here. Any persistent failures would point to a new hypothesis space, and therefore to the possibility of still better predictions.

In image reconstruction, the present solutions seem to be based on uniform independent weighting to all values of luminance for each pixel. I have a suspicion—perhaps shared by John Skilling, although he expresses himself in very different terms—that a deeper hypothesis space that to some degree "anticipates" correlations of adjacent pixels may be still better. Of course, we would have to accumulate a great deal of further experience before we could be sure that we were at step (3).

We hope that entropy considerations will be brought to bear on other problems of optimal strategy, and perhaps with enough experience we shall learn how to define our hypothesis space for such problems, just as confidently as physicists now do in statistical mechanics.

References

Barker, William H. (1977) Information theory and the optimal detection search, Oper. Res. **25**, 304-314.

Jaynes, E. T. (1978) Where do we stand on maximum entropy? in R. Levine and M. Tribus, eds., The Maximum Entropy Formalism (Cambridge, Mass.: MIT Press), pp. 15-118.

Pierce, John G. (1978) A new look at the relation between information theory and search theory, in R. Levine and M. Tribus, eds., The Maximum Entropy Formalism (Cambridge, Mass.: MIT Press), pp. 339-402.

MAXIMAL ENTROPY IN FLUID AND PLASMA TURBULENCE: A REVIEW

David Montgomery

Los Alamos Scientific Laboratory, Los Alamos, New Mexico

Permanent address: Physics Department, College of William and Mary, Williamsburg, Virginia 23185

Three recent applications of maximal entropy procedures to models of turbulence in plasmas are described. They are (1) the calculation of "most probable" states in guiding-center plasmas and vortex fluids, (2) the calculation of "most probable" magnetohydrodynamic equilibrium profiles, and (3) the prediction of absolute equilibrium Gibbs ensemble spectra for Navier-Stokes fluids. Further conjectures on the role of entropies in turbulence theories are offered.

C. Ray Smith and W. T. Grandy, Jr. (eds.),
Maximum-Entropy and Bayesian Methods in Inverse Problems, 455–468.

1. Introduction

Turbulence theory is only slowly being admitted to the status of a full-fledged branch of mainstream statistical mechanics. Several reasons for this may be suggested. (1) Nearly all of mainstream statistical mechanics deals with thermal equilibrium systems or systems that are approaching thermal equilibrium, while turbulence by its very nature never gets close to thermal equilibrium and is, therefore, hard to fit into the categories. (2) There is nothing exotic or difficult to achieve about turbulence, and in varying degrees it characterizes virtually all the fluids we come into contact with, such as air, water, and most soups. The diagnosis of plasmas is more difficult than that of fluids, so plasma turbulence is a little easier to remain unaware of, but it seems to be just as prevalent in interesting cases. (3) The detailed manifestations of turbulence are very diverse and situation-dependent, and it has been harder to abstract a core of "universal" behavior from the phenomena than it has been for many other statistical mechanical situations.

In keeping with the theme of this conference, I want to discuss some examples of ways in which information theory has recently impinged on the discussion of nonlinear behavior of turbulent fluids and plasmas. This has occurred in some cases without the investigators' full awareness of their debt to information theory—somewhat like Moliere's gentleman who discovered that he had been speaking in prose all his life without knowing it.

There is an unmistakable tendency in ordinary fluid turbulence for the system to try to increase its disorder, and to evolve toward a state that, if we perceive it in any common-sense way, appears to be more scrambled and chaotic than the state out of which it may have developed. At the same time, there are circumstances in which common-sense intuitions are clearly inadequate, and when a sort of macroscopic orderliness seems to emerge, rather eerily, from what appears to be microscopic disorder. Entropies that we may define for various mathematical models of these phenomena can be useful in predicting what happens. Their degree of utility depends simply upon the extent to which their predictions are confirmed by computer simulations and experiments. I consciously mention computer simulations first because at present they provide most of the "data" we have to test theoretical predictions against. The kinds of experiments needed to test the theories are hard to do, and so far they have not intrigued experimentalists or funding agencies enough to command the resources to do them. (The explanation for this is more socio-economic than scientific, and is no longer even very interesting.)

This is not the place to engage the larger questions and controversies that have surrounded the infancy and adolescence of information theory. Rather, I want to introduce, in a somewhat limited framework, three ad hoc applications in which maximal entropy formulations appear to have been useful in predicting the results of large-scale computer simulations of plasma and fluid phenomena, and even some experiments. The three applications are different, all require some specialized jargon, and none can be described as fully explored. They concern (1) the agglomeration of line vortices in a two-dimensional line vortex model of an ideal fluid, (2) the definition of an entropy for magnetofluid equilibria, to determine which profiles in a controlled fusion confinement geometry are "most probable," and (3) the relaxation of Navier-Stokes or guiding-center plasma flows to the absolute equilibrium values predicted by the Gibbs ensemble for Fourier coefficients. In each case, I will provide only the minimal background introduction necessary to make the nature of the problem comprehensible. What I hope to do is to persuade a few investigators not now working in these areas to take a look at what has happened, and to try to improve upon the results.

2. Collections of Line Vortices and/or Guiding Center Plasmas

There are intelligible arguments for making a zeroth-order approximation of the large-scale atmospheric or oceanic circulations as two-dimensional Navier-Stokes flows, in which the fluid variables are functions of only two spatial coordinates, and the velocity fields are in the plane of variation. There are also reasons for similarly idealizing the electrostatically-produced motions of a plasma across a strong dc magnetic field. In either case, the crudest possible model for the dynamics is the Lin-Onsager one [1,2], in which only two coordinates are required to locate the position of the ith straight-line vortex, which moves with its alignment always parallel to the z-axis. In the plasma interpretation, the line vortex is replaced by the excess electrical charge on a tube of magnetic field lines parallel to the z-axis [3,4]. For simplicity, the magnetic field is taken to be uniform and constant, and in the z-direction.

In either case, the velocity field $\mathbf{v} = [v_x(x,y,t), v_y(x,y,t), 0]$ is related to the stream function ψ and the vorticity $\boldsymbol{\omega} = \omega\hat{e}_z$ by

$$\mathbf{v} = \nabla \times \psi\hat{e}_z$$

$$\boldsymbol{\omega} = \nabla \times \mathbf{v} = -\nabla^2\psi\hat{e}_z \qquad (1)$$

and the vorticity $\boldsymbol{\omega} = \omega\hat{e}_z$ is given by

$$\omega = \sum_i \kappa_i \, \delta(\mathbf{x} - \mathbf{x}_i) ,$$

where κ_i is the strength of the ith vortex and $\mathbf{x}_i$ is its (two-dimensional) location. If we ignore the dissipation, each vortex is convected undiminished with the local flow velocity $\mathbf{v}$ so that the dynamics simplifies to

$$\left[\frac{\partial}{\partial t} + \mathbf{v} \cdot \nabla\right]\omega = 0 . \qquad (2)$$

The dynamics described by Eq. (2) can be cast in canonical form by introducing canonical variables

$$(q_i, p_i) = |\kappa_i|^{1/2} (x_i, y_i \operatorname{sgn} \kappa_i) , \tag{3}$$

where κ_i is either positive or negative, and the Hamiltonian function H is

$$H = -\frac{1}{2\pi} \sum_{i,j} \kappa_i \kappa_j \ln|\mathbf{x}_i - \mathbf{x}_j| . \tag{4}$$

The equations $\dot{q}_i = \partial H/\partial p_i$, $\dot{p}_i = -\partial H/\partial q_i$ are equivalent to Eq. (2).

For simplicity we take N vortices of strength $+\kappa$ and N more of strength $-\kappa$ in a square two-dimensional volume $L^2 = V$, and assume either periodic boundary conditions or an ideal, reflecting, free-slip boundary. This causes the logarithms in Eq. (3) to be replaced by the solutions [5,6] of $\nabla^2\psi = -\kappa_i\delta(\mathbf{x}-\mathbf{x}_i)$, subject to the chosen boundary condition, but that is not a serious modification.

The system so described is a peculiar one, in that the total phase space is bounded. There is no infinite momentum space the way there usually is; all the q_i and p_i are just bounded configuration space coordinates. This turns out to have some extraordinary implications for the states that the system can get itself into.

Our usual instincts would lead us to calculate expectations of phase functions using the canonical ensemble

$$D_{eq} = \exp\left[\frac{F-H}{K_B T}\right] , \tag{5}$$

where F is a free energy, K_B is Boltzmann's constant, and T is the temperature. This can be and has been done [4], with the result that the $\langle H \rangle$ one calculates is in general negative, for all $T > 0$. Clearly H, which is a constant of the motion for a single system, can be chosen to be positive by rearranging the initial positions of the vortices, however their configurations evolve. There is a question, then, as to what the states with high values of the energy (Hamiltonian) look like. Can we characterize any of their features by statistical mechanical methods?

The answer, which has now been reached by several means, is that the vortices segregate themselves, more positives on one side of the box and more negatives on the other [7,8]. This is in contrast with the uniform-density predictions for both species that emerge for the $\langle H \rangle < 0$ states.

We now show a maximal-entropy method of describing these high-energy states [7]. A direct ancestor of the calculation is Van Kampen's treatment [9] of the condensation of a Van der Waals gas (see also Lynden-Bell [10]). Assuming $N \gg 1$, we use a cellular or "occupation number" representation for the density of positive and negative vortices in the ith cell of a partition of the volume V into cells that are small compared to V but still large enough to contain many vortices. We call these densities n_i^+, n_i^-; giving them for all i determines the state of the system with an accuracy that becomes better and better as N and the number of cells get larger. Boltzmann combinatorics can be invoked to assert that the entropy of a particular configuration may be defined as

$$S = -\sum_i (n_i^+ \ln n_i^+ + n_i^- \ln n_i^-) ,$$

which, if we shrink the cell size and vortex strength to zero and let $N \to \infty$ in such a way that $N\kappa$ remains finite, becomes

$$S = -\int n^+ \ln n^+ \, d^2x - \int n^- \ln n^- \, d^2x \, . \tag{6}$$

It is not hard to show that if this S is now maximized subject to the constancy of the total vorticity of both signs and constant total energy, we arrive at an exponential dependence of the vorticity densities on ψ. However, ψ and the vorticity are still related by Poisson's equation, so

$$\nabla^2\psi = -\kappa n_+ + \kappa n_- = -\exp(-\alpha_+ - \beta\psi) + \exp(-\alpha_- + \beta\psi) \, . \tag{7}$$

The $\alpha_\pm$ and β are Lagrange multipliers associated with the constancy of the vorticity and energies respectively. For symmetric situations (ψ as much positive as negative), which can be achieved by either picking $\psi = 0$ on the boundary or using periodic boundary conditions, we have $\alpha_+ = \alpha_-$. It can then be readily shown that no solutions exist unless β is negative. If we let $|\beta|\psi \equiv \phi$, we then have

$$\nabla^2\phi + \lambda^2 \sinh\phi = 0 , \tag{8}$$

where $\lambda^2 \equiv 2\exp(-\alpha_+)|\beta|$, as the equation whose solution gives the most probable, or maximum-entropy, stream function. Since β plays the role of an inverse temperature, it is natural to describe what we have as a "negative temperature state." λ, β, and $\alpha_\pm$ are determined by requiring that the energy and total positive and negative vorticity calculated from $n^\pm$, ψ match prescribed initial values. The states with $\langle H \rangle < 0$ and $T > 0$, obtained from Eq. (5), all go away when we take the "Vlasov" limit $N \to \infty$, $\kappa \to 0$.

The solutions to Eq. (8) have been explored in considerable detail [11,12,13], resulting in a fairly complete understanding of what the states described by Eq. (8) look like. They consist of a large maximum and a large minimum, corresponding to a pair of counter-rotating vortices. There are also solutions with more maxima and minima, but they always have lower entropy [11,14]. Numerical solutions of the dynamics of several hundred vortices have established qualitatively [7,8] that their configuration evolves toward states similar to the solutions of Eq. (8), but quantitative agreement has not been achieved.

One significant point about this phenomenon and its magnetohydrodynamic analogs is that higher entropies can correspond to states that are macroscopically ordered and seem at first glance "less probable" than those from which they evolve. One might not have understood this without the attempt to define an entropy functional quantitatively.

3. Maximum-Entropy Magnetohydrodynamic Equilibrium Profiles

A central preoccupation in controlled thermonuclear research (CTR) has been the achievement of magnetohydrodynamic (MHD) equilibria inside a bounded region, often a cylinder or torus [15]. It is desired to confine a plasma at thermonuclear temperatures and to hold the hottest, densest part of it away from the container walls. Whether any such ideally quiescent state can be achieved is doubtful, but there is no doubt that some states are much more nearly quiescent than others: the turbulent components of the fields can be and have been made small compared to the mean fields. It is often far from clear why confined plasmas choose a particular spatial profile, rather than another, for the mean magnetic field **B**, the mean current density **j**, the mean pressure p, and so on. It may be that maximal entropy formalisms can have something to say on that subject.

A brief and grossly oversimplified introduction to MHD equilibrium theory is necessary to motivate the development. The basic equations for the mean fields are [15]

$$\mathbf{j} \times \mathbf{B} = \nabla p \tag{9}$$

and

$$\nabla \times \mathbf{B} = \mathbf{j}\,, \tag{10}$$

where Eq. (9) expresses balance of the pressure gradient ∇p for the confined magnetofluid and the magnetic $\mathbf{j} \times \mathbf{B}$ force associated with the plasma current density **j**. Equation (10) is just the constitutive (Biot-Savart) connection between **j** and **B**. Equations (9) and (10) are to be supplemented by the condition $\nabla \cdot \mathbf{B} = 0$. It is useful to introduce a vector potential **A** for which $\mathbf{B} = \nabla \times \mathbf{A}$ and $\nabla \cdot \mathbf{A} = 0$, and it also turns out to be useful to introduce another auxiliary field $\boldsymbol{\pi}$ such that $\nabla \times \boldsymbol{\pi} = \mathbf{A}$.

If we seek axisymmetric equilibria in cylindrical coordinates (r,θ,z), we can pick virtually arbitrary current profiles of the type $\mathbf{j} = [0,j_\theta(r),j_z(r)]$ and be certain of being able to find **B**'s and p's to go with them. The boundary can be imagined to be an infinitely long cylinder with radius $r = a$, for present purposes. The **B**-field, **A**-field, and $\boldsymbol{\pi}$-field can readily be calculated, given a set of boundary conditions, and also have nonvanishing θ and z components that are functions of r.

The only nonvanishing component of Eq. (9) is

$$\frac{\partial p(r)}{\partial r} = j_\theta B_z - j_z B_\theta , \tag{11}$$

which can be integrated subject only to the very weak requirement that p(r) be nonnegative.

The game, in much of CTR theory, has been to investigate the linear stability of solutions to the full set of time-dependent MHD equations, with equilibrium profiles of the type we have been discussing as the zeroth order [15]. Some fancier effects such as toroidal curvature and resistive dissipation may be built in. The difficulty is that there are literally infinite numbers of profiles whose stability may be investigated, and even slight changes in the details greatly alter the relevant linear mathematics, just as they do in hydrodynamics. The game can go on forever, and there is clearly a need for a basis for a decision as to which profiles to select for analysis, if one can be found.

Often the experimenter has only a few parameters at his disposal: the total z-current, say, or the total z-flux of the magnetic field, or perhaps the total magnetic stored energy or helicity. This information is always far less than is required for a pointwise tailoring of the dependence of **j** and **B** on r. The plasma finds its own profile by processes that are very poorly understood. It is natural to inquire whether perhaps some of the profiles are not in some sense "more probable" than others, given a few simple constraints [16].

We are required to introduce a quantitative characterization of the "probability" of a continuous classical field. There are no microscopic "particles" to pre-discretize it for us the way there were for the vortex system. Some kind of postulate is required to assign probabilities to the particular profiles.

Since z-current and z-magnetic flux are at the disposal of the experimenter, it seems natural to adopt $j_z(r)$ and $B_z(r)$ as the two independent field variables. Together with boundary conditions, $j_z(r)$ and $B_z(r)$ determine all the other components of all the fields. Originally, to motivate the combinatorics, we imagined a field line representation [16] for j_z and B_z, cutting a cross section of the cylinder up into small areas labeled with an index i, and letting n_i^B and n_i^J be the number of magnetic field lines and current field lines piercing the ith cell. This leads to a discrete entropy

$$S = - \sum_i n_i^B \ln n_i^B - \sum_i n_i^J \ln n_i^J \tag{12}$$

as a measure of the logarithm of the probability of a given configuration; then we pass back to the continuum limit by letting the cell size shrink to zero and the number of field lines go to infinity, getting

$$S = -\int B_Z \ln B_Z \, d^2x - \int j_Z \ln j_Z \, d^2x \tag{13}$$

as the entropy functional to be maximized.

The prescription leaves one with the disadvantage of an uncertain weighting factor for the two partial entropies. (What is the ratio of the number of current field lines to the number of magnetic field lines?) This has led Ambrosiano and Vahala [17] to introduce dimensionless versions of the fields to be discretized, arriving at a slightly different entropy,

$$S = -\int B_Z \ln \frac{B_Z}{\Phi} \, d^2x - \int j_Z \ln \frac{j_Z}{I} \, d^2x \, , \tag{14}$$

instead of Eq. (13), where $\Phi \equiv \int B_Z \, d^2x$ is the z-flux of the magnetic field and $I \equiv \int j_Z \, d^2x$ is the total current down the cylinder.

This S can then be maximized subject to such constraints as

$$\Phi = \int B_Z \, d^2x = \text{const.}$$

$$I = \int j_Z \, d^2x = \text{const.}$$

$$E = \int (B_Z^2 + B_\theta^2) \, d^2x = \text{const.}$$

$$H_m = \int (A_\theta B_\theta + A_Z B_Z) \, d^2x = \text{const.}$$

The constraints are taken into account in the usual way with Lagrange multipliers, and various boundary conditions on the fields and potentials are imposed. As in the vortex-fluid case, there results an exponential dependence of the j_Z and B_Z on the other fields:

$$B_Z(r) = \Phi \exp[-\alpha_B - \beta B_Z(r) - \gamma A_Z(r)] \, , \tag{15}$$

$$j_Z(r) = I \exp[-\alpha_J - \beta A_Z(r) - \gamma \pi_Z(r)] \, . \tag{16}$$

The α_B, α_J, β, and γ are the Lagrange multipliers.

However, there are the constitutive relationships to be satisfied, just as in the vortex case:

$$\nabla^2 A_z = -j_z , \tag{17}$$

$$\nabla^2 \pi_z = -B_z . \tag{18}$$

When Eqs. (15) and (16) are inserted on the right-hand sides of Eqs. (17) and (18), a highly nonlinear set of second-order ordinary differentio-integral equations results for the field variables. Except in very special cases, these require numerical integration. Considerable effort has been invested in this problem by Ambrosiano and Vahala, with some quite satisfactory results. Two limits have been explored in some detail [17]: the "tokamak" regime, in which the ratio of Φ/I is large, and the "pinch" regime, in which Φ/I is not large and the B_z and B_θ are comparable. The reader is referred to the paper [17] for details of the resulting profiles.

Lacking in the MHD-profile case, as in the line vortex fluid case, is any dynamical argument or kinetic equation (such as Boltzmann's equation) which predicts temporal evolution toward the maximum entropy states that have been calculated. Whether or not such H-theorems can in some sense be proved, and under what assumptions, is as far as I am concerned a completely open question.

A second rather uncertain point concerns the assignment of probability measures for fields that can change sign. In the preceding development of the expression for $S = S(j_z, B_z)$, it was implicitly assumed that all the field lines pointed in the same direction. Physically there is nothing to prohibit changes in sign of $j_z(r), B_z(r)$, and indeed in the case of "reversed field" z-pinches, B_z does sometimes pass through zero. This raises the question of how to define an entropy for such a state.

A possibility is to split the field into an "up" and a "down" part:

$$B_z(r) = B_z^+(r) - B_z^-(r) , \tag{19}$$

where $B_z^\pm(r)$ are both nonnegative. Then the previous definition of an entropy can be generalized to

$$S = -\int B_z^+ \ln\left[\frac{B_z^+}{\Phi^+}\right] d^2x - \int B_z^- \ln\left[\frac{B_z^-}{\Phi^-}\right] d^2x - \int j_z \ln\left[\frac{j_z}{I}\right] d^2x . \tag{20}$$

Maximizing this S subject to the previous constraints and treating $B_z^\pm$ as independent gives

$$B_z = \Phi^+ \exp(-\alpha_B - \beta B_z - \gamma A_z) - \Phi^- \exp(\alpha_B + \beta B_z + \gamma A_z) \tag{21}$$

and

$$j_z = I \exp(-\alpha_J - \beta A_z - \gamma \pi) \tag{22}$$

to be substituted into Eqs. (17) and (18).

The division of the given flux Φ into a positive part Φ^+ and a negative part $\Phi^- \equiv \Phi - \Phi^+$ clearly is not unique and must somehow be inferred. If the division is regarded as arbitrary, then S becomes a function not only of E, H_m, I, and Φ, but of Φ^+/Φ^- as well. If, without any very persuasive arguments for doing so, we maximize S with respect to Φ^+/Φ^-, then the partition becomes unique. The numerical results of Ambrosiano and Vahala [17] to date indicate that the maximal entropy is achieved when $|\Phi^\pm| \to \infty$, with $\Phi/|\Phi^\pm| \to 0$, with the "up" field and "down" field being very large, but with a finite difference.

It is hard for me to know what to make of this result. The idea of treating a classical field as a probability distribution and defining an entropy for it is perhaps strange enough, without trying as well to define an entropy for a field that can take on both signs. Rather deep conceptual questions are involved, and any attempt at a definitive resolution is probably premature at this point.

4. Relaxation of Fourier Moments for Navier-Stokes Turbulence; Carnevale's H-Theorem

There is another way of discretizing the two-dimensional flow of a uniform density Navier-Stokes fluid besides the Lin-Onsager delta-function line vortices mentioned in Section 2. We can instead work with a continuous, nonsingular vorticity distribution $\omega(x,y,t)$, where

$$\frac{\partial\omega}{\partial t} + \mathbf{v} \cdot \nabla\omega = \nu\nabla^2\omega , \tag{23}$$

and where the velocity field $\mathbf{v} = \nabla\psi \times \hat{e}_z$ is related to the vorticity again by $\nabla \times \mathbf{v} = \omega\hat{e}_z$. The continuous vorticity distribution has the advantage that we can incorporate a finite value of the coefficient of kinematic viscosity ν; any kind of dissipative process is hard to work into the line vortex model.

Equation (23) is particularly easy to deal with if we assume rectangular periodic boundary conditions and decompose ω and $\mathbf{v}$ into Fourier series, such as

$$\omega = \sum_{\mathbf{k}} \omega(\mathbf{k},t) \exp(i\mathbf{k} \cdot \mathbf{x}) . \tag{24}$$

Then Eq. (23) becomes

$$\left[\frac{\partial}{\partial t} + \nu k^2\right] \omega(\mathbf{k},t) = \sum_{\mathbf{p}+\mathbf{r}=\mathbf{k}} M(\mathbf{r},\mathbf{p})\, \omega(\mathbf{r},t)\, \omega(\mathbf{p},t) , \tag{25}$$

where the coupling coefficient M is

$$M(\mathbf{r},\mathbf{p}) = \frac{\hat{e}_z}{2} \cdot (\mathbf{r} \times \mathbf{p}) \left[p^{-2} - r^{-2}\right] .$$

If we like, we may think of the real and imaginary parts of the $\omega(\mathbf{k},t)$ as a set of phase space coordinates, analogous to canonical coordinates and momenta for gas molecules, which can provide the basis for a statistical mechanics of turbulence. (This way of considering turbulence statistically was presented by Lee [18] and has been developed by Kraichnan [6,19].)

If we neglect the viscosity ($\nu = 0$), we throw away a significant physical effect, but with what is left of Eq. (25), we are able to prove a Liouville theorem. If we further truncate the description of a finite (but arbitrarily large) number of Fourier coefficients, we are left with a somewhat mutilated model, but one to which all the paraphernalia of equilibrium statistical mechanics should apply. In particular, time averages of functions of the Fourier coefficients should be calculable as phase space expectations with respect to the probability distribution given by the Gibbs canonical ensemble [19].

To write down the canonical ensemble (which is, of course, obtained by maximizing the entropy of the ensemble as a whole, or else is derived from the microcanonical ensemble assuming some sort of ergodic or mixing hypothesis), we need to know the constants of the motion (isolating integrals) for the motion described by Eq. (26) with $\nu = 0$. Before the truncation in Fourier space, there are many such constants; but after the truncation, it appears that only two of them are still constant. They are the energy E,

$$E = \sum_{\mathbf{k}} \frac{|\omega(\mathbf{k},t)|^2}{k^2} ,$$

and the mean square vorticity or "enstrophy" Ω,

$$\Omega = \sum_{\mathbf{k}} |\omega(\mathbf{k},t)|^2 .$$

The canonical ensemble D_{eq} = const. x $\exp[-\alpha E - \beta\Omega]$, where the Lagrange multipliers α and β are the reciprocal "temperatures" that go with the constancy of E and Ω, respectively. Expectations of modal energies and enstrophies are straightforward to calculate using the two-temperature canonical ensemble D_{eq}, and we find (indicating ensemble averages by brackets $\langle\ \rangle$):

$$\langle|\mathbf{v}(\mathbf{k})|^2\rangle = \frac{\langle|\omega(\mathbf{k})|^2\rangle}{k^2} = \frac{1}{\alpha + \beta k^2} . \tag{26}$$

The α and β are determined by requiring that

$$\langle E \rangle = \sum_{\mathbf{k}} \frac{1}{\alpha + \beta k^2}, \qquad \langle \Omega \rangle = \sum_{\mathbf{k}} \frac{k^2}{\alpha + \beta k^2} \tag{27}$$

for a given $\langle E \rangle$ and $\langle \Omega \rangle$.

This is the first step in what has become a rather thoroughly elaborated theory of two-dimensional Navier-Stokes turbulence [6]. Time and space are lacking even to outline this theory here. We shall consider only the very limited question of the kinetic theory of the approach to thermal equilibrium of the kind described above.

The classical statistical mechanical approach to nonequilibrium kinetic theory has been through the BBGKY hierarchy or something equivalent [20]. Something like

$$S = -\sum_{\mathbf{k}} \int f_{\mathbf{k}} \ln f_{\mathbf{k}} \, dX_{\mathbf{k}} \tag{28}$$

would be the appropriate entropy, where $X_{\mathbf{k}}$ stands symbolically for the real or imaginary part of the $\omega(\mathbf{k},t)$, and $f_{\mathbf{k}} = f_{\mathbf{k}}(X_{\mathbf{k}},t)$ is the single "particle" phase space distribution for the $X_{\mathbf{k}}$. What one would expect would be that, under certain approximations, $f_{\mathbf{k}}(X_{\mathbf{k}})$ should tend toward a Maxwell distribution as $t \to \infty$, with a temperature that will yield the moments (26).

Most of the literature of nonequilibrium turbulence theory has not developed along these routes, however [21]. Instead, most of the theories have dealt with the modal energies $\langle |\mathbf{v}(\mathbf{k}, t)|^2 \rangle$ ("moments") rather than with the probability distributions, which typically never appear in the theory explicitly.

An ingenious marriage of these two points of view was performed by Carnevale [22]. His result is perhaps the most elegant application that has been made of the information theory perspective in the theory of turbulence. What he did, in essence, was to minimize (28) subject only to the constraints that the $\langle |\mathbf{v}(\mathbf{k},t)|^2 \rangle$ are instantaneously known, but with no assumptions about their ratios. The result of this variational calculation is Maxwellians for the $f_{\mathbf{k}}$ that contain the moments as temperatures. If then the S is evaluated using these Maxwellians, the resulting expression (up to unimportant additive constants) is [22]

$$S = \sum_{\mathbf{k}} \ln \langle |\mathbf{v}(\mathbf{k}, t)|^2 \rangle, \tag{29}$$

a time-dependent function. This new "reduced" entropy is then, in turn, maximized by the moments given in Eq. (26)!

It is a remarkably pleasant consequence of some of the mainstream attempts at a time-dependent theory of evolving turbulence in terms of the moments that dS/dt turns out to be ≥ 0, with the equality sign applying only for the case of the absolute thermal equilibrium moments (26). This is particularly true for the so-called eddy-damped quasi-normal Markovian approximation (EDQNMA) [21], which has now been shown to contain a previously unknown H-theorem. This is, to my knowledge, the first H-theorem to have been deduced that refers, not to the relevant probability distribution, but only to its moments. The result is being extended and elaborated by Carnevale in a sequence of papers [22] that cannot be done justice to here. The $d[\Sigma_{\mathbf{k}} \ln |\mathbf{v}(\mathbf{k},t)|^2]/dt$ has also been evaluated numerically for actual solutions of the 2-D Navier-Stokes equations, with and without viscosity. The result is a monotonically decaying function of time plus a small amount of noise [23].

A rather wide-open area concerns how the entropy enters in a forced, dissipative calculation where the presence of forcing and dissipation keeps the entropy forever away from its minimum value. In particular, the question of whether a Kolmogoroff-Obukhov power-law cascade can be related to a rate of entropy production is an open one, and is of interest. Many interesting directions present themselves, so it is hard to know which way to go next.

5. Closing Remarks

We have offered a quick glimpse of three areas in which problems involving fluids or magnetofluids have to some extent yielded to procedures motivated by information theory [24, 25]. I am not sure how respectable, from the point of view of the mainstream information theorists, these applications are. Certainly the physical dynamics folds into the loss of information in a fundamental way, and I have not regarded any of the applications as an exercise in pure statistics or probability theory. It has always been in the nature of the history of statistical mechanics that it has acquired new computational tools, which require many decades of elucidation before the practitioners are satisfied with their understanding of them. I personally find using the new tools more enjoyable than elucidating them, but I am glad there are people who take the opposite viewpoint. There can, at this point, be no doubt that information theory has provided some new tools that we are far from realizing all the implications of.

Many pressing questions for the future could be enumerated, but I will mention only two. There is first the philosophical basis of treating classical turbulent fluid field distributions as probability distributions, particularly when they can change sign, already alluded to. Second, there is a question of the extent to which the "most probable" distributions are overwhelmingly the most probable. Statistical mechanics is a useful predictor largely because almost all systems in the ensemble are in many senses almost identical: the fluctuations (or noise) are fractionally small, compared with the most probable or mean values of the phase functions. So far we have virtually no information on such questions in the cases addressed here.

Acknowledgments

This work was supported in part by the National Aeronautics and Space Administration under Grant NSG-7416 and in part by the U.S. Department of Energy.

References

1. C. C. Lin (1943) On the Motion of Vortices in Two Dimensions (Toronto: University of Toronto Press).
2. L. Onsager (1949) Nuovo Cimento Suppl. **6**, 279.
3. J. B. Taylor and B. McNamara (1971) Phys. Fluids **14**, 1492.
4. D. Montgomery, 1975, in C. de Witt and J. Peyraud, eds., Plasma Physics: Les Houches 1972 (New York: Gordon and Breach), pp. 425-535.
5. C. E. Seyler (1976) Phys. Fluids **19**, 1336.
6. R. H. Kraichnan and D. Montgomery (1980) Rep. Prog. Phys. **43**, 547.
7. G. Joyce and D. Montgomery (1973) J. Plasma Phys. **10**, 107.
8. D. Montgomery and G. Joyce (1974) Phys. Fluids **17**, 1139.
9. N. G. Van Kampen (1964) Phys. Rev. **A135**, 362.
10. D. Lynden-Bell (1967) Mon. Not. R. Astron. Soc. **136**, 101.
11. D. L. Book, B. E. McDonald, and S. Fisher (1975) Phys. Rev. Lett. **34**, 4.
12. B. E. McDonald (1974) J. Comput. Phys. **16**, 360.
13. G. A. Kriegsmann and E. L. Reiss (1978) Phys. Fluids **21**, 258.
14. A term in S was omitted in Ref. [11], where this conclusion was reached. It has since been put in and the entropy recomputed, but the conclusion did not change (B. E. McDonald, private communication).
15. See, for example, G. Batemann (1978) MHD Instabilities (Cambridge, Mass.: MIT Press).
16. D. Montgomery, L. Turner, and G. Vahala (1979) J. Plasma Phys. **21**, 239.
17. J. Ambrosiano and G. Vahala (1981) Phys. Fluids **24**, 2253.
18. T. D. Lee (1952) Q. Appl. Math. **10**, 69.
19. R. H. Kraichnan (1967) Phys. Fluids **10**, 1417.
20. D. Montgomery (1976) Phys. Fluids **19**, 802.
21. See, for example, D. C. Leslie (1973) Development in the Theory of Turbulence (Oxford: Oxford University Press) or R. H. Kraichnan (1975) Adv. Math. **16**, 305.
22. G. Carnevale (1979) Ph.D. Thesis (Harvard), and (1982) J. Fluid Mech. **122**, 143; G. Carnevale and G. Holloway (1981) Information decay and the predictability of turbulent flows, NCAR Preprint; G. Carnevale, U. Frisch, and R. Salmon (1981) J. Phys. A **14**, 1701.
23. That Eq. (26) is a good prediction for the time averages of the solutions to Eq. (25) with $\nu = 0$ had been previously established by C. E. Seyler, Y. Salu, D. Montgomery, and G. Knorr (1975) Phys. Fluids **18**, 803.
24. E. T. Jaynes (1957) Phys. Rev. **106**, 620, and **108**, 171.
25. E. T. Jaynes (1963) in K. W. Ford, ed., Statistical Physics, Vol. III (New York: W. A. Benjamin), pp. 181-218.

ESTIMATION OF THE BOUNDARY OF AN INCLUSION OF KNOWN MATERIAL FROM SCATTERING DATA

J. M. Richardson

Rockwell International Science Center, 1049 Camino Dos Rios, Thousand Oaks, California 91360

An interesting type of probabilistic inverse problem associated with the scattering of elastic waves from inhomogeneities in solids is that in which the prior statistics of the inhomogeneity is non-Gaussian. Here we consider a relatively simple example of such a problem in which the inhomogeneity is an inclusion with known material but unknown boundary. The appropriate measurement model for longitudinal-to-longitudinal backscatter has the typical signal-plus-noise form in which the noise is assumed to be Gaussian and the signal is given by the application of the Born approximation to an inclusion described by a characteristic function for a random domain. The so-called signal also incorporates the transfer function characterizing the response of the measurement system to a fictitious point scatterer. Assuming a discretized physical space represented by a lattice of points, the prior random properties of the characteristic function are described by a set of independent random variables, each of which has the values 0 and 1 at each point with prescribed probabilities. Thus, in a prior sense, all possible domains (including sets of separate domains) are represented (at least within the resolution implied by the lattice). Our problem is to find the most probable (in the posterior sense) characteristic function given the results of the scattering measurements. We have found that the direct maximization of the posterior probability with respect to the characteristic function is computationally intractable. We have devised a computationally convenient method obtained by maximizing the appropriate probability function of the characteristic function and the additive noise while regarding the measurement model as a continuous set of constraints to be handled by the conjugate vector method. With appropriate analytical manipulations we finally obtained a convex function of the Lagrange multiplier vector to be minimized by computational means. Using synthetic input test data, we have estimated the inclusion boundaries for a number of cases. Compared with conventional imaging techniques, the estimations of the boundary geometries are surprisingly good even for rather sparse sets of input data.

C. Ray Smith and W. T. Grandy, Jr. (eds.),
Maximum-Entropy and Bayesian Methods in Inverse Problems, 469–474.

1. Introduction

The probabilistic approach to the inverse problem associated with the scattering of elastic waves from an unknown flaw involves a stochastic measurement model that contains assumptions concerning the a priori statistics of both the measurement error and the possible flaws. Then the output of the inversion procedure is the most probable flaw, given the measurements. It is clear that the performance of the inversion procedure for a given set of scattering data improves with the amount of correct a priori information restricting the flaw statistics. Here we consider a highly restricted case in which it is assumed that the flaw is an inclusion of known material (isotropic) but with an unknown boundary. The a priori statistical model of the inclusion is nonparametric in the sense that all possible boundaries are represented, including the boundaries of two or more separate inclusions. Each of these possible geometries is confined to a specified rectangular localization domain.

The above is an example of non-Gaussian flaw statistics. An investigation of probabilistic inversion involving several types of flaw statistics has been conducted by Richardson and Gysbers (1977) using a relatively conventional approach entailing the minimization of "many-valley" functions, or, alternatively, the maximization of "many-mountain" functions. The computational difficulties arising from this situation were severe. The present treatment, to be discussed in the ensuing sections, obviates the many-valley difficulties by a procedure that leads ultimately to the computational minimization of a convex function which inherently has the "single-valley" property.

2. Formulation

The appropriate measurement model for L → L pulse-echo scattering from an inclusion is given by the expression

$$f(t,\mathbf{e}) = \alpha \sum_{\mathbf{r}} \delta\mathbf{r}\, p''(t - 2c^{-1}\mathbf{e}\cdot\mathbf{r})\, \Gamma(\mathbf{r}) + \nu(t,\mathbf{e}) , \tag{1}$$

where $f(t,\mathbf{e})$ is the received waveform at time t with an incident propagation direction $\mathbf{e}$ and $\nu(t,\mathbf{e})$ is the associated error. The function $p''(t)$ is the second derivative of $p(t)$, the so-called reference pulse; $\Gamma(\mathbf{r})$ is the characteristic function of the inclusion at the position $\mathbf{r}$; c is the velocity of L waves; and α is a parameter dependent on the material properties of the inclusion and the host medium both assumed to be known a priori. Here we assume that t, $\mathbf{e}$, and $\mathbf{r}$ are discretely valued. In particular, $\mathbf{r}$ takes vector values on a finite cubic lattice where $\delta\mathbf{r}$ is the volume of a unit cell.

To complete the description of the measurement model we must define the a priori statistical properties of ν and Γ. We assume that ν and Γ are statistically independent of each other. The error ν is assumed to be a set of Gaussian random variables with the properties

$$E[\nu(t,\mathbf{e})] = 0 \tag{2a}$$

$$E[\nu(t,\mathbf{e})\ \nu(t',\mathbf{e}')] = \delta_{\mathbf{ee}'}\delta_{tt'}\ \sigma^2 . \tag{2b}$$

The values of the characteristic function at two positions are assumed to be statistically independent, with $\Gamma(\mathbf{r})$ taking the values 0 and 1 with probabilities 1-P and P, respectively. In the present treatment we assume that P is independent of $\mathbf{r}$ and thus $\Gamma(\mathbf{r})$ is an example of a stationary non-Gaussian random process.

3. General Solution

The general problem is to find the best estimate of $\Gamma(\mathbf{r})$ given the measured values of $f(t,\mathbf{e})$. Here we define the best estimates to be the most probable (in the a posteriori sense) function $\Gamma(\mathbf{r})$; that is, we find the set of values of $\Gamma(\mathbf{r})$ at the various points $\mathbf{r}$ that maximizes the conditional probability density $P[\{\Gamma(\mathbf{r})\}|\{f(t,\mathbf{e})\}]$ (or conditional probability in case (3) where the $\Gamma(\mathbf{r})$ are discrete valued). This is equivalent to maximizing $P[\{f(t,\mathbf{e})\}|\{\Gamma(\mathbf{r})\}]P[\{\Gamma(\mathbf{r})\}]$ since $P[\{f(t,\mathbf{e})\}]$ is regarded as fixed in the maximization process.

It is easy to see that

$$\log P[\{f(t,\mathbf{e})\}|\{\Gamma(\mathbf{r})\}]$$

$$= \frac{1}{2\sigma^2} \sum_{t,\mathbf{e}} \left[f(t,\mathbf{e}) - \alpha \sum_{\mathbf{r}} \delta\mathbf{r}\ p''(t - 2c^{-1}\mathbf{e}\cdot\mathbf{r})\ \Gamma(\mathbf{r}) \right]^2 + \text{const.} \tag{3}$$

Using the well-known identity

$$-\frac{1}{2\sigma^2}\nu^2 = \underset{w}{\mathrm{Min}} \left[-w\nu + \frac{1}{2}\sigma^2 w^2 \right], \tag{4}$$

where $w = w(t,\mathbf{e})$ will be called the conjugate vector, we now obtain

$$\underset{\{\zeta\}}{\mathrm{Max}} \log P[\{f\},\{\Gamma\}] = \underset{\{\zeta\}}{\mathrm{Max}} [\log P[\{f\}|\{\Gamma\}] + \log P[\{\Gamma\}]]$$

$$= \underset{\{\zeta\}}{\mathrm{Max}}\ \underset{w}{\mathrm{Min}}\ \Phi , \tag{5}$$

where

$$\Phi = \sum_{t,\mathbf{e}} \left[-w(t,\mathbf{e})\ f(t,\mathbf{e}) + \frac{1}{2}\sigma^2\ w(t,\mathbf{e})^2\right]$$

$$+ \sum_{\mathbf{r}} [\lambda(\mathbf{r})\ \Gamma(\mathbf{r}) + \log P(\Gamma(\mathbf{r}))]\ , \tag{6}$$

in which

$$\lambda(\mathbf{r}) = \alpha \sum_{t,\mathbf{e}} \delta\mathbf{r}\ p''(t - 2c^{-1}\mathbf{e}\cdot\mathbf{r})\ w(t,\mathbf{e})\ . \tag{7}$$

Our procedure now is to reverse the order of the maximization and minimization operations (assuming this is permissible) and then to carry out the maximization operation analytically. This can be done in the case considered here. Let us define the function $g[\lambda(\mathbf{r})]$ by the relationship

$$\underset{\Gamma(\mathbf{r})}{\text{Max}}\ \{\lambda(\mathbf{r})\ \Gamma(\mathbf{r}) + \log P[\Gamma(\mathbf{r})]\} = g[\lambda(\mathbf{r})]\ , \tag{8}$$

where the value of $\Gamma(\mathbf{r})$ giving the maximum is denoted by

$$\Gamma^*(\mathbf{r}) = h[\lambda(\mathbf{r})]\ . \tag{9}$$

We obtain

$$\underset{\{\zeta\}}{\text{Max}}\ \Phi \equiv \psi = \sum_{t,\mathbf{e}} \left[-w(t,\mathbf{e})f(t,\mathbf{e}) + \frac{1}{2}\sigma^2 w(t,\mathbf{e})^2\right] + \sum_{\mathbf{r}} g[\lambda(\mathbf{r})]\ . \tag{10}$$

The next step is to minimize ψ with respect to w, yielding $\hat{w}$. The best estimate $\hat{\Gamma}$ is then given by

$$\hat{\Gamma}(\mathbf{r}) = h[\hat{\lambda}(\mathbf{r})]\ , \tag{11}$$

where $\hat{\lambda}(\mathbf{r})$ is obtained from Eq. (7) by substituting $\hat{w}$ for w.

The minimization of ψ must be carried out by computational means. This minimization process is greatly facilitated by the fact that ψ is a convex function of w, as we have already mentioned in the introduction. Another favorable factor is that the dimensionality of the vector w is the same as that of the measurement vector f, which in turn is usually far less than that of the state vector Γ. These comments constitute the motivation behind the present procedure.

Explicit expressions for the functions g and h are given below:

$$\min_{\Gamma} [\lambda\Gamma + \log P(\Gamma)] = g(\lambda) = \frac{1}{2}[\Gamma + \log P + \log(1-P)] + \frac{1}{2}|\Gamma + \log P - \log(1-P)| , \quad (12)$$

$$\Gamma^*(\mathbf{r}) = 1[\lambda(\mathbf{r}) + \log P - \log(1-P)] = h(\lambda) , \quad (13)$$

in which P is the a priori probability that $\Gamma(\mathbf{r}) = 1$.

4. Computational Examples

A series of test runs was made for a two-dimensional system using a noiseless set of test waveforms $\tilde{f}(t,\mathbf{e})$ derived on the assumption that the actual acoustical (longitudinal) impedance deviation corresponds to a circular inclusion of radius 4 (in dimensionless units). The assumed reference waveform p(t) was a typical sinusoid modulated by a Gaussian envelope. Two incident directions (each defined by a vector $\mathbf{e}$) were chosen, one orthogonal to the other, that is, one in the x-direction and one in the y-direction. In the inversion algorithm a typical level of noise was assumed.

The results are illustrated in Fig. 1. The assumed boundary is indicated by the solid line and the estimated boundary by the dots. Owing to the

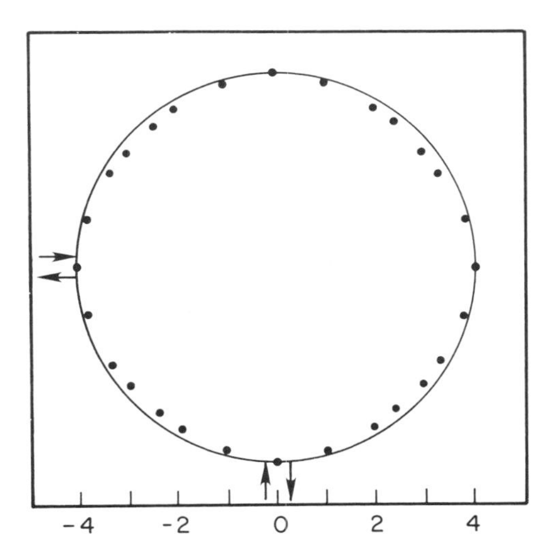

Figure 1. Inversion results for the case where scatterer is assumed a priori to be a known inclusion but with an unknown boundary. Estimated boundary (•) of inclusion vs. assumed boundary (—). Incident and scattered directions of elastic waves are indicated by arrows.

highly restrictive nature of the a priori statistical assumptions, the results show that the set of simulated measurements is almost adequate for the present case. The results are perhaps accidentally better than they ought to be, in view of the rather coarse discretization involved. In Fig. 2 we illustrate the nature of the auxiliary function $\hat{\lambda}(\mathbf{r}) = \hat{\lambda}(x,y)$ for the present case.

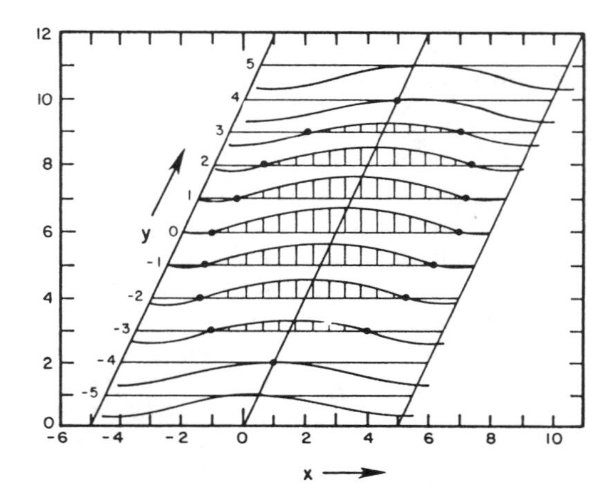

Figure 2. The nature of the function $\hat{\lambda}(\mathbf{r}) = \hat{\lambda}(x,y)$.

Acknowledgment

This work was supported by Independent Research and Development funds of Rockwell International.

Reference

Richardson, J. M., and Gysbers, J. C. (1977) Application of estimation theory to image improvement, 1977 Ultrasonic Symposium Proceedings, IEEE Cat. No. 77 CH 1264-1SU (1980).

VERTICES OF ENTROPY IN ECONOMIC MODELING

Hillard H. Howard and John C. Allred

Los Alamos National Laboratory, Los Alamos, New Mexico 87545

The economic process can be modeled in terms of eight elementary vertices that take account of entropy generation and energy and mass flows in specific processes. This representation emphasizes aspects of the economic process that are frequently neglected in other models, particularly the role played by materials in facilitating specific processes.

C. Ray Smith and W. T. Grandy, Jr. (eds.),
Maximum-Entropy and Bayesian Methods in Inverse Problems, 475–488.

1. Introduction

The relationship, if there is one, between entropy and economics has been a subject of speculation since the time of Sadi Carnot (1796-1832), who has been called the first resource economist. At the time, entropy had been neither defined nor named, but the concept that Carnot used in establishing the efficiency of engines was in fact that of entropy [1].

Since Carnot's time, many important advances have been made in the understanding of entropy. The names of Clausius, Clapeyron, Boltzmann, and Gibbs come to mind. More recently Shannon and Brillouin have explicated the relationship between information and entropy.

2. Entropy and Economics

Because entropy is a measure of disorder, and because objects of value frequently are ordered, the possible relationship between low entropy, or order, and economic value has occurred to many people, particularly scientists and engineers, independently. However, the introduction of the concept to economics was done by Nicholas Georgescu-Roegen [2], first in essays, then in his book The Entropy Law and the Economic Process. Unfortunately, Georgescu-Roegen has not been able to formulate a recipe for the application of entropic concepts to economic analysis in a quantitative way, though he deserves credit for his continuing insight into possible uses of entropic analysis. In this paper we work toward quantitative application of these ideas.

In his monumental treatise on economics, Human Action [3], the late Ludwig von Mises gave a definition of economics that is still acceptable to many economists today. He defined economics as the science of ends and means, and asserted that economics does not make value judgments as to goals, but analyzes means as to their relative appropriateness, given a particular goal. Thus a means is either "good" or "bad" as it does or does not lead in an efficient way to a goal, without reference to the desirability of the goal. By denying value judgments in this definition, von Mises argued that economics partakes of the nature of physical science, which attempts likewise to be value-independent.

We believe this definition may indicate a fruitful approach to the use of entropic analysis in economics, in the following way. Suppose we call the present state of the world A. We wish somehow to alter that state to B. There are many possible paths from A to B, and if physical quantities only are involved, there are path integrals of entropy generation along each possible path. In the real, irreversible, world it is not possible to make a state transformation without entropy generation. Query: Is the path of least entropy generation, subject perhaps to other constraints, the most desirable path economically? With Georgescu-Roegen, we hypothesize that it is. If this hypothesis is correct, accounting for entropy in processes ought to serve some use in economic analysis.

At first glance it may appear that the proposition that the optimum economic path is the path of least entropic increase is to suggest that the only scarce resource is available energy, or, roughly, "free" energy. Such a hypothesis appears to ignore the Marxian elements of production (land, labor, and capital), and to ascribe scarcity only to available energy. The apparent implication is that land, labor, and capital are not scarce, but this implication is only superficial. For the fact is that land, labor, and capital are gained by the expenditure of available energy: land is acquired by exploration and capture, or by purchase, the expenditure of capital; labor is possible through the use of the negentropic material products of solar energy; and capital is nothing more nor less than the store of the fruits of the use of available energy.

Economic modeling that depends almost entirely upon physical parameters is indeed a radical departure. Economic models cannot be divorced from human preferences, which are analogous to the driving forces of classical mechanics. Human preferences determine final goals—the "ends," goods and services. Preferences are arrived at by many processes not included in economic models—politics, ethics, aesthetics, traditions. Economic models are used to study the various means by which a given set of goals (ends) may be achieved. We suggest a break from traditional economics. In our approach, labor and capital, including land, can be incorporated into the models using their entropic contents if we can calculate their "embodied energies" [4]. Some entropy contents are well understood, whereas others must be studied at a most basic level before we can begin to make calculations.

Economic modeling is, of course, not the decision-making process. Rather, it is a tool for the decision maker. If human preferences are removed during analysis, then judgments with respect to human preferences are left to the decision maker. "Shadow prices"* are not hidden in the analysis, but are applied during the decision-making activity. Economic pathways that are optimized from an entropic viewpoint are then translated to a monetary system to express values in terms of price.

3. The Connection Between Energy and Matter

The First Law of Thermodynamics states the conservation of energy and matter. With the addition of the Second Law, the energy (matter) balance of the First Law has an added aspect, namely, that for a real time process in an isolated system some energy (matter) is lost to a form that cannot be used to do work except for the very special case of reversible processes. An increase in the entropy is said to have occurred. Entropy is a measure of

---

*Shadow prices reflect the actual social value of goods and services as opposed to the prices observed in the marketplace, which may be distorted by intervention by a third party, for example, government.

the quality of the energy (matter) in a system. One can imagine that the true effect of the Second Law of Thermodynamics is to bound the First Law. Even though energy (matter) is conserved, some of it is lost for further use. The amount lost is related to the increase in entropy.

We are accustomed to calculating entropies for chemical reactions. In these cases we have concerned ourselves only with the energy and matter actually participating in the reaction. We have ignored until now the vessels and other equipment that contain the reacting energy and/or matter. Although we have concentrated on the energetics of our systems, we can also measure the quality of matter in process equipment by calculating the entropies for the material components.

4. Forms of Entropy

We are acquainted with the various manifestations of energy: potential, kinetic, nuclear, electromagnetic, ... However, most people are not so familiar with the manifestations of entropy. We identify five forms of entropy that a physical system may display. Within any given system a particular form may predominate. The five forms are considered distinct because of their distinct applications. All forms of entropy are additive, based on their statistical nature, and must be summed if we are to arrive at the total entropy.

First is the standard entropy, S, of thermodynamics, $dS = \delta Q/T$, first used by Carnot, named by Clausius, and explicated as a statistical property of the states of a system by Boltzmann. Q is the thermal energy of transformation and T the absolute temperature.

A second form of entropy is encountered in the study of chemical reaction rates. The equation for the velocity constant leads to the identification of a free energy of activation and the entropy of activation analogous to the thermodynamic equation for free energy. The entropy of activation is again a manifestation of the statistical nature of entropy. Entropy of activation is a measure of the probability that the reacting species will form an activated complex capable of decomposing into the products of the reaction or equally that the partition functions for the electronic states are favorable for the reaction to take place. An interesting aspect of this form of entropy is that in chemical reactions the entropy and free energy factors are always in opposition. The actual reaction path is a compromise between the two.

Third is the entropy of mixtures, again defined statistically, but measuring the purity, or lack of it, of a substance. A substance composed of but a single kind of atom is statistically in a unique state, with mixing entropy equal to zero.

Fourth is entropy of structure—of architecture of a crystal, for example. If a crystal lattice has the correct atom in each of its defined lattice sites, its structural entropy is zero. Dislocations, vacancies, and other defects introduce disorder and increase the entropy, just as impurities do in the previous instance.

The fifth form of entropy has been discussed very little previously. So far as we are aware, R. S. Berry first wrote about it in a discussion of the entropy of machined surfaces in automobiles [5]. One of us (JCA) independently arrived at the same concept in the attempt to evaluate the entropy of machined surfaces in energy-converting devices [6]. We call this form "entropy of shape." It is a measure of the preciseness of definition of a surface. If it were possible to produce a surface of perfect flatness, say within a tolerance of a single lattice spacing, thus producing a unique statistical state, the entropy of shape of the surface would be zero. Looser tolerances give larger entropies of shape, but the entropy of shape per unit area of a typical machined surface, such as that of a piston or bearing, appears to be small in any case.

It may appear that the entropy of surface is a special case of the entropy of structure. We make a distinction here to emphasize the anthropomorphic nature of entropic analysis. In many machines it is surface properties that dominate performance of a given task. In some applications the internal structure of a device may be substituted freely, provided the surface properties are kept constant. In the entropic analysis of such a device one would make the distinction obvious.

Table 1 lists the five forms of entropy and gives examples of entities having low entropy for each case.

Table 1. Low-entropy entities for the five forms of entropy

Form of entropy	Entity of low entropy
Thermal	Any system at equilibrium at extremely low temperature ($\rightarrow$ 0 K)
Activation	$2\ HI = H_2 + I_2$ (lower entropy) $HI = H + I$ (higher entropy) (but $2\ HI = H_2 + I_2$ prevails because of energy economy)
Mixture	Extremely pure material; chemicals of analytical grade; pure metals
Structure	Crystals of high regularity; diamond, sapphire, quartz (also often of high purity)
Shape	Machined surfaces, superfinishes, surfaces of optical mirrors and lenses

5. Entropy Densities in the Economic Process

In discussing possible relationships between entropy and economics it is helpful to have a system that makes accounting of entropy generation both easy and accurate. The basic element of the system is called a vertex. A vertex represents a device or process through which energy and/or material flows, within which an energy-material interaction may (but not necessarily does) occur. A vertex is an eight-terminal device with four inputs and four outputs (see Fig. 1). Not all terminals need be activated for a particular vertex. However, since mass and energy flows are conserved through a vertex, at least in steady state, if an energy (mass) input terminal is activated, at least one of two energy (mass) output terminals must also be activated, and vice versa.

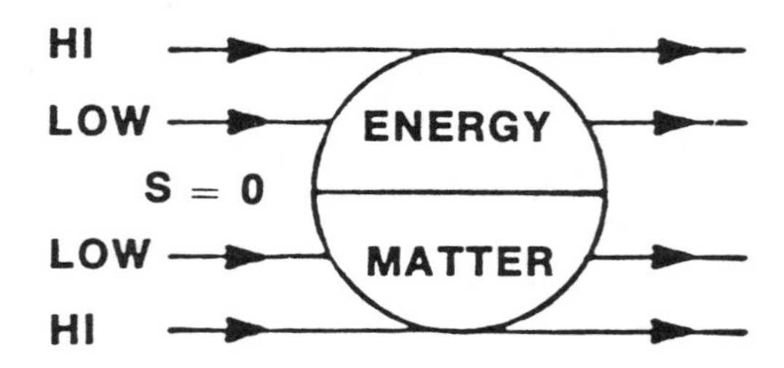

Figure 1. General entropy vertex.

The vertex pictured in Fig. 1 labels energy and matter as either "high" or "low" in entropy content. This labeling illustrates a salient point. Any economic discussion of relative values must, at bottom, assume an anthropomorphic view. The label "low" is attached to energy or matter that has economic value or that we find useful as desirable input to the next vertex. Matter or energy of little or no use to us at the next vertex we label "high." Grades of matter are not necessarily attached to some absolute value of intrinsic entropy, as can be illustrated by example. In the plant-animal cycle

$$CO_2 + H_2O + nh\nu \rightleftarrows \text{sugars} + O_2$$

the components on the right-hand side of the equation are demonstrably of lower entropy than those on the left since, in an isolated system, the reaction proceeds very slowly from right to left, oxidizing the sugars to CO_2 and H_2O. In the photosynthetic reaction with chlorophyll as catalyst, going left to right, a plant finds itself in the steady state far from equilibrium first described by Prigogine and co-workers [7], which can occur in an open system having a continuing flow of energy ($nh\nu$).

Thus it would appear that, while the anthropomorphic preference is for sugars and oxygen, in some sense a plant prefers the left-side components! This conclusion seems to us to emphasize the interesting idea that the entropy concept, as it may bear on economics, has its anthropomorphic side. That is, the economic value of an object or system of low entropy is enhanced precisely because humans define those structures, compositions, and states that they regard as unlikely, and therefore scarce, as being of low entropy.

At any vertex we can rate the energy/matter on an entropy scale as is often done in energy flow analysis. However, in entropic analysis we must keep track of the entropy for all grades of energy/matter. Before studying the algebra of entropy vertices in more detail, let us examine the single general vertex more closely. If we are clever, in principle we can calculate exactly the entropy of all the inputs and outputs of a vertex. The vertex itself is the black box that converts the inputs to outputs and in so doing changes the entropy of the closed system that contains the vertex and all of the vertex's inputs and outputs. From the Second Law of Thermodynamics we know that the total change in entropy for our closed system is greater than or equal to zero. Note that any given input may be changed by the vertex into an output with entropy lower than that of the given input, but when the system is taken as a whole, including sources and sinks of matter and energy, the net increase in entropy must be zero or positive. Vertices become more useful to us if we move the boundaries of our closed system to the boundaries of our black box, the vertex. We now see that the change in entropy at the vertex is again positive or zero and is bounded from below by the difference of the entropies of the inputs and outputs of the vertex:

$$\Delta S_V \geq \sum_i S_i^0 - \sum_j S_j^I \, . \tag{1}$$

We can extend our argument to see that the increase in entropy of the universe ΔS_u (or any closed system containing the vertex) is bounded from below by the right-hand side of Eq. (1). That is, we know the minimum amount by which the entropy of a closed system containing the vertex must be increased by the presence of the vertex. We now have the relationships

$$\Delta S_u \geq \Delta S_V \geq \sum_i S_i^0 - \sum_j S_j^I \, . \tag{2}$$

To get a lower bound on the increase in entropy to our closed system (or the universe), we need not know the details of the vertex, but only its inputs and outputs.

Every vertex represents either a machine or a process, or a combination of such elements. Hence every vertex is a source of entropy. In global steady state (which is optimal and also unrealizable), one can calculate dS/dt at every vertex. Then the sum of dS/dt over all vertices gives the rate of entropy generation for a system that we can represent by a diagram of interconnected vertices. If dS/dt is in fact subject to an extremum principle, rate analysis of entropy generation is a tool of considerable importance for achieving economic designs. An urgent need in entropic analysis is to establish the existence or nonexistence of an extremum principle. Is it truly the case, as many believe, that minimizing the rate of entropy production indeed maximizes the overall efficiency in terms of "minimizing our regrets"?

All vertices can be represented as compositions of eight elementary vertex types. Figure 2 shows these vertex types, names them, and cites real-world examples of each. As already noted, energy (mass) inputs imply energy (mass) outputs. Figure 2 shows only the "low-entropy" inputs and outputs, and only those that typify the vertex. Every vertex has in actuality high-entropy outputs of both energy and matter, if only because of friction and wear in the vertex. Most vertices have substantial high-entropy

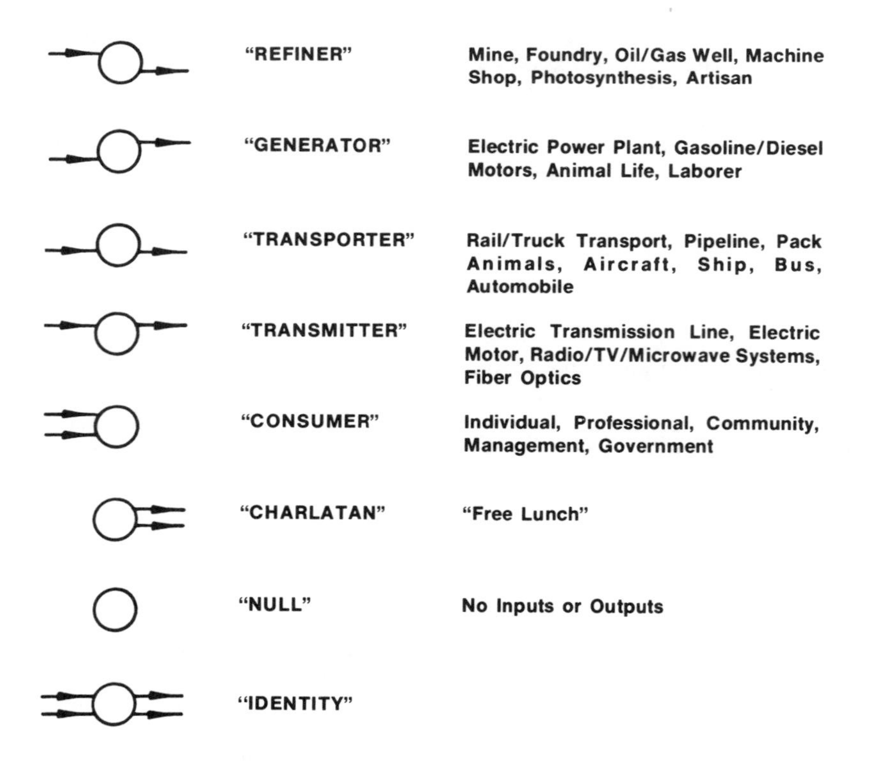

Figure 2. Eight basic entropy vertices.

inputs as well. The vertices of Fig. 2, except for the null and identity vertices, share the common feature that some form of regulation, monitoring, or control is implicit in each. Outputs must be measured and controlled, else the vertex will cause some difficulty—perhaps self-destruction in the case of a machine, perhaps a tumor or death in a living organism.

Two vertices, the "consumer" vertex and the "charlatan" vertex, have interesting properties that deserve further discussion.

6. The "Consumer" Vertex

If the purpose of production is consumption, Adam Smith's proposition is admirably demonstrated by the "consumer" vertex. The "consumer" takes in "low-entropy" energy and matter and exhausts "high-entropy" heat and waste material. But note, first, that if the consumer is to maintain even quasi-steady state, the supplies of matter and energy that he requires must be adequate. Second, it is precisely his sort of wastes that can injure or render inoperative the feedback and control circuits of the other vertices. Finally, unlike the other vertices, the "consumer" has no way to sense and regulate on a low-entropy energy or matter output. It may happen that the "consumer" generates information, which Shannon and Brillouin have shown has negentropic content. His sensing must evaluate undesirable alternatives and choose among them, the risk being that an improper choice may destroy not only other vertices essential to the economic process, but the consumer vertex as well.

Any entity or mechanism that acquires low-entropy energy and/or matter and that produces no low-entropy output of either is called a "consumer" in this scheme. Thus, professionals, communities (in the organized sense), managements, and governments are included. These latter entities produce, or try to produce, information. To be effective, information must be communicated, understood, and acted upon. It is also clear that the purpose of the production of information by these organizations is to alter the control mechanisms of the other vertices. Even if the intent is to make control changes for the "consumer's" benefit, such may not occur in the absence of correct understanding of the operation of the other vertices. Likewise, incorrect or garbled transmission of the information will produce an undesired result if the misinformation is acted upon.

7. The "Charlatan" Vertex

"Charlatan" proposes to produce low-entropy matter and/or energy using only high-entropy inputs. The entropy law does not allow such a process. Thus proposals to produce gasoline from water or high-quality fuel from garbage are "charlatan" vertices. "Charlatan" also occurs in hidden form, such as when "refiner" and "generator" are interconnected, purportedly producing a net output of low-entropy matter or energy. Because "charlatan" cannot perform as represented, this vertex is virtual; it does not in fact exist.

8. Some Specific Examples

Figures 3, 4, and 5 show specific examples of the usefulness of vertices in describing physical processes. In this formalism it appears to be quite simple and natural to include changes in entropy both of energy and matter; the concept of "embodied energy" arises quite naturally as well.

For example, Fig. 3 illustrates the tempering of steel. Thermal energy is added (upper left), embodied in part in the revised carbon-iron solution, which lowers the entropy of the matter affected. In quenching, the thermal energy is rejected to the quenching fluid. There is a net gain in thermal entropy.

Figure 4 demonstrates entropic changes in the electrolysis of, say, water to hydrogen and oxygen. This process can be, ideally, taken to be reversible as to the energy balance. Thus the figure shows energy at very low entropy entering the system just above the horizontal diameter, with water entering at lower left. The electrical energy can in principle be embodied in the separated products of electrolysis, and in principle can be recovered completely by the inverse process. In fact, fuel cells do perform with remarkably high efficiency although not, of course, with efficiency of 100%. However, the third input, matter at low entropy, represents the electrolytic or fuel cell itself: special electrode material, container, valving and piping, and so on. All of this material is degraded entropically during the operation of the cell; at least part of the material is lost irreversibly during this operation. It is this happening that Georgescu-Roegen wishes to draw to our attention to as an inescapable part of the economic process.

Finally, Fig. 5 represents schematically the process of machining, as by a lathe, mill, or shaper. For specificity, suppose we are describing the turning of screw threads by a die applied to a metal rod in a lathe. Energy at low entropy is supplied at left. Most of this energy is degraded to thermal energy in the cutting process, but a part of it is embodied into the screw threads because the surface area—hence the surface free energy—is increased. This embodied energy together with the work of cutting lowers the entropy of the matter as it passes through the vertex. Once again, however, the vertex diagram permits us to take into account the entropic degradation of the machine, as well as the unrecoverable loss of a part of the matter being worked, as it passes through the vertex.

9. Production, Consumption, and Economic Pathways

From the viewpoint of physical science, the economic terms "production" and "consumption" cannot be taken literally. Neither matter nor energy can be created nor destroyed (except by their mutual interconversion—and then it is their sum, the mass-energy, that is conserved). The economic processes "production" and "consumption" are actually transformations of the physical states of the components (both energetic and material) of the goods appearing in the economic process. The different physical states manifest themselves in qualitative changes in matter and energy:

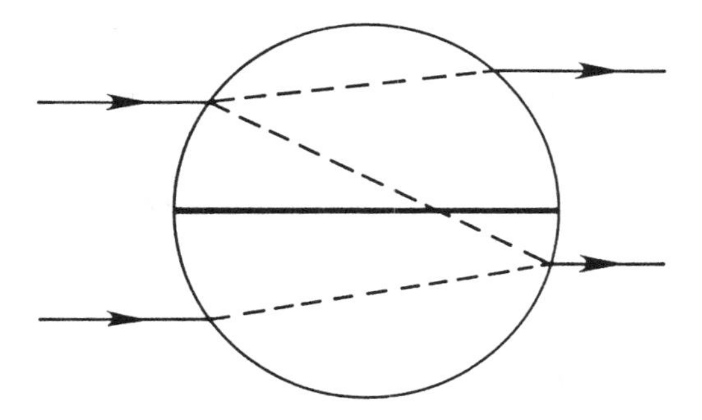

Figure 3. Tempering steel.

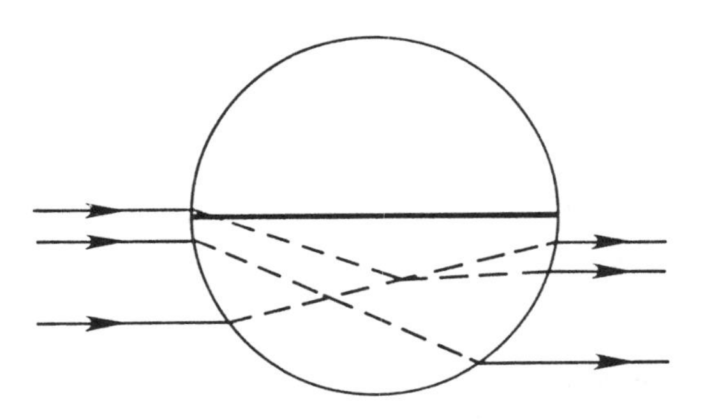

Figure 4. Electrolysis.

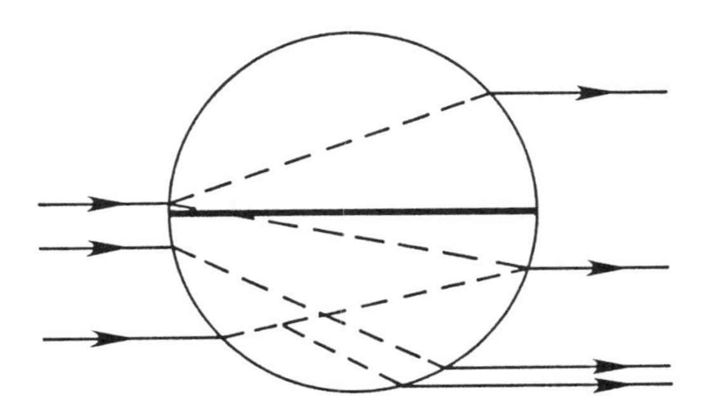

Figure 5. Machining.

their forms, shapes, purities, and availabilities. Entropy is a physical "state function." That is, entropy is a property of a body that depends only on its "state," not on the path or method by which it achieves that state. Since entropy is a state function of the system, if one changes the entropy of a system, one changes the state of the system. However, there are many states of a given system that are acceptable in an anthropomorphic sense. As Feynman aptly put it, entropy measures the number of ways one can rearrange the insides (of a system or object) without being able to tell the difference from the outside. Experience tells us that some changes in state of a system make the system no longer acceptable for our desired use. Our proposition is that entropy is a good measure of that transition between acceptable and unacceptable. Systems undergoing sufficient increases in entropy certainly pass from acceptable to unacceptable. Can the converse be established? From thermodynamics we know that reversible processes, including isentropic ones, require an infinite amount of time to occur; thus they are not truly representative of real-world processes. Because real-world processes occur in finite times, the entropy must change. Changes from acceptable to unacceptable states (or vice versa) must entail changes in entropy. We thus realize that entropy may be a "quantitative measure of quality," and thus an apt candidate for an economic variable.

Let us examine this possibility. Economic changes are path functions, not state functions. When one considers only the goods "produced" and "consumed," the entropies of the goods are not path functions because the entropy of the goods in the final state is the same no matter how it went from the initial to the final state. However, the entropy generated in the surroundings may be highly dependent upon the path. Entropic analysis of the economic pathways may be much more enlightening than the study of the entropies of the initial and final states of goods. Initial and final state entropies of the raw materials and finished products certainly give lower bounds on the generation of entropy by the economic process. At each level of sophistication in an entropic analysis the "economic pathways" can be divided into steps with inputs and outputs. Thus we can define a total entropy production (or negentropy expenditure) for a pathway.

Economics and the physical sciences attempt to develop models, concepts simpler than the complex systems that exist in the real world, abstractions that retain the essential features of the real-world systems. Some descriptions of physical systems can ignore dissipative or frictional processes and still give useful results as, for example, the models of atomic quantum mechanics. The mechanics of particle motion sometimes describe the motions of real objects in a useful way even when dissipative processes are neglected. But if we wish to have even an approximately correct representation of the motion of a golf ball or a bullet or an aircraft, dissipative forces mut be included. However, efforts to calculate from first principles such phenomena as friction between solid surfaces and fluid turbulence have generally not met with success.

The construction of economic models requires the consideration of dissipative forces and events. Surely dissipative processes need to be considered and taken into account in economic modeling if the model is to approximate the real world in a useful way.

Although the economic process is not reducible solely to matter and energy, it cannot violate the laws to which these last elements are subject. For example, it is unfortunate that often, in arguments concerning the constraints of the economic process, there is a failure to recognize the qualitative differences between kinds of energy (for example, electric and thermal), as when the energy requirement of the United States is stated as 75 quads without further qualification. Further, the interrelatedness of matter and energy in processes occurring in finite time has not usually been either understood or taken into account.

In his paper "Embodied Energy and Economic Valuation" [4] Costanza demonstrates a remarkable correlation between the energy embodied in various goods and services in many sectors of the U.S. economy and the dollar value of unit output. For almost all sectors of the economy the ratio of embodied energy to dollar value is nearly constant. Within the analysis are notable exceptions in the sectors dealing with primary energy inputs to the economy. Costanza speculates that the degree of departure from other sectors is related to the net energy yield or available free energy added to the economy. It appears, however, that what Costanza regards as embodied energy is actually input of free energy. Nor is it always the case that the energy remains "embodied" within the output; in many cases (it appears to us), what actually happens is that free energy is applied in the fabrication of a good, becomes entropically degraded to heat, and is then radiated to the surroundings, eventually to space. We cite this example as illustrating the comprehensive nature of the entropy concept in describing what actually occurs in economic processes.

A purpose of this paper is to improve the dialog and mutual understanding between economists and physical scientists. We believe the addition of the entropic coordinate to economic systems clarifies such problems as the appearance of externalities (entities produced in the economic process which the market ignores because of lack of value), scarcity of material resources, valuation of material goods and energies, and material and thermal pollution.

If economic theory is to guide political and social planning, aims, and goals, it is surely necessary that economics "get its physics right." To the extent that physicists can help in getting the physics right, some of our economic problems may be better understood and hence more correctly solved.

Acknowledgment

This work was performed under the auspices of the U.S. Department of Energy.

References

1. Sadi Carnot (1824) Reflections on the Motive Power of Fire, reprinted 1960, E. Mendoza, ed. (New York: Dover).

2. Nicholas Georgescu-Roegen (1966) Analytical Economics: Issues and Problems (Cambridge, Mass.: Harvard University Press); (1971) The Entropy Law and the Economic Process (Cambridge, Mass.: Harvard University Press); (1977) The steady state and ecological salvation: a thermodynamic analysis, BioScience **27**, 266-270; (1978) Technology assessment: the case of solar energy, Atl. Econ. J. **6**, 15-21; (1978/79) Energy and matter in mankind's technological circuit, J. Bus. Admin. (Univ. of B.C.) **10**, 107-125; (1979) Energy analysis and economic valuation, South. Econ. J. **45**, 1023-1058; (1980) Matter: a resource ignored by thermodynamics, in L. E. St.-Pierre and G. R. Brown, eds., Chemrawn I (invited lectures presented at the World Conference on Future Resources of Organic Raw Materials, Toronto, July 10-13, 1978) (Oxford: Pergamon Press), pp. 79-87.
3. Ludwig von Mises (1966) Human Action, 3rd rev. ed. (Chicago: Contemporary Books, Inc.).
4. R. Costanza (1980) Embodied energy and economic valuation, Science **210**, 1219-1224.
5. R. S. Berry (1972) Recycling, thermodynamics and environmental thrift, Bull. At. Sci. **27**, 8-15.
6. John C. Allred and Hillard H. Howard (1977) Second Law Concepts and Energy Converters (Houston: Biophysical Research Corporation), pp. 27-30.
7. I. Prigogine, G. Nicolis, and A. Babloyantz (1972) Phys. Today **25** (Nov.), 23-28.

SUBJECT INDEX